Principles of Evolution

Systems, Species, and the History of Life

Jonathan Bard

Vice President: Denise Schanck
Senior Editor: Elizabeth Owen
Editorial Assistant: Jordan Wearing
Production Editor: Deepa Divakaran
Illustrator: Nigel Orme
Layout: Nova Techset Ltd
Cover Designer: AM Design
Copyeditor: Jayne MacArthur
Proofreader: Sally Livitt
Indexer: Jonathan Bard

ISBN 978-0-8153-4539-8

Library of Congress Cataloging-in-Publication Data
Names: Bard, Jonathan.
Title: Principles of evolution : system, species, and the history of life /
Jonathan Bard.
Description: New York, NY : Garland Science, 2016.
Identifiers: LCCN 2016025611 | ISBN 9780815345398 (pbk. : alk. paper)
Subjects: LCSH: Evolution (Biology)
Classification: LCC QH366.2 .B367 2016 | DDC 576.8--dc23
LC record available at https://lccn.loc.gov/2016025611

Published by Garland Science, Taylor & Francis Group, LLC, an informa business,
711 Third Avenue, New York, NY, 10017, USA, and 3 Park Square, Milton Park, Abingdon, OX14 4RN, UK.

Printed in the United Kingdom
15 14 13 12 11 10 9 8 7 6 5 4 3 2 1

Visit our website www.garlandscience.com

Diagram on front cover: *The first three billion years of life.* A stylized diagram showing a likely paths by which the descendants of the last universal common ancestor (LUCA) of all subsequent taxa diversified to give the major groupings of today's organisms. First, the various types of bacteria evolved (the Gram-positive and Gram-negative eubacteria and the archaebacteria). A long time later, it is likely that a Gram-negative eubacterium, perhaps an early ancestor of *Gemmata obscuriglobus*, acquired a nucleus through endosymbiosis of an archaebacterial cell. This first eukaryotic common ancestor (FECA) later acquired a mitochondrion through endosymbiosis, and this last eukaryotic common ancestor (LECA) diversified to give single-celled eukaryotes (protists), fungi, and animals. An early descendant of the LECA acquired a chloroplast from a cyanobacterium, again through endosymbiosis; its descendants evolved to become algae and plants. These four kingdoms are distinguished today by the number of their cell flagella and by the different compositions of their cell membranes and polysaccharide coats (Figure 9.4 provides more detail).

About the author
Professor Jonathan Bard has published widely in the fields of developmental, theoretical, and systems biology. He was a member of the MRC Human Genetics Unit and the University of Edinburgh, and is now attached to Balliol College, Oxford.

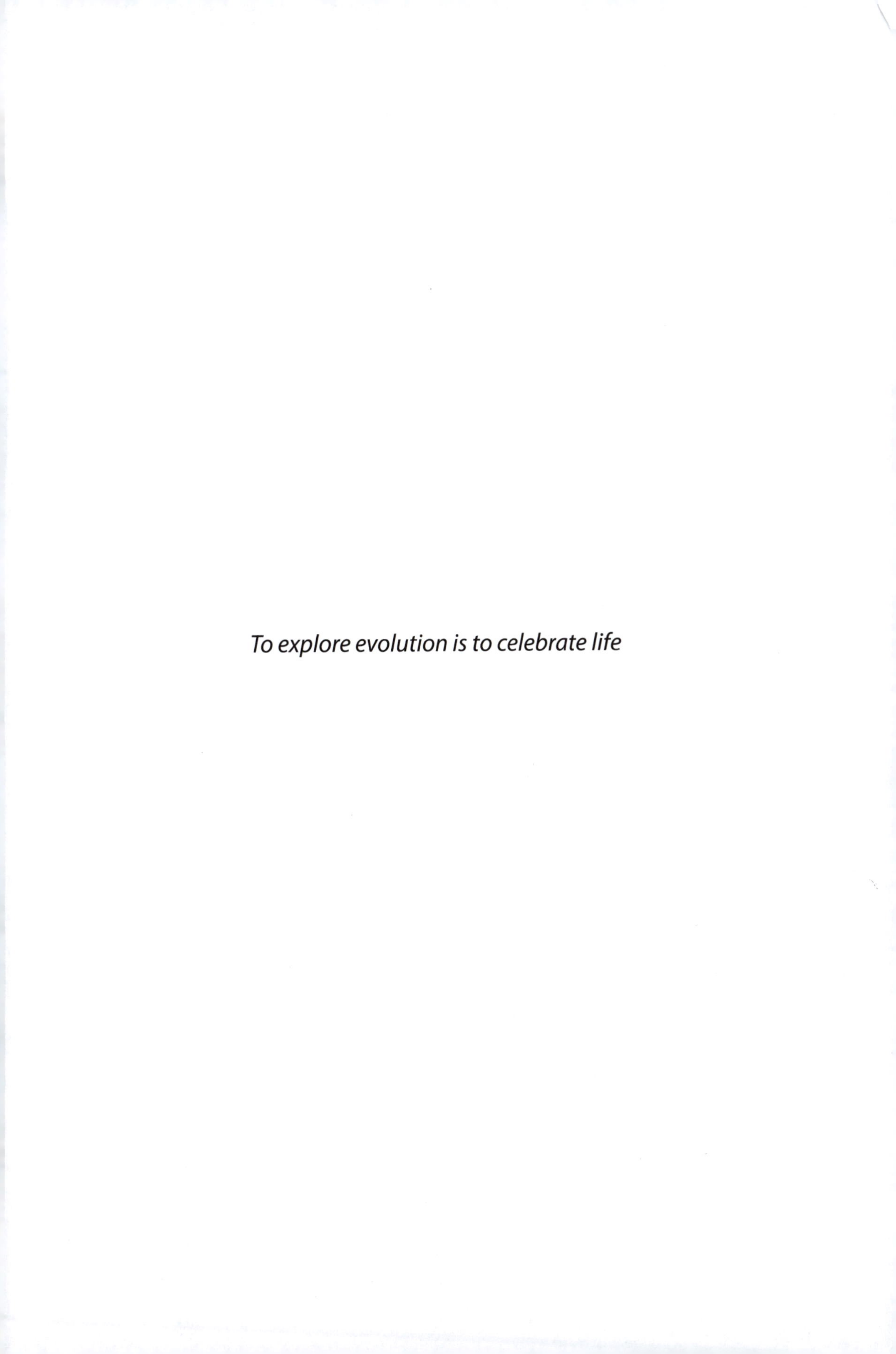

To explore evolution is to celebrate life

Preface

Evolution is the most wonderful area of biology: not only does it have an overarching grandeur, but it provides a perspective that unifies everything that we know about life. In 1973, Theodosius Dobzhansky wrote that 'Nothing in biology makes sense except in the light of evolution.' He was right then and he is right now. The history of evolution is however hard to untangle because what we see today is a contemporary snapshot of a rich and complex biosphere of more than a million species, together with an equally rich and complex, but incomplete fossil record. This record extends back more than three billion years with ever-decreasing clarity to a very simple bacterium that was the first universal common ancestor of all subsequent species. Any author writing a book about evolution thus has the joint tasks of giving that long history together with its supporting evidence and of explaining how change happened and, as Darwin noted, is still happening.

As there are already many books about evolution, any potential reader naturally wonders what this one offers that others do not. It has four aims. The first is to be terse, focusing particularly on the principles of evolution – it assumes that readers know a fair amount about biology, and want neither secondary material repeated nor excessive amounts of detail. The second is to explain the two major themes of the scientific study of evolution, the evidence for evolution and the mechanism by which novel speciation happens – the problem that so interested Darwin. The third is to give the core features of the history of life. The final aim of the book, and one that particularly distinguishes it from others, is to consider the subject from the perspective of systems biology.

This systems biology approach to biological thinking is relatively recent and is based on two ideas. The first is that most activity in living organisms is driven by complex protein networks rather than by individual proteins. This realization particularly helps understand evolutionary developmental biology (evo-devo) and the origins of variation. The second is that causality in living organisms is widely distributed; this can be seen, for example, in considering the relationship between the genotype and the phenotype or in how selection operates. Systems biology provides a natural context for integrating such complex evolutionary knowledge. The systems approach also has a third use: its formal language is helpful in constructing and understanding phylogenetic trees, both cladograms and phylograms (Chapter 5).

The first section of the book gives some necessary background. Section two includes the evidence for evolution and the history of life (Chapters 4–11). This evidence comes from analysis of the fossil record, from phylogenetics, and from evo-devo, with each contributing its own insights. As a result, this history is distributed across the section and so has its own table of contents. The third section (Chapters 12–16) covers the core mechanism or theory of evolution: that new species arise from existing ones through selection acting on variants in small, reproductively isolated populations. The section concludes with a chapter on the evolution of *Homo sapiens*, partly because so much is known and partly because of its particular interest to us. The book ends with a series of appendices that cover some wider aspects of evolution; the fourth of these considers the different viewpoints of evolution and creationism.

There are two areas of evolution where the treatment is light. The first is plant evolution, and this is partly because of the thinness of the fossil evidence, partly because the genetic evidence is not simple, but mainly because of the limitations of the author. The second is the mathematical theory of evolutionary population genetics. This is because the theory mainly focuses on only a single step in the complex mechanism of speciation, how gene distributions in a population can change under random drift, mutation, and selection. There are other reasons: a proper treatment of the theory requires far more mathematics than most biologists have, while the amount of space needed to discuss the theory in adequate depth would have been disproportionate to its importance in the wider scheme of evolution. The solution that I have adopted for handling this theory is to use its important ideas and results where appropriate, particularly in the contexts of genetic inheritance within small groups, of selection, and of coalescent analysis, and to refer the interested reader to the relevant textbooks for mathematical detail.

There is also an interesting difficulty with the theory of evolutionary population genetics, and this is with its view of genes, which dates from the 1940s. Then, a gene was viewed as the heritable component of a phenotypic trait, a view that derived from Mendel's work on peas. He worked with traits that bred reliably, such as seed shape and color, and assumed that the organism had factors whose activity underpinned all such traits. Now, of course, we know that most such trait-defined genes rarely breed true and also have little molecular meaning. Genes are of course DNA sequences coding for particular functions, often detected through mutations that affect an organism's phenotype. Unravelling the complex relationship between trait-defined genes and DNA-sequence genes is not straightforward, but an attempt has been made in the last parts of Chapters 12 and 13 using the framework of systems biology.

Two areas that are treated in a more detailed and integrated way here than in other textbooks are evo-devo and the mutational basis of anatomical variation (Chapters 10–13). A systems-based introduction to the principles of development leads on to an analysis of the functional homologies across embryos (evo-devo). This is followed by detailed discussion about the origins of variation as reflected in changes to embryological development, which is of course where adult anatomical variation usually originates. For Darwin, variation was the most difficult evolutionary problem to solve and he just had to take its occurrence for granted. Now, of course, it is obvious that anatomical and physiological variants usually result from germline mutations, but the idea that a change in a DNA base has a simple or direct effect on the phenotype is rarely correct or even helpful. The view taken in this book is that many of the important and immediate effects of such mutations are on those proteins operating in the networks that drive embryogenesis. Such mutations lead to changes in the functional outputs of these networks, which in turn lead to the broad spectrum of heritable anatomical variants seen within the adult populations. It is on this distribution across a population that selection acts. Without variation there can be no selection: it is the origin of speciation.

A note on the literature. Only a small amount of the vast literature on evolution can be discussed here. A few core publications are given at the end of each chapter, but most are cited in the text and detailed in the final reference list. These many papers, books, and websites point readers to important experiments and areas that may only be briefly discussed. The level of the book is such that undergraduate students should be able to read it without needing to refer to these references, although I hope that they will want to. Final year undergraduate and graduate students are encouraged to read more deeply: one only gets to grips with the science of evolution by seeing in detail how nature has been made to yield its secrets.

Timing convention. There are several abbreviations for historic time in the literature. Those used here are By and Bya, My and Mya, and Ky and Kya for billions, millions, and thousand years, respectively, with the additional 'a' meaning 'ago.'

Jonathan Bard
Oxford, 2016

Acknowledgments

I thank Susan Offner, Stephen Oppenheimer, Jordi Paps, Sebastian Shimeld, and Steven Zimmer for their helpful comments on individual chapters, and the many authors who gave me permission to use their photographs and drawings. I am particularly grateful to Vernon French who read the whole draft and saved me from many errors and infelicities. Many of the ideas in this book were developed in the course of a seminar on systems biology that ran for several years in Balliol College, Oxford, and I would like to express my appreciation to the other organisers, Denis Noble, Tom Melham, and Eric Werner, together with the regular attendees, for all the discussions that we had. I also thank Liz Owen, my editor, Denise Schanck, my publisher, and the production team at Garland for their help and support. Finally, I am indebted to Gillian Morriss-Kay, my wife, who read the whole manuscript from the harsh perspective of someone who has edited a scientific journal for a decade. It is a fortunate author of a biological text whose wife not only has a wide and deep knowledge about anatomy and zoology, but is happy to share it and discuss difficult points, even at breakfast. Finally, any book in an area as broad as evolution will inevitably include errors, and they are all my own responsibility.

The author and publisher would like to thank external advisers and reviewers for their suggestions and advice in preparing the text and figures.

Ting-Fung Chan, The Chinese University of Hong Kong, Hong Kong; Olivier Hanotte, The University of Nottingham, UK; Janet Hoole, Keele University, UK; A. Edward Salgado, Christian Brothers University, USA; Erik P. Scully, Towson University, USA.

Contents

Contents II: The History of Life

The focus of this book is on the science underlying our understanding of evolution, but in doing this, it of course covers the history of life over the last four billion years. As this information is distributed across the chapters in which the various aspects naturally emerge, the following table allows the interested reader to follow the history sequentially.

SECTION 1
AN INTRODUCTION TO EVOLUTION

This first section aims to provide sufficient background for the reader to understand the science behind the evidence for evolutionary change (Section 2) and the work that has unraveled the mechanisms by which that change occurs (Section 3). Many readers will know some of this material and should feel free to browse through these early chapters or just use them for reference.

Chapter 1 is an introduction that attempts to do no more than set the context for approaching evolution and the perspective from which this book is written. A particular focus here is systems biology; this is a new area of biology, one of whose aims is to explore the properties and roles of the complex protein networks that drive much of development in the embryo and physiology in the adult. It is important in any study of evolution because variation and change usually derive from mutations that affect the properties of these networks. More information on systems biology is given in Appendix 1.

Some basic history (Chapter 2) is required to provide the context for understanding how our current knowledge of evolution was achieved, why different problems were worth working on at different times, and how one area of the subject built on others. This chapter is very short, and interested readers can find more detail on this fascinating subject in Appendix 2.

The key problem in evolution, as Darwin realized, is how new species evolve from existing ones. Exploring this requires appreciating how species are recognized and how they are organized. Chapter 3 discusses today's current diverse range of organisms and how they are taxonomically grouped on the basis of their properties. In a sense, this chapter asks the questions that the rest of the book has to answer.

CHAPTER 1
APPROACHING EVOLUTION

This first short chapter discusses the approach that this book takes to investigating the detail of evolution.

Our planet teems with life. In the sky, on the land, in the sea, and in between, there are some millions of species of plants, animals, fungi, bacteria, and viruses, and almost all thrive as they develop and reproduce using energy directly or indirectly derived from the sun. This is possible because each species is appropriately adapted to the environment in which it lives: its ecosystem provides it with food, and it, in turn, provides nourishment and a context for other organisms, sometimes while it lives and always when it dies. Unicellular organisms, be they prokaryotic or eukaryotic, can reproduce by simple division, whereas multicellular organisms mainly reproduce by mating – no multicellular individual survives or reproduces in the long term unless it is part of a population.

The key purpose of the science of evolution is to make sense of this richness, to understand how it arose, to track the diversity of life backwards in time to its origins, and to study the mechanisms by which that diversity evolved and continues to evolve, always ensuring that species adapt to their environment. The data required for this enterprise come from across the whole of biology, and include information from contemporary and fossilized organisms, molecular biology and cytogenetics, comparative embryology, comparative genomics, and even from the comparison of the two gene sequences in a particular pair of chromosomes in a diploid individual. It is worth noting that, other than this last example, evidence from an individual organism or even the most wonderful of fossils, rarely represents more than just an interesting fact in this context. The science of evolution mainly deals with interpreting comparisons of data from and relationships among species.

Such comparisons are, however, just the start: understanding evolutionary change requires a much broader approach because all organisms live and evolve in a rich and complex biological and physical environment, and have always done so. Whether a species thrives or dies out depends on its interactions with that environment. What we know about that species, in particular, and evolutionary biology, in general, depends on ecology and even the climate as much as on genomics.

We would, of course, love to study evolution directly, but we cannot: eukaryotic speciation generally requires perhaps a million generations and, for larger species, even more millions of years for mutation to generate and for the environment to select a new population that is recognized as a novel and distinct species. Heritable change to an organism's phenotype is usually far too slow to be appreciated in the here and now; therefore, research work has to be indirect.

It is sometimes claimed that, unlike physics, biology does not have any profound theories. Physics, however, requires four forces and many, many laws about how they operate to explain the world of particles and inanimate objects. What is unusual about biology, compared with physics and, indeed, all other areas of science, is that it has a single underlying concept that integrates so much of its subject matter. This concept is that novel species evolve from existing ones through environmental or natural selection acting on heritable variants, with the result being that new species form through descent with modification - this is evolution. Darwin had this deep insight into how evolutionary change happens around the 1840s but only published it some 15 years later in his masterwork *On the Origin of Species*, an 1859 classic that is still worth reading. Darwin's single insight provides the framework within which it is possible to discuss the whole of life, from its simple beginnings to now.

A systems biology perspective

This is a book whose main underlying focus is on speciation, or the mechanisms by which new species form from existing ones. As Darwin realized more than 150 years ago, this is the key problem in evolution and this is because the reproductive isolation of a new species is a one-way route to further diversification. At the macroscopic level, a great deal is known about speciation, but the molecular details are still opaque because it is a long way from a mutation in the DNA that affects the structure or expression details of a protein to a downstream change in anatomy or behavior that eventually helps to produce reproductive isolation. While a fair amount is understood about the mechanisms that form normal tissues, knowledge of the effects of mutations that have minor effects on these mechanisms, rather than just blocking them, is very limited. This is because a minor change in a protein sequence rarely has an obvious advantageous effect on the rich and complex phenotype of an organism.

Perhaps the major advances in biological thinking over the last decade have been in the area of systems biology (Noble, 2008), a subject whose roots come from the realization that many proteins do not work alone but co-operate within networks that produce a single functional output, one that could rarely have been predicted on the basis of the role of any of its individual proteins. This realization has changed how we think about the relationship between the genotype and the phenotype, particularly with respect to the effects of mutation (Capra & Luisi, 2014). Systems biologists are also grappling with the nature of biological complexity, particularly through the ways in which events at different levels interact through feedback and hence how causality is distributed. The analysis of protein networks reflects a narrow view of systems biology, while the study of complexity across levels and systems requires a broader approach (Bard, 2013a). Some of the basic ideas of systems biology and its formal language are discussed in Appendix 1.

The most obvious area of evolution where systems biology in the narrow sense matters is in evo-devo, the area that investigates molecular similarities in the development of very different organisms. As far back as the 1970s, it had become clear that proteins such as cytochrome C, which has similar functions in very different organisms, had amino acid sequences that could be seen as deriving from a common ancestor (they were homologs). By the 1980s, many other examples had been discovered where homologous proteins played the same key role in the development of very different organisms. A good example is the Pax6 protein: this transcription factor lies at the base of the hierarchy of mechanisms responsible for eye development across the phyla to the extent that the Pax6 protein, whose activity initiates the development of the camera eye in the mouse, can substitute for its homolog in *Drosophila* and initiate a compound eye in the fly (Gehring, 1996).

More recently, it has become apparent that such homologies extend upwards to the next level of molecular organization, that of protein networks, where very similar networks have much the same roles in very different organisms (Gilbert, 2014). Across development, homologous pathways and networks often control signaling, patterning, differentiation, cell division, and cell death, as well as morphogenesis in different phyla (Chapters 10 & 11). Knowledge of the programmed cell death (apoptosis) pathway in mammals, for example, derived from identifying homologs of proteins first discovered in the developing nematode worm, *Caenorhabditis elegans*. Such networks are important in two evolutionary contexts: first, the fact that there are such homologies across the phyla is strong evidence for very different species sharing a last common ancestor; second, anatomical variation, the basic driver of evolutionary change, comes from mutations that modulate the output of these networks, particularly those that control patterning and growth. Systems biology, in the narrow sense of understanding how protein networks generate their outputs, is at the very heart of evolutionary thinking.

While the effect of a mutation on the function of an individual protein can often be straightforward and even measurable *in vitro*, the effect of that mutated protein on the output of a complex network in which it participates is usually impossible to predict, unless, of course, the mutation causes that network to fail. It is even harder to predict the downstream effect of the original mutation on that organism's phenotype, other than that it may produce an anatomical or physiological variant in the organism that should be heritable. Such phenotypic change is, however, only the beginning of speciation. Whether that change is worth maintaining depends on selection and this, in turn, depends on the local environment, an umbrella term that covers the effects of population pressure and competition from the original and other organisms, the climate, food availability, and, indeed, anything that affects reproductive success. Not only are the events within an individual level complicated, but there are also interactions between levels: population numbers depend on climate and food, while phenotypic variants need to be able to reproduce with other members of their species.

This sense of levels and the interactions between them is captured in the theory of evolutionary population genetics (the modern evolutionary synthesis), which was developed around 1925–45. This theory describes how the frequencies of subpopulations of variants within a population change due to to natural selection, genetic drift, and other factors (Chapter 14). The most recent incarnation of population genetics, known as coalescent theory, also incorporates variation in known gene sequences, and so includes data and concepts from levels extending from the genome to populations (Chapter 8). The models, though complex, make predictions about the sequence possessed by the most recent common ancestor of the population and how population numbers have varied since this organism lived. Coalescent theory captures a great amount of complexity and is a prime example of broad systems biology, although it is not usually viewed in this context.

Complexity is ubiquitous across evolutionary studies. Although the concept of descent with modification is straightforward in principle, it is not in practice. While the fitness of a variant in a given environment can sometimes be measured as a numerical constant, rarely, if ever, can that constant be calculated or predicted. This is because the extent to which one variant organism leaves more or fewer offspring in a particular environment than its peers depends on many factors, some of which may interact. It is premature to expect that systems biology can yet solve the difficult problems of the science of evolution, but, without some means of dealing with the intrinsic complexity of variation and selection, they will remain unsolved.

Nevertheless, and in spite of the fact that systems biology is a new and incomplete subject, it is to be hoped that its ideas will help illuminate some of the questions in evolutionary science that remain unanswered. A medieval rabbi remarked in a different context that "it is not your task to complete the work, but neither are you free to avoid it," and so it is with evolution: every generation adds their perspective to a great tradition but there will always be further evolutionary problems to solve.

Further reading

Bard J (2013) Systems biology - the broader perspective. *Cells* 19:413–431.

Darwin C (1859) On the Origin of Species by Means of Natural Selection, or the Preservation of Favoured Races in the Struggle for Life. http://darwin-online.org.uk/converted/pdf/1859_Origin_F373.pdf

Gilbert SF (2014) Developmental Biology, 10th ed. Sinauer Press.

Noble D (2008) The Music of Life: Biology Beyond Genes. Oxford University Press.

CHAPTER 2
A POTTED HISTORY OF EVOLUTIONARY SCIENCE

This chapter briefly summarizes the key steps and the major figures over the last 200 or so years who were responsible for the modern understanding of evolution. It provides the context for appreciating the many strands that make up our contemporary understanding of the subject. A much fuller history of this fascinating story, together with references, is given in Appendix 2.

The pre-Darwinian era

Until the end of the eighteenth century, the general belief across Europe was that the Bible provided a true account of the origins of life, although a few people had had their doubts. The first serious step that led the way to modern views of evolution came from Charles Bonnet, a Swiss biologist, who originally trained and practiced as a lawyer. Around 1770, he suggested that there had been progress up the great chain of being (or the ladder of life) from basal molds through plants, insects, worms, shellfish on to fish, birds, and quadrupeds to the ultimate perfection of man. Other naturalists such as Erasmus Darwin also put forward ideas that life had evolved from primitive origins, but all their arguments were based on logic rather than evidence.

There were actually quite good reasons for believing the Bible then: the biological world seemed constant and static, animals were perfectly suited to their environment, one species could not breed with another, and there was little or no evidence in favor of any other account. The Bible story was credible because there was no alternative rational explanation (albeit that the Bible gives different creation stories in Genesis 1 and 2). The one oddity was the presence of fossils, but these were mainly viewed as organisms either lost in or displaced by Noah's flood.

The first evolutionary scientist in any modern sense was Jean-Baptiste Lamarck (1744–1829): he realized that, on the basis of tissue geometry, annelid and parasitic worms could not be treated as being on the same step of the ladder, as Bonnet had suggested: the anatomy of the former was far more complicated than that of the latter, and, in particular, had a coelom or body cavity. In 1809, he suggested that evolution occurred, not by ascending a ladder, but through successive descending branches from a common root. Lamarck was thus the first person to put forward the idea of descent with modification and did so on the basis of scientific evidence. He did not, of course, know how change happened or was inherited, but suggested that organisms had two intrinsic abilities which were, in essence, to become more complicated and to adapt, with the resultant changes being inherited. These ideas were accepted for the best part of the next century by most biologists, including Darwin.

Lamarck's dreadful reputation today derives from the obituary written by George Cuvier, his fellow professor at the *Muséum National d'Histoire Naturelle* in Paris, who was the major anatomist and paleontologist of his day. Cuvier did not believe in evolution (or transmutation, the term then used) because, as he interpreted the fossil evidence seen in the different rock layers, the data pointed to successive extinctions followed by new bouts of creation. He used his grossly unfair and misleading obituary in 1829 to say that Lamarck was no more than a foolish philosopher, and to destroy Lamarck's work and reputation.

In spite of Cuvier's attempts, the idea that life had evolved rather than had been created was becoming widely discussed in scientific circles in the 1830s. One obvious reason for this was the increasing number of vertebrate fossils being discovered that obviously bore little relationship to modern animals. While it was not possible to date them, it was clear as early as the 1820s that the deeper, and hence older, were rock strata, the more primitive were the fossils that they contained. The biblical stories thus became rather less credible than they had been. Paleontology and geology soon became sister subjects and the insights that derived from them were a driving force for establishing that evolution could be responsible for the diversity of modern life. What was, of course, missing was any idea of how change could happen.

The Darwinian era

Around 1837, Charles Darwin (1809–1882) came to realize that Thomas Malthus's idea, that exponential growth in human populations would be held in check by food availability, also applied to the rest of the living world: natural competition among variants would lead to the success of the fittest and this was the key to understanding how evolution occurred. Darwin then spent more than 20 years assembling data on variation and selection, but did not publish his work because he did not want a public argument (his health was poor) or to upset his wife, a devout Christian. However, his hand was forced by Alfred Russel Wallace (1823–1913), who had had similar ideas while working in the Malayan jungle on what is now called bio-geography, an area of study that he invented. In 1858, Wallace set down his ideas in an essay on how species form and sent it to Darwin to pass on for publication. After much soul searching and discussion with friends, Darwin then abstracted an older, unpublished paper of his own, and in the same year both were published in the *Journal of the Linnean Society*. Darwin then rapidly wrote *On The Origin of Species* (1859), which looks at how new species form, a shrewd way of discussing evolution without having to be too explicit. The book itself focuses on a single question: What are the implications of the variation that any observer can see within a species?

His answer, buttressed by a wide range of examples from across the then-known biological spectrum, starts with the fact that, left to themselves, populations grow exponentially but such growth is constrained by pressures from the environment such as food availability. Darwin then saw new species as arising from the variants that naturally formed within the populations and suggested that these would disappear or flourish even to the extent of replacing their parent population as selection pressures from that environment dictated (the *struggle for existence*). Such Malthusian competition, now interpreted as natural selection leading to survival of the fittest, would, if the new variant flourished, result in descent with modification, which is, of course, evolution. The balance of variation and selection, the two ideas that underpin the book, are the key to understanding the origin of new species. In 1871, Darwin produced his second book on evolution, *The Descent of Man, and Selection in Relation to Sex*, in which he discussed sexual selection and human evolution. A year later, he published *The Expression of the Emotions in Man and Animals* in which he sought to show that what might seem special to man had, in fact, a basis in animal behavior.

Darwin's wonderful work left several questions unanswered, the most important being about the mechanisms that created variants and their inheritance. There were also questions about the relative importance of hard inheritance, the effect of intrinsic variation, and of soft inheritance (often called Lamarckian inheritance), due to environmentally determined change that became heritable, and whether variation proceeded by slow continuous change or substantial discontinuous jumps (saltations). The basic question of inheritance was solved in the 1880s, mainly by August Weismann (1889), who proposed the theory of the continuity of the germplasm: this stated that inheritance is carried by sperm and eggs in their nuclei and that the key factor in generating variation is crossing over of chromosomes during meiosis. His view was that soft inheritance was impossible because there was no mechanism by which it could work, and even today there is little evidence to support this possibility. Darwinian evolution combined with Weismannian cell biology formed the basis of neo-Darwinism.

The era of evolutionary genetics

A deeper understanding of variation became possible when Gregor Mendel's work on genetic inheritance, which had been published in 1865 and then forgotten, was rediscovered around 1900. Soon afterwards, Hugo de Vries proposed his mutation theory of evolution. Over the next three decades, the modern evolutionary synthesis was put together by a series of major figures that included Sergei Chetverikov, Ronald Fisher, Theodosius Dobzhansky, John Haldane, Sewall Wright, Julian Huxley, and Ernst Mayr (Chapter 12). This combined neo-Darwinist ideas on evolution with models of population genetics to produce a mathematical theory of how trait and gene frequencies could change under the influence of selection. The strength of the synthesis was that it integrated evolutionary change and population genetics in a formal mathematical framework that was originally based on two key points:

- Within a population, there is slow accumulation of mutations that have small effects and these lead to genetic and phenotypic (trait) variation.
- Natural selection (Malthusian competition) acts on these variants, particularly when a subpopulation in which they are common becomes isolated in some way. If the novel environment of this subpopulation favors these variants and they do well (by producing more fertile offspring, the technical meaning of survival of the fittest), they will come to dominate the population, and continue varying to the extent that they may eventually form a new species, which will not breed with the original one (this, incidentally, is the simplest and best definition of a species).

It was not, however, until the 1940s that Mayr and others realized that the mechanisms by which new species formed was through small groups of organisms becoming reproductively isolated from their parent populations. If this happened in an environment with different selection pressures from those of its parents, then a different set of existing alleles would come to dominate the subpopulation, and that further mutation would eventually lead to reproductive isolation and a new species. In the late 1960s, Motoo Kimura pointed out that genetic drift also played a part in the evolutionary genetics of speciation.

In a population where there are several alleles of a gene, genetic drift, the result of random breeding, plays an important role in their distribution through the population. This realization had two important implications. First, that genetic drift is as important as mutation in creating phenotypic variants, and often more important in the short term as mutation rates are low. Second, in small segregated groups of organisms, the asymmetric gene distributions resulting from genetic drift are key to producing new variants rapidly. It is from these that new species eventually arise.

A further point implicit in this synthesis is that organisms are engaged in a battle of survival and are intrinsically selfish. This idea, summarized in Tennyson's phrase "Nature, red in tooth and claw" from his 1850 poem *In memoriam* had been argued against as early as 1902 by the Russian biologist (and anarchist) Pyotr Kropotkin in his book *Mutual Aid* about how organisms depend on each other to survive and flourish, but it took until the 1960s before a wider view, known as kin selection, which incorporated altruism (acting in a way that was beneficial to others, particularly relatives, at the expense of one's own interests), was accepted and explained within the modern evolutionary synthesis. This synthesis has since been extremely useful in analyzing population changes, in working through how selection can operate, and in seeing how speciation can occur.

These ideas were based on the idea that speciation takes place at a slow and uniform rate, depending on the gradual incorporation into a population of mutations that survived selection. In the early 1970s, however, Niles Eldridge and Stephen Gould suggested that novel speciation did not simply occur through the slow accumulation of small changes in a species (this is phyletic gradualism or anagenesis) but could proceed through periods of stasis alternating with short periods of rapid change, a process they called punctuated equilibrium. This was not to say that change occurred through a single mutation that had a major effect (saltation), but that the weight of accumulated mutation led to a sudden change. As ever, the answer lay in observation and, by this time, the fossil record was very rich. Detailed examination of long-lived families of fossilized organisms, extending from small bryozoans (coral-like sea organisms) to large dinosaurs, revealed examples of punctuated equilibrium that reflected changes in the environment and hence changes in selection pressures. Other organism sequences, however, showed phyletic gradualism. In other words, there turned out to be a spectrum of change dynamics.

Meanwhile, a major change was taking place in ideas on how best to group species for discussions of evolution. Traditionally, this had been on the basis of embedded classes with common or shared properties (Linnaean taxonomy; see Chapter 5 and Appendix 1) so that a species was a member of a family, families were grouped in orders that were grouped in classes, and so on, irrespective of when that species lived. There had, however, been some dissatisfaction with such a hierarchy because it failed to capture evolutionary relationships in any coherent way. In the 1950s, Willi Hennig produced a far better system that he called phylogenetic systematics, and that is now called cladistics. Here, species are linked by the relationship *descends with modification* on the basis of shared and derived characteristics (Chapter 5), and relationships are traced back to last common ancestors. So successful has this approach been in integrating living and extinct organisms that Linnaean taxonomy has had to be modified to include its results (Chapter 7).

The molecular era

These classical ideas were all based on observable traits and behaviors (the *phenotype*), which, of course, could say nothing tangible about their then unknown heritable underpinnings (the *genotype*). The discovery of DNA structure and the molecular apparatus for turning DNA sequences into proteins in the 1950s and 1960s was a landmark in this context, as it was across the whole of biology. It not only demonstrated that the genotype was coded in specific DNA sequences, but also showed that variation was achieved through mutation of those sequences. Work over the last 30 years, in particular, has explained a great deal about what is in the genome, the types of mutation that have occurred, and their evolutionary implications (Chapter 8).

One fruitful area of research here has been the use of computational approaches to analyze similar DNA sequences in very different organisms whose respective proteins turn out to have very similar functions: comparative analysis of the various DNA sequences of a widely distributed gene, such as that coding the cytochrome C protein, can generate a hierarchy of the most likely paths of descent from an original candidate sequence possessed by a last common ancestor on the basis of inherited mutations. The result has been that the coding sequences of a range of homologous proteins in a wide range of organisms can be integrated into an evolutionary hierarchy of descent with mutational modification. Such phylogenetic trees or phylograms not only show how contemporary organisms should be grouped, but also do so with a greater precision than is possible with standard taxonomy and so allow this record to be strengthened (Chapter 8). Computational genetics has also allowed us to analyze early events in evolution for which there are virtually no fossil data such as the likely nature of early prokaryotic organisms, the archaebacteria, and eubacteria, and the ways in which eukaryotic life evolved from them (Chapter 9).

Phylogenetic analysis can now be extended using coalescent theory to model the mutational changes of current DNA variants backwards in time back to a last common ancestral sequence. The technique for doing this takes, for example, the two sequences of the same gene on an individual's pair of chromosomes and conducts a phylogenetic analysis within the context of a model of population genetics and a likely mutation rate. The analysis produces estimates of the original sequence, how many generations back it was present, and the size of the population over time since that original sequence started to vary (Chapters 12 & 13). Such approaches have highlighted the various population bottlenecks that occurred as early groups of *Homo sapiens* colonized the world (Chapter 16).

Further information came from the application of molecular genetics to investigating an old insight, that the generation of novel or changed anatomical features in adults reflects the effect of changes taking place during embryogenesis (it seems that this obvious point was first pointed out to Darwin by Thomas Huxley). There had always been some interest in similarities in the development of different embryos, but it was not until the 1980s that it became possible to compare the activity of homologous proteins in very different organisms. This work opened up the field now known as evo-devo (Chapters 10 & 11), and it soon became clear that very similar genes, which clearly had an early common ancestor and so were homologous, had very similar roles in a wide range of organisms. One well-known example here is the set of *Hox* genes that organize body patterns in animals as different as mouse and *Drosophila*. Evo-devo has provided some of the most powerful evidence on how evolutionary change occurs.

The most recent insights here have come from the realization that much of embryonic development results from the outputs of the complex protein networks that drive growth, differentiation, and movement. It has now been shown that there is a great deal of protein homology in the networks that drive similar events in very different organisms and this is clarifying the evolutionary relationships among very different organisms (Chapter 11).

abilities for detailed speciation in living organisms, and on niche adaptation for exploring the early stages of separation (Coyne & Orr, 2004). What does need to be remembered in cases where the interbreeding test is not possible is that separate species can only be viewed as being closely related if their differences clearly derive from shared features that were possessed by a recent common ancestor (Chapter 5).

The numbers of species today

The numbers of living species so far identified is very large: the catalog of life database (www.catalogueoflife.org) includes information on about 1.1 million living species of which the very great majority (~920,000) are arthropods. These databases are still being added to and there are certainly many more species that are known informally but have yet to be officially cataloged. What is clear, however, is that we are a long way off having a full list of species.

The world-wide investigations by naturalists over the last few centuries have certainly led to the identification of the very great majority of the larger species (**Table 3.1**), and many of the smaller ones, but, as one goes down in scale to the small invertebrates, tiny plants, and fungi that are barely visible to the naked eye, there will inevitably be many yet to be discovered. A recent estimate of biodiversity, or the number of eukaryotic species on earth, is 8.7 ± 1.3 million; if this is correct, less than 20% of the distinct organisms with which we share the planet have been identified (Mora et al., 2011). One reason why this number is so large is that we have little idea of the number of parasites that are specific to each species. There is a standard joke that goes as follows: "Which phylum has the most species?" The answer is the nematodes (round worms). This is because every other species has a nematode parasite, although most have yet to be cataloged (hence the wide range seen in Table 3.1). There are, however, many, many other species-specific parasitic organisms that remain unknown and their identification will increase the total. Another reason is that we are unlikely to have cataloged all the single-celled and other very small eukaryotic organisms; indeed, it is unlikely that we ever will: there will always be odd niches that we will have neither the resources nor the imagination to investigate. Furthermore, evolution has not ceased and groups of existing variants are always making the transition to becoming new species.

Such reasons also help explain the unexpectedly low number of prokaryotic species so far identified: the PubMed taxonomy database includes only ~18,000 bacterial and ~800 archaeal species. The prokaryotic populations of most habitats and organisms, other than

Table 3.1: Numbers of cataloged species in various groups of eukaryotes (numbers are approximate because different sources give different numbers)

Species	Number
Monocotyledonous plants	~5 K
Flowering plants	~300 K
Nematode worms	80–1000 K
Beetles	~450 K
Fishes	~28 K
Birds	~10 K
Mammals	~5 K

humans, have not been systematically investigated, so there are bound to be many more prokaryotic species to be discovered. It is, however, harder to define a prokaryotic than a eukaryotic species. Because bacteria divide asexually, a morphological rather than a breeding criterion is normally used, but there is a limited number of possible shapes for a single-cell organism, so individual species in each morphological group have to be distinguished partly by their DNA differences and partly on the basis of the hosts to which they are restricted (Wayne, 1988; Oren & Garrity, 2014). We may have to accept that the differences within groups of bacteria reflect a continuum rather than a set of distinct species (note that a similar problem occurs in eukaryotes in what are known as ring species; see Chapter 15). As to the viruses, their effects are usually organism-specific and can only be identified if they lead to some disease. As we care little about the diseases that affect the great majority of organisms, the ~3400 virus species cataloged in 2014 is bound to be a gross underestimate of the full number.

Scale: from microns to meters

One of the most remarkable features of the biosphere is the range of sizes of multicellular organisms. Living vertebrates, for example, extend from *Paedophryne amauensis*, a small frog less than 8 mm long, to the blue whale, *Balaenoptera musculus*, which is up to 30 m long (the volume ratio is about $1:10^{10}$). Flowering plants extend from less than 1 mm (*Wolffia globosa*, a duckweed that weighs about 150 mg) to almost 100 m in height (for example, the giant sequoia, *Sequoiadendron giganteum*, and the fig, *Ficus benghalensis*, which both weigh well over 1000 tons). The length ratio of the two trees to *Wolffia* is about $1:10^7$ and the weight ratio about $1:10^{20}$. Going down to the level of a bacterium that is about 2×10^{-3} mm in diameter and weighs about 10^{-11} g or a single-cell eukaryotic yeast, which can be as little as 3×10^{-3} mm across and weighs about 3×10^{-11} g, it is clear that the scale of life is almost inconceivably large.

When one compares the smallest with the largest, the differences are more than obvious, but the similarities should not be ignored. The organ systems in the frog and the whale are remarkably similar, and the range of cell types in the two is thus much the same. This is, of course, because their last common ancestor was a late amphibian with the same set of organ systems. The most important types of cellular organization in vertebrates are epithelia, which cover external surfaces and line internal tubes and are essentially two-dimensional sheets; mesenchyme, which forms most tissue three-dimensional masses (such as dermis, muscles, bones, tendons, cartilage); neurons and their support cells (such as glia and astrocytes); and the many subtypes of blood cells. Other, more specialized cells are subtypes of these main classes. At this level of discussion, the main difference between vertebrates and invertebrates is that the latter have a more restricted set of mesenchymal cells: they lack bones, fascia, and a sophisticated set of blood-cell types.

As to the plants, the similarities between *W. globosa* and the fig tree are particularly apparent in their respective reproductive systems as both have the flowers, pistils, stamens, and seeds characteristic of flowering plants. The differences are in the supporting tissues. Comparison of the fig or sequoia with a small woody shrub called *Salix herbacea*, which is only a few centimeters high, shows that it is hard to discover any differences in the anatomical entities or cell types that make up the plant – the differences are merely of scale and minor morphological detail. Naturally, our eye looks for differences, and inevitably focuses on the exterior rather than on the organism as a whole. Most of an organism's tissues are on the inside and, if scale is set aside, are usually similar to most other organisms in the same phylum.

Variation within a species

If one excludes the minority of eukaryotic species that reproduce by parthenogenesis (asexual reproduction, for example water fleas such as *Daphnia* and many single-cell eukaryotes) or are essentially hermaphroditic (organisms carrying both male and female reproductive cells, a group that includes many examples of plants, slugs, mollusks, and nematodes), reproduction involves mating, and this implies the presence of members of the opposite sex. In other words, when one speaks of a species, one is implicitly talking about a population, and the individual members of that population are never identical – variation is normal. Even identical twins, which start off the same and with the same DNA, soon develop small differences. This is owing to the fact that, for all the DNA error-correcting molecular machinery in cells, a few such somatic mutations are incorporated into the genome of the daughters every time a cell divides. However, such mutations are not inherited as they are not in the germline.

The major, immediate sources of variation within a population come from the essentially random combinations of genes that occur during reproduction (genetic drift). This has two sources: first, the homologous combination between pairs of chromosome that occurs during meiosis arbitrarily assorts the genes, after allowing for linkage effects; second, mating brings together two arbitrarily chosen sets of genes. Any additional copying errors that occur during meiosis or development just add a further small amount of randomness, albeit that they are inherited, together with any future mutations that appear in the next generation of sperm and eggs.

Such resulting phenotypic variation within a population reflects the effects of the mutations that make minor changes to protein function and tend to have minimal effects on individuals and their reproductive abilities. Cases where a heritable or germline mutation has a more dramatic effect fall into two classes: those whose effects are deleterious to the individual are likely to be lost, particularly if they generate congenital abnormalities; those that enhance the ability of the organism to thrive and reproduce in a particular environment will be kept and may slowly spread through the population. They may even be candidate genes for facilitating novel speciation, although these are just about impossible to identify.

Within any population there is thus a broad spectrum of unique genotypes with the phenotype of each individual, other than of very young identical twins, being different. If we just look at humans, the major areas where we immediately notice the distinctiveness of an individual are in overall size, pigmentation patterns (skin, hair), and minor differences in the relative size of particular features, notably in the face; in terms of the various anatomical bits and pieces, all humans are essentially identical. The molecular basis for minor differences in phenotype between individuals is discussed in Chapter 11.

The anatomical differences between species

Before considering how organisms are classified into the various taxonomic levels on the basis of their anatomy, it is worth taking a brief look at the origin of the differences among them. In taxonomies, more basic differences are reflected in the criteria used for high-level taxa, such as phyla and classes, and smaller differences for low-level ones, such as species and subspecies. Nevertheless, the grouping of cetaceans and ungulates shows that taxonomies can sometimes have an odd feel about them in terms of the underlying biology. It is therefore necessary to take a slightly deeper look at how the differences across groups of similar organisms arise and why the hierarchies are less arbitrary than they sometimes seem.

The fundamental point here is that differences among organisms reflect the different ways that embryos and larvae develop and so produce the adults; very few distinguishing features reflect events taking place during the growth phase that follows morphological development in the embryo and larva. The key insight that helps in making sense of these events came from von Baer almost 200 years ago (see Chapter 10 for details), and seems obvious today. This was that basic features, such as the body plan, emerge early in embryonic development, while more minor interspecies differences appear much later, building on what is already there. Although embryological stages are not precisely linked to taxonomic classifications, they provide the basis for them as early events reflect higher taxonomic levels and later events, lower ones (Davidson & Erwin, 2009).

The key distinctions between high-level taxa lie in the basic body patterns, and reflect distinct patterning events in the early embryo. These define the ~60 eukaryotic phyla; examples are mollusks (an external mantle and a radula for eating), arthropods (invertebrates with a segmented body, a chitin exoskeleton, and jointed limbs), and gymnosperms (seeds develop from leaves or leaf derivatives such as cones). Within a phylum, the next stages of development of most members are often similar, with interspecies differences appearing later. If one considers the vertebrate subphylum: all go through a series of early stages that lead to an elongated bilateral embryo with a notochord, a rod just ventral to the neural tube, and a segmented mesodermal body musculature. The latter separate into those that do have a proper head and those that do not (these are the lancelets which lack neural crest cells - Chapter 10).

Those with heads then develop backbones, their shared feature, and go on to produce paired lateral outgrowths. If these become fins, the animals are fish, but if these outgrowths develop into limbs, the result is usually a land vertebrate, with subsequent developmental features distinguishing amphibia, reptiles, birds, and mammals. In each of these taxa, subsequent changes in, say, head morphology produce further subdivisions of classes and families. Later in development, minor differences appear that distinguish the various species within a family. Typical areas where such terminal differences occur include size, tooth morphology, and the details of skin patterns.

The taxon of Eutherian (placental) mammals illustrates this. The essential disparities among them are not particularly impressive in the greater scheme of things: they all have, to a close approximation, the same body form and the same cell types. Their primary differences reflect their evolutionary adaptations to different environments. Comparisons between mouse and man show that the obvious differences are in size, hair coverage, a tail, and brain complexity, but these are as nothing compared with the internal similarities. Both have very similar cardiovascular, liver, reproductive, sensory, alimentary (gut), neuronal, and other organs systems, and it is very hard to tell under the microscope whether a cell is from a mouse or a man. This is, of course, why the mouse is such a good model system for man. This is not to underplay the differences but to show where they lie and what they are. In considering lines of evolutionary descent, it turns out that similarities are just as important as differences.

It has already been noted that all multicellular organisms are made of a very similar set of cell types that include epithelial cells that form sheets and tubes, neurons, muscle cells, pigment cells, reproductive cells, and the like. Similarly, there is a short list of cells for plants that includes support cells, transporting cells, storage cells, and reproductive cells.

What is wonderful about nature is that small permutations on a limited number of themes can lead to such a wonderful range of variants.

Taxonomy and the grouping of species

It is immediately obvious that the diverse species of life today can be grouped into families or taxa (a taxon is a collection of organisms grouped on the basis of shared characteristics and can include one or more levels). Similar taxa can be grouped into broader taxa on the basis of shared properties, and so on upwards (**Box 3.1**). Membership of a particular taxon at a particular level is defined by the possession of a set of properties, with higher ranks having a smaller number of defining properties than lower ones and so including more organisms. Thus, for example, the taxon of marsupial mammals sustains the later development of their early-born offspring in a pouch, the taxon of placental mammals sustains the fetus to birth through a placenta, and the taxon of monotremes produces eggs whose yolk provides food for their embryos. These three taxa are members of a higher-level taxon, the mammalia, whose membership is defined by the possession of milk-producing mammary glands.

A classification based on such similarities is known as a taxonomy, and the essential feature of the current taxonomy is a graded hierarchy of classes whose highest level is generally considered to have three members: eukaryotes and the two prokaryotic taxa, eubacteria and archaebacteria. Below these are kingdoms and these reflect the expansion of the various classes of eukaryotes, although the classification of single-celled eukaryotes is still contentious. A level lower are the phyla, a taxon based on body plan. Within each phylum, there are lower ranks of taxa that include class, order, family, genus, species, and subspecies, together with other intervening levels that are sometimes inserted because they have been deemed convenient (see Box 3.1). Although such classifications go back to Aristotle, the credit for modern approaches to the subject is given to Linnaeus through his book *Systema Naturae*, published in 1735. Species are, following Linnaeus's practice, given a double name written in italics: the former is the genus or family name, the latter the species name; thus *Pan paniscus* and *Homo sapiens* are the species names for bonobos and humans, respectively. Subspecies have a third name so that, for example, the Sumatran tiger is *Panthera tigris sumatrae*.

In systems terminology, such a taxonomy is a series of embedded classes linked by the relationship <A> *<is a subclass of>* <B>, normally shortened to <A> *<is a>* <B>, and can be

Box 3.1: Classic taxonomic description of the domestic cat

According to www.ncbi.nlm.nih.gov/Taxonomy, the full taxonomic hierarchy (with *taxonomic* levels) for the domestic cat is:

cellular organisms; Eukaryota (*domain*); Opisthokonta; Metazoa (*kingdom*); Eumetazoa; Bilateria (*subkingdom*); Deuterostomia (*infrakingdom*); Chordata (*phylum*); Craniata; Vertebrata (*subphylum*); Gnathostomata (*infraphylum*); Teleostomi; Euteleostomi; Sarcopterygii; Dipnotetrapodomorpha; Tetrapoda (*superclass*); Amniota; Mammalia (*class*); Theria (*subclass*); Eutheria (*infraclass*); Laurasiatheria (*superorder*); Carnivora (*order*); Feliformia (*suborder*); Felidae (*family*); Felinae (*subfamily*); *Felis* (*genus*); *Felis catus* (*species*).

In more formal, systems terms, this 27-level hierarchy is a series of embedded classes or sets linked by the relationship *<is a subclass of>* (abbreviation: *is a*). Thus, <my cat, Coco> *<is a>* *<Felis catus>* *<is a>* *<Felis>* *<is a>* <Felinae> *<is a>* <Felidae> *<is a>* <Feliformia>, etc.

visualized as a hierarchical tree or as sets of embedded circles (Chapter 5, Appendix 1). It is important to realize that such a hierarchy, even where it is based on the interbreeding abilities of living organisms, says nothing about evolutionary descent. Even where taxa include extinct, fossilized organisms with timings, they still say nothing explicit about evolution. One reason is that the dates associated with a particular fossilized species are rarely precise as to when that species evolved, what were its descendants, and when it became extinct, unless the latter date coincided with a major extinction. The more important reason, however, is that time is not a link that describes evolutionary descent; for this, one needs the details of anatomical change.

Evolutionary hierarchies or cladograms, which will be discussed in subsequent chapters, are different from taxonomic hierarchies for two important reasons. First, the link that connects terms is *<descended with modification from>* rather than *<is a>* (Chapter 5). Second, in evolutionary hierarchies, every link connects, in principle at least, two species of organisms, the parent and the child, although it can be hard to identify the original parent species. In Linnaean classifications, all higher levels reflect groups of organisms, alive or extinct, that share features; such classifications carry only an implication of common descent. For Linnaean classifications even to be consistent with evolutionary hierarchies, further information needs to be added.

It is worth noting that, because a pair of taxa may share some common features, this does not necessarily imply that they are closely related evolutionarily: the common features could just reflect a common adaptation to a particular environment. Thus, for example, although cactuses and desert euphorbias can look very similar because both are succulents with spikes and thick layers of cuticle, they are only distantly related taxonomically. Similarly, sabre teeth have arisen independently many times (Chapter 5). In fact, there are many examples of very similar morphological features that have evolved independently as adaptations to a particular environment rather than through common descent. These are known as homoplasies and reflect what is known as convergent evolution (McGhee, 2011); their existence emphasizes the importance of using a broad range of phenotypic characteristics in making taxon assignments.

It used to be difficult to assign organisms that looked very similar to a particular taxon as it was not always clear whether the similarity derived from convergence or close inheritance. DNA analysis usually solves such problems today, and is widely used to confirm the validity of Linnaean classifications. Comparative DNA sequencing of a particular gene across a group of organisms shows which of its members are closely related and which only distantly. Thus, analysis of the various 18S ribosome RNA genes was important in showing that there was an early and high-level subdivision of the animal kingdom into protostomes and deuterostomes (Peterson & Eernisse, 2001; Giribet, 2008). It was also used to confirm that the subdivision of the protostomes into molting Ecdysozoa and non-molting Lophotrochozoa reflected a high-level, and hence a very old, evolutionary separation. The origins of the phyla are discussed briefly in Chapter 4 and more thoroughly in Chapter 9.

A difficulty with today's standard class taxonomy is that it tries to solve two problems. It aims to group species on the basis of physical similarity and to include within particular subhierarchies species with a common evolutionary ancestor, and the latter requirement can lead to some surprising results. Thus, for example, the ungulates were originally mammals that walked on their toes and whose nails evolved into hooves; it has now become clear on the basis of DNA analysis that not only have hooves evolved more than once, but also that the cetaceans (whales, dolphins, and other sea mammals) have the same common ancestor as modern even-toed ungulates such as pigs, sheep, and camels. Although they were originally considered as unrelated groups, the class taxonomy had to be changed so as to include both within a single class. Today, the cetaceans are now grouped with these even-toed ungulates in a higher-level class, the Cetartiodactyla (Chapter 7). Although there are occasional such changes, the current taxonomy can now be viewed as fairly stable.

Classifying life today

The domains at the top level of the current taxonomy of living organisms are the eubacteria (anuclear prokaryotes with a membrane containing glycerol-ester lipids), the archaebacteria (anuclear prokaryotes with glycerol-ether lipids), and the eukaryotes (with nucleated cells). Below the domains is the taxon of kingdoms, which particularly subdivides the eukaryotes. The two major ones are the Archaeplastida (subkingdoms: red algae, green algae, and plants, all of which have chloroplasts with double membranes), and the opisthokonts (subkingdoms: animals, fungi, and some single-celled eukaryotes, all of whose motile cells such as sperm have a single flagellum); there are also several minor kingdoms, most of which are single-celled eukaryotes (see Chapter 9).

The next important level down is the phylum and it has an important biological meaning: each phylum includes all species possessing the same body plan; the taxon includes 35 animal, six fungal, 11 plant, one moss, 52 bacterial, and five archaeal members. Well-known examples are the arthropods and the flowering plants (note: plant phyla are sometimes call divisions). **Table 3.2** shows 22 of the main animal phyla (the omitted 15 or so minor phyla are protostomes). Four are radiata, which have rotational symmetry and are diploblastic; this means that the embryos have two layers, an external epithelial ectoderm and an inner epithelial gut, with any muscles forming later in development. The rest are bilatera, which have basic mirror-image symmetry and are all triploblasts, having an intervening layer of mesoderm between the ectoderm and endoderm that makes muscles and other cell types not seen in diploblasts. The bilaterial phyla are divided into two major groups on the basis of embryonic development: most are protostomes (the first cavity in the early embryo becomes the mouth and the second the anus) and only three are deuterostomes (the roles of these cavities are reversed, other than in amniotes where the evolution of the large yolk changed the way that the gut formed in fishes and their descendants). The chordates are the major deuterostome taxon and their most important subgroup is the vertebrates.

A phylum was initially defined solely on the basis of morphology, but this can be ambiguous, as the example of the echinoderms shows. Adult echinoderms such as starfish have essentially radial symmetry, normally with five arms, but their larvae, which are triploblastic, initially show bilateral symmetry. Later in their development, these larvae undergo a major and very rapid reorganization as internal rudiments in the larva rapidly restructure it, producing limbs and even rebuilding the gut (Wray, 1997). For the purposes of classification, the larval form with its three layers and bilateral morphology is a better indication of the nature of the organism than the adult morphology with its radial symmetry. Echinoderms are therefore included in the bilateria.

Today, the most successful phylum on the basis of cataloged numbers is that of the arthropods (see Table 3.1); these invertebrates, which possess a chitin exoskeleton and jointed limbs, have flourished from the Cambrian Period onwards. The famous evolutionary biologist J.B.S Haldane is said to have replied to a question from a theologian about what the

Table 3.2: The main living animal phyla: most of those excluded are worm protostomic phyla with few species

Superphylum			Phylum	Common name	Distinguishing characteristic	Species described
Radiata			Cnidaria	Coelenterates	Nematocysts (stinging cells)	~11,000
			Ctenophora	Comb jellies	Eight "comb rows" of fused cilia	~100
			Placozoa		Small (1 mm), flat, and featureless	1
			Porifera	Sponges	Perforated inner wall	5000+
Bilateria	Deuterostomes		Chordata	Chordates	Hollow dorsal nerve cord, notochord, pharyngeal slits, endostyle, postanal tail	~100,000+
			Echinodermata	Echinoderms	Fivefold radial symmetry in living forms, mesodermal calcified spines	7000
			Hemichordata	Acorn worms, pterobranchs	Stomochord in collar, pharyngeal slits	~100
	Proto-stomes	Ecdysozoa animals that shed their exoskeleton	Arthropoda	Arthropods	Chitin exoskeleton	1,134,000+
			Kinorhyncha	Mud dragons	Eleven segments, each with a dorsal plate	~150
			Loricifera	Brush heads	Umbrella-like scale at each end	~122
			Nematoda	Round worms	Round cross section, keratin cuticle	80–1000 K
			Nematomorpha	Horsehair worms	Nonfunctional gut	~320
			Onychophora	Velvet worms	Legs tipped by chitinous claws	~200
			Priapulida	Penis worms	Eversible proboscis	16
			Tardigrada	Water bears	Head and body with four segments	1000+
		Lophotrochozoa Ciliated larvae	Annelida	Segmented worms	Multiple circular segments	17,000+
			Brachiopoda	Lamp shells	Lophophore and pedicle	300–500
			Bryozoa	Moss animals, sea mats	Lophophore, no pedicle, ciliated tentacles	5000
			Entoprocta	Goblet worms	Anus inside ring of cilia	~150
			Mollusca	Mollusks/molluscs	Muscular foot and mantle round	112,000
			Nemertea	Ribbon worms		~1200
			Phoronida	Horseshoe worms	U-shaped gut	11

living world tells us about God by saying, "If one could conclude as to the nature of the Creator from a study of creation, it would appear that God has an inordinate fondness for stars and beetles."

As to the prokaryotes, bacteria are subdivided into about 50 phyla on the basis of morphology; they mainly have outer membranes that contain glycerol-ester lipids and peptidoglycan (Cyanobacteria, an important bacterial phylum, contain cellulose rather than peptidoglycan). The rarer Archaea have cell membranes that contain glycerol-ether lipids and are subdivided into four phyla. Archaea were originally known as extremophiles because they were found in what were considered unfriendly environments, such as those with high salt concentrations (the halophiles), high temperatures (up to 105°C; the thermophiles), high acidity or alkalinity, and those lacking oxygen (the methanogens). More recently, however, and because the search has been widened, Archaea have been found in almost every environment that has been studied. The relationship between Archaea and bacteria is discussed in more detail in Chapter 9.

The taxonomic levels between the phylum, defined by body plan, and the species, defined by the interbreeding criterion, are essentially artificial constructs based on morphological criteria that indirectly reflect the development of the organisms within a phylum; in general, higher taxonomic levels reflect early developmental events and lower ones much later events. The Integrated Taxonomic Information System (www.itis.gov/) shows that the domestic cat hierarchy has 27, the honey bee has 18, and the garden daffodil has 12 levels between kingdom and species. There is little that is contentious in these levels between the phylum and the species, albeit that names occasionally change (for example, the family of legumes used to be called the *Leguminosae* but are now called the *Fabaceae*).

In considering whether to assign different living organisms to distinct species when breeding experiments are impractical, the main criteria are the external characteristics, such as size, skin patterns, feather details, and coloring. An example is Darwin's classic work on how specific finches flourished on individual islands in the Galapagos: it is now clear that a key distinguishing feature is beak morphology and its adaptation to particular island-specific food resources (**Figure 3.1**). As the finches are still on their respective islands, this example of speciation is still accessible for study (Abzhanov et al., 2006; Abzhanov, 2010; Rands et al., 2013). Extinct taxa can, however, only be studied on the basis of fossilized tissues: these mainly reflect the tough exoskeleton of invertebrates, the hard tissues of plants, and the bones and teeth of vertebrates. Paleontologists have to know anatomy in very great detail and be able to recognize quite small differences between sets of fossilized remains if they are to produce convincing arguments that similar tissues represent different species.

The full current taxonomy is now so large and complicated (see Box 3.1) that it can only be held in a database and viewed online, and there are several versions (these include www.itis.gov/ and www.ncbi.nlm.nih.gov/taxonomy/). While the higher levels represent abstract groupings and are inevitably generalizations, they do try to capture a structural, an evolutionary, and, implicitly, a developmental relationship. Each species represents a group of real organisms and the numbers in some of the phyla are very large (see Table 3.1). The key reason for this is that natural variation allows organisms to explore new habitats. Every small region on the planet affords organisms an opportunity to invade it and, once there, not only change themselves and form a new species, but also to change that habitat (for example, it may deplete one source of food while offering itself as another). This may allow further organisms to invade the territory, to adapt and make a living there – and so further speciation occurs!

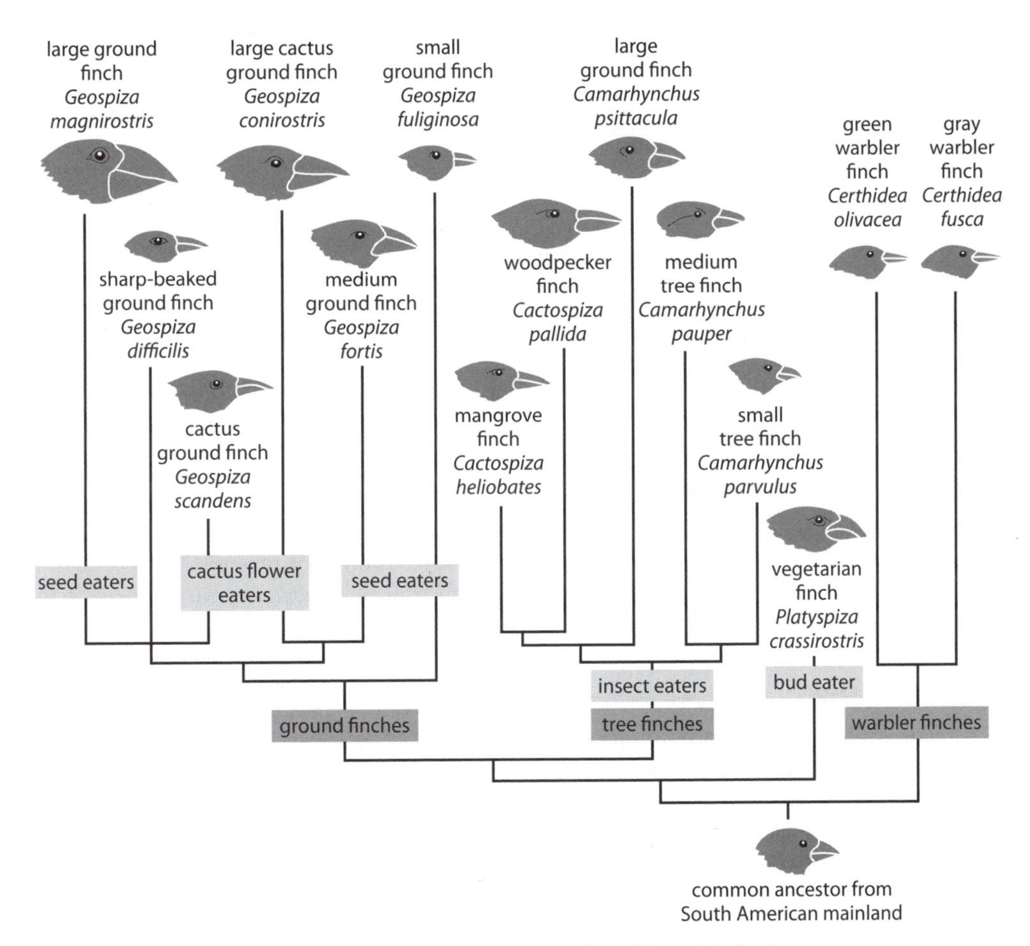

Figure 3.1 A taxonomic representation of Darwin's finches.

The mechanisms of speciation – a summary

The diversity of contemporary species discussed in this chapter raises two important and related problems for the evolutionary biologist. The first concerns the historical origins of this diversity; the second is the mechanism by which change happened. The solution to the first comes from the fossil record, where it exists, and from the analysis of homologous DNA sequences in related organisms, as this shows how they are grouped and hence when they last shared a common ancestor. The second is more difficult because it highlights an apparent contradiction. If a species is defined as a population of organisms that is reproductively isolated, it does not seem possible that it can generate another population that is also reproductively isolated not only from all other species, but also from the parent species.

There is a good solution to the problem of how species originate, but it was not easy to discover. The problem was initially posed by Darwin (1859), and his answer was that natural selection acted on variation, but he knew nothing of the origins of variation nor anything about genes and inheritance. Some of the answers become clear in the 1950s with the exploration of the implications of population genetics, although a full understanding of the actual role of genes awaited the discovery of DNA coding and molecular genetics in the 1950s–1970s, while it is only recently that the ways in which mutation affects the phenotype have become clear. Providing the data to explain the mechanism by which new species form occupies the major part of Section 3 (Chapters 12–15). Here, a summary of the four stages that lead to speciation is given.

Stage 1: Reproductive isolation

The first active step on the way to novel speciation occurs when a small subpopulation becomes reproductively isolated in some new habitat different from that in which its parent population lived, and there are many ways in which this can happen. Because this new population is small, its distribution of genotypes and phenotypes will be slightly different from those of the parent population, owing to genetic drift (random sampling).

Stage 2: Novel phenotypes appear

The effect of this reduced and unbalanced gene profile is the appearance of novel phenotypes, partly due to the Wahlund effect that increases the proportion of recessive phenotypes (Chapter 14), and partly due to the increased frequency of rare alleles because of genetic drift. Over time, these effects together with that of continual mutation leads to further variation, with change being relatively rapid.

Stage 3: Selection

As a result of this different phenotype distribution and because the selection pressures in this new habitat are different from those in the original one, individuals in the new population will do better or worse in the new habitat than those in its parent population might have done. The subpopulation that, as a whole, does worse in the sense that it cannot maintain its numbers will eventually be lost. If, however, there are individuals in the subpopulation with appropriate phenotypes that allow them to produce more fertile offspring, then they will flourish (this is the definition of fitness).

Underpinning these events at the phenotypic level is a different set of events at the genotypic level. The gene pools of the parent and new populations start by being slightly different because of genetic drift. As further mutations and gene distributions accumulate over time, changes to existing traits appear in the phenotype and are available for further positive or negative selection. As a result, the population becomes ever better suited to its new habitat.

At this stage, changes in both the phenotype and genotype distributions in the new population have amplified its original differences from the parent population. Because the two populations are reproductively isolated, there can be no equalizing of genes between them and, as time passes, they become increasingly different. At this stage, however, there is unlikely to be any genetic barrier to interbreeding because the accumulating genetic differences represent minor mutational changes – the two populations are now variants (this is essentially what happens in artificial breeding).

Stage 4: Speciation

Speciation is the process that leads to genetic incompatibility between the two populations, and occurs as a result of three types of further genetic change. The first and second are relatively fast and can overlap; the third takes much longer. The first affects the phenotype and results in individuals in one population becoming unwilling to mate with those in the other (prezygotic isolation). The second derives from mutation that decreases the likelihood of pregnancy and the fertility of any hybrid that results from such a mating through the build-up of genetic incompatibilities.

The third type of change results from the slow accumulation of structural changes to the genome (e.g. chromosomal rearrangements). As these accumulate, they make hybrids increasingly less fertile because their chromosomes become unable to pair up during meiosis. Thus, although horses and donkeys are distinct species with 64 and 62 chromosomes respectively, they are close enough to be able to interbreed. But their offspring, mules and hinnies with 63 chromosomes, are normally infertile, there being a single published exception (Ryder et al., 1985). Eventually, and this may take millions of years, these structural changes accumulate to the extent that fertilization and mitosis fail through nondisjunction of chromosomes. At this stage, the two populations have finally become separate species. As a result, the new species cannot be absorbed back into its parent community; it can only maintain itself and generate further novel species. Thus, speciation represents a one-way mechanism to future diversity.

One reason why evolution is so hard to study is that the genetic changes that lead to speciation are of very low probability and can take thousands, even millions of generations to occur first in groups of individuals and then to spread throughout a population. The next chapter discusses the history of life, the origins of today's diversity, and the glacial speed with which evolutionary change occurs.

Key points

- There are more than a million known species distributed across about 50 phyla (basic body forms) with vertebrates varying in size from less than 1 cm (a frog) to more than 30 m long (the blue whale). Plants likewise show a similar size range.
- There is no definition of a species that is appropriate in all contexts, but, where appropriate, the best is that a species is a group of organisms that cannot breed with other groups.
- Where a definition of species based on breeding cannot be used (for example, for extinct organisms), species are usually distinguished on the basis of their morphology, habitat or DNA profile.
- Groups of organisms can be arranged in hierarchical trees on the basis of anatomical features with broader features having higher levels than more specific ones (a Linnaean taxonomy). This hierarchy roughly reflects the order in which the events of development take place.
- Within any population, variation is normal and mainly derives from gene assortment during meiosis and mating.
- Novel speciation builds on this variation.

Further reading

Gaston KJ & Spicer JI (2004) Biodiversity: An Introduction. Blackwell Science Ltd.

Polaszek A (ed.) (2010) Systema Naturae 250 – The Linnaean Ark. CRC Press.

Ruse M & Travis J (eds) (2009) Evolution: The First Four Billion Years. Harvard University Press.

SECTION 2
THE EVIDENCE FOR EVOLUTION AND THE HISTORY OF LIFE

This section considers the evidence for evolution: this derives from careful comparisons of comparable features in different species, some living and some extinct. The fossil record is the most obvious source of this evidence; it demonstrates the major lines of species radiations and extinctions, particularly over the last ~600 My (Chapter 4). Although this record provides a historical context for evolution, fossil descriptions are only a start: the evidence for evolutionary descent comes from a close comparison of the anatomy of similar fossils to identify features indicating descent with modification. The formal approach for doing this is known as cladistics and will be discussed in Chapter 5. The most complete set of fossil data comes from the anatomical record of vertebrate evolution; this is examined in Chapter 6 (from fish to birds) and Chapter 7 (mammals). Although it can never be shown to be complete and there will never be certainty that all the relevant fossils are known, detailed analysis of the anatomical changes in closely related species has shown how the major transitions occurred, such as those from fish to amphibians and from dinosaurs to birds.

An alternative means of elucidating the evolutionary relationships among living species comes from computational analysis of their DNA sequences. This generates a hierarchy known as a phylogram, one of whose strengths is that it shows the degree of closeness among groups of contemporary species (Chapter 8). As such approaches are based on the existence of last-common-ancestor sequences from which more recent sequences derive through mutation, computational and statistical methods can also be used to model random mutation backwards in time and so suggest the most likely common ancestral sequences from which they derived. A third use of computational methods is in exploring deep time, the period for which there is no fossil record: such analyses identify the extent to which apparently similar genes in organisms from, for example, different phyla are related. This work has uncovered the origin of eukaryotes and the roots of the major groups of organisms; Chapter 9 discusses this hidden history of the first three billion years of life.

The third line of evidence comes from identifying organisms that are very different yet share homologous proteins and protein networks that have very similar functions during embryogenesis; this is the area known as evo-devo (Chapter 11). Understanding the results of this work requires some background knowledge about embryonic development; this is provided in Chapter 10. One of the successes of evo-devo methodology has been to identify a set of proteins, protein networks, and likely phenotypic properties that would have been possessed by the last common ancestor of all animal species with basic bilateral (mirror) symmetry. This organism is known as urbilateria and lived at the base of the Cambrian Period, or perhaps a little earlier.

It is worth noting that these three lines of evidence, from anatomical comparisons, from DNA sequence analysis, and from comparative molecular embryology or evo-devo are completely independent and it is a natural prediction that their conclusions should support rather than contradict one another. Each, of course, probes its own area with different degrees of temporal and species resolution, but, where they overlap, they turn out to be entirely compatible. This is not surprising as each examines a specific facet of a single history, and detailing it is the secondary purpose of this section.

CHAPTER 4 THE FOSSIL RECORD

The clearest evidence about the history of life and the origins of contemporary diversity comes from the fossil record. This chapter first summarizes what the rocks tell us of the early years, from the origins of life more than 3 Bya up to the Cambrian explosion. There are then brief histories of the invertebrates, vertebrates, and plants, each of which aims to show how and when the major subgroups evolved; fuller details of vertebrate evolution are given in Chapters 6 and 7. The dates given here are based on rock ages; readers who require information on how rocks and fossils form and how they are dated should consult Appendix 3.

Both the record of fossils and our ability to date them are now impressive, and improving as new examples and new dating techniques are discovered. Although the preservation of soft tissue from early, small life forms is inevitably limited to a few examples, there is some information about the organisms that lived during the first 3 billion years of life, before the Cambrian explosion when the diversity of animal life evolved from a few types of primitive organism. After that, there is an abundance of fossils, although never as many species as one might like and often too few of any single species to be able to get a sense of its development, size distribution, and sexual dimorphism, or even its dates of appearance and extinction. Nevertheless, our knowledge of fossils and their dates is good enough to describe much of the history of life.

The fossil record of rock-dated organisms is important for two reasons. First, it gives deep insight into the anatomical steps by which life evolved from a primitive bacterium to the rich diversity of today – it is the direct evidence of evolution. Second, it can indicate some of the details of how transitions between very different organisms occurred. Important examples include the transition from fish to amphibian and from dinosaur to bird. The former involved the evolution of two major novel features: limbs and their supporting girdles (the shoulder and pelvic bones) from fins, and the lung apparatus from the esophagus, while the latter just required a lightening of the skeleton and improvements in feather morphology (Chapter 6; Clack, 2012). Only a few such transitions can be touched on here, but the interested reader is particularly referred to the wiki that gives more than 200 fossils to illustrate 24 transitions that led to the evolution of a wide variety of genera, such as insects, land vertebrates, birds, horses, and humans (en.wikipedia.org/wiki/List_of_transitional_fossils).

The fossil record does, of course, have its limitations. The most obvious reason is that fossilization is a random and rare process that depends on dead organisms degrading slowly in conditions where mineralization can preserve their organization before it is lost. This means that organisms lacking a hard internal or external skeleton will rarely have been preserved, and even here most soft tissues will have been degraded before they will have been mineralized. It is only in the most fortuitous of circumstances that small, soft organisms will have been preserved. This means that not much evidence of animal life should be expected to be found before the Cambrian Period (541 Mya) when skeletons first appeared, or of plant life before the mid-Ordovician Period (~470 Mya) when bryophytes were able to live on land because a waxy cuticle had evolved to protect these primitive organisms from desiccation. The fossil record thus represents a series of almost random snapshots, few of which represent early evolution, that mainly focuses on the hard tissues of physically robust organisms, many of which were relatively large.

Even when a fossilized organism can be identified and dated, however, this information does not indicate when it first appeared and when it died out, unless this happened during a major extinction. Furthermore, that fossil alone gives limited insight into exactly which species the organism evolved from unless the details of its anatomy can be analyzed in the context of the anatomy of other similar organisms. Such analysis should at least show which features are primitive and which derived, and Chapters 5–7 will consider how to explore relationships among extinct organisms, particularly vertebrates. The process involves a detailed analysis of comparative vertebrate anatomy and the use of cladistics to identify hard evidence of lines of evolutionary descent.

The discussion of the fossil record given below only covers the higher-level taxa and is inevitably terse and broad-brush. It is not possible to do more than summarize here the main features of the many tens of thousands of fossilized organisms that have been discovered and dated. The purposes of this chapter are to provide a general history of much of life, as shown by the fossil record, and to provide a context for the detailed analyses that follow.

The beginnings

The earth formed some 4.5 Bya. The oldest rocks date to 3.8 Bya, and the first clear evidence of life comes from stromatolites, some of which have been dated back to ~3.5 Bya (Chapter 9). These are fossilized mats or piles of sediment and marine eubacteria, some of which may have been early cyanobacteria capable of photosynthesis; it is noteworthy that a few stromatolites are still present in today's hypersaline lakes and marine lagoons. There are also 20–60-µm microfossils from around that time that may well have included prokaryotes (Knoll, 2015). There is no direct fossil evidence for archaebacteria, the other major class of early prokaryotes, in this early period, but they may have left traces in rocks aged about 2.7 Bya in the form of hydrocarbon biomarkers that derive from the unique lipids such as phytane found in their membranes (Ventura et al., 2007). As archaebacteria and eubacteria share some basic molecular biology and metabolic pathways, they must also share a common ancestor dating to an earlier time (Valas & Bourne, 2011). While the early fossil record is very limited, it seems clear that the activity of cyanobacteria led to the oxygenation of the earth's atmosphere about 2.4 Bya.

The first unicellular eukaryotes seem to have evolved around 2 Bya with the simplest multicellular organisms probably appearing very soon after. A few primitive fossils have been dated back to this time that are almost certainly eukaryotic. They include the enigmatic and disk-like Gabon macrofossils from ~2.1 Bya: one example was about 12 cm wide (El Albani et al., 2014); another had a spirally coiled ribbon-like structure that resembles *Grypania spiralis*, a similar organism that lived about 1.4–1.3 Bya and was up to 50 cm long and 0.2 mm wide (Han & Runnegar, 1992). Other such organisms date to about 1.65 Bya and include phosphatized multicellular organisms from India that resemble today's filamentous and coccoidal cyanobacteria and filamentous eukaryotic algae (Bengtson et al., 2009). By 1.2–1.0 Bya, small multicellular cyst-like organisms that derived from freshwater environments can be seen in phosphatic nodules (Strother et al., 2011).

It seems that cell differentiation evolved around 1.2 Bya. A very early multicellular organism showing evidence of several cell types is *Bangiomorpha pubescens*, a filamentous red alga that lived then and whose descendants still survive (Butterfield, 2000; Javaux, 2007). It had a holdfast for adhering to rocks, with different examples having one of two spore types that are typical of today's *Bangiomorpha*. It thus seems that sexual dimorphism was present in those early times. The origins of diploidy are unknown but probably date to before the earliest eukaryotes since bacteria today can exchange DNA. The advantages of diploidy are obvious: having two copies of a gene not only enables an organism to survive after one copy has become dysfunctional through mutation, but also facilitates variation. The ways in which the apparatus of sexual reproduction evolved do, however, remain unclear, mainly of course, because soft tissue is rarely fossilized.

By 800 Mya, there was clearly a rich, marine biota whose individuals were less than a centimeter in size, unless they had an extended ribbon morphology (Knoll, 2015). The first larger-scale organisms for which there is fossil evidence are from the Ediacaran Period (635–542 Mya, originally called the Vendian Period) and the best-known sites are in Australia. These show a substantial radiation some 575 Mya (the Avalon explosion) that led to an increase in complexity (Martin et al., 2000). Today, their morphologies are mainly seen as compression fossils: these are impressions of the organism's surface made in sand or mud and maintained when they lithified to become sandstone or shale.

Ediacaran fossils such as *Dickinsonia costata* and *Spriggina floundensi* (**Figure 4.1**A and B) are often very beautiful but enigmatic since their exact relationship to almost all current organisms remains unclear. A few Ediacaran organisms may have had radial symmetry and hence were probably diploblastic, with some having a holdfast to glue them to rocks. Some of the later Ediacaran organisms were clearly bilateral and were probably early triploblastic organisms; if so, they were early examples of the great majority of current animals. Examples here are *Kimberella*, which seems to have been a primitive mollusk (Fedonkin & Waggoner, 1997), and *S. floundensi*, which had aspects of trilobite morphology. There are also burrows in

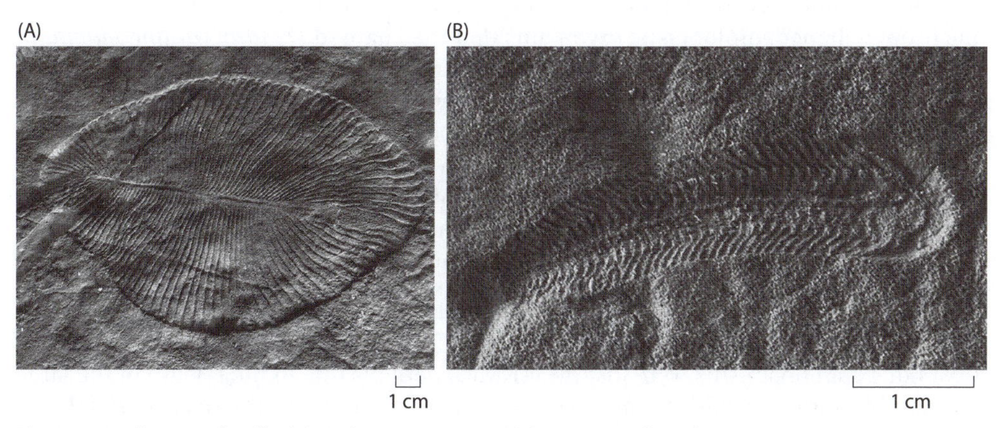

Figure 4.1 Ediacaran fossils. (A) *Dickinsonia costata*; (B) *Spriggina floundensi*.

Figure 4.2 *Dendrogramma enigmata* and *discoides* (*) may be surviving organisms from the Ediacaran Period.

Ediacaran rocks (trace fossils) and these suggest that primitive worms had evolved by the end of the period. While it is generally thought that all Ediacaran life forms were marine, there is evidence that some lived on land (Retallack, 2013). Droser and Gehling (2015) have recently examined the Ediacaran biota and concluded that "aspects of the ecology, such as trace fossils, taphonomy, and morphology, reveal that these fossils show characteristics of modern taxa." In short, modern fauna have their ancestral roots in the Ediacaran organisms.

A recent intriguing observation has suggested that one family at least of these Ediacaran organisms may still survive, hidden in the depths of the ocean. Just et al. (2014) re-examined material from off south-east Australia that had been dredged up in 1986 and fixed in ethanol and formaldehyde. They identified two species of small (2–5 mm) mushroom-shaped, diploblastic organisms that they named *Dendrogramma enigmata* and *D. discoides* (**Figure 4.2**). While these organisms are superficially like Ctenophorans (comb jellies) or Cnidaria (such as jelly fish), they lack many of their cell types and tissues, particularly their gonads, although this may be through immaturity. Just et al. (2014) also point out that the *Dendrogramma* have a strong morphological similarity to Ediacaran organisms and may be descendants of an early ancestor of the Ctenophora and Cnidaria. Not surprisingly, they end their paper with the suggestion that a further search for living examples of *Dendrogramma* is needed.

The Cambrian explosion

Most Ediacaran organisms were lost by 541 Mya, perhaps due to the release of the large amounts of hydrogen sulfide known to have occurred around then (Wille et al., 2008), and this extinction was followed by the Cambrian Period. A great deal is known about the organisms of the Early (or Lower) and Middle Cambrian Periods because the fossil record is extensive and of very high quality showing a great deal of soft-tissue preservation, more than for any other period. The standard explanation for this is that the organisms died in soft sand under anoxic conditions and thus decayed slowly enough for the soft tissues to be fossilized. An additional factor may have been that annelid and other worm phyla only evolved from their primitive Ediacaran ancestors during the later Cambrian; once there in sufficient quantities, they ate and so removed many of the dead soft-bodied animals, initially on the seabed and later on land (Rota-Stabelli et al., 2013), so lessening the chance of post-Cambrian soft tissue being available for fossilization (Brasier, 2010).

The Cambrian record of marine organisms makes clear that the very great majority of the current animal phyla evolved over a relatively short period in evolutionary terms. The Chengjiang Biota in Yunan, China (Lower Cambrian, 525–520 Mya) includes 16 phyla, with the great majority being triploblastic bilateria (organisms with predominantly mirror-image symmetry); amongst these are the first chordates and semi-chordates. The Burgess Shale biota (c. 505 Mya, Middle Cambrian) in Canada includes at least 15 known phyla, one of which seems to have included an early chordate known as *Pikaia gracilens* (**Figure 4.3**A). Most of these phyla represent extant invertebrates such as the mollusks and arthropods, although others were lost. Two examples from phyla that became extinct are *Opabinia*, with five eyes and a small trunk, and *Anomalocaris*, a large predator with a lobed body, two large lobed limbs emerging from the head and a radial mouth (**Figure 4.3**B and D**)**. Other groups that were originally thought to have represented phyla that became extinct are now having their taxonomic status reconsidered. An example here is *Hallucigenia sparsa*, which had a long thin body with limbs and dorsal spikes (**Figure 4.3**C); it is now considered to be a lopopodian worm and an early ancestor of today's velvet worms, members of the Onychophoran phylum (Smith & Ortega-Hernández, 2014).

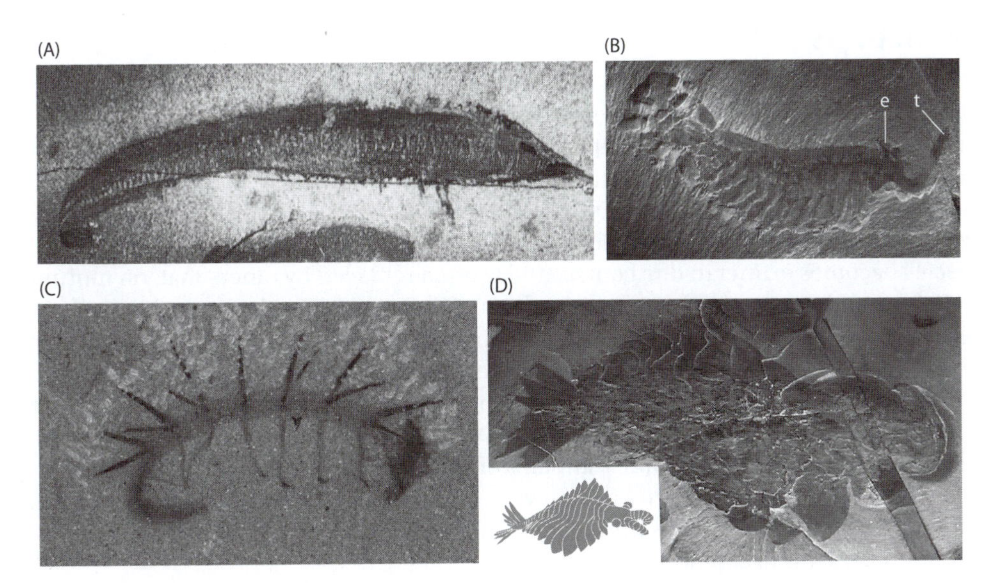

Figure 4.3 Burgess Shale organisms. (A) *Pikaia gracilens*, a chordate several centimeters long. (B) *Opabinia regalis*. This strange invertebrate had five eyes (e) and a trunk (t) or proboscis that terminated in a claw-like appendage. (C) *Hallucigenia*, probably a lopopodian worm. (D) *Anomalocaris*. This large invertebrate (1–2 m long) had flexible lobes for swimming, a disk-like mouth made of plates, compound eyes, and two large limbs protruding from the head (see drawing).

The soft-tissue structures seen in fossils of the Lower and Middle Cambrian show the key functional features of animal anatomy to have evolved fairly rapidly. The fossils of *Ottoia prolifica*, for example, indicate that this Middle Cambrian worm had a gut with a mouth, gizzard and anus, muscles, a proboscis, a dorsal nerve cord, and spines and hooks in its epidermis (Conway Morris, 1977). Perhaps the most surprising feature of the Cambrian organisms is that so many had eyes, with the number ranging from two to at least five (most of today's spiders have eight eyes). Some organisms had simple eyes, but several had compound ones with many ommatidia, the earliest known example being *Cindarella eucalla*, a Lower Cambrian trilobite (c. 520 Mya; Zhao et al., 2013) from the Chengjiang Lagerstätte, a site possessing fossils of exceptionally high quality. Each of its eyes seems to have had about 2000 ommatidia and it is reasonable to assume that integrating and interpreting all these optical inputs would have required a significant amount of sophisticated neural processing by its small brain. It is clear that, by the end of the Cambrian Period, extensive cell differentiation and tissue diversity had evolved.

The Cambrian biota evolved from Ediacaran ancestors, but the reasons why the Cambrian environment favored the generation of so much additional novelty are still not clear. The simplest explanation is that, towards the end of the Ediacaran Period, muscle, a late-forming epithelial derivative in radiata such as jellyfish, acquired its own germ layer, the mesoderm (Seipel & Schmid, 2005). For reasons that are not understood, this change was accompanied by a basic change in early embryo geometry as radial symmetry became bilateral. Once in place, of course, mesoderm was available to form nonepithelial cell types other than muscle and so opened up a new range of anatomical options and hence novel species. From these, natural variation and the absence of predators enabled these novelties to colonize hitherto empty ecological niches.

Extinctions

The bio-history of the last 500 My is one of radiations followed by extinctions followed by further radiations of the surviving organisms (**Table 4.1**), with different groups of organism being predominant in different periods. The reasons for the major period-changing extinctions are varied: some were climate-based; others derived from the effects of asteroids hitting the earth with devastating results (see Table 4.1). Extinctions are not, however, limited to those caused by global catastrophes: the fossil record shows that every species becomes extinct in due course, and is often replaced by others that, on morphological grounds, seem to derive from it, although a few, such as the horseshoe crab and velvet worms, have changed relatively little over very long periods (Fortey, 2011). The fossil record is thus one of continuous novelty and extinction. The so-called natural extinction rate, or, more accurately, the typical species lifespan, has been measured (Lawton & May, 2005), and seems to depend on the class of organism. Allowing for the vagaries of the fossil record, some estimates of these lifespans are 4–5 My for marine animals, 1–2 My for mammals, ~10 My for bivalves, and ~2 My for graptolites (index fossils). Extinction has to be viewed as normal and has the function of freeing ecological niches for new organisms to exploit.

Invertebrate evolution

Both embryological and genomic evidence show that animals fall into two basal classes distinguished on the basis of how their guts form (Gilbert, 2014): protostomes in which

Table 4.1: Some major extinctions

Name	Mya	Likely cause	Genera lost	Main results
Ordovician–Silurian	c. 445	Gondwana moved south, global cooling	60%	Loss of most brachiopods, bivalves, corals, and echinoderms
Late Devonian	374	Unclear	~25% marine	Marine losses: most trilobites, brachiopods, and corals. Land insects and plants seem to have been unaffected
Permian–Triassic	252.2	Volcanism and complex environmental effects, perhaps amplified by an asteroid collision	~83%	Every group was affected. On land, most mammal-like reptiles, labyrinthodont amphibians, and large insects were lost. At sea, all spiny sharks, trilobites, sea scorpions, and most snails, echinoderms, brachiopods, bivalves, ammonites, and crinoids were lost
Triassic–Jurassic	201.3	Volcanism + global warming	~20% marine	Loss of conodonts in sea. On land, loss of many large nondinosaur archosaurs, some therapsids, and large amphibia
Cretaceous–Paleogene (C-T)	66	Asteroid collision and perhaps other factors	~75%	All dinosaurs (except birds) and ~75% of other animal and plant species were lost on land; the giant sea lizards were also lost

the mouth derives from the first cavity in the embryo and the anus from the second (>30 phyla), and the less common deuterostomes (four phyla) in which cavity development is reversed. It is worth noting that in amniote deuterostomes, the evolution of first the yolk and later the placenta changes the early geometry and the central part of the gut tube forms first to consume yolk and the mouth and anus form later (Chapter 10). There are two other core differences between the groups. First, the location of the main nerve cord is different, being ventral in protostomes (under the gut) and dorsal in deuterostomes. Second, most deuterostomes apart from echinoderms (starfish) have, in their early stages, at least, a notochord extending along much of their length (Holland, 2005).

The invertebrates encompass the organisms from some 30 protostome phyla, all of which have a nervous system and a gut; the term is thus an umbrella word for what are likely to be many millions of species whose oldest may have evolved well before the Cambrian Period. The invertebrate fossil record is, of course, highly biased towards those organisms with a robust exoskeleton, such as arthropods with chitin exoskeletons and mollusks with calcium carbonate and sometimes chitin shells. As the great majority of invertebrates is either soft-bodied or has an exoskeleton that is fragile, most extinct species are likely to have left little trace.

Such limited evidence as there is for the post-Cambrian softer invertebrates mainly comes from impression fossils, for example large damselflies shown in **Figure 4.4**, from organisms caught and preserved in amber from tree resin, from trace fossils that record animal activity (such as movement tracks and boreholes), and from those few Lagerstätte where conditions were such that soft-bodied organisms were encased in sediment before allowing the preservation of wonderful detail (see www.fieldmuseum.org/science/blog/mazon-creek-fossil-invertebrates and en.wikipedia.org/wiki/Lagerst%C3%A4tte).

Figure 4.4 Cast of a *Meganeura* fossil whose wingspan is about 40 cm. It is on display at the Evolution Gallery of the Museum des Sciences Naturelles, Brussels.

Although multicellular life extends back to at least 1.65 Bya (Bengtson et al., 2009), evidence of invertebrate fauna is minimal until the Late Ediacaran Period. The fossil record then expanded rapidly (the Cambrian explosion) so that, towards the end of the Lower Cambrian Period (525–520 Mya), there were many distinct animal phyla, all being protostome invertebrates apart from the few species in the deuterostome taxon. By the end of the Cambrian Period, at least 15 or 16 invertebrate phyla had evolved, with many being various forms of worms and their derivatives, several of which have failed to survive. For reasons probably associated with the diversity of habitats and the paucity of predators, the Cambrian was a period when many variants of a basic triploblastic worm evolved and, as it were, tried their luck.

The widest morphological diversity is seen in the most common Cambrian phylum, the arthropods. These are invertebrates with an exoskeleton and a segmented body, with most segments possessing jointed appendages (Giribet & Edgecombe, 2012). One major group of arthropods that was present in the Cambrian and finally lost in the Permian extinction was the trilobites; trilobites had a three-lobed body, 2–4 pairs of head limbs, and many body segments, each of which had a pair of limbs and so were biramous: the upper was an outer walking limb and the lower an inner gill, a morphology that is maintained today in, for example, crabs. Thousands of trilobite species have been distinguished between the Early Cambrian Period (c. 515 Mya), when they first appeared, and the end of the Permian Period (201 Mya), when they died out. Today, there are only four major arthropods subphyla: the Insecta with three pairs of legs, the Myrapoda that include centipedes with one pair of legs per segment and millipedes with two, the Chelicerata such as spiders with 8 pairs of legs, and the Crustacea such as crayfish and crabs with five or more pairs of legs and often many swimmerets.

The early invertebrates were all marine animals; two major anatomical changes were required for them to colonize land. The first was a skeleton that was both impervious to water so as to avoid desiccation and strong enough to bear the animal's weight; the chitin exoskeleton satisfied both of these criteria. The second was a means of absorbing oxygen from the air, which was achieved through the evolution of trachea: these are epithelial tubules that extend inwards from openings in the larva's external ectoderm known as spiracles, and through which oxygen diffuses. One of the earliest fossilized animals showing such trachea is *Pneumodesmus*, a millipede from the mid-Silurian (428 Mya), but there is evidence from trace fossils that arthopods may have colonized land as much as 490 Mya (MacNaughton et al., 2002; Pisani et al., 2004).

Fossils of flying insects date from the later Carboniferous Period (~320 Mya) and it seems that they had two pairs of wings (like contemporary butterflies, dragonflies, and damselflies). From these evolved a wide variety of flying insects: in some the forewing remained but the hindwing became a balancing organ or haltere, with examples being diptera such as *Drosophila*; in others, the hindwing remained and the forewings became either halteres as in the strepsiptera, an order that are mainly parasites, or wing covers (elytra) as in the Coleoptera (beetles). The aerial radiation was so successful that the great majority of contemporary insects can fly - even cockroaches, whose modern form appeared in the Early Cretaceous (c. 140 Mya).

Fossilized organisms not only reflect evolutionary history, but can also be an important measure of the environment in which they flourished. Today, most flying insects are small, with the largest living butterfly, the female Queen Alexandra's birdwing, having a wingspan of 250 mm and a body 80 mm long. Early on in the evolution of flight (c. 285 Mya), however, trace fossils show that *Meganeuropsis permiana*, a genus related to today's dragonflies, had wingspans of up to 700 mm, a body length of about 300 mm, and an estimated mass of several hundred grams (Mitchell & Lasswell, 2005; see Figure 4.4). Theoretical analysis suggests that such insects would not be able to take in enough oxygen through their spiracles to fly in today's atmosphere (Gauthier & Peck, 1999), which itself would not have been dense enough to support their flight. Some thought has therefore gone into working out how these enormous insects were able to fly. The most likely answer seems to be in two parts: first, they were probably able to increase tracheal efficiency through "breathing" cycles of thoracic compression and relaxation (Westneat et al., 2003); second, the atmosphere then probably had more oxygen than now (up to 35% compared with today's 20%) and so would have been denser (Berner, 1999).

The invertebrate phyla of today represent those descendants that survived the various extinctions. Most radiated widely, with the mollusks, nematodes, and arthropods being particularly speciate; almost a million contemporary arthropod species have now been cataloged. In contrast, there are a few invertebrate phyla, such as the Placazoa, the Cycliophora, the Loricera, and the Phoronida for each of which 10 or fewer living species have so far been discovered; these eke out a living in isolated and specialized habitats.

Invertebrate evolution is too broad a subject to be more than briefly summarized here. Readers interested in the general area should consult Clarkson et al. (2014), while details of the evolution of insects, the most speciate arthropod subphylum, are given in Grimaldi and Engel (2005), who focus on the phenotypic data, and Misof et al. (2014), who summarize the phylogenomic relationships.

Chordate and vertebrate evolution

The deuterostomes include the chordates (~65,000 species), the echinoderms (~7,000 species), the hemichordates and the tunicates (~120 species), and perhaps the Xenacoelomorpha; this last phylum includes a very few flatworms whose taxonomic position is defined by genetic similarities or phylogenetics rather than by developmental anatomy (Philippe et al., 2011). Hemichordates such as the acorn worm have a larva with a notochord, while the elongated adult has gill slits in its pharyngeal region. The tunicates initially have a larval form very like a tadpole with a head/body region that includes primitive eyes, a tail, and a notochord. They then undergo metamorphosis, dissolve their brain, and form a tunic shape that cements itself to a rock. In both cases, the role of the notochord is to strengthen the larva.

The chordates are divided into two subphyla, the vertebrates and the cephalochordates, with *Branchiostoma floridae* (commonly known as amphioxus or the lancelet) being the sole living species in the latter subphylum. Its substantial notochord persists throughout life, just providing rigidity for its segmented muscles, unlike the thin notochord of vertebrates whose role is mainly developmental (it first secretes sonic hedgehog, a signaling protein that initiates the development of neurons and vertebrae, and eventually becomes part of the intervertebral disc; Holland, 2005). Amphioxus has little in the way of the vertebrate brain structures and sensory organs (eyes and ears) seen in vertebrates and which derive from cells migrating away from the neural crest that forms as the neural tube closes (Chapter 10). Jandzik et al. (2015) have shown that, in amphioxus, the tissue that produces neural crest cells in vertebrate embryos is transitory and no neural crest cells form.

Paleozoic era

There are, owing to fortuitous fossilization processes in the Early Cambrian Period, some very early aquatic chordates whose soft tissues have been preserved. The most primitive in its tissue anatomy is *Pikaia gracilens* (505 Mya; see Figure 4.3A), found among soft tissue fossils from the Middle Cambrian Burgess Shale. It had myotomes (segmented muscle blocks), a notochord, and probably gills and a mouth but no jaws, teeth, or eyes. It may thus be close to stem-group chordates (Morris & Caron, 2012), and if it has a contemporary descendant, it is likely to be amphioxus.

Earlier chordates, but with more complex anatomy, have been found in the Chengjiang biota from the Lower Cambrian Maotianshan Shales and include nine species with well-preserved, soft-tissue structures. The oldest among them is probably *Myllokunmingia* (524 Mya), which seems to have had a cartilaginous skull and skeletal structures. *Haikouella jianshanensis* (Shu et al., 2003) was a chordate dating to a little later (520–515 Mya), and analysis of several hundred examples shows that it had gills, a notochord, chevron-shaped muscle segments, a heart and blood vessels, eyes, pharyngeal teeth, a nerve cord, and reproductive organs (Shu et al., 2003; Chen, 2009; **Figure 4.5**). Another fossil, *Haikouichthys* from the Chengjiang shales (also from the Lower Cambrian, 520–515 Mya), seems much more like a primitive jawless fish than *Haikouella* because it had cranial features and fins, albeit having segmented gonads. Other primitive chordates from these shales include *Yunnanozoon*, which was probably a semichordate from the same period (~520 Mya; Shu et al., 1996), and *Shankouclava anningense*, a tunicate.

The early chordates of the Cambrian Period diversified rapidly with key vertebrate features appearing almost immediately (on an evolutionary timescale), producing a taxon of ever-increasing richness (**Figure 4.6**). From the Late Cambrian until the end of the Triassic, the most common chordate remains are those of conodonts. These small, segmented eel-like creatures (Donoghue et al., 2000) lacked vertebrae and jaws and so were agnathans, like contemporary lampreys and hagfish. The best-preserved evidence of conodonts are their sets of mineralized tooth elements, and these are now used as index fossils. Conodonts became extinct a little before the end of the Triassic, and their immediate successors were the small jawless and toothless ostracoderms characterized by a cartilaginous skeleton and plates of dermal bone that formed in the mesenchyme of the skin and protected their head.

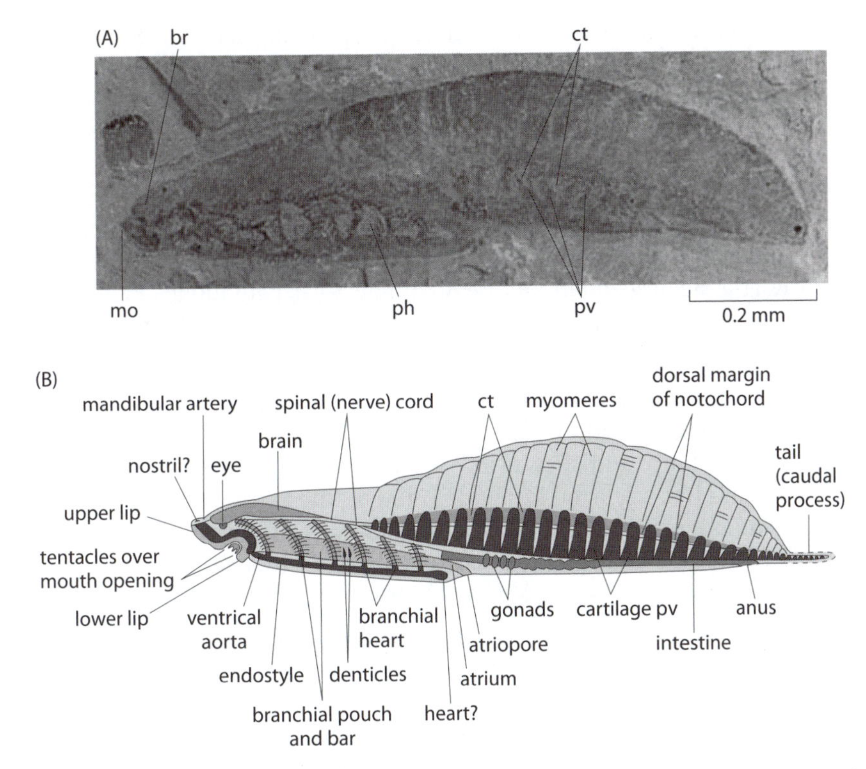

Figure 4.5 *Haikouella*, an early chordate from the Cambrian Period (~515 Mya). (A) An example of the fossil (br, brain; ct, connecting tissue; mo, mouth; ph, pharynx; pv, protovertebrae). (B) A drawing showing the anatomy.

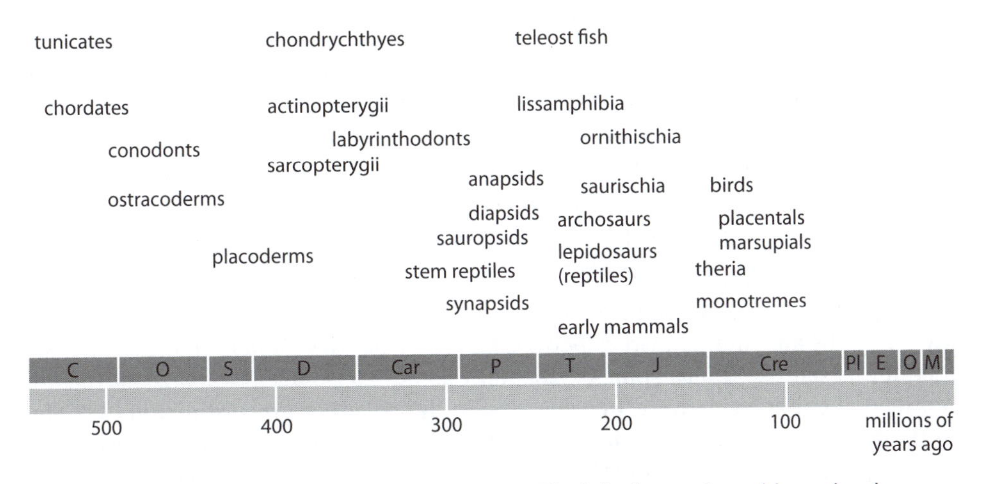

Figure 4.6 The fossil record data for the chordate taxon. The left of a word roughly marks when the group first appears in the fossil record. No formal attempt has been made to link groups other than through proximity because the times when fossils are first observed do not provide adequate evidence for lines of descent.

The ostracoderms gave rise to the first jawed fish, the placoderms (Early Silurian, 440 Mya, to Late Devonian, 360 Mya) that also had bony dermal plates protecting their heads. By the mid-Devonian, three major families of fishes had evolved, the Chondrichthyes that had cartilaginous skeletons, and the Actinopterygii and the Sarcopterygii, both of which had bony skeletons. The Actinopterygii, which were the most common taxon, had simple rayed fins, while the Sarcopterygii had fleshy lobed fins with the pectoral and pelvic ones containing terminal rays linked to articulating bones, much like the proximal parts of the limbs of tetrapods; it is from this group that land animals evolved. Today, most fish are teleosts (Early Triassic descendants of the Actinopterygii), with lungfish and coelacanths being among the very few surviving fish descendants of the Sarcopterygii.

The full progression from sarcopterygian fish to land animal took some 30 million years (390–360 Mya) and required a wide variety of adaptations that included major respiratory changes, as well as the transition from fins to limbs (Chapter 6). The final steps from a marine lungfish with limbs and lungs to an early amphibian able to clamber onto land took less than 10 My and had occurred by about 374 Mya. The first fully terrestrial vertebrates were the labyrinthodonts (360–150 Mya), a rich family of amphibian-like organisms with complex teeth (hence their name) and massive skulls that gave rise to modern amphibians, known as Lissamphibia, around 250 Mya or a little earlier (Marjanović & Laurin, 2007).

The appearance of reptiles can be recognized by the lightening of the skeleton, but the key step in the transition away from amphibians was the evolution of the shelled, amniotic egg that could develop away from water, early examples of which are not preserved in the fossil record. This occurred in the Late Carboniferous Period (315–310 Mya), or more than 40 My after the earliest amphibians. The first preserved reptilian footprints date to about 315 Mya (Falcon-Lang et al., 2007), the oldest known amniote is *Hylonomus lyelli* (~312 Mya), and the earliest taxon with an unequivocally reptile-like skeleton seems to be the sauropsid captorhinids (Müller & Reisz, 2005) from around the same time.

Hylonomus lyelli and other very early reptiles lacked fenestrae or spaces between the skull bones, but such features soon appeared, with the reptiles splitting into two main groups that can be distinguished by their skull morphology. Synapsids had a single skull opening on each side of the head behind the eye (a temporal fenestra), and saursopsids had two such fenestrae (**Figure 4.7**). In due course, the sauropsid clade gave rise to diapsids that maintained two fenestrae, anapsids that lost their fenestrae, and euryapsids with a single upper fenestra (these became, for example, sea reptiles and all were lost in the Cretacean extinction). Today's descendants of these early reptiles fall into one of three groups: the synapsids are mammals; the diapsids are birds and reptiles such as crocodiles, lizards, and snakes; the third group, the anapsids, which have no fenestrae, are tortoises and turtles.

The synapsid clade soon radiated to became the dominant reptiles in the Permian Period: early synapsids are often known as pelycosaurs (for example, *Dimetrodon,* 295–272 Mya, famed for its sail-like dorsal spines - Chapter 6), and later ones as therapsids. The great majority of the therapsids were, however, lost in the great Permian extinction (252 Mya), and the subsequent and much diminished diapsid taxon was drastically reduced at the beginning of the Mesozoic Period to a small group of rodent-sized creatures. These started to radiate during the Late Triassic, evolving to become mammals during the Jurassic Period (Chapter 7). For this reason, the Permian synapsids are often called the mammal-like reptiles or, more accurately, proto-mammals. In contrast, the diapsids such as the early *Petrolacosaurus* (c. 302 Mya, a reptile about 40 cm long) seem to have played a minor role in the Permian fauna, although the earliest archosaurs such as *Protorosaurus* were first seen then.

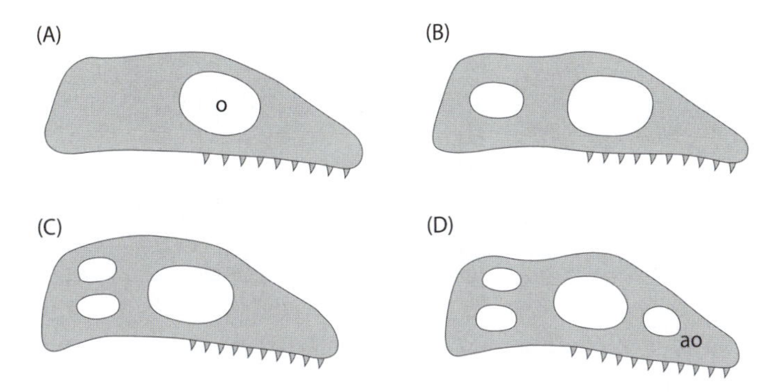

Figure 4.7 Skull fenestrae of the main reptile clades. (A) An anapsid skull with an orbit for the eye (o) but no postorbital fenestrae – a characteristic of turtles. (B) A synapsid skull with the low (infratemporal) opening of the mammal-like reptile clade. (C) A diapsid skull with the lower and upper (supratemporal) fenestrae characteristic of most other reptiles. (D) A diapsid skull with the additional antorbital fenestra or fossa (ao) that was a defining synapomorphy of the Archosaur clade. Not shown is the euryapsid skull, which has a single upper postorbital fenestra; this was a characteristic of plesiosaurs and ichthyosaurs (Mesozoic sea reptiles).

Mesozoic era

During the Early Triassic Period, the diapsid reptiles radiated within two main clades, the Archosauria and the Lepidosauria, which included lizards and probably ichthyosaurs, a taxon of large marine Mesozoic reptiles, and eventually snakes and tuataras. Archosaurs can be distinguished from other reptiles by their socketed teeth, additional openings in the skull and jaws, particularly the antorbital fossa (see Figure 4.7D), and a fourth trochanter on the femur (a knob for muscle attachment). The origin of plesiosaurs, the other taxon of large marine Mesozoic reptiles, is in a group of diapsid reptiles known as the sauropterygians that may be related to turtles.

The early archosaur radiation included dinosaurs, crocodiles, and pterosaurs, together with a range of other forms. The dinosaurs soon formed two distinct taxa: the Saurischia, who retained the original lizard-type hip, and the herbivorous Ornithischia, characterized by bird-hipped morphology and a new lower-jaw bone (the predentary). Some archosaurs, together with many amphibian and non-archosaur reptile taxa, were lost in the extinction at the end of the Triassic (c. 200 Mya). There was, however, a major expansion of the dinosaurs soon after the beginning of the Jurassic Period with the appearance of the saurischian-derived sauropods (large, pillar-limbed herbivores) and theropods (mainly carnivores who walked on two legs) and by a large radiation of ornithiscian dinosaurs. The broad taxon of dinosaurs dominated land for the whole of the Jurassic and Cretaceous Periods, or for almost 140 My.

One further important taxon of dinosaurs evolved from saurischian theropods; these were the birds, and there is a good fossil record of the evolutionary steps leading to this new taxon (Chapter 6). The first feathers appeared in the Middle–Late Jurassic theropods (c. 160 Mya), for example *Aurornis xui* (Godefroit et al., 2013), and seem to have been a common feature, probably as an adaptation to cold. The best-known feathered dinosaur with bird features, *Archaeopteryx*, appeared ~10 My later (**Figure 4.8**A), and seems to have lived in trees and on the ground, only being able to glide. Flight requires more than feathers, and, during the Cretaceous Period, this clade of therapods underwent loss of weight

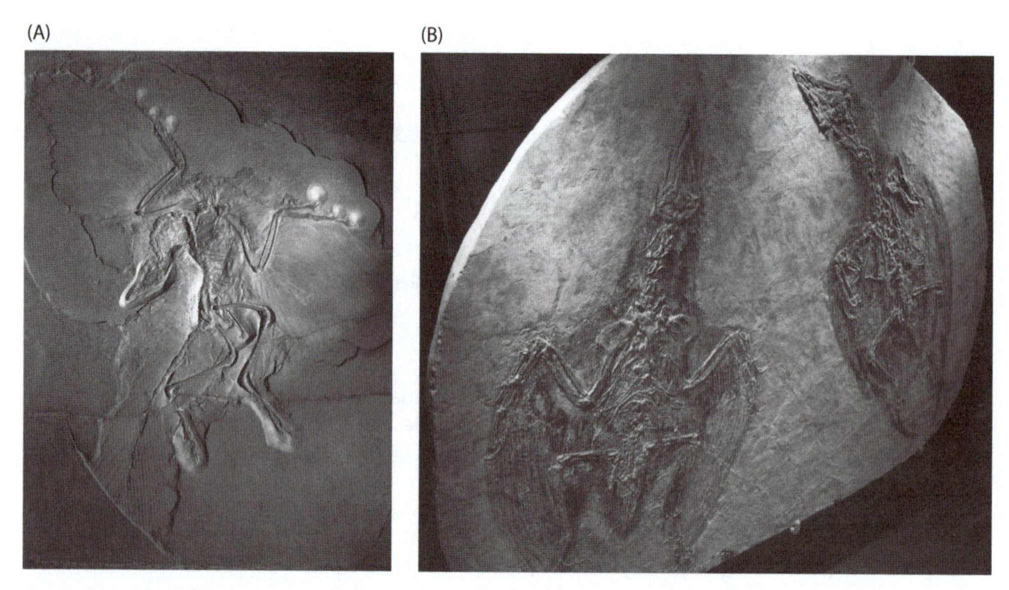

Figure 4.8 (A) The fossil of *Archaeopteryx* in the Berlin Museum für Naturkunde. The bone structure shows that the animal was a dinosaur, with clear evidence of long feathers on its forelimbs, and so was a transitional species with many bird features. (B) Fossil of *Confuciusornis sanctus*, an early bird with feathers and a beak.

through lightening of bones, together with replacement of the tail with a much smaller pygostyle, and of the jaws and teeth with a beak and gizzard (Louchart & Viriot, 2011). An example of an Early Cretaceous bird with these features is *Confuciusornis* (c. 125 Mya); it could probably fly but not very well because it had a low keel for anchoring pectoral muscle, while its forelimbs maintained claws (Wang et al., 2012; **Figure 4.8**B).

Synapsid evolution continued very slowly during the Mesozoic, with the progression from mammal-like reptiles to mammals gathering pace in the Jurassic. The fossil record does not, of course, show when mammary glands first evolved, but there are other key markers of mammals of which the most important are the malleus and incus, two novel bones used for hearing; it is convenient to use the presence of these two middle-ear bones as the defining characteristic of true mammals (Chapter 7). An early organism showing an intermediate form between the reptilian and mammalian jaw structures is *Morganucodon,* which first appeared during the Late Triassic Period (c. 205 Mya).

The synapsid clade played a very minor role during the Mesozoic era, although the fossil evidence suggests that the three families of mammals alive today, which include the Eutheria (placentals), the Metatheria (marsupials), and the Prototheria (monotremes), first became distinct in the mid-Jurassic (~175 Mya). It seems that the initial division was between the early Prototheria (for example, *Ambondo*, c. 167 Mya) that lay eggs and the Theria, such as *Dryolestes,* a line that led to the Eutheria and Metatheria (von Koenigswald, 2000), both of which are placental.

Cenozoic era

The C-T extinction led to the loss of all marine reptiles and archosaurs except the crocodiles and birds, and the ecological space so liberated allowed the small taxon of remaining synapsids to radiate. Their history since the C-T extinction is one of amplification and radiation and perhaps the most interesting evolutionary novelties are associated with the mammalian colonizations of the air (the bats) and the sea (for example, whales and walruses). A considerable amount is now known about the return of the mammals to the sea ~50 Mya (Chapter 7), but very little on how mammals evolved to fly as the earliest known fossilized bats (52 Mya) have a morphology similar to present ones (Teeling et al., 2005).

The vertebrates have dominated the world for over 400 My because their skeleton of mineralized bone enabled motile organisms to adapt to almost any environment, be it land, sea, or air. Bone has the second advantage that it lends itself to fossilization, as a visit to any museum of natural history shows. The vertebrate fossil record is thus far more complete than that of any other taxon. The major gap in the record is the lack of skeletal details of the early proto-mammals that survived the Permian extinction and slowly evolved to give rodent-like early mammals. Their most common remains are very small teeth (Chapter 7). A diagrammatic summary of when the major vertebrate taxa each made their appearance is given in Figure 4.6, while Chapters 6 and 7 provides a fuller and more formal analysis of vertebrate evolution.

Plant evolution

Plants evolved from unicellular cyanobacteria (green algae) that first appeared some 2.9 Bya. These were the first organisms able to capture light energy from the sun and use it to synthesize sugars from carbon dioxide and water, so releasing oxygen; with some, at least, soon being able to synthesize cellulose (Nobles et al., 2001). Unicellular, eukaryotic algae such as Dasycladales were present in the Cambrian seas and evolved into the various types of algae seen today. They are the ancestors of land plants and the first good evidence for their existence is spores typical of today's liverworts that date to the mid-Ordovician Period (~470 Mya). These, together with hornworts and mosses, are bryophytes and the most primitive plant form with a waxy cuticle that enables them to live out of water (Ligrone et al., 2012). They do, however, lack vascularized tissue for transporting water and are therefore all small. The earliest fossilized bryophytes date rather later to the Early Silurian Period (c. 440 Mya). The spores of primitive vascular plants appear in the Upper Ordovician (443–417 Mya), while full vascularization evolved during the Silurian Period. Wood and roots appeared c. 400 Mya (Raven & Edwards, 2014) and, soon after, the oldest known seed plant *Elkinsia polymorpha*, evolved; this was a seed fern with seed-bearing shoots.

The evolution of lignin, which combines with and strengthens cellulose to form wood, allowed vascularized plants to increase dramatically in size, and early examples of this were tree ferns such as *Wattieza* (mid-Devonian) that were up to 8 m tall. Around then, progymnosperm trees that still produced spores rather than seeds were common. It was not until the Late Carboniferous (c. 320 Mya) that true gymnosperms that produced seeds from leaves appear in the fossil record. These are mainly conifers whose seeds are made by cones, which are a leaf derivative, but the clade also includes cycads and ginkgoes. The evolution of leaves is poorly understood but they were certainly present at c. 360 Mya (Late Devonian Period). With leaves trapping energy and wood providing mechanical strength, trees soon grew to considerable heights, dominating the land wherever the climate allowed. The resulting woodlands were very extensive and much of our current oil and coal derives from the forests of the Carboniferous Period (354–290 Mya).

Angiosperms (flowering plants whose petals are modified leaves) had, on the basis of the morphology of fossilized pollen, evolved by the early Cretaceous (~120 Mya), although genomic analysis dates the root of extant angiosperms to at least 160 Mya (Hochuli & Feist-Burkhardt, 2013; Amborella genome project, 2013). Such angiosperms fall into two main classes, dicotyledons and monocotyledons, which are named for the number of cotyledon or seed leaves that initially develop from the seed. It is worth noting that the grasses, the major source of human calories, are monocotyledons that only evolved about 40 Mya. **Figure 4.9** summarizes the dates when various plant types first appear in the fossil record, and it is clear that most features of plants apart from flowers seem to have evolved around 400–360 Mya, a period known as the Devonian explosion. What the fossil record does not make clear is the detailed sequence of inheritance here; this is discussed in Chapter 5.

Finally, viewing the evolution of the complexity of living organisms as a whole, it seems that anatomical richness developed exponentially, beginning with a very slow period during which prokaryotic then primitive eukaryotic organisms evolved. This was followed by a period of increasing organism complexity but generally little increase in size between the Ediacaran and Cambrian Periods (635–485 Mya). It was only then that first animals and then land plants started to increase in size and diversity. The speed of evolution was such that, well before the time of the Permian extinction (252 Mya), there were rich fauna and flora. At an anatomical level, the great majority of the tissues that plants and animals now possess had evolved; all the general features of life were in place. The last 250 My have mainly been a period of consolidation, marked only by minor new features such as hairs, feathers, and flowers, together with one substantial and very recent advance over the last 7 My, the increasing sophistication of the hominin brain.

Within this arc of time, different important events have taken very different amounts of time. It took, for instance, little more than 5 My or perhaps the same number of generations (Table 6.2) for a tetrapod lungfish to give rise to the first amphibian able to clamber onto land. In contrast, it took at least 40 My and perhaps as many generations (Chapter 6) for the amphibian egg that developed under aquatic conditions to evolve into the amniote egg whose complex external and internal membranes allowed development to take place under dry conditions. Likewise, speciation can be very slow: the ~4 My since wolves, coyotes, and the many varieties of dogs separately evolved from a common ancestor has not been long enough for them to become distinct species unable to interbreed and produce fertile offspring. Human evolution is also interesting here: if human generation time is

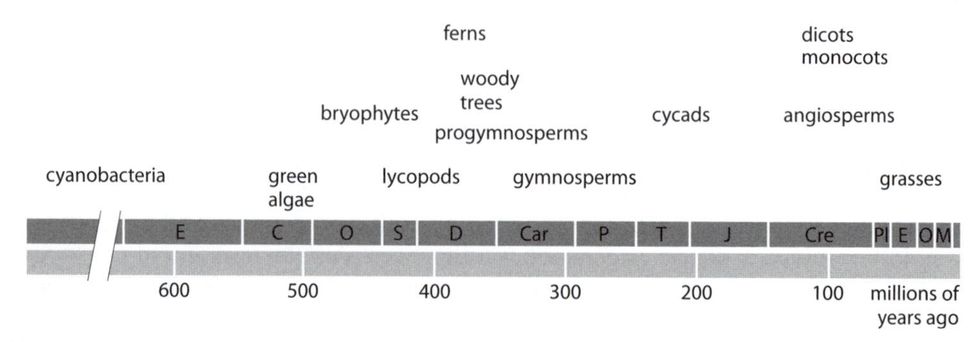

Figure 4.9 Diagram showing the emergence of new families in the evolution of plants, based on the fossil record.

about 25 years or a little less, the 200 Ky since the earliest known fossils of *Homo sapiens* were laid down represents 10,000 generations at most. Our species seems to have changed relatively little during this period.

Key points

- Prokaryotic life evolved more than 3 Bya, with unicellular eukaryotes appearing around 2 Bya and multicellular ones very soon after.
- Although the first large organisms evolved during the Ediacaran Period (635–542 Mya), the great majority were lost before the Cambrian Period.
- The core bilaterian phyla, both protostomes and deuterostomes, can be traced back to the Early Cambrian Period (540 Mya), but there is little evidence on their early divergence from simple worms.
- Plants derive from algae that, in turn, derive from very early cyanobacteria which undertook photosynthesis and possessed a cellulose-strengthened cell membrane.
- All early vertebrates were fish, with amphibia evolving from a sarcopterygian ancestor whose fins had bones homologous to those of the proximal parts of limbs.
- Land was colonized around 374 Mya. Amphibia evolved rapidly but it took at least 40 My for the amniote egg to evolve and so allow early development to take place away from water.
- Synapsid reptiles dominated earth during the later part of the Paleozoic era, but were mainly lost during the Permian extinction, surviving only as small rodent-sized reptiles.
- Diapsid reptiles, particularly dinosaurs, flourished during the Mesozoic era giving rise to birds, their major survivor from the C-T extinction.
- Mesozoic proto-mammal synapsids evolved to become mammals during the Mesozoic era, survived the C-T extinction, and flourished in the subsequent Cenozoic era.

Further reading

Benton J & Harper DAT (2009) Basic Palaeontology: Introduction to Paleobiology and the Fossil Record, 3rd ed. Wiley-Blackwell.

Bottjer DJ, Etter W, Hagadorn JW & Tang CM (eds) (2002) Exceptional Fossil Preservation: A Unique View on the Evolution of Marine Life. Columbia University Press.

Knoll AH (2004) Life on a Young Planet: the First Three Billion Years of Evolution on Earth. Princeton University Press.

Willis K & McElwain J (2013) The Evolution of Plants. Oxford University Press.

CHAPTER 5
DARWINIAN DESCENT WITH MODIFICATIONS: EVOLUTIONARY TAXONOMY AND CLADISTICS

This chapter provides the formal framework for analyzing evolutionary evidence in the context of Darwin's perception that evolution reflects "descent with modification." This view has a key implication: if species B shares some features with another species, A, but has other features that are modifications of those in species A, then species B is likely to have descended from a line whose origin included species A. This chapter shows how such an idea can be formalized in cladograms: these are hierarchies or evolutionary trees that link species through the relationship *<derives with modification from>*. Such cladograms can be used to explore the detail of evolutionary relationships, and plant evolution is used as a case study.

The meaning of evolution is precise: new species start to evolve as a result of novel, heritable changes occurring in members of an existing species. A new species becomes established if these changes are sufficient to ensure two results: first, that the new group can reproduce and thrive in some novel environment, and, second, that the genes and chromosomes of the new group eventually undergo sufficient genetic change that individuals can no longer interbreed with those of their parent species to produce fertile offspring. If successful interbreeding between the two groups can occur and results in fertile offspring, the two are just variants or subspecies. These changes imply, of course, that any new species carries most of the properties of its parent species but has at least one novel and heritable feature, usually recognized because it ensures the adaptation of the new species to the environment in which it evolved.

This is, of course, Darwin's "descent with modification," and it can be expressed more formally. Suppose that an ancestral species **A** has a set of anatomical properties or traits (**ijk..pqr..**), the expectation is that any descendant, but different, species **B** has most of these ancestral properties (known as plesiomorphies) but with at least one novel feature that has resulted from the loss or change of an existing property (for example, property **p** is modified to become property **x**) so that **B** now has the properties (**ijk..qr..x**). This process can be repeated with **B** giving rise to a new species, **C**, characterized by a further novelty, **y**, which derives from **q** and so has properties (**ijk..r..xy**), and perhaps again a little later to another species, **D**, with novel feature **z** from **r** and hence with properties (**ijk..xyz**). If there is a contemporary descendant, **A**', of the original ancestor species **A**, it would be expected to lack the **x y z** innovations, although other changes in **A**' might have evolved since **A** lived. The key point is that evolutionary change can only be recognized through examining the relationships among species; a single species on its own says nothing in this context.

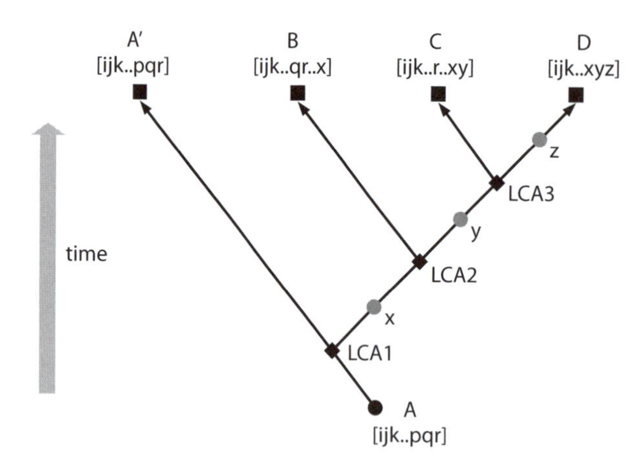

Figure 5.1 A cladogram illustrating Darwinian descent with modification. This has been drawn to illustrate how an ancestral organism, A, has given rise to four different organisms, A′, B, C, and D, any of which could be alive or extinct today. Grey circles indicate the novel modification that characterizes downstream species. Note also that lines of descent are not shown as stemming directly from known species, but from last common ancestors (LCA); this is because the exact species from which a descendant evolved is not known.

All this phenotypic information can be expressed as a diagram known as a cladogram (from the Greek *klados* = branch). This is a model of the evolutionary history of a set of species (**Figure 5.1**) constructed on the basis of simple triplet facts of the form:

<species F> *<derives with modification from>* <species G>

Such a statement immediately raises a problem because, for a fossil species, one can never be sure about its precise ancestor, only its approximate location on the diagram. In the case of species B in the example above, all one can say is that its parent species had to be a species whose phenotype includes property x that separated it from the inheritance line from species A. This further species is considered to be the last common ancestor (LCA) of all organisms with property x, and the historical facts are rewritten in the form:

<species F> *<derives with modification from>* <LCA with property x>

The cladogram for this example is shown in Figure 5.1 and the terms are of the form:

<species C> *<derives with modification from>* <LCA2>

<species D> *<derives with modification from>* <LCA3>

<species LCA3> *<derives with modification from>* <LCA2>

Such a cladogram can, of course, be extended to include further species (or, indeed, any taxonomic class or *taxon*) as the data demand. The key point, however, is that every taxon included in a branch of the hierarchy should share with the others on that branch a single LCA that is at the base of the branch (species A in this case). A hierarchy or tree that includes only the descendants of a single, root species is known as a monophyletic group.

Figure 5.1 includes an important convention: the gray circle labeled **x** signifies that all species on its far side possess property **x**, unless there is a further downstream species in which **x** has been lost. In this case, all descendant species will also lack **x**, but these species should maintain a sufficient number of ancestral features that their classification is not ambiguous. They will thus still share the same LCA with descendant species that have retained **x**. Thus, early flying insects such as the ancestors of dragonflies had two pairs of wings and evolved from unwinged ancestors. In the Diptera (gnats, mosquitoes, and flies such as *Drosophila melanogaster*), the posterior wings evolved to become halteres (small, vibrating balancing organs). In contrast, the Strepsiptera (small "twisted wing" parasitic insects) retain their posterior flying wings, and it was the original anterior wings that evolved to become halteres. Because both groups share many other properties, such as limb number, they are all insects with the same common ancestor.

Figure 5.1 also clarifies the two sorts of prediction that descent with modification makes. First, whenever a novel property arises as the result of change to an ancestral property (for example, **x**, which derives from **p**), it should normally be present in all subsequent downstream taxa; an example of this is the amniote egg produced by the first reptiles and maintained in all reptile-derived vertebrates. Such derived characters that are shared by all downstream taxa are called synapomorphies and represent strong and clear evidence for evolutionary descent (a summary of cladistics terminology is given in **Box 5.1**). In cases where a property is lost (for example, humans, unlike other primates, have minimal amounts of thick body hair), the cladogram can be viewed as accurate provided that the "difficult" species possesses sufficient number of other synapomorphies and plesiomorphies to sustain its position in the tree.

Second, very different and widely separated species that share a common ancestor (such as the different vertebrates) would still be expected to maintain many common, ancestral properties or plesiomorphies (these are the **ijk** tissues in Figure 5.1). To repeat the example from Chapter 3, the tiny frog *Paedophryne amauensis* and the enormous blue whale *Balaenoptera musculus* both have standard vertebrate eyes. Indeed, the fact that nerve cells in squid and humans are essentially the same in both structure and function indicates that this cell type is a plesiomorphy and hence that neurons only evolved once, and that vertebrates and mollusks share a very ancient common ancestor that had such cells. Such ancient shared properties are particularly studied in evo-devo approaches (Chapter 11).

It is important to realize that the formalism represented in Figure 5.1 says nothing about time. One might immediately think that the framework applies to fossilized organisms where timings are known. The timing is, however, irrelevant: even for two closely related fossilized organisms, the dates when each appeared and became extinct are rarely precise enough to confirm that one derived from the other. The only properties considered in making a cladogram based on anatomy are the structural features inherent within related organisms, whether they are living or extinct. In fact, timing can be misleading as many organisms carry on with little change long after other species evolved from their line.

Organizing species on the basis of descent with modification is thus fundamentally different from the classical Linnaean approach of organizing species on the basis of sets of shared properties (Chapter 3), with higher levels having more general properties than lower ones. Thus, for example, low-level Linnaean categories, such as families, can be defined using details such as tooth number (Canidae, for instance, have 42 teeth and Felidae have 30), while higher-level categories such as phyla are defined by body plans. The hierarchies of traditional taxonomy are, like cladograms, also mathematical graphs, but the structure of the triplets is different: instead of *<derives with modification from>*, the relationship is *<is a member of the class of>*. A Linnaean taxonomy is a set of nested classes based on morphology, where species are grouped within genera, which are grouped within families, orders, classes, phyla, and so forth. In principle, the grouping of terminal (or leaf) species in both Linnaean and cladistics hierarchies should be the same and only the higher classes should be different. This is because the former represents a set of ever-more-inclusive groups and the latter a hierarchy of LCA species. This is generally so in practice, but Linnaean taxonomy has occasionally had to be revised to ensure that the groupings are consistent with the evolutionary history (Chapters 3 and 7).

Cladistics

The idea of grouping organisms on the basis of shared, derived characteristics had long been thought about, but it was not until the 1950s that it was formalized by Willi Hennig, who called his approach "phylogenetic systematics" (Hennig, 1966). However, it is now known as cladistics (Kitching et al., 1998; Lipscombe, 1998) and has become the standard way of analyzing evolutionary change and interspecies relationships, although its use is not always straightforward. Cladistic analyses can, for instance, give ambiguous results in cases where there are many characteristics to consider, some of which may be of secondary significance. In addition, as in all attempts to formalize differences, it has difficulties with parameters that vary continuously (such as size). Furthermore, the terminology, summarized in Box 5.1, is not intuitively obvious. Nevertheless, because it captures the nature of evolutionary relationships so well, cladistics is universally used to describe evolutionary descent, and it will provide the language of all subsequent chapters.

Constructing a cladogram for a set of related species is generally straightforward, in principle at least. The process starts by listing a set of characters for each species: some will be species-specific and others will be shared by two or more of the species, with those that are shared by all being viewed as plesiomorphies or ancestral. These characters are then integrated into a taxon matrix of species and their properties; this is then used to construct a tree that describes the evolutionary relationships among all the species in the most parsimonious way (originally known as a Wagner Tree). In essence, the first step is to identify a set of LCAs in which innovations (or apomorphies) first appear and to identify the species that carry them; if there are several such species, they are all in the same clade and share what is now a synapomorphy. The next step is to group species on the basis of such shared properties. Eventually, the analysis produces a set of statements of the general form:

<LCA P> *<derives with modification x from>* <species B>

<species A> *<carries the modification x and derives from>* <LCA P>

where all modifications are explicit. This information is sufficient to construct a tree.

For simple cases, producing this tree can be done by hand, but, for more complicated examples where species have many characters, the tree needs to be constructed computationally and there are readily available algorithms and programs for this. One such program is PHYLIP (http://evolution.gs.washington.edu/phylip/general.html): this tries out all combinations of possible lines of evolution to work out the tree that requires the least number of assumptions and hence is the most likely to represent the evolutionary history of the clade. Cladistics cannot, of course, say what actually happened, only suggest the most probable way that it did happen. In many cases, however, the most likely version of a cladogram is very much more likely than any other version, and in simple cases is the only plausible version.

It is worth repeating that the one aspect of evolution that cladograms do not capture is time, and this is for two reasons. First, time is not a property of a species that gives any clue to inheritance or phenotypic change. Second, the precise time when a new species arose and when it died out, except where there was a global extinction, is very rarely known because of limitations in the fossil record. This means that the length of the branch in a cladogram carries no implication either about the degree of differences or the time between species. The most that can be done is assign a date when the earliest fossil showing an apomorphy first lived, but this date, of course, is provisional, and to be revised in the light of any future data.

Box 5.1: The main cladistics terms

Clade: This is the complete group of species that share a last common ancestor (LCA) and its visualization is a cladogram; it is also known as a monophyletic clade, to distinguish it from a paraphyletic clade in which some of the taxa are missing (for example, a dinosaur clade that excludes birds). A **cladogram** has at its base a last common root ancestor species that gives rise to a spread of intermediate species and terminates in a set of leaf species.

Taxon: Any named group of related organisms, sometimes of the same taxonomic rank (for example, the bryophyte taxon includes mosses, hornworts, and liverworts).

Cladistics: The inheritance of properties. As evolution proceeds, a pre-existing or ancestral property (a plesiomorphy) in a species changes in some heritable way to produce an innovation or a novel, derived property (an apomorphy). If this species then gives rise to a group of further organisms, each should inherit that property; the original apomorphy thus becomes a shared, derived property (a synapomorphy – "syn" means together). A novel feature that is specific to a terminal leaf of a branch (often a living species) is known as an autapomorphy (**Figure 5.2**).

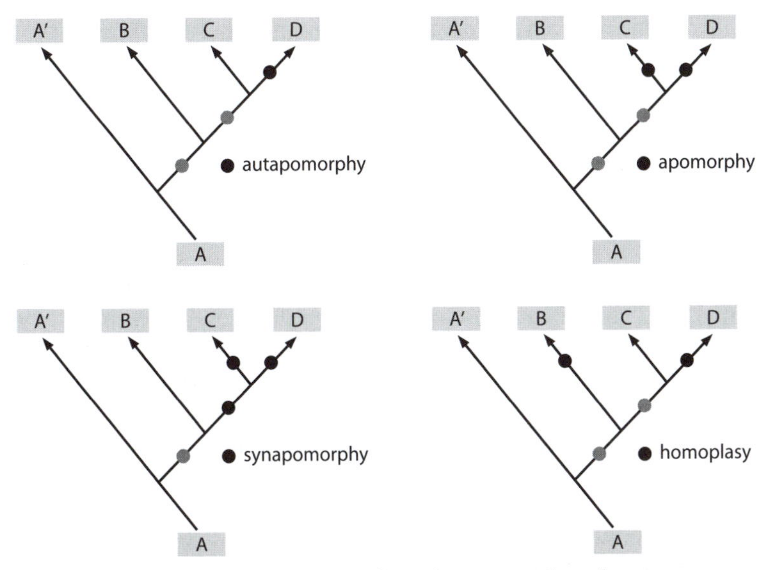

Figure 5.2 Diagram illustrating the various types of novel property.

Crown group: After the evolution of some apomorphy, the crown group (**Figure 5.3**) is the LCA with that apomorphy together with all its descendants, both extinct and alive. The crown group is thus a subclade of a wider and earlier clade of which the parent species was once a terminal species or leaf. A single synapomorphy is the minimal requirement for specifying a crown group clade.

Stem group: This reflects an earlier clade than the crown group: the stem group (see Figure 5.3) is the clade minus the crown group; it thus includes species related to but not part of the crown group.

Basal group: This is the earliest diverging subclade within a clade (see Figure 5.3), and is thus a sister clade to all other subclades.

Box 5.1: The main cladistics terms (*Continued*)

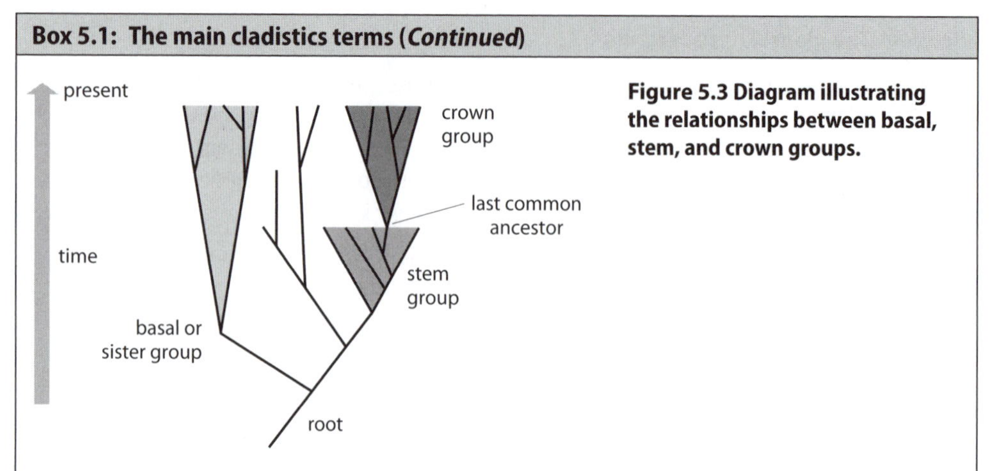

Figure 5.3 Diagram illustrating the relationships between basal, stem, and crown groups.

If the synapomorphies in different organisms later evolve to give different structures, these structures are homologs (for example, the forewings of early flying insects evolved to give halteres in Strepsiptera and elytra or wing covers in beetles). Properties that appear similar, but arise independently and not from plesiomorphies (for example, bat and bird wings), are known as homoplasies and the taxa that possess it form a grade. This is sometime called homoplastic or convergent evolution

The term LCA is used in the context of groups of species, while the term most recent common ancestor is used for discussing individuals within a species.

The use of the *<descends with modification>* link also enables properties other than anatomical ones in living organisms to be examined to see if they reveal evolutionary relationships. The terminology of cladistics with its various "morphies" indicates that it was intended for use with anatomical structures, the most obvious feature of the phenotype and the only one consistently preserved in the fossil record. The idea of studying evolution through analyzing inherited changes is not, in principle, limited to anatomy, and an obvious alternative is the inheritance of mutations as part of the genotype. Chapter 8 discusses the analysis of variants in DNA sequences and how gene-sequence relationships among the genomes of different living organism allow inheritance patterns to be inferred over both short and long periods of time. Hierarchies derived from sequence data are known as phylograms; they are usually much more precise than cladograms as they are based on much more detailed information. They can be used to probe deep time and provide information about the early stages of evolution where there are no fossil data (Chapter 9). An important difference between cladograms and phylograms is that, while the length of the former's branch arms carry no explicit meaning, the length of phylogram branch arms indicates, qualitatively at least, the extent of mutational difference between parent and daughter sequences. It is worth noting that any tree composed of such triplets is an example of a mathematical graph, and its more general use is discussed in Appendix 1.

The third area in which this type of analysis is relevant is in the inheritance of protein function as part of the molecular phenotype. As with DNA analysis, one can only analyze living species, but the exploration of how homologous proteins and protein networks in very different organism have very similar functions, particularly during embryogenesis, provides strong indications of shared inheritance and identifies very early plesiomorphies. Exploring the light that this approach casts on evolutionary history, particularly in the context of development (evo-devo), is the major topic of Chapter 11.

These graphical approaches thus give three ways of constructing evolutionary history on the basis of completely independent types of data: anatomical tissues, DNA sequences, and the function of homologous proteins and networks, often during embryogenesis. It is an obvious prediction that each of these should group living species in the same way; this is because they represent different views of the same evolutionary events. In turn, these hierarchies should mesh with what is known about the evolutionary timescale as inferred from the fossils record. In cases where the amounts of data are not the same, lower-resolution groupings should be compatible with higher-resolution ones. It is hard to identify even a single case where this prediction is not met.

Convergent evolution and homoplasies

Just because two groups of organisms share a novel feature, it does not necessarily mean that they share a recent evolutionary relationship: that particular feature could have evolved independently. Consider the vertebrate wing of pterodactyls and birds with diapsid skulls, and bats with synapsid skulls: the fossil record shows that each wing has its own characteristic skeletal details and each evolved separately from vertebrate forelimbs originally used for walking (bats and pterodactyls) or grasping (birds). Each, of course, also reflects an adaptation that originally enabled the species to colonize a new environment (for a more detailed discussion, see Chapter 13). It is thus clear that the vertebrate wing is not a helpful property for analyzing evolutionary descent.

In traditional evolutionary analysis, a similar character that has evolved separately in different species (that is, it is shared, but not derived from, the same common ancestor) is described as resulting from convergent evolution. In cladistics, such a character is called a homoplasy and the group of species that display it form a grade, not a clade (see Box 5.1). One well-known example of a homoplasy is the sabre tooth: although this is only present in one extant species of tiger, the fossil record shows that this dental feature occurred in many extinct species, only some of which were closely related to the cat family (Van Valkenburgh, 2007; Goswami et al., 2011). The earliest were Permian mammal-like reptile species that were mainly carnivores, although the Anomodontia were, on the basis of the structures of their other teeth, vegetarians that probably used their sabre teeth for defense (Cisneros et al., 2011; this paper includes what is probably the first recorded example of tooth decay). A more recent animal with sabre teeth was Thylacosmilus, a marsupial mammal that lived in South America. The other examples were members of Felidae-related taxa, but not closely related as they do not share an LCA possessing a sabre tooth.

The camera-type eye is another classic example of a homoplasy: this evolved completely separately in the Vertebrata and the Cephalopoda, taxa whose LCA lived at the beginning of the Cambrian Period, or possibly earlier. These eyes are identical in function and similar in form but differ in the ways that they develop and in their structural detail (**Figure 5.4**; Harris, 1997). Not only do these two types of eye have very different lens crystallins, but the retinal nerve fibers in vertebrates migrate over the photoreceptors (hence the blind spot where the nerves coalesce to form the optic nerve, which migrates through the retina), whereas the cephalopod fibers migrate behind the retina (hence there is no blind spot). It is intriguing that these two very different eye types use homologs of some of the

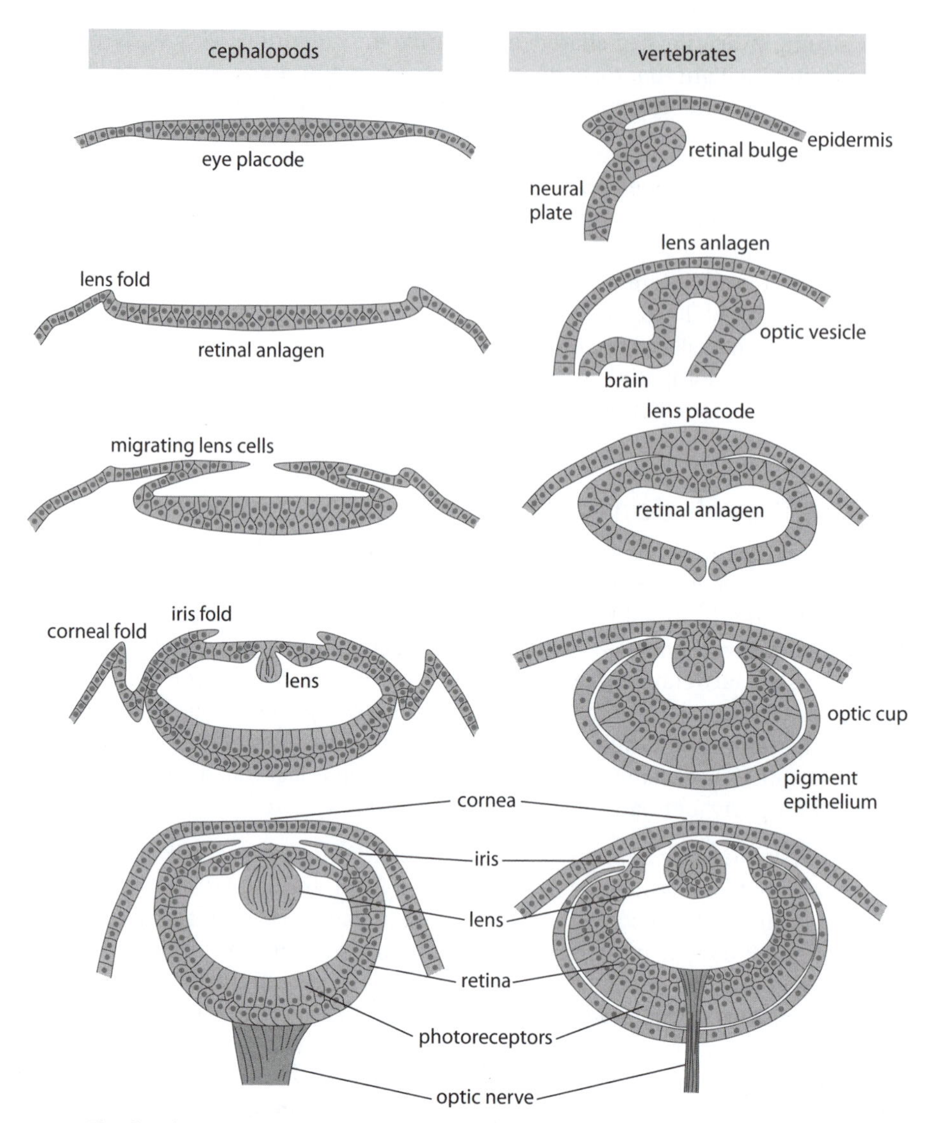

Figure 5.4 The development of the cephalopod and vertebrate camera eyes. The former develops almost completely from an epithelial placode in head ectoderm. The latter forms from two regions of head ectoderm: a neuronal part that will form the retina and iris, and a dermal part that will form the lens and cornea. The vertebrate eye also has a scleral component (not shown) that surrounds and strengthens the eyeball and which derives from neural crest (from Harris, 1997).

same genes such as those for the rhodopsins, which are also found in the compound eyes of arthropods; this may reflect parallel evolution at the molecular level where proteins from simpler-eyed ancestors have been retained (Yoshida & Ogura, 2011).

Plant cladistics

A good taxon of species that illustrates how cladistics works is the plants. There are, on the basis of obvious morphology and increasing degrees of complexity, only six major groups of chlorophyll-bearing eukaryotic organisms: green algae, mosses, lycopods (for example,

Table 5.1: Plant properties – all possess chloroplasts	
Class	**Some distinguishing properties**
Green algae (charophytes) such as pondweeds	Freshwater organisms containing cellulose and possessing spores and green spurs
Moss (bryophytes)	Land organisms with spores, cuticle, non-vascular stems, and primitive leaves that show apical growth
Club moss (lycopodiophytes)	Land organisms with roots, vascular tissue, spores, and microphyll (single vein) leaves
Ferns (monilophytes)	Novelty: megaphyll leaves (complex veins)
Gymnosperms	Novelty: seeds that develop from leaf stems and not spores
Angiosperms	Novelty: flowers and seeds that form in an ovary

clubmosses), ferns, gymnosperms, and flowering plants. **Table 5.1** shows a few of their distinguishing properties. Before constructing the cladogram, two points need be emphasized. First, the analysis only looks at these major groups together with only a few core features: the purpose of the exercise is not to provide a full analysis of the evolution of plants, but to illustrate in as simple a way as possible how cladistics examines taxon data. Second, no one today would construct a cladogram of a set of contemporary organisms on the basis of morphological features alone; it would always be checked, supplemented, and refined by phylogenetic analysis of a range of homologous DNA sequences from these organisms (Chapter 8).

Table 5.1 shows first how the structures in successive classes increase in complexity through apomorphies (innovations) and second that, once a taxon has acquired new properties, more complex plants maintain and share these derived properties (synapomorphies). Green algae have very little structure beyond spurs, but, like all other plants, contain cellulose (Mikkelsen et al., 2014). Mosses have anchoring filaments and very primitive leaves that can both be seen as modifications of these spurs, with a cuticle that reflects differentiation of the original spur epithelium and regulates evaporation. Club mosses show more complex tissues than other mosses: they have single-veined leaves, roots, and vascularized stems to facilitate water transport that, respectively, derive from primitive leaves, anchoring filaments, and the epithelia of the moss stem. Ferns and all further classes have megaphyll (multiveined) leaves that are modifications of single-veined leaves. In gymnosperms, the seeds derive from earlier spores but develop from leaf stems; their characteristic cones reflect leaf modifications. The most sophisticated plants are angiosperms with flowers that derive from leaves and that produce seeds in ovaries that form in the basal part of the flower. It is clear that plant progression exactly meets the formal demand of Darwinian descent with modification.

The best way to look at this information is to express it visually. It is straightforward to make a cladogram in which those organisms with fewer synapomorphies and more novelties deriving from those features are further from the root (**Figure 5.5**). If the dates of the earliest known fossils for each taxon (see Figure 4.9) are included in this figure, it is clear that, as expected, time meshes with evolutionary progression. Note also that charophyte algae form the basal group and are sisters to every other clade (McCourt et al., 2004).

There is an interesting problem that the data so far given cannot solve and that is the detail of the evolution of flowering plants (angiosperms). First Threophrastus (c. 370 BC,

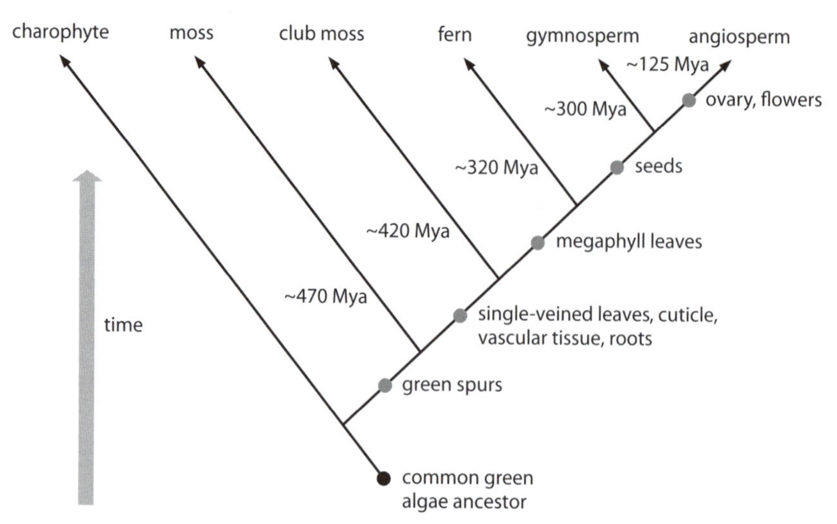

Figure 5.5 Cladogram showing the evolution of some major plant taxa on the basis of their possession of novel derived properties (shown as large gray dots) that separate the classes. The numbers indicate the time of the oldest known fossils of these classes.

Appendix 1) and then John Ray (1682) showed that the flowering plants divide into two obvious classes: those whose embryo develops a single cotyledon (a primitive or seed leaf) and those that develop two cotyledons. The adult plants can easily be distinguished on the basis of the adult leaves: the monocotyledons have parallel veins (orchids, grasses, and palms, to name a few), while the dicotyledons have branching veins (all other trees and most flowering plants). There is no obvious rationale for deciding which came first. Not even fossil data are helpful, partly because we lack appropriate fossils and partly because, as Figure 4.9 shows, the various types evolved too closely in time to provide temporal clues as to what happened.

There is, however, another property that angiosperms share with gymnosperms. They both have pollen: the small, rounded particles whose hard coats contain the male gamete (reproductive cell) and some support cells. Pollen morphology is highly plant-dependent but one feature is common: they all have furrows with apertures that swell in the presence of moisture, enabling male gametes to escape the coat. All gymnosperms and monocots have single-furrowed (monocolpate) pollen together with a few dicot angiosperms (an example being the Magnoliids); the great majority of angiosperms, however, have three-furrowed (tricolpate) pollen, and are known as the eudicots or true dicots (**Figure 5.6**). These two properties - cotyledon number and colpate number - are enough to work out the most parsimonious line of evolutionary descent (**Figure 5.7**). It is clear that the monocolpate dicotyledonous angiosperms were more basal and a common ancestor gave rise separately to monocolpate monocots that lost a cotyledon, and tricolpate dicots that gained two furrows (Soltis et al., 2008).

The evolution of the major plant taxa can be dissected out on the basis of a few well-defined characters that are shared and derived (synapomorphies). This simplicity does not extend to the minor taxa; for example, classifying trees on the basis of their anatomic characters alone can be particularly difficult because so many species look similar. The solution for living species, of course, is to use DNA analysis to discover how closely related they are.

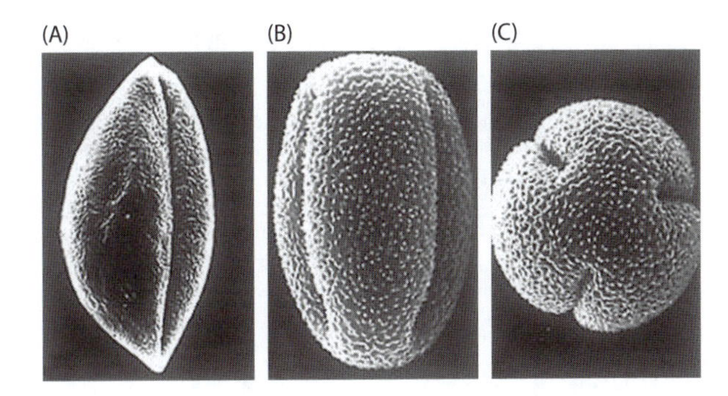

Figure 5.6 (A) Monocolpate pollen of *Convallaria majalis*. (B, C) Tricolpate pollen of *Aspazoma amplectens*. The pollen grains are 40–60 μm in length.

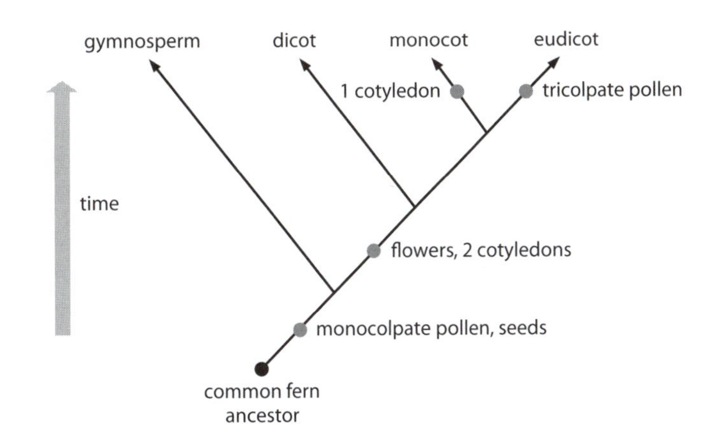

Figure 5.7 A cladogram showing how the various types of angiosperms evolved from a common fern ancestor.

An example of a hidden plesiomorphy

This chapter has covered the essentials of how the details of evolutionary history and relationships can be discovered using cladistics. In summary, a member of a species carries its deep, historical features (plesiomorphies), as well as its clade-specific (synapomorphies) and species-specific novelties (autapomorphies). The interesting question arises when the chain is broken because, say, a synapomorphy is lost, and an illustration of this comes from snakes that lack limbs. All the fossil, morphological, and molecular evidence points to the snakes being a reptilian clade whose most obvious defining feature today is their lack of limbs (an apomorphy for the current taxon). Although limb loss seems to have started about 100 Mya and occurred progressively (Martill et al., 2015), the fossil data cannot, of course, prove that the ancestors of today's snakes had normal limbs, only suggest it.

Careful examination of pythons shows that they do, however, have tiny hindlimbs, but not forelimbs, and it has required experimental and molecular embryology (evo-devo work) by Cohn and Tickle (1999) to unravel the details of limb formation in snakes. They first examined the patterning of Hox genes along the anterior–posterior axes of early chick and python embryos (**Figure 5.8**A), because there is now a considerable body of evidence (Chapters 8 and 11) showing that the expression pattern of these genes is responsible for the later organization of the axial tissues. The pattern in the chick shows how the arrangement of three *Hox* expression domains, in particular, specifies the cervical and thoracic regions, with the forelimb domain being just anterior to the expression regions *HoxC8* and

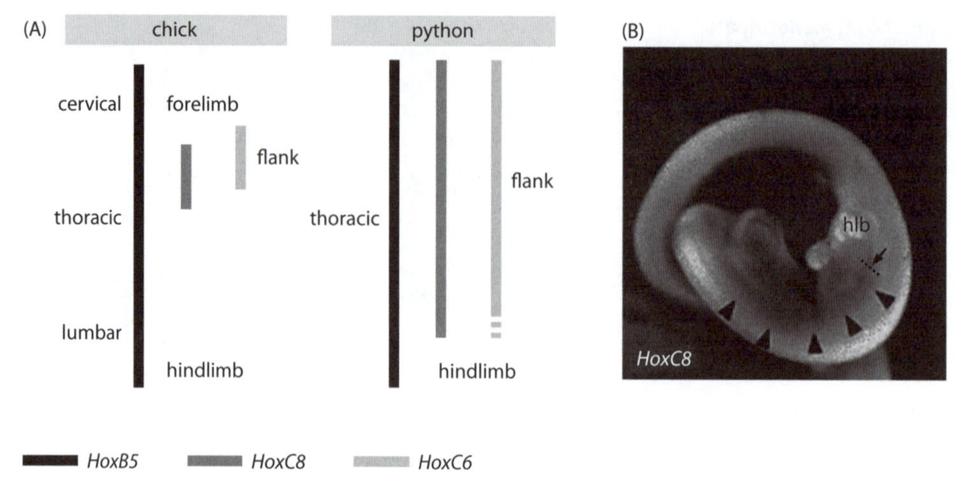

Figure 5.8 Hindlimb development in pythons. (A) Diagram comparing expression domains of *HoxB5*, *HoxC8*, and *HoxC6* (the broken end indicates lack of certainty about precise expression limits) in chick and python embryos. The expansion of *HoxC8* and *HoxC6* expression domains in python correlates with expansion of thoracic identity in the axial skeleton and the flank identity in lateral plate mesoderm. (B) *HoxC8* expression (arrowheads) extends posteriorly to the level of the hindlimb bud (hlb). The dashed line and arrow mark the posterior expression boundary of gene expression.

HoxC6; there is then a domain in the posterior cervical region where the forelimb can arise. In the python, the whole of the anterior region carries the Hox patterning associated with thoracic tissue (there are ribs all the way up to the head region); there is no cervical region and hence no forelimb.

The situation in the hindlimb is different. The Hox patterning is such that the early hindlimb bud forms in the python as it does in the chick (**Figure 5.8**B). Cohn and Tickle (1999) showed that the reason why the python bud does not develop further is that the signaling systems that activate proliferation and patterning in the early limb mesenchyme of the chick are not present in the snake. Python mesenchymal cells do, however, retain their ancestral abilities: if their limb bud mesenchyme is grown in the presence of chick signaling tissue, the mesenchyme can be activated to participate in normal proliferation and patterning.

The implication is clear: modern snakes carry their evolutionary history in their genes, still maintaining those needed for limb formation, but their expression is repressed by an evolutionary modification in the signaling system. The historical line of synapomorphies has been tinkered with but not lost. It is also intriguing that a similar, but reciprocal, effect has been found in Cetacea (whales and dolphins): here, forelimbs form while hindlimb buds cease development shortly after they have formed (Chapter 7). The implications of such patterning changes in the context of variation are discussed in Chapter 13.

Key points

- The evidence on evolutionary relationships can be collated in hierarchical trees, known as cladograms, which reflect descent with modification. Cladistics provides the formalism for constructing and studying such trees.

- Cladograms are mathematical graphs used to organize relationships among groups of species on the basis of shared and derived features. If the fossil record is used, their features reflect detailed anatomy. A typical relationship is:

  <species B> *<descends with modification from>* <species A>

  <species C> *<descends with modification from>* <species B>

- Plesiomorphies represent ancestral characters (for instance, one found in species **A**, **B**, and **C**). Synapomorphies represent shared and derived characteristics (species **B** and **C** share a property, **R**, that is a modification of property **Q** in species **A**). **C** differs from **B** because it has an autapomorphy or new, derived character, **T**, which derives from character **S** possessed by species **B**.
- Using these three types of feature, a cladogram can be constructed and used to sort out lines of descent.
- The presence of similar properties in two species is not of itself unequivocal evidence of the species being related as the property may have arisen independently in each. Such properties are known as homoplasies and examples are wings and sabre teeth.

Further reading

Kitching IJ, Forey PL, Humphries CJ & Williams DM (1997) Cladistics: Theory and Practice of Parsimony Analysis, 2nd ed. Oxford University Press.

Willis K & McElwain J (2013) The Evolution of Plants. Oxford University Press.

CHAPTER 6 THE ANATOMICAL EVIDENCE FOR VERTEBRATE EVOLUTION: FROM FISH TO BIRDS

This and the next chapter analyze the best-known evidence for evolution, that there are shared and derived anatomical relationships (synapomorphies) among living vertebrates that can be followed through a fossil record of specimens extending back in time more than 500 My. The aim of these chapters is to use this evidence to construct the history of vertebrate life. Such is the level of detail in the fossil record that it is simply not practical to be comprehensive. This chapter covers only the major taxa and transitions in the evolution of fishes, amphibians, reptiles, and birds, while Chapter 7 discusses the mammals. Together, the two chapters show that the fossil evidence is sufficient to enable all the vertebrates to be included within a single clade that links today's rich vertebrate fauna back to the earliest chordates of the Lower Cambrian (>529 Mya). Of particular importance here is the quality of the fossil record in showing how major transitions, such as that from a fish's fin to a vertebrate limb, occurred.

It is impossible to make full evolutionary sense of the vertebrates on the basis of those species that are alive today. This is because the structural differences between the major classes of vertebrate are profound, and there are no contemporary intermediates. Elucidating the history of eukaryotic evolution has thus depended heavily on the remarkable fossil record. However, because soft tissues rarely lasted long enough to be fossilized, this evidence comes almost entirely from their fossilized bones and teeth, tissues that were originally hardened by the presence of hydroxyapatite. Teeth also contain dentine and enamel, the latter being ~96% mineral and the hardest substance made by chordates.

Analysis of the many vertebrate organisms that have been fossilized has provided more detail about their evolution and classification than have the fossils of any other taxon (**Figure 6.1, Table 6.1**). For all of its breadth and depth, however, this knowledge is not complete: there are still insufficient fossils from the early stages of vertebrate life to fill in the details of their evolution or those of some of the transitions from one major group to another. Thus, for example, there is an important set of fossils showing how fins evolved to become limbs over a period of about 30 My, but there little information about how mammalian forelimbs became adapted for flying in the small animals that became bats, or how early archosaur reptiles adapted to become the flying pterosaurs, and only a little more on how later ones became birds. The reason is that the transition from land to air required a substantial lightening of the bones to facilitate flight, and light bones are fragile. The transition from sea to land, however, required strengthening and extending the skeleton of large fishes so that the limb bones and girdles became strong enough to enable them to live on land. Large heavy bones are obviously far more likely to be fossilized than small

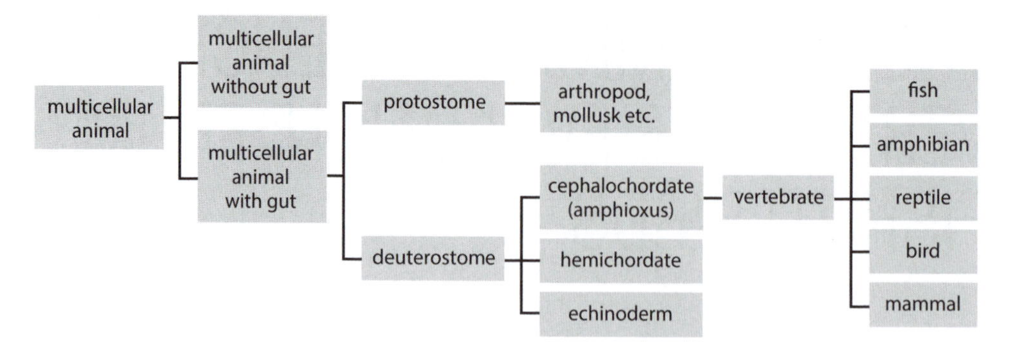

Figure 6.1 A Linnaean taxonomy of the multicellular animals locating the vertebrates. Note that this is not a phylogeny based on "descent with modification" but a taxonomic classification where the relationship is "is a member of the class of" (sometimes simplified to "is a").

light ones - the vertebrate fossil record is essentially a random sample of life with a strong bias towards larger and more recent organisms.

Deciphering the fossil record means finding the evidence to support the view that one taxon evolved from another. This, in turn, means closely examining the anatomy of sets of similar fossilized organisms to identify which of their features can be seen as shared and derived (synapomorphies), so implying descent, and which are novel (apomorphies). Note that the age of the fossilized organism plays no role in these assignments; this is because species can continue to exist long after others have evolved from them. Thus, *Haikouella*, a sophisticated chordate is found in Lower Cambrian rocks (~515 Mya; Chapter 4), while *Pikaia*, a much simpler chordate, was found in rock from the Middle Cambrian (~505 Mya) some 10 My years later. Such are the vagaries of the fossil record. Individual fossil dates indicate only when that organism died; they say nothing about when its species evolved and when it became extinct. Cumulatively, of course, fossil dates indicate the direction of time, but such dates cannot be part of the criteria for determining evolutionary relationships.

Fish

Contemporary fish

This is the taxon in which bone evolved and it is made in two ways in embryos. Dermal (or intramembranous) bone is laid down directly by skin dermal mesenchyme and was first used to form armored plates in the body and by head mesenchyme to form most of the skull; it is also the origin of the dermal rays present in fins. Endochondral bone, which comprises the rest of the skeleton, forms through ossification of a cartilage model laid down by other mesenchyme throughout the body. One of the transitions from marine to land-based vertebrates was the replacement of dermal rays in fins with endochondral digits in limbs.

The major current fish taxa include three with cartilaginous skeletons and two with bony skeletons. The first two of the former group are agnathans - they lack jawbones (gnatha) and paired fins but are clearly an advance on *Amphioxus*, a cephalochordate that lacks eyes, a skeleton, and embryonic neural crest cells. The lampreys have cartilaginous skulls and vertebral elements (called arcualia), complex eyes, and a dorsal fin. The slime-producing hagfish are much more primitive: they have cartilaginous skulls with eyespots but

Table 6.1: Selected properties of some major vertebrate taxa

Major vertebrate taxa	Some selected properties
Anamniotes	
Cephalochordate	Mouth, no eyes or fins, segmented muscles, pharyngeal slits, notochord (for example, amphioxus – also known as the lancelet)
Agnathan fishes	Jawless mouth, fins, cartilaginous skeleton, eyes, neural crest (for example, lamprey)
Chondrichthyan fish	Jaws, cartilaginous skeleton (for example, sharks)
Osteichthyan fishes	Jaws, bony skeleton
Ray-finned fishes (teleosts)	Fins with bony rays (for example, salmon)
Lobe-finned fishes	Pectoral and pelvic lobe fins with 1 + 2 bones and terminal rays (for example, lungfish)
Labyrinthodonts (early amphibians)	Four limbs, large tail with fin, heavy skull roof, complex teeth, (extinct; for example, *Ichthyostega*)
Lissamphibians (frogs, newts)	Four limbs, normal teeth, aquatic larva
Amniotes	
Anapsid reptiles	Skulls lacking temporal fenestrae (for example, turtles)
Diapsid reptiles	Skulls with two temporal fenestrae (for example, lizards)
Archosaur reptiles	Diapsid skulls with antorbital and mandibular fenestrae
Pseudosuchia	"Lizard-hips," quadrupedal gait, crurotarsal ankle joint (for example, crocodiles)
Avemetatarsalia	Mesotarsal ankle joint (for example, theropods and birds)
Theropod dinosaurs	Bipedal, "lizard hips" (extinct; for example, *Tyrannosaurus rex*)
Sauropod dinosaurs	"Lizard hips," quadrapedal gait with massive legs with minimal digits, (extinct; for example, *Brachiosaurus*)
Ornithischian dinosaurs	"Bird hips," predentary bone on lower jaw, mesotarsal ankle joint (extinct; for example, *Stegosaurus*, hadrosaurs)
Birds	Skulls with two temporal and antorbital fenestrae, "bird hips," bipedal, three-digit wing, no predentary bone (for example, chickens)
Pterosaurs	Keel, light bones, mesotarsal ankle joint, five-digit wing (extinct; for example, pteranodon)
Synapsid reptiles	Skulls with a single temporal fenestra, limbs (extinct; for example, pelycosaur)
Mammals	Synapsid-derived skull, warm-blooded, hair, mammary glands, mainly placental (see Chapter 7)

neither fins nor vertebrae, although the latter may represent a secondary loss (Ota et al., 2011). Incidentally, the hagfish has a curious eating habit: it bites the flesh of other fish then ties a knot in itself, which it slides to its head; it next pushes the knot over its head and pulls its mouth and a morsel of flesh away from the source of its dinner. Other cartilaginous fish (the chondrichthyes), such as the sharks and rays, have vertebrae, jaws, and paired fins but lack swim bladders.

The current bony fish taxon (the Osteichthyes) has two subtaxa: the ray-finned fishes (teleosts, which evolved from the Actinopterygii), which includes the great majority of fish species; and the lobe-finned fishes (the Sarcopterygii), both of which have swim bladders. The latter group now includes lungfish and coelacanths, whose lobed fins are notable for including bones and joints that are roughly characteristic of the fore- and hindlimb bones of amphibians, although they end in bony rays rather than digits. It is worth noting that a feature of sarcopterygians is teeth with complex foldings, a feature that also characterizes the earliest amphibians, the labyrinthodonts.

The fish fossil record

The earliest evidence of chordate hard tissue is the tiny fossilized teeth (~0.1 mm) of *Haikouella* (Lower Cambrian; see Figure 4.5) that were located in its pharynx and used to break up food already engulfed rather than for biting. In some Cambrian fossils, soft tissue is ossified and, in such early chordates, segmented muscle, another synapomorphy of the early vertebrata, is more often preserved than evidence of any early axial skeleton (see Figures 4.3A, 4.5). This is because muscle is composed of dense protein, while the notochord and any early skeletal support tissues were made of nonmineralized cartilage, which was hydrated and so easily lost. An exception, however, is the Lower Cambrian chordate *Haikouichthys ercaicunensis*; its fossils show traces of vertebral cartilaginous elements (Shu et al., 2003).

Although early segmented Cambrian chordates such as *Pikaia* and *Haikouella* (Chapter 4) were marine organisms, the earliest vertebrates that were unequivocally fish were the late Cambrian conodonts (~515 Mya). These were eel-like agnathans with a complex feeding apparatus of dental elements rather than teeth, although they probably contained dentine and enamel (**Figure 6.2**; Donoghue & Sansom, 2002). The conodont taxon survived until the end of the Triassic Period (~200 Mya).

From early but unknown ancestors of the conodonts, the various groups of ostracoderm agnathan fishes evolved at the end of the Ordovician Period (c. 488 Mya). They are subdivided into a broad group of fish taxa characterized by head shields. A particularly important group is the heterostracans; these had head shields of dermal bone, a body covered with trilayered scales, and a tail (Sansom et al., 2005). There were also

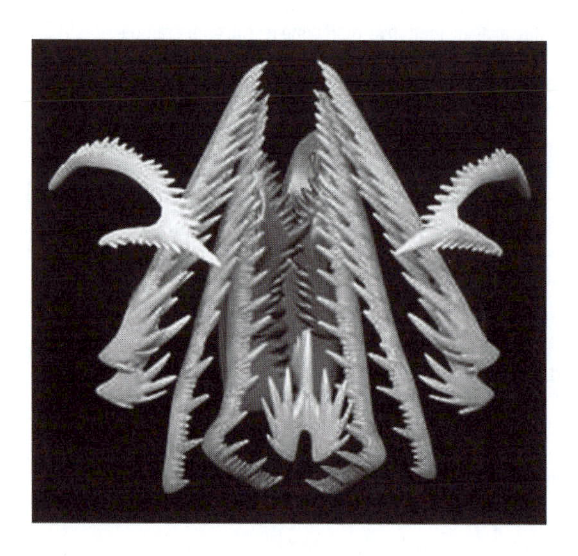

Figure 6.2 Rostral view of a model of the dental elements of an ozarkodinid conodont. Each element is ~1–2 mm long.

ostracoderm fishes with other features, for example *Sacabambaspis janvieri*, a heterostracan that possessed a posterior notochordal lobe rather than a tail (Pradel et al., 2007). Its fossils are sufficiently well preserved that the bone structure can be analyzed: its head shield was made of dermal bones whose layered structures included dentine, acellular bone, and enamelloid with pulp and spongy spaces so that, in some cases, it appears as if they are made of compacted teeth. In no ostracoderm fish has an axial skeleton been preserved in the fossil record; any vertebral elements must have been cartilaginous and probably small.

By the mid-Silurian Period (c. 430 Mya), primitive jaws were present in two groups of fishes, the acanthodians and the placoderms, although their line of descent is not yet precisely known because the fossil record is thin here (Brazeau & Friedman, 2014). By 409 Mya, jaw bones had evolved in a placoderm, *Gnathostomata gegenbaur* (Zhu et al., 2013). These jaws formed from the mesenchyme of the first pharyngeal arch, an embryonic tissue that had originally given rise to the anterior skeletal part of the first gill opening (**Figure 6.3**). Both groups had paired fins and a cartilaginous endoskeleton with a layer of perichondral bone, and these or their common ancestor may have been the earliest organisms with vertebrae. The acanthodians, however, show the first evidence of endochondral bone, which permeates the vertebral cartilage, while the placoderms maintained a dermal skeleton of bony plates. Both taxa probably had swim or gas bladders (these develop as pouches off the gut anterior to the stomach and are homologous with lungs), and also showed evidence of internal fertilization and development. Of immense interest but unclear significance is a fossil placoderm with an umbilical cord connected to an internal embryo (Long et al., 2008).

The forerunners of today's cartilaginous fish, such as rays and sharks, were probably members of the acanthodian taxon: Burrow and Rudkin (2014) report that *Nerepisacanthus denisoni*, a beautifully preserved acanthodian fish from the Late Silurian Period (c. 420 Mya) had scales whose detailed structures are typical of those seen in later chondrichthyan organisms. These cartilaginous organisms appear in the fossil record at about 409 Mya: Miller et al. (2003) have reported an early fossilized shark with a braincase and paired pectoral fin-spines typical of chondrichthyans. Placoderms are, however, more likely to have been the direct ancestors of the Osteichthyes (modern bony fish) with a key apomorphy being the development of modern dentition (Rücklin et al., 2012). The oldest osteichthyan fossils date from the end of the Silurian (~420 Mya) and all are actinopterygian with rayed fins. The earliest known sarcopterygian fish with lobed fins were a little later, dating back to the early Devonian (~410 Mya); although most lived in the sea, some lived in fresh water.

The details of the early arborization of the fishes is still not entirely clear and it seems that early forerunners of the line that eventually led to the Osteichthyes (bony) and the Chondrichthyes (cartilaginous) fish clades survived for quite a long time. Thus, *Janusiscus schultzei*, an early Devonian gnathostome (~415 Mya), had features of both taxa (Giles et al., 2015).

The information given in the paragraphs above is sufficient to produce a low-resolution cladogram that shows the likely lines of descent from Cambrian chordates to the various major taxa of modern fishes (**Figure 6.4**). A few key species are included from the many that are known (Janvier, 1996). The different relative speeds of change during the Middle Devonian period (~410 Mya) emphasize the importance of depending on the details of organism anatomy rather than on time for elucidating evolutionary relationships.

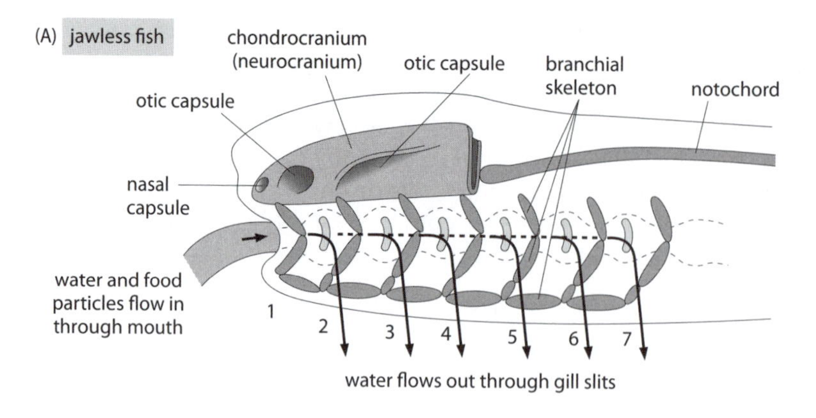

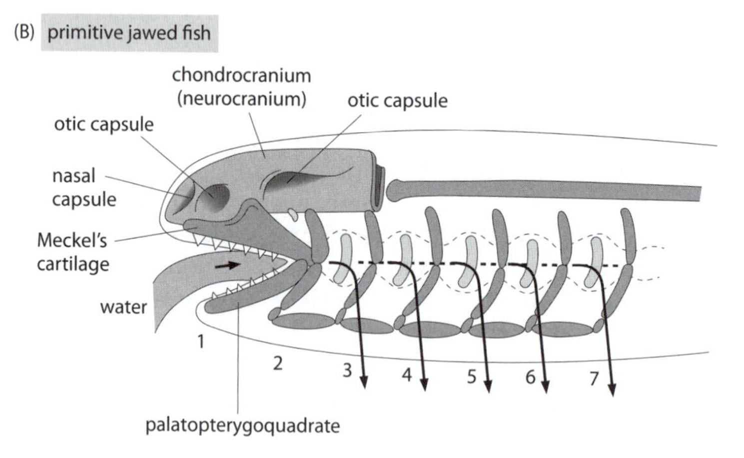

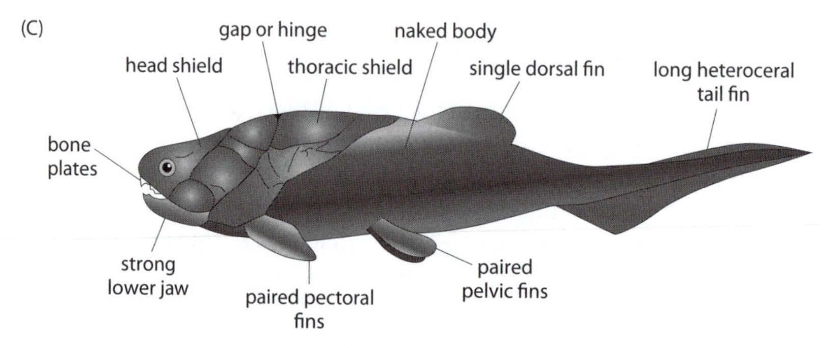

Figure 6.3 The jaw bones of fish derive from the mesenchymal cartilages of the first branchial arch of agnathan fish. (A) The cranial morphology of an early agnathan fish. (B) An early jawed fish. (C) The placoderm *Coccosteus decipiens*, an early jawed fish (gnathostome).

From sea to land

From a human viewpoint, the two most important events in the history of evolution were the colonization of land through fish evolving into amphibians and the evolution of the amniote egg, and both were complex. The former demanded the development of fore- and hindlimbs for movement, and of complete pelvic and pectoral girdles to anchor the

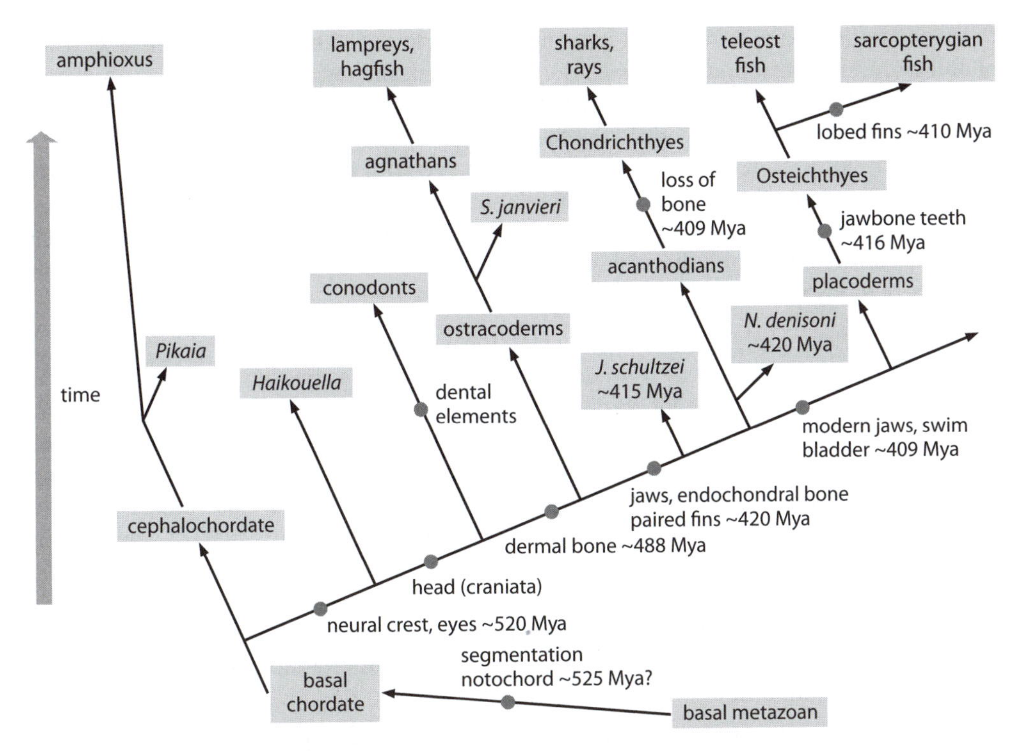

Figure 6.4 A cladogram showing the evolution of the major fish families from a basal chordate. Synapomorphies, major taxa and selected species (italics) are shown, together with the dates of appearance in the fossil record. Branch points represent unknown last common ancestors.

limbs and to support the body (Clack, 2012). It also involved changing the oxygen-uptake system from a gill system to one based on lungs. This needed the evolution of lungs from the gas-bladder pouches off the esophagus, together with modifications to the heart and body musculature. It also required a sophisticated metamorphosis that mediates the change from a fish-like larva to a land-based quadruped.

For several reasons, it is clear that the ancestor of amphibians was a sarcopterygian rather than a teleost fish, the most obvious being that the former has pectoral fins whose proximal skeleton includes a humerus, an ulna, and a radius, and whose pelvic fins include a tibia, a femur, and a fibula; teleost fins lack these limb bones, having only rays of dermal bone. Extant sarcopterygian families include the Dipnoi (lungfish) and the Actinistia (whose only living member is the coelacanth), and both anatomical and genetic evidence suggests that the likely ancestor of tetrapods was a member of the former (Brinkmann et al., 2004). In addition, not only do Dipnoi have lungs, but the early development of modern lungfishes and of amphibians is almost indistinguishable and very different from that of contemporary teleost fish (Gilbert, 2014; Semon, 1901). Today, lungfish fall into two classes: the Australian ones that have working gills and a single lung, and all the others that have two lungs and atrophied gills; the latter are more derived and hence the more likely ancestor of amphibia.

Over the last decade or so, the anatomy of a series of late-Devonian fossils, some recently discovered and others re-examined, have illuminated the nature of the transition from sea to land that took place during the relatively short period of 380–374 Mya (**Table 6.2**). They all fall within the Tetrapodomorph clade: these are all the tetrapods and those ancestors that diverged from the Dipnoi clade. One of these, the lobed-fin *Eusthenopteron* (385 Mya), makes a very plausible early ancestor for the amphibia: apart from its fin bones, it had the labyrinthodont teeth, nostrils that connected with the oral cavity and an intracranial joint that allows the head to turn, all of which are seen in the earliest amphibian fossils. Although only hard tissues have been preserved, *Eusthenopteron* would, on the basis of these features, probably have had the typical primitive paired lungs and breathing capacity of today's lungfish.

Clack (2012) has emphasized that the selection pressure that favored the evolution of limbs was not the potential to walk on land. Primitive limbs had lengthened before the evolution of pectoral and pelvic girdles with sufficient strength to support the weight of the organism on land. In addition, digits, which formed from endochondral bone rather than the dermal bone of rays, started to evolve before the transition to land. *Sauripterus*, a fossilized lobe-finned fish with well-developed girdles, ~370 Mya, had radials of endochondral bone, as well as rays of dermal bone in its fin (Davis et al., 2004). It seems likely that the evolution of limbs ending in autopods rather than rays was to facilitate movement through shallows with large amounts of seaweed; it was, however, exactly the exaptation required for walking on land. This term means an adaptation that was suited for one function had the potential to evolve later and so lead to a further adaptation (this was previously called preadaptation).

The nine species of fossils mentioned in Table 6.2, some of which are shown in **Figure 6.5**, have been ranked on the basis of their autapomorphies, but these do not give a clear line of descent and so do not provide sufficient evidence to construct a cladogram - further

Table 6.2: The transition from sea to land – the fossil evidence

Name	Mya	Family	Apomorphies
Kenichthys	c. 395	Sarcopterygia	Earliest fish with nostrils linking to the oral cavity (Zhu & Ahlberg, 2004)
Eusthenopteron	c. 385	Sarcopterygia	Aquatic, internal nasal adaptations, labyrinthodont teeth; the bones of pectoral and pelvic fins had growth plates for lengthening (Meunier & Laurin, 2012)
Gogonasus	c. 380	Sarcopterygia	Aquatic; middle ear typical of a land animal (Long et al., 2006)
Panderichthys	c. 380	Sarcopterygia	Four unjointed digit-like bones, tetrapod cranium (Boisvert et al., 2008)
Tiktaalik	c. 375	Sarcopterygia	Basic wrist with rays, flexible neck, lungs, basic pectoral and pelvic girdles (Shubin et al., 2014).
Ichthyostega	c. 374	Labyrinthodontia	Amphibian-type lungs, strong ribs, and fore- and hindlimbs (seven jointed toes) + fish gills and tail – able to clamber on land (Pierce et al., 2012)
Acanthostega	c. 365	Labyrinthodontia	Eight forelimb and seven hindlimb jointed digits, nonweight-bearing forelimbs, complete pelvic girdle (Coates, 1994)
Tulerpeton	c. 365	Labyrinthodontia	Six jointed digits, powerful "wading" limbs, pectoral girdle, lungs, no gills (Lebedev & Coates, 1995)
Pederpes	c. 348	First land tetrapod	Five (+ one?) digits. Land-adapted feet (Clack, 2002)

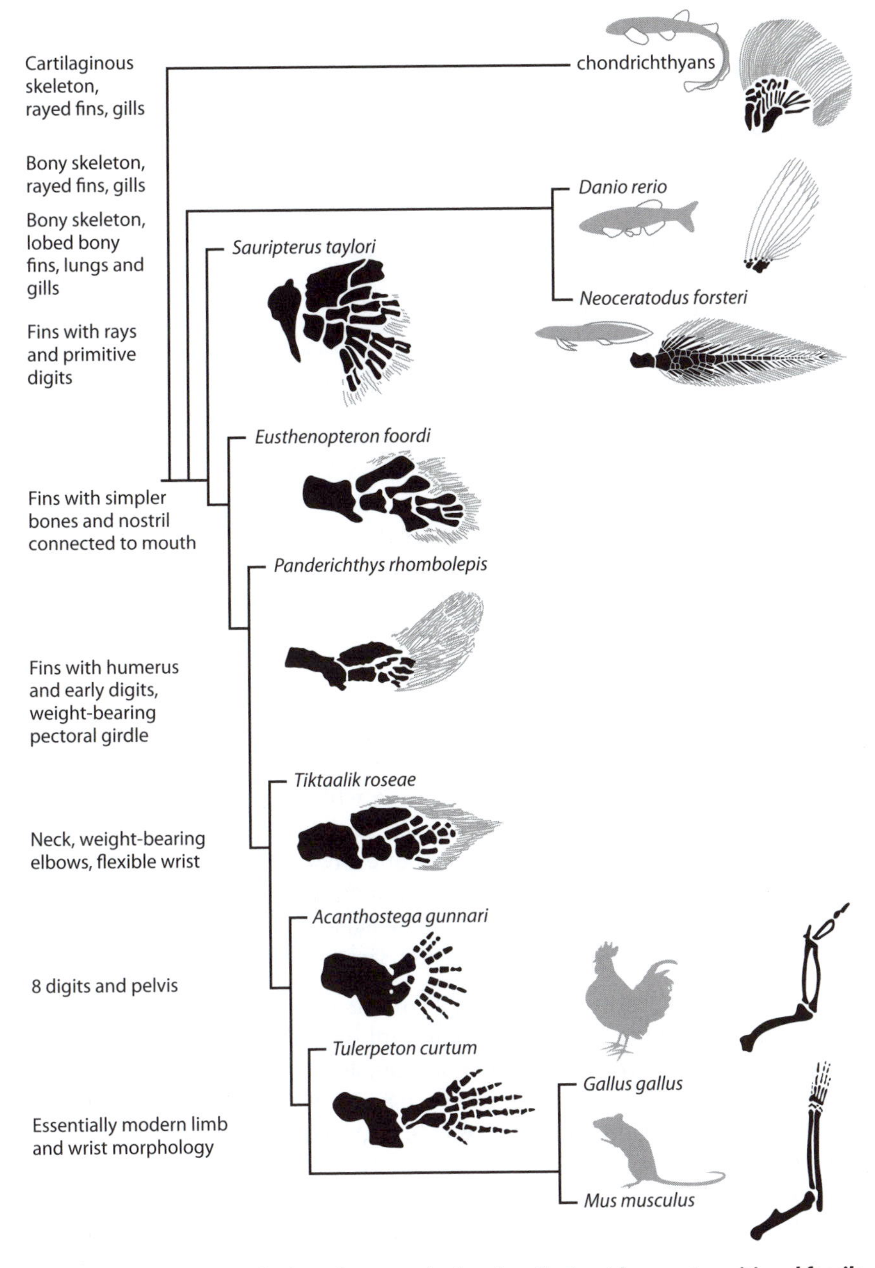

Figure 6.5 A diagram of the evolution of tetrapods showing the best-known transitional fossils.

species will be required. Thus, for example, *Panderichthys* (c. 380 Mya) has bony digits but a rigid neck, whereas *Tiktaalik* (c. 375 Mya) has rayed digits and a mobile neck. Similarly, a comparison of *Ichthyostega* (374 Mya) and *Acanthostega* (365 Mya) shows that the former had stronger forelimbs, while the latter had stronger hindlimbs. Such evidence suggests that, within the early tetrapodomorpha, there was a broad range of marine taxa living near the coastline that were evolving to exploit the opportunities that this niche offered. If so, more such transitional fossils may well be discovered in the coming years.

Limb evolution

Because the transition from sea to land was so important, it is worth considering each of the various changes in a little more detail. The lobed fins of early lungfish already had homologs of the tetrapod long bones, albeit that these were short and needed to lengthen (growth plates, the prerequisite for this, are seen in *Eusthenopteron*), so the major innovation required was the evolution of autopods (wrists and digits).

Twenty years ago, this seemed to be a difficult evolutionary problem because the terminal fin rays were thought to derive from neural crest cells, although these normally only make bone cells in the head. Although it has now been shown that local dermal mesenchyme from the lateral trunk forms these rays (Lee et al., 2013), all the endochondral long bones of the limbs, including those of the digits, form from mesenchyme derived from lateral mesoderm (Gilbert, 2014). The transition from rays to digits thus required migration of such mesenchyme into the distal parts of the early fin bud, together with a re-patterning of the development of this region so as to suppress ray formation and activate digit formation. The morphology of *Sauripterus* (Figure 6.5) demonstrates that this change was a gradual process (Davis et al., 2004), while the range of digit numbers in the various transitional species (see Table 6.2) shows that the mechanism for number specification took some millions of years to stabilize (Gilbert, 2014).

Girdle evolution

The pectoral girdle today includes the scapula (shoulder blade), the clavicle, and the coracoid bones. Comparative anatomy has shown that the tetrapod pectoral girdle derives from the nonweight-bearing partial pectoral girdle in fish, which supports the pectoral fin and is attached to the skull base. Part of the scapula, for example, derives from the scapulocoracoid and the remainder from the cleithrum, the bone linking the cranium to the pectoral fin (Holland, 2013). Equally important to the development of the tetrapod pectoral girdle was its change in position from the base of fish skull in fish to the cervical vertebrae. This separation of the girdle from the cranium allowed land vertebrates to have a mobile neck.

Evolution of the weight-bearing pelvic girdle with its three bones, the ilium, pubis, and ischium, was more complex. Lungfish have a minimal pelvis comprising the two pubic bones, which link to the pelvic fins but do not articulate with the vertebrae. It is therefore surprising that a fairly early transitional fish such as *Tiktaalik* (375 Mya) had a substantial girdle with broad iliac processes and acetabulae, but lacked the ischial bones that link to the vertebrae (Shubin et al., 2014). It is clear that this girdle was not used for walking but for enhanced propulsion through water. Boisvert et al. (2013) have analyzed the steps needed for the ischium and ilium to evolve and suggest that the most important was the dorsal migration of precartilage cells of the pubis so as to provide the raw material for the other bones; some re-patterning of their differentiation was also necessary to ensure that three bones formed rather than one.

Breathing

The other major step needed for life on land was an ability to breathe air rather than to absorb dissolved oxygen from water. As fish nostrils are blind-ending sacs lined with olfactory epithelium whose sole purpose is to provide a sense of smell, a key step was the extension of the nostrils through to the oral cavity, a feature first seen in *Kenichthys* (see Table 6.2), a small sarcopterygian fish. As none of the transitional fossils includes soft tissues, comparisons between modern lungfish and amphibians are needed to understand the physiological changes associated with the move to land. It is likely that early lungfish had a cardiovascular system much like that of their modern descendants: this includes primitive lungs in addition to gills, together with a heart with a single ventricle and a partially divided atrium separating most of the pulmonary and systemic flows. This heart is a simplified version of the three-chambered heart of extant amphibians.

Lungfish breath through buccal pumping (using their cheeks) as do today's amphibians, although the latter's breathing is strengthened by the additional force provided by the transverse abdominal muscles (Simons & Brainerd, 1999). In addition, amphibians have moist skin that can exchange gases (cutaneous respiration). Neither lungfish nor amphibians have a diaphragm; indeed, the only nonmammalian taxon to have a diaphragm is the crocodile and this muscular sheet seems neither to play a major role in normal breathing nor to be homologous with the mammalian diaphragm (Munns et al., 2012). Taken together, the evolutionary steps (see Figure 6.5) required to progress from the lungfish cardiovascular and pulmonary systems to those of the early amphibian do not seem as great as once they did.

Amphibians

Today, there are relatively few amphibian species, and the taxon (the lissamphibians) only includes toads, frogs, newts, salamanders, and the limbless caecilians. All have larvae with fins, gills, and a tail structure that reflects their evolutionary history. Most larvae develop in water, although some are internally incubated (in the mouth or in pouches) or develop on land in eggs whose external membrane allows gas exchange but is not permeable to water. All larvae, other than those of the axolotl, eventually metamorphose: they lose their tails and develop lungs that allow the adult animals to breathe air and so live on land.

The early amphibian fossil record is extensive: because there were no competitors when the earliest taxa colonized land, the amphibians thrived unchallenged for many tens of millions of years, giving rise to a host of species, some of which were very large, before declining during the Permian period, with most becoming extinct during the Early Cretaceous. The main amphibian taxon was the Temnospondyli, whose defining feature was the structure of its vertebrae, each of which was in several parts; many also had armor and clawed digits. Their size range was broad with the largest known being *Temnospondyli prionosuchus* (~270 Mya) that was ~10 m long. Perhaps the most important and certainly the best known of its early subtaxa were the labyrinthodonts (360–150 Mya) characterized by relatively massive skulls and teeth with internal foldings, hence their name (**Figure 6.6**). Of the other amphibian taxa, the best known are the lepospondyli; these first appeared in the Early Carboniferous ~350 Mya and were much smaller and lighter than the labyrinthodonts, as well as being characterized by simple vertebrae. They are the most likely ancestors of both modern lissamphibians and reptiles.

Unfortunately, the fossil record of the early lissamphibians is poor because they were relatively small and had fragile bones. The oldest lissamphibian fossil is from the Early Triassic (c. 250 Mya), but molecular dating has suggested that they may have evolved earlier (Marjanović & Laurin, 2007; San Mauro, 2010). The origin of the lissamphibians is still controversial. The fossil evidence is compatible with three hypotheses: that they are polyphyletic, that they descended from the Temnospondyli, and that they descended from

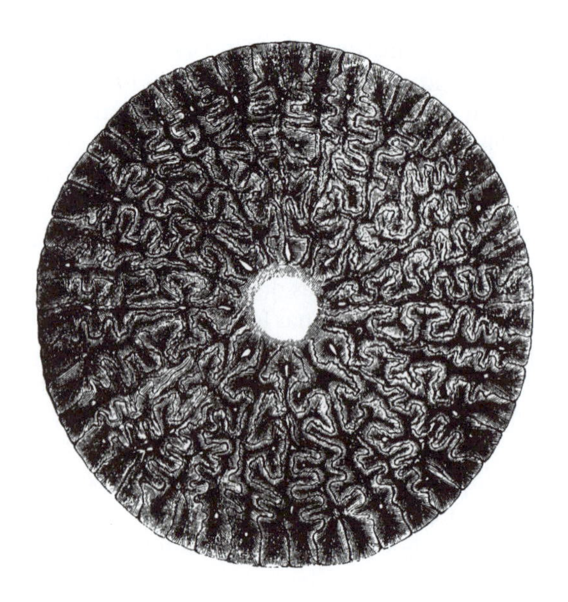

Figure 6.6 Drawing of a labyrinthodont tooth showing details of the convolutions.

the Lepospondyli, themselves descendants of the Temnospondyli. Current evidence suggests that the third, lespondyl, hypothesis is the most convincing (Marjanović & Laurin, 2009; Pyron, 2011).

The transition to reptiles and the evolution of the amniote egg

The first unambiguous reptile so far identified is probably *Hylonomus* (~312 Mya), recognized by its typical reptilian footprints. The captorhinids, reptiles with the head morphology of late amphibians, followed soon after. If *Hylonomus* really was among the initial reptiles, the transition from the first amphibian (374 Mya) to the first reptile seems to have taken some 60 My. This seems a surprisingly long time, given that there were vast areas of noncoastal land with a host of unoccupied ecological niches available for animals to colonize. Furthermore, amphibians with some early reptile features such as small legs and short tails emerged by about 355 Mya; these Microsauria were a clade of Lepospondyli, but their exact taxonomic status is still unclear.

The obvious reason for the delay was the time required for an egg to evolve that did not need water in which to develop; this was the essential apomorphy of the first reptile. The obvious route to achieve this would have been to build on the reproductive strategy shown by extant lungfish and lissamphibians: in both of these, the egg includes yolk granules which, as the embryo develops, mainly become located in the cells that will form the gut, and there are no external support membranes around the embryo. The end result for both lungfish and lissamphibians is a small larva that continues to develop in an aqueous environment. An alternative possibility was to build on a solution adopted much earlier by some chondrichthyan fishes, that of internal fertilization, simple placental development, and live birth of a small fish. This route is similar to that adopted much later by Eutherian mammals (Chapter 7).

In fact, the origins of the amniote egg reflect a third approach, that of teleost fish whose eggs contain a zygote with a separate yolk. Here, a series of cell divisions follows fertilization and, in due course, the peripheral cells of the embryo cells migrate over the yolk surface (a process known as epiboly, https://zfin.org/zf_info/movies/movies.html). Some of these cells help become the embryo, while the remainder forms part of the gut and consumes the yolk, so nourishing the embryo which soon becomes a freely swimming fry. Adopting this solution for the amniote egg required the yolk granules to accumulate separately within a yolk with its own (vitelline) membrane on which is located the zygote. Such an egg was much smaller than, but essentially similar to, a normal chicken egg.

It thus seems that the developmental route that led to the amniote egg is based on the reinvention of a mode of egg formation apparently lost in the sarcopterygian–amphibian–lissamphibian line. The difficulty in reconstructing this evolutionary history, however, is that nothing is known about the eggs of Devonian vertebrates. An alternative solution would have been for both lungfish and early amphibia to have had teleost-type eggs that eventually evolved to give the amniote eggs of early reptiles. If so, then one has to assume that modern Dipnoi yolkless eggs reflect a post-Devonian apomorphy with the similar eggs of modern lissamphibians separately evolving and so being a homoplasy. Other than through the fortuitous discovery of appropriate fossilized eggs in which soft tissues have been preserved, it is hard to see how this aspect of amniote evolution will be resolved.

The teleost-type approach required many other changes before an amniote egg was able to develop on land. The most obvious of these was the production of a protective shell that blocked evaporation; this allowed the embryo to develop away from water without drying out. The other major changes were internal with the early embryo making four extra-embryonic membranes: a yolk sac that absorbs nutrients from the yolk (like teleosts), an outer chorion that allows oxygen exchange, an amnion that retains fluid around the embryo, and an allantois that is a reservoir for nitrogenous waste (**Figure 6.7**). Only when these membranes were in place could the embryo develop beyond gastrulation. Once a modern reptile or avian egg is laid, the embryo continues its development and only leaves the egg when it is capable of independent life on land - there is no aqueous stage. These

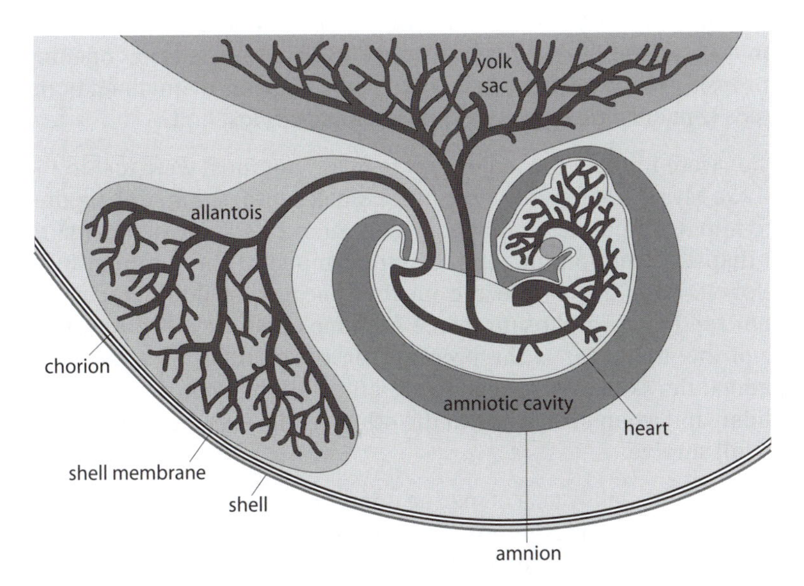

Figure 6.7 The amniote egg of a mammal, showing the internal membranes and associated cardiovascular system.

extraembryonic membranes are a synapomorphy for all reptiles and their descendants, including placental mammals whose internally developing eggs lack a shell and a yolk.

The evolution of the amniote embryo and egg thus required a complete reorganization of almost every event in the early development of both teleosts and amphibians, although very little is known about how the transition from the amphibian to the amniote egg was achieved. One early suggestion (Needham, 1931) was that the allantois evolved from a diverticulum off the larval bladder. Such was the complexity of the steps leading to the amniote egg that it is not surprising that it took 50 My or more. Understanding how the amniote egg evolved remains a major unsolved problem. Indeed, unless soft-tissue fossils are discovered that show the necessary structural detail, it is unlikely that the exact sequences of events that resulted in the amniote egg will ever be discovered.

Reptiles

The late Carboniferous period is marked by the appearance of tetrapods that seem to have had both amphibian and reptilian features. Amongst these was *Casineria* (c. 335 Mya; Albardi, 2008; Paton et al., 1999), a lizard-like organism with pentadactyl claws adapted to land but whose exact taxonomic status remains unclear because the sole fossil lacks hindlimbs and most of its head. Unambiguous reptiles, marked by lightened skeletons and pentadactyl limbs, together with footprint trace fossils, appear in the Late Carboniferous (315–310 Mya; Falcon-Lang et al., 2007). A well-known example is *Hylonomus lyelli* (~312 Mya), a small lizard-like organism that left footprints, while the earliest well-populated taxon seem to be the captorhinids (Müller & Reisz, 2005).

Both amphibians and early reptiles had skulls that lacked openings (temporal fenestrae) on each side behind the eye (see Figure 4.7). By about 304 Mya, however, araeoscelidian reptiles appear in the fossil record with two such openings on either side; they were probably used for attaching muscles whose particular role was to strengthen biting. Soon after, the reptile clade split into two main groups that can be distinguished on the basis of skull morphology: sauropsids such as *Petrolacosaurus* (302 Mya; Reisz, 1977) had two temporal fenestrae on each side of the skull, while synapsids had a single lower opening. It seems that the sauropsids soon split into three groups: the diapsids maintained both fenestrae, the euryapsids kept only the upper one, and the anapsids lost both.

The synapsids were initially the major reptile taxon but most were lost in the Permian extinction (252 Mya). The few survivors were small reptiles that, during the Mesozoic era, evolved into mammals and radiated widely after the K-T extinction (66 Mya; Chapter 7). The diapsids were minor reptiles during much of the Permian but started to flourish before its end. In the Mesozoic era, they became the dominant clade, comprising the major reptile and archosaur taxa, which included the birds. A well-known plesiomorphy of the archosaur is the presence of a preorbital fossa (see Figure 4.7D), which lightened the skull. After the K-T extinction, all the archosaurs were lost except the crocodilian and avian taxa, while of the other reptile taxa, only snakes, lizards, and the tuatara still survive.

The euryapsidae were mainly taxa of marine nondinosaur reptiles that contained the ichthyosaurs, placodonts, and plesiosaurs, and were all lost in the K-T extinction. Analysis of euryapsid skeletons as a whole showed that they almost certainly evolved from diapsid reptiles and, as this seems to have happened several times, the euryapsidae are viewed as a grade rather than a clade within the sauropsids. The ichthyosaurs looked very much like modern dolphins; this shared morphology is a good example of convergent or homoplastic evolution.

The anapsids survived the K-T extinction and today they are the Testudines (turtles, tortoises, and terrapins). The details of their evolution remain unclear because the fossil data are thin: the earliest known fossil turtles lived about 260 Mya, long after the first reptiles (Lyson et al., 2013). Even the genomic data are ambiguous. Mitochondrial DNA sequencing (Zardoya & Meyer, 1998) points to the testudines being closer to the archosaurs than today's Lepidosauria (reptiles with overlapping scales such as the lizards and snakes), while the presence and absence of various unique microRNAs give the opposite view (Lyson et al., 2012).

The exact timings of when these various groups of reptiles evolved and whether the appearances of the fenestrae were unique events are still contentious as the early reptile fossil record is thin. Nevertheless, the details of these fenestrae are the clearest synapomorphies for defining the major reptilian clades and their descendants. It should be emphasized that in extant taxa the temporal fenestrae are often missing or have altered morphology. There is, for example, only one fenestra in modern lizards and none in snakes or most extant mammals. Nevertheless, the details of the fossil record together with the presence of other shared derived characteristics confirm that the lizards and snakes had diapsid ancestors. In contemporary mammals, the bones surrounding the original lower temporal fenestra of synapsid reptiles now form the zygomatic arch (Chapter 7). Because the temporal fenestrae are such a key synapomorphy in distinguishing the main land animal clades, relatively few other anatomical features are required to establish the core cladogram describing the evolution of the taxa of limbed vertebrates that evolved from lungfish ancestors. A simple such cladogram is given in **Figure 6.8**, although not all the dates are as precise as one might like.

The Permian reptiles

The synapsid clade radiated to become the dominant reptiles of the Permian period: early examples are known as pelycosaurs and later ones as therapsids. The earliest-known pelycosaur is *Archaeothyris* (~306 Mya) and the best known is *Dimetrodon* (295–272 Mya), famous for its sail of elongated neural spines whose role may have been regulating temperature (**Figure 6.9**). Curiously, a similar sail formed on *Spinosaurus*, a later diapsid dinosaur (112–97 Mya) – this is clearly an example of homoplastic evolution (Chapter 5). The great majority of therapsids were lost in the Great Permian Extinction (252 Mya), leaving just a small group of rodent-sized creatures. During the Mesozoic, these few synapsid taxa slowly evolved into rodent-like mammals and, for this reason, their Permian ancestor synapsids are often called the mammal-like reptiles or proto-mammals, a better name because they had no overt mammalian features. After the K-T extinction, they radiated into the niches previously occupied by archosaurs and flourished to the extent that they (which includes us) have come to dominate the land (Chapter 7).

When they first evolved, diapsids such as the early *Petrolacosaurus* (c. 302 Mya), a reptile about 40 cm long), seem to have played a minor role in the Permian fauna. The early diapsids gave rise to two main groups, the lepidosaurs and the archosaurs. The former radiated more widely in the Mesozoic era, giving rise to such oddities as the Late Triassic *Kuehneosaurus latus* (c. 210 Mya), a lizard with winged forelimbs that it probably used to attenuate falling speeds through gliding (McGuire & Dudley, 2011). Today, there are relatively few

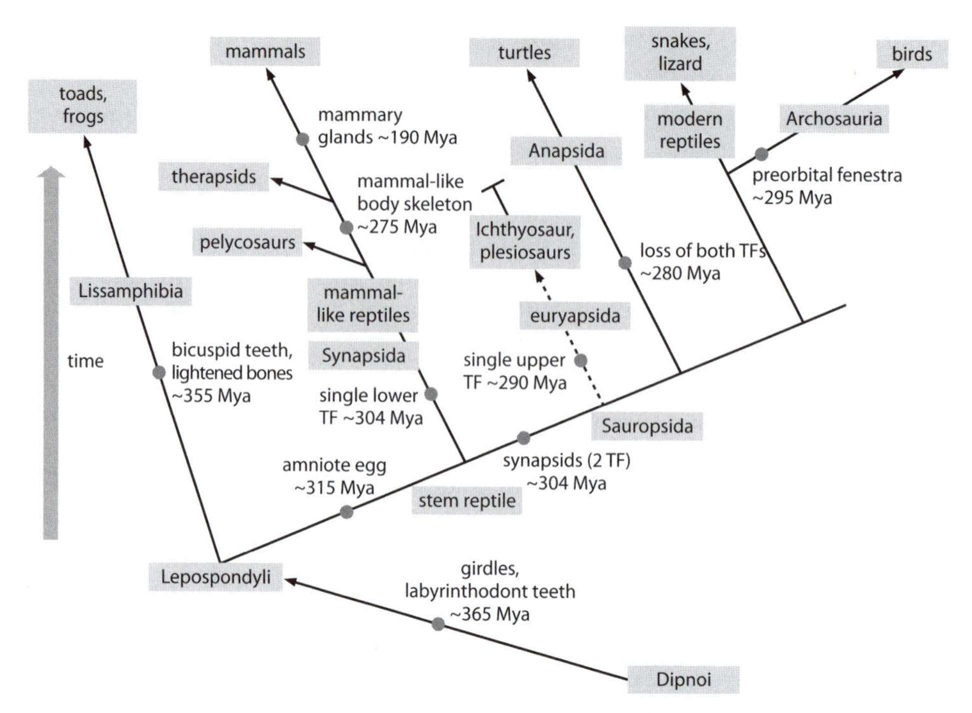

Figure 6.8 A cladogram showing the likely steps leading to the evolution of the major reptile taxa from amphibians, which, in turn, stem from Dipnoi (lungfish). Key apomorphies are shown together with the approximate date that they became established. The evolution of the euryapsida is shown dashed as it may represent several events (TF: temporal fenestra).

major *Lepidosaur* taxa, just the squamates (snakes and lizards), turtles, and the tuatara, a lizard-like animal found only in New Zealand. The exact timing of the first archosauromorphs (the most basal members of taxon) is unclear: the earliest known examples date to about 250 Mya (Ezcurra, 2014; Marsicano et al., 2016).

Archosaurs can be distinguished from other reptiles by a series of synapomorphies. The best known of these are the preorbital and mandibular fenestrae, the former an opening between the bones anterior to the eye (see Figure 4.7) and the latter an opening in the lateral part of the jawbone, both of which lightened the skull. It is at first sight odd that developing what seem such minor properties as these were key steps on the evolutionary journey of a group that probably emerged very early in the Triassic and went on to dominate the Mesozoic world. The key role of the preorbital fenestrae is still unclear, but it seems to have housed a diverticulum off the nasal cavity (or perhaps a gland or muscle attachment; Witmer, 1987) or may just have been an early autapomorphy of the basal taxon whose role was to lighten the skull. Other synapomorphies included socketed teeth that were less likely to be pulled out and a ridge on the femur, the fourth trochanter, which served as a muscle attachment and facilitated bipedality; this feature was lost in some later tetrapod archosaurs such as the crocodiles and sauropod dinosaurs.

The mesozoic reptiles

The main radiation of the archosaurs (**Figure 6.10**) started during the Triassic with the evolution of dinosaurs, Pseudosuchia, and pterosaurs, together with a range of other

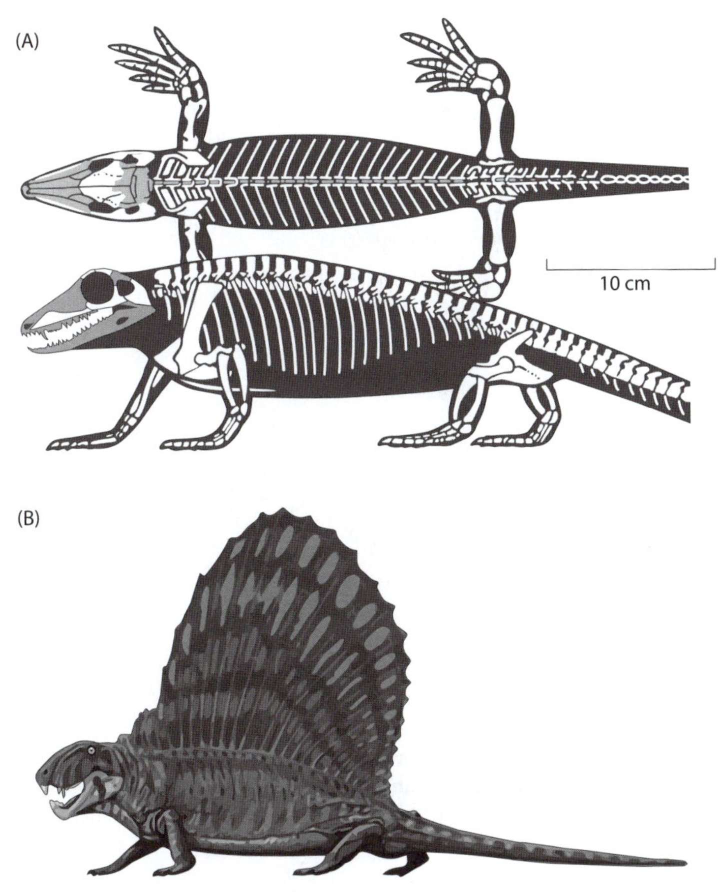

Figure 6.9 Synapsid (proto-mammal) pelycosaurs. (A) A drawing of *Archaeothyris*, a small lizard-like reptile. (B) A reconstruction of *Dimetrodon grandis*.

forms. Many of these, together with many amphibian and nonarchosaur reptile taxa, were lost in the extinction at the end of the Triassic (c. 200 Mya). This clade again divided into two clades. The avemetatarsalia (originally the ornithosuchia), which includes all archosaurs that are nearer the birds than the crocodiles, contained the dinosaurs characterized by hip morphology and more than 10 other anatomical features (Nesbit, 2011), together with the pterosaurs that were flying reptiles having very light air-filled bones and a keeled breastbone for flight-muscle attachment (these are a homoplasies shared with birds). The Crurotarsi included the Pseudosuchia, whose best-known taxon is the crocodiles with a massive skull, two sets of dorsal protective plates, together with several minor clades. Thousands of archosaur species are now known and it is only possible to mention the major clades here.

A defining archosaur feature is the ankle joint (**Figure 6.11**): the early Triassic archosaurs, which had stiff ankle joints (for example, *Eodromaeus murphy* c. 240 Mya; Martinez et al., 2011), soon gave rise to two major clades whose ankle joints evolved differently (Thulborn, 1980). The Crurotarsi (which means lower-leg foot bones) have a specialized ankle where, amongst other features, the astragalus and the calcaneum form a hinge because the joints with the tarsals is inflexible. The Avemetarsalia, which means bird foot bones,

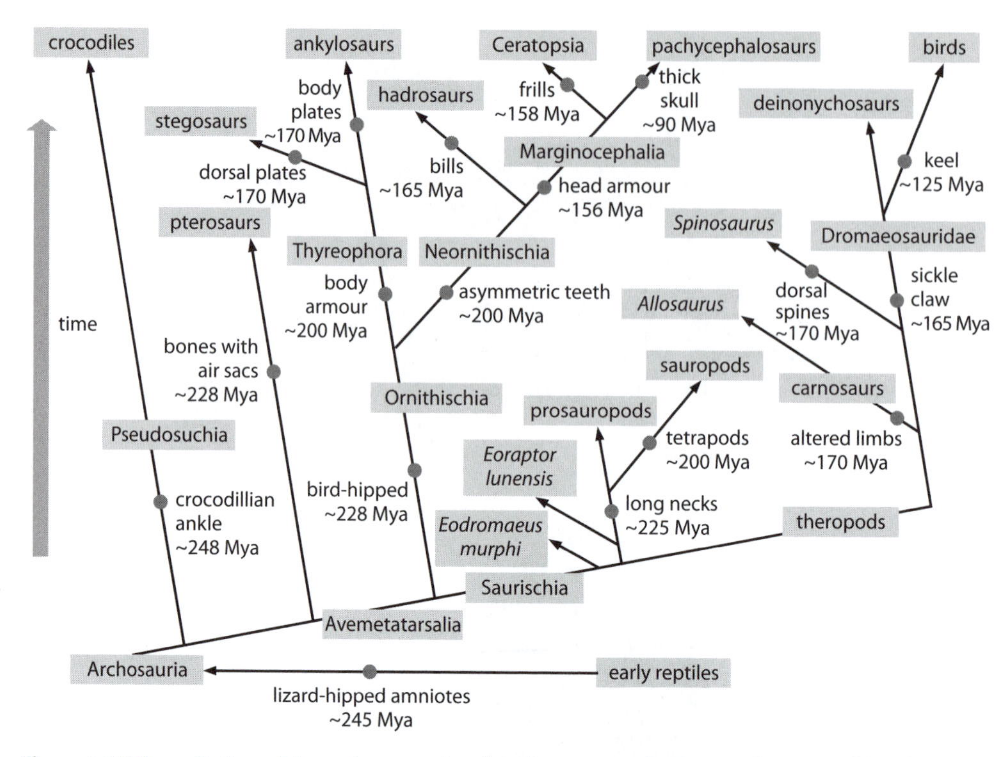

Figure 6.10 The radiation of the archosaurs showing the many major taxa and apamorphies that define them, together with the approximate times when they first appear in the fossil record. There is still some ambiguity in the exact details of the dinosaur cladistics and some allocations of early taxa should be viewed as provisional.

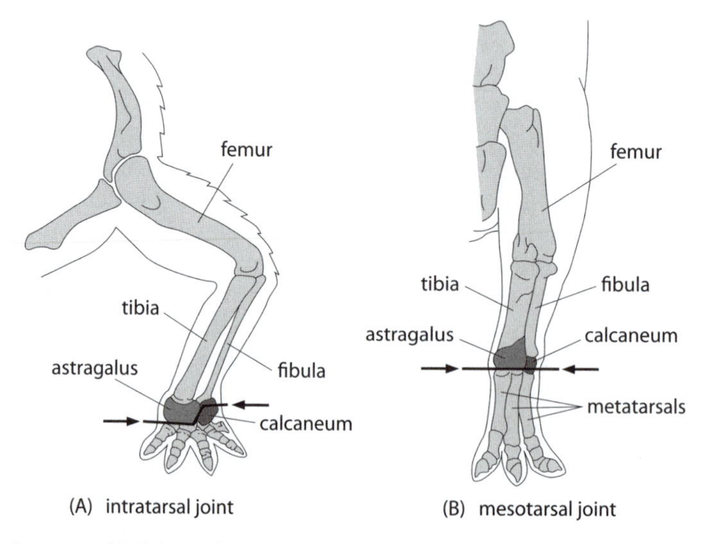

Figure 6.11 Archosaur ankle joints with the line of flexion in heavy black. (A) The intratarsal ankle joint shown by members of the Crurotarsi clade, which includes the crocodiles and other related taxa, is mainly between the calcaneum and the astragalus, the two bones of the ankle. (B) The mesotarsal joint shown by all members of the avemetatarsalia. This is between the ankle bones and the metatarsals and is more flexible.

had mesotarsal ankle joints where the astragalus and the calcaneum form a joint with roller capacity because the joints with the tarsals are flexible.

The expansion of the Avemetarsalia in the Triassic is marked by the appearance of the Saurischian and Ornithischian clades. The saurischians included the sauropods, large, pillar-limbed herbivores that walked on four legs, and the theropods that were mainly carnivores that walked on two legs and were the ancestor of the birds. The Ornithischian clade includes a large radiation of the herbivorous dinosaurs that is characterized by the presence of a predentary bone that extends the lower jaw and facilitates the eating of plants; all were, on the basis of their teeth, herbivorous.

The Saurischian and Ornithischian clades can be distinguished on the basis of the geometry of their pelvis (**Figure 6.12**), with its three linked bones, the upper ileum, which attaches to the sacrum of the backbone, the backward-facing ischium, and the pubis. In the basal lizard and all early archosaurs, the pubis points downwards and forwards, and this arrangement, known as lizard-hipped, also characterizes all Saurischia, except for the birds. The Ornithischia, in contrast, have a pubis that is almost horizontal with the anterior part extending forwards and a posterior point that aligns with the ischium, an arrangement seen in birds and therefore described as bird-hipped.

The saurischian radiation was particularly diverse. Early small carnivorous dinosaurs such as *E. murphy* soon gave rise to two taxa, the long-necked prosauropods such as *Eoraptor* and the therapods. The former gave rise to the eventually enormous sauropods (such as *Apatasaurus* and *Supersaurus* whose estimated weight was 35–40 tons), which were mainly herbivorous; the latter evolved into a wide variety of mainly carnivorous dinosaurs ranging in size from the Late-Jurassic *Compsognathus* of about 2.5 kg to the semiaquatic *Spinosaurus*, which weighed more than 10 tons and was larger than *T. rex*. A cladogram showing the evolution of the best-known archosaur taxa is given in Figure 6.10 together with core synapomorphies and the dates when they are first seen in the fossil record.

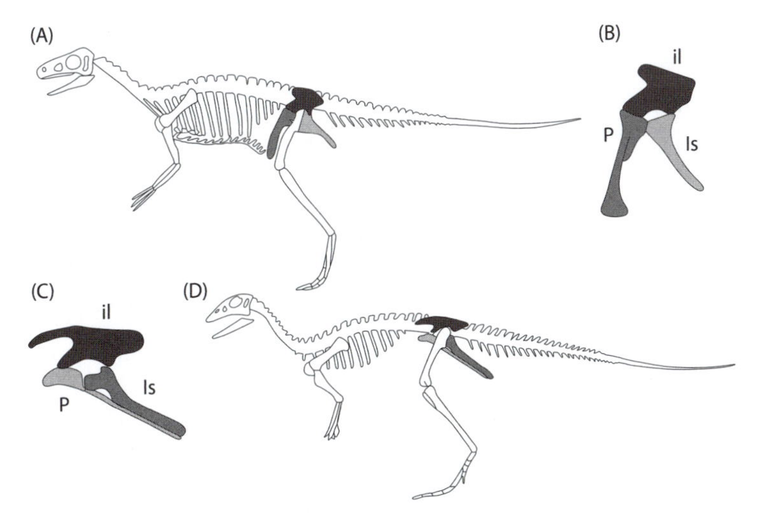

Figure 6.12 (A, B) Drawings of *Eoraptor*, a basal Saurischian, with a lizard-hipped, forward-directed pubis. (C, D) *Lesothosaurus*, a basal Ornithischian, with its bird-hipped, rear-directed pubis (P: pubis; il: ischium; il: ileum bones).

The Ornithischia (see Figure 6.10) are a complex group, and their detailed cladistic analysis is still under discussion. The most recent analysis has two major taxa, the Thyreophora, or armored dinosaurs (for example, *Stegosaurus* and *Ankylosaurus*), and the Neornithischia, characterized by having the enamel on their lower teeth laid down so as to facilitate eating tough plants. One major taxon here is the Marginocephalia (fringeheads): it includes the heavy-skulled bipedal Pachycephalosauria and the quadrapedal Ceratopsia that had a synapomorphic rostral upper-jaw bone and often displayed a horny frill around the bill (for example, *Triceratops*). A second major clade is the bipedal Ornithopoda, characterized by a horny beak and an elongated pubis but lacking an armored head. This clade includes the iguanadonts, the first dinosaurs to be discovered, and the later hadrosaurs, many of which had large hollow cranial crests (for example, *Charonosaurus jiayinensis*; **Figure 6.13**).

The archosaur radiation was so successful that the clade came to dominate the earth throughout the Mesozoic era and some thousands of species have so far been cataloged; there were probably many more that are either yet to be found or that have left no fossil record. Such was their ubiquity that there was barely any ecological space for the other, major clade of vertebrates, the synapsids, which, by the end of the Cretaceous had evolved

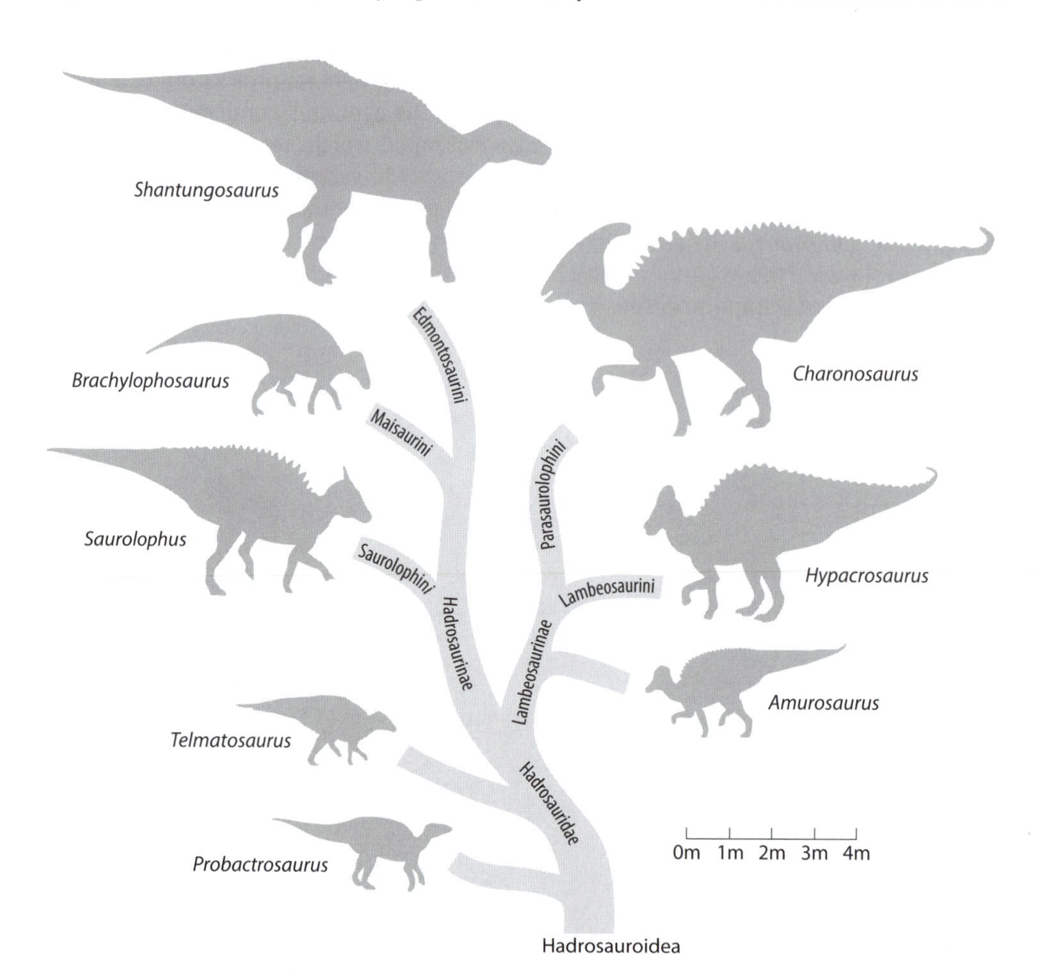

Figure 6.13 The hadrosaur radiation.

into the various mammalian taxa, all small-bodied. The archosaurs ruled the land and air for almost 200 My until the K-T extinction at the end of the Cretaceous Period (66 Mya), after which the only archosaurs that survived were the birds and crocodiles. Finally, there was space for the mammals to radiate.

The origin of birds

The anatomy of living birds does not at first sight fit neatly into any obvious category: they have, for example, the antorbital and diapsid fenestrae of archosaurs, the bird-hipped pelvis of Ornithischian dinosaurs, the omnivorous abilities of Saurischian dinosaurs, and the breast keel and light, hollowed bones of pterosaurs. They are, alone among contemporary diapsids, warm-blooded and feathered (Chen et al., 2015). Although Huxley in the nineteenth century had felt that the skeletons of small carnivorous theropods such as *Compsognathus* were similar to those of birds, it was commonly held for a long time that the clade was an early spur off the diapsid line that developed independently of other taxa. These views were challenged in the 1960s by John Ostrom and Robert Bakker on the basis that the fossil and other evidence meshed better with Huxley's view than any alternative, and more recent evidence confirms this view. Although the light bones of birds do not fossilize well, there is a clear record of the transition from theropod dinosaur to bird (Xu et al., 2014; see also en.wikipedia.org/wiki/List_of_fossil_bird_genera).

The origins of feathers may be rather earlier than was once thought: filamentous skin structures have been noted in both theropods and Ornithischian heterodontosaurs and it is thus possible that these were present early in dinosaur evolution (Alibardi et al., 2009; Zheng et al., 2009). Simple feathers date back to about 170–165 Mya (Zhang et al., 2008) and their presence on many theropods seems initially to have been for insulation and perhaps display rather than flight (Alibardi et al., 2009; Martin & Czerkas, 2000; Xu, 2012). Later, fully feathered animals such as *Anchiornis huxleyi* (~155 Mya) and *Archaeopteryx* (~150 Mya; see Figure 4.8A) were clearly transitional organisms between dinosaurs and birds, albeit that they were almost certainly incapable of flight (Longrich et al., 2012). Both are important: the former has sufficient detail in its melanosomes for its feather colors to be identified (red, black, and white; Li et al., 2010) – the latter not only possessed the dinosaur features of lizard-hipped pelvis and jaws with sharp teeth, but also had bird-like features, such as a wing-like forelimb with three clawed digits. If the original function of feathers was for insulation, then it was probably this that protected birds from the cold associated with the K-T extinction (66 Mya). Feather evolution is clearly another example of an exaptation.

Flight does however require more than feathers, and, during the Cretaceous Period, this clade of proto-bird therapods underwent a series of major anatomical changes. These included a loss of weight through the lightening of bones and the formation of a deep sternal keel for the attachment of enlarged pectoral flight muscles (Lee et al., 2014). In addition, the tail was replaced by a much smaller pigostyle of fused vertebrae, the jaw bones and teeth by a beak and gizzard (Louchart & Viriot, 2011), forelimbs by wings, and a lizard-hipped pelvis by a bird-hipped pelvis. This last feature incidentally mirrors the morphology of the Ornithischian dinosaur hips, but is a homoplasy and not a synapomorphy. An example of an Early Cretaceous bird with these features is *Confuciusornis* (c. 125 Mya; see Figure 4.8B); it could probably fly but not very well (Wang et al., 2012). The evolutionary line from sauriscian therapod to modern bird is now unambiguous and clear.

The question of whether homeothermy (constant body temperature maintained, for example, by endothermic metabolism) evolved in birds or whether they inherited it from parent dinosaurs has yet to be answered unequivocally. There is, however, good evidence that some of the anatomical features associated with endothermy were present in various dinosaurs, and even very early archosaurs (Seymour et al., 2004): these include the upright posture associated with fast movement, and the likely cardiac function and bone morphology associated with endothermy in birds and mammals. In addition, large dinosaurs would have been endothermic to some extent because of their small surface:volume ratio, while small, fast dinosaurs like the theropod ancestors of the birds could only have maintained the speed appropriate to their morphology if they had some means of keeping their bodies warm; feathers would have helped, but mesothermy at least may also have been required (Grady et al., 2014). Feathers would of course have been counterproductive for cold-blooded animals: the insulation would have slowed their warming in the sun.

In summary, the evidence clearly shows that, from the first chordates of the Cambrian period to the reptiles and birds of the late Cretaceous period, the vertebrates form a monophyletic clade. There is an impressive amount of fossil detail to help understand two major transitions, from fish to amphibians and from therapsid dinosaur to bird. The one important gap in our knowledge of this period is how, over ~50 My, an amphibian egg that developed in an aqueous environment evolved to become an amniote egg with a set of complex membranes and a shell that sustained its development on dry land.

Key points

- The lines of descent from the earliest chordates to contemporary vertebrates can be described within a complex cladogram on the basis of synapomorphies.
- The key synapomorphies that distinguish today's mammalian and reptilian clades are the former having a single and the latter having a double temporal fenestra in the cranium. This division occurred some 300 Mya.
- The transition from lungfish to amphibian is detailed in the fossil record through a succession of organisms over the period 395–348 Mya, with the first tetrapod walking on land some 374 Mya.
- Perhaps the most complex and least understood evolutionary event in vertebrate evolution was the development of the amniote egg that allowed embryogenesis to take place away from water. This seems not to have been in place until a little before ~315 Mya or some 60 My after the first amphibians.
- Birds evolved from theropod dinosaurs that already had feathers.

Further reading

Clack J (2012) Gaining Ground: The Origin and Evolution of the Tetrapods, 2nd ed. Indiana University Press.

Diogo R (2008) The Origin of Higher Clades: Osteology, Myology, Phylogeny and Evolution of Bony Fishes and the Rise of Tetrapods. CRC Press.

Kardong KV (2014) Vertebrates: Comparative Anatomy, Function, Evolution, 7th ed. McGraw-Hill.

CHAPTER 7
THE ANATOMICAL EVIDENCE FOR VERTEBRATE EVOLUTION: MAMMALS

There is a great deal of information about the anatomical changes that characterized the evolution of the mammalian clade. These mainly derive from the considerable amount of fossilized skeletal material, while the origins of some of the soft tissue anatomical synapomorphies that define mammals are beginning to be understood. These include the evolution of hair and mammary glands, changes to the jaw and the middle ear, the expansion of the neocortex, and, particularly, the adaptations required for internal fertilization and placental development. This chapter starts by exploring the fossil evidence; it then examines the ways in which the various mammal-specific anatomical features evolved through changes to embryonic development. The last part of the chapter is a case study that uses evidence from anatomy, embryology, and genomics to show how a group of land animals returned to the sea.

For all their diversity, the ~4000 walking, swimming, and flying extant mammalian species fall into only three high-level taxa: the Prototheria (monotremes), the Metatheria (marsupials), and the Eutheria (placental mammals). Although each has its own features (**Table 7.1**), there are a series of synapomorphies that distinguish them all from their reptilian ancestors. These include the malleus and incus bones of the inner ear, whose evolution came from a remodeling of the reptilian jaw joint, the transitioning of the temporal fenestra to become the zygomatic arch, the formation of a secondary palate to facilitate breathing while eating, and other skeletal and dental changes, including the formation of molars. There are also soft tissue synapomorphies that include mammary glands, a large brain with a particularly increased neocortex, hair, endothermy, and, for the Metatheria and Eutheria, a coiled cochlea and major changes to the reproductive system. Of these innovations, only those that lead to skeletal changes can generally be picked up in the fossil record, although there are occasional examples where a fossil includes hair and skin.

Prototheria (monotremes)

The very few monotremes (the duck-billed platypus and the four species of echidnas or spiny anteater) are the most primitive in terms of their reproductive systems. Males have a penis for reproduction but not excretion; females lay eggs and have a reproductive system close to that of their reptilian ancestors; both possess a single internal cloaca into which the gut, ureter, and, in the case of females, the oviduct terminate and the single external cloacal exit is used for defecation, excretion, and, for females, reproduction. Adult monotremes differ from reptiles in having hair or quills and a degree of endothermy; adults also have a leathery bill rather than teeth, although fossil forms and young platypuses have teeth.

Table 7.1: Some synapomorphies of current mammalian taxa

Property	Synapsid reptile	Prototheria (monotremes)	Metatheria (marsupials)	Eutherians
Middle-ear bones	1	3	3	3
Tricuspid molars per half jawbone	No cusped teeth	Australosphenida tricuspid molars in juveniles	Four tribosphenic tricuspid molars	Three tribosphenic tricuspid molars
Neocortex	No	Yes, small	Yes	Yes
Interhemispheric communication	Anterior commissure	Anterior commissure	Anterior commissure	Corpus callosum + anterior commissure
Cochlea	Straight	Straight	Coiled	Coiled
Hair	No – scales	Yes	Yes	Yes
Whiskers with touch sensitivity	No	No	Yes	Yes
Endothermy	No	Incomplete	Yes	Yes
Mammary glands	No	Skin patches	With nipple	With nipple
Pouches	Unlikely	Platypus – no Echidna – yes	In most cases	No
Reproductive system, excretion, and defecation	F: oviduct M + F: common cloaca M: penis for reproduction	F: oviduct M + F: common cloaca M: penis for reproduction	F: Two uteruses, a complex vagina, and a single birth canal whose terminal region meets the anal canal to give a very short common cloaca M: Bladder voids into the anal canal and the penis is only used for reproduction	F: uterus, vagina, and three separate exits M: dual-purpose penis + anus
Placenta (egg size)	No (lizard eggs: ~1 cm)	Bilaminar, within egg for uterine secretions (~4 mm)	Choriovitelline (130–350 μm)	Chorioallantoic (~70 μ)

F: female; M: male.

Monotremes, like their reptilian ancestors, lay eggs, but only about 10 days after fertilization: early embryonic development up to the fetal stage takes place inside the mother. Once the small quantities of yolk have been absorbed, the embryo gets its nutrition from maternal secretions into the oviduct that pass through its parchment-like shell (Cruz, 1997; Renfree, 2010). After the eggs are laid, the mother keeps them warm for a further 10 or so days until they hatch at about the same state of development as those of newborn Metatheria. Monotremes lack nipples: their mammary glands secrete milk directly to the skin where the young puggles can lick it. Echidnas have a pouch for rearing their young, so providing protection for embryos; platypuses lack a pouch and their young have to cling to their mothers.

Metatheria (marsupials)

The Metatheria, or marsupials, such as the kangaroo, have four molars on each side of their upper and lower jaw, and only replace a few of their original (milk) teeth. Metatheria display other features that include the formation of whiskers (a hair specialization), a range of pouch types (Tyndale-Biscoe & Renfree, 1987), and a coiled cochlea in the inner ear rather than the straight cochlea of monotremes. The evolution of this coiled cochlea enabled the range and frequency resolution of sound to be increased without requiring additional space within the temporal bone, and was presumably a response to the enhanced abilities of the three-boned middle ear. These features together required an increased functional ability for the organ of Corti, which converts vibrations into nerve impulses. These properties together required an enhanced neuronal processing ability, particularly in the neocortex.

The Metatheria have an external cloacal opening and the male reproductive system is much the same as that of monotremes. The female internal reproductive tract is, however, very different: each oviduct has changed so that it includes a uterus within which basic placental and embryonic development occurs, and it terminates in a vagina whose entrance is within the cloaca. Pregnant metatherian females develop a simple, yolk sac placenta that nourishes the embryo until soon after its forelimbs have developed to become functional. The embryo then detaches from the placenta, enters the birth canal, and is born. The neonate then uses its prematurely large forelimbs to climb to the pouch situated in the belly area. This premature forelimb development in marsupials, as compared with those in birds and eutherian mammals, is one of the best-known examples of heterochrony and the way in which this change occurs is under investigation (Keyte & Smith, 2010). Within the pouch there are mammary glands with nipples, and the mouth of the neonate is shaped for plugging onto one of them; the resulting milk flow enables the neonate to complete development.

After a rich history of evolution and diversification (Williamson et al., 2014), metatherian mammals are now mainly found in Australasia. There are still many species of opossums and shrew opossums in South America, but only opossums survive in North America.

Eutheria (placental mammals)

Placental mammals replace all their original milk teeth and have three or fewer molars on each side of their upper and lower jaws. These teeth have different enamel morphologies from those in metatherian teeth (Von Koengswald, 2000; Davis, 2011); such dental features are useful in identifying fossil material. They lack the small epipubic bones of the pelvis of other mammalian taxa, having lost them during the Cretaceous Period (Novacek et al., 1997). Their most important synapomorphies, however, are in the brain and the reproductive systems of both males and females. The eutherian brain is marked by a particularly large neocortex and the presence of the corpus callosum, a large region below the cortex, whose fibers link the two halves of the brain (Suárez et al., 2014). There are also changes to the external reproductive systems of both males and females as compared with those seen in Metatheria. Males have a dual-purpose penis for urination and reproduction, and an anus rather than a cloaca. Females have separate orifices for defecation, excretion, and reproduction, and, when pregnant, produce a chorioallantoic placenta. This allows the embryo to complete much of its development before birth so that, other than their need for milk, most neonates are capable of an immediate and considerable degree of independence.

Today's mammals include a rich variety of animals that today extend in size from shrews to elephants on land, and include bats and flying (gliding) squirrels in the air, together with cetaceans (such as whales) and pinnipeds (seals) in the sea. The richness of the clade reflects the extensive radiation of the early eutherians after the K-T extinction as they colonized the wide range of habitats that had been freed up by the extinction of the dinosaurs. Readers interested in how this happened and the associated fossil record are referred to the books listed at the end of the chapter, while there is a wiki that details how the horse evolved (en.wikipedia.org/wiki/Evolution_of_the_horse).

The transition from synapsid reptile to early mammal

The initial radiation of the synapsid reptiles started towards the end of the Carboniferous Period, and they soon became the dominant animal group. Although pre-Triassic synapsids are informally known as the mammal-like reptiles, there was nothing mammalian about them and proto-mammals, their other name, is probably preferable. The dominant taxon then was the Pelycosauria, (for example, *Dimetrodon*, with its spiny dorsal sail; see Figure 6.9), but it was supplanted in the middle Permian (c. 270 Mya) by the Therapsida, a clade that included reptiles of all sizes from rat to cow. The most noticeable difference between the two taxa was that the former were more sprawling and the latter more upright. During the Permian Period, the synapsid reptiles flourished, dominating the great land masses (Kemp, 2005; Kardong, 2014).

Most synapsid taxa were lost during the Permian extinction (c. 252 Mya) with only three surviving into the Triassic. The dicynodonts (beaked herbivores) and therocephalians (carnivores) eventually died out, but the cynodonts, the only taxon with a secondary palate, flourished, with one becoming the last common ancestor of today's mammals. The cynodonts initially included omnivores and herbivores with a range of sizes but, as archosaur hegemony increased on land during the Triassic, the clade seems to have been reduced to a group of small nocturnal shrew-like insectivores that lived in burrows from which they emerged at night. This niche diminution led to what is known as the nocturnal bottleneck (Gerkema et al., 2013), with only those few taxa that evolved apomorphies enabling them to cope with these conditions being able survive.

There are two reasons for accepting that this is what happened. First, all the fossilized synapsid material from the Triassic and Early Jurassic Periods comes from very small insectivores with proportionately small teeth; there is no skeletal material from larger synapsids that would have been much more likely to have been preserved (Ungar, 2010). Second, many common mammalian properties today reflect the life of ancestral animals that were small, nocturnal, and needed to work hard to maintain their body temperature. These include hair, endothermy, and the presence of brown fat that can be metabolized quickly to raise heat (even today, shrews have to spend inordinate amounts of time eating just to maintain body temperature), together with the strengthening of those senses that facilitate living in the dark. These include acute hearing (and an improved inner ear), sensitive touch (particularly whiskers), relatively large eyes, and a strong sense of smell (shown by well-developed turbinate bones in the nasal area).

As the fossil evidence that remains from these small animals (usually less than 10 cm long, excluding the tail) is, of course, very limited, often being just teeth and only sometimes jaws, the important changes to soft tissues have to be inferred on the basis of often very limited skeletal data. Three features, in particular, characterize early mammals. First was the presence of the two new inner-ear ossicles, the malleus and the incus, that evolved from two of the reptilian jawbones, the lower articular and the upper quadrate (Anthwal et al., 2013). These ossicles amplify sound beyond the level transmitted by the stapes alone and so indicate enhanced hearing ability (Gill et al., 2014; Rowe et al., 2011). Second was the changed jawbones and third was the enhanced cusp organization of the molar teeth, which indicate the nature of mastication and hence diet (Gill et al., 2014).

There was thus a period between the end of the Permian (252 Mya) and the Early Jurassic during (c. 190 Mya) which the cynodonts slowly started to acquire mammalian features (**Figure 7.1**). The synapsids living in this period are known as the mammaliaforms (or mammaliaformes); earlier taxa are sometimes grouped into the mammaliamorphs. An example of a large mid-Triassic (~230 Mya) mammaliaform is *Trirachodon*, a social animal some 50 cm in length that possessed a secondary palate (Groenwald et al., 2001). Two early organisms showing transitional forms between the reptilian and mammalian jaw structures are *Morganucodon*, which first appeared during the Late Triassic Period (~205 Mya) or some 10 My before *Hadrocodium wui*, and *Sinoconodon rigneyi*, a carnivorous mammaliaform from the Early Jurassic Period (~195 Mya). Both possessed two jaw joints: the original one connecting the articular and quadrate bones, and a new jaw joint between the dentary and squamosal bones (**Figure 7.2**). The immediate likely function of this change was that it enhanced mastication, but it also paved the way for the later detachment of the bones of the early joint to become the future ossicles, with consequent improvements to the auditory system.

It is impossible to know exactly when the first animal evolved that nourished its offspring with milk and this is, of course, because mammary glands did not fossilize. The generally accepted criterion for determining whether an animal was a mammal is therefore the possession of detached ossicles. The earliest synapsid so far identified as having them is *H. wui*, a very small animal (~3.2 cm long) with a relatively large brain cavity that lived ~195 Mya (Early Jurassic; Anthwal et al., 2013). Kemp (2005), however, points out that there are other definitions for a mammal, and these include the structure of the jaw hinge and the mode of dental activity. Each, of course, represents a point on the spectrum of morphological change, and, in the absence of the key soft tissue data, the choice of definition is always going to have a subjective element. Given that we only have skeletal data, the choice really reduces to which of the early taxa should be included in the mammaliaforms and which in the proto-mammals.

The early evolution of mammals

It used to be thought that the acquisition of mammalian features was a linear process slowly taking place during the Jurassic. However, it is now clear that there was a wide

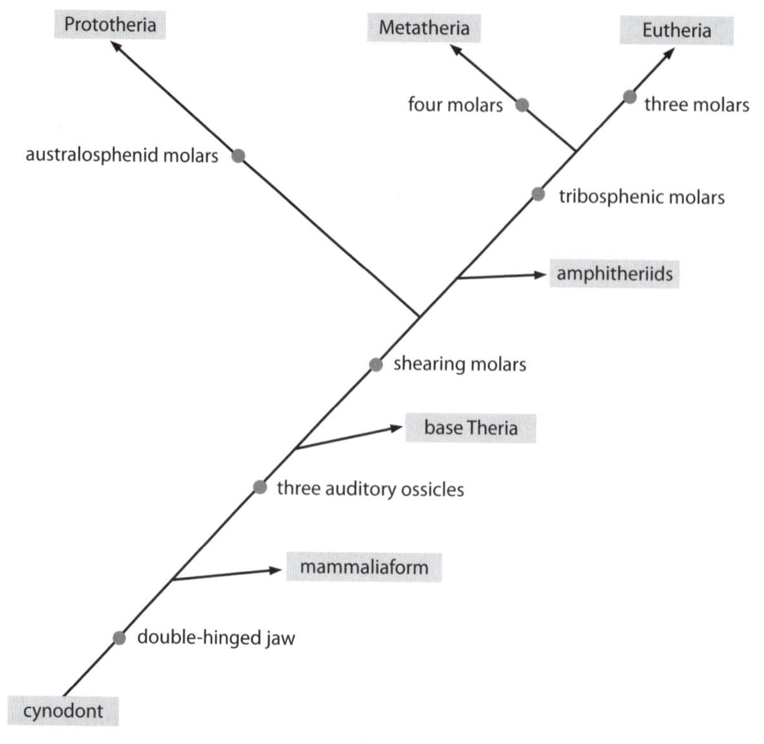

Figure 7.1 Simplified cladogram showing the main lines leading from a Triassic cynodont to the modern mammalian taxa, together with their defining skeletal synapomorphies. Most genera that have become extinct have been omitted, but the amphitheriids have been included to emphasize that the two sorts of tricuspid molars, both of which could be used for shearing and grinding, represent homoplasies rather than synapomorphies.

variety of species with mixes of mammalian features; the Mesozoic synapsid clade now has 300 or so known mammalian genera (Luo, 2007). These represent a wide diversity of species, albeit that they all remained relatively small, so avoiding the attention of dinosaurs. From its rodent-like beginnings c. 195 Mya, the taxon diversified widely with the fossil evidence clearly showing, for example, that, by about 164 Mya, the early mammals included both gliding and swimming species such as *Volaticotherium antiquum* and *Castorocauda lutrasimilis,* which were sister taxa to the monotremes and probably near the line to the other early mammalian clades.

Volaticotherium antiquum was an insectivorous mammal with a gliding membrane similar to that of a modern-day flying squirrel (Meng et al., 2006). Its teeth were highly specialized for eating insects, and its limbs were adapted to living in trees. The gliding membrane (or patagium) was insulated by a thick covering of fur, and was supported by the tail, as well as the limbs. The discovery of *Volaticotherium* provided the first record of a gliding mammal and it is ~70 My older than the next to have been found.

Castorocauda lutrasimilis, although it lived at about the same time as *Volaticotherium,* was very different: it showed features seen in modern semi-aquatic mammals such as beavers, otters, and the platypus. Fossilized impressions indicate that there was some webbing present between the toes (Ji et al., 2006) and that the caudal vertebrae were flattened dorsoventrally, while the tail was broad with scales interspersed with hairs that grew less frequent

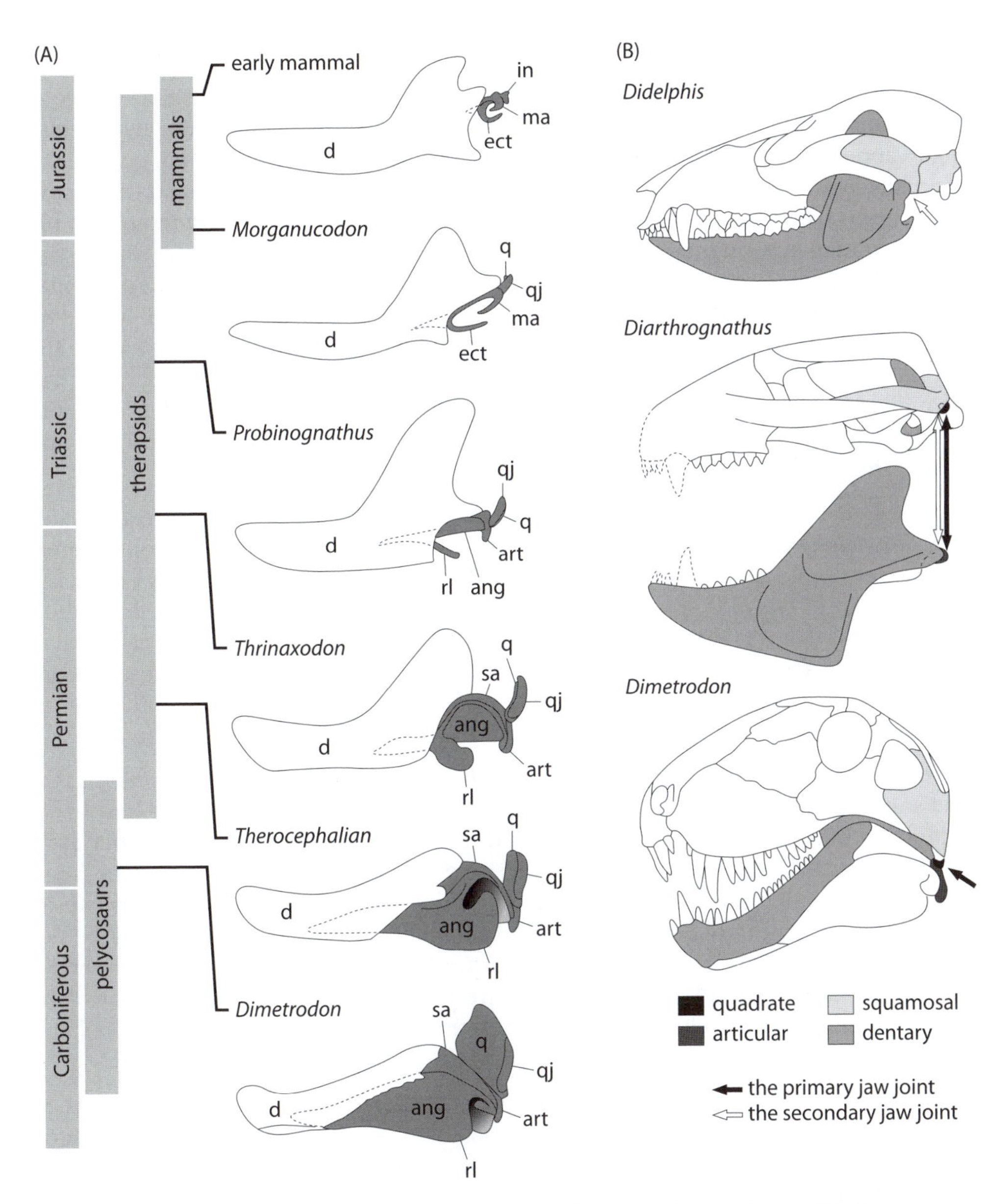

Figure 7.2 Paleontological evidence for mammalian middle ear and jaw evolution. (A) Diagrams of lateral views of jaw skeletal elements showing the transitions that led to the mammalian morphology. The geological record and occurrence of each animal are indicated on the left. For clarity of comparison, no teeth are shown. Note that the postdentary elements (articular [art], surangular [sa], and angular [ang]) and the upper jaw elements (quadrate [q] and quadratojugal [qj]), indicated in gray, became separated from the dentary [d] and reduced in size during the transition from pelycosaurs to mammals. The sequence of changes in the fossil record does not represent a true ancestor–descendant relationship, only structural grades. (B) Changes in jaw articulation during mammalian evolution. In the pelycosaur, *Dimetrodon* (bottom), the quadrate and articular formed a functional jaw joint (black arrow). In the "advanced" cynodont, *Diarthrognathus* (middle), an additional jaw joint is present between the squamosal and dentary (white arrow). In an extant marsupial, *Didelphis* (top), the functional jaw joint has been taken over only by the squamosal and dentary. (ect: ectotympanic bone; in: incus; ma: malleus; rl: reflected lamina).

toward the tip. As the tail was very similar to that of modern beavers, *Castorocauda* presumably used it for swimming in the same way as its modern counterparts.

It seems to have been around 170–160 Mya that modern mammalian dentition was beginning to appear. Most mammaliaforms had molars adapted to shearing food, but species eventually evolved whose molars also had a grinding ability; these are australosphenid teeth (three cusps) now seen only in monotremes. The earliest is *Shuotherium* (c. 167 Mya), identified only on the basis of its jaw and lower teeth; one of its molars was australosphenid-like, indicating that *Shuotherium* was a sister taxon to the early monotremes (Kielan-Jaworowska et al., 2004; Davis, 2011). This latter clade diversified widely, and an interesting example of an ancestral mammal from the Late Jurassic is *Dryolestes* (~160 Mya). Although it lacked a coiled cochlea, the internal skull anatomy indicates that it had the auditory innervation seen in the Metatheria and Eutheria. *Dryolestes* thus seems to have been an intermediate between the monotremes with essentially straight cochlea and modern therians with coiled ones (Luo et al., 2011a).

The oldest metatherian fossils so far identified date to around 125 Mya, and an early example is *Sinodelphys*, a furry metatherian ancestor with four molars adapted to an arboreal lifestyle (Luo et al., 2003; Williamson et al., 2014). It was found in what is now China, but was then part of Pangaea, a supercontinent that originally included most of the world's landmass, but that was then in the process of breaking up. The details of the dispersal of the Metatheria across the various land masses remains unclear because the fossil record is poor, but subsequent metatherian fossils have been found in North and South America and eventually in Australasia (~50 Mya), which they reached by crossing Antarctica.

The earliest current evidence of a eutherian animal is the fossil of *Juramaia sinensis* that lived some 160 Mya, with skeletal analysis showing that it possessed three tribosphenic molars and other more minor eutherian characteristics (Luo et al., 2011b). Other early eutherians have continued to be found and an example is *Sasayamamylos* from Japan (~125 Mya), identified by its fossilized jaw with three molars and four premolars (Kusuhashi, 2013). The eutherian mammals started as small, mouse-sized animals and did not exceed the size of a small dog before the K-T extinction (66 Mya). The skeletal differences between the Eutheria and Metatheria are relatively minor and include the reduction of the number of molars in each half jaw from four to three, the loss of the epipubic bone, and the gain, in some cases, of a baculum (penis bone).

It does, however, seem odd, particularly as the eutherian reproductive system builds on that of the metatherians, that the oldest eutherian organisms so far identified are so much earlier (~167 Mya) than the oldest metatherian ones (~125 Mya). It would therefore not be surprising if there were older metatherian fossils still to be found. The fossil evidence as a whole shows that, by the end of the Cretaceous Period (~66 Mya), there was a wide variety of mammalian species, but that much of this rich prototherian, metatherian, and eutherian fauna was lost in the K-T extinction. The much greater loss of all marine reptiles and archosaurs, except the crocodiles and birds, emptied a wide variety of ecological niches and the absence of large predators facilitated the radiation of the three small mammalian taxa to fill them.

The major evolutionary feature of the mammalian radiation since the K-T extinction has been the success of the eutherians. The monotremes seem never to have been particularly successful, and only the platypuses and spiny anteaters remain today, restricted to Australia and New Guinea. The Metatheria originally spread widely, being particularly successful in South America; there they filled many of the niches now occupied by eutherian taxa and showed similar anatomical adaptations (all are examples of homoplasies). More recently, however, they have been almost completely displaced by Eutheria, as relatively few metatherian taxa were apparently able to compete with them, an exception being the opossum, which is still common in North America. Today, their major habitat is Australasia.

Since the K-T extinction (the Cenozoic Period), eutherian mammals have come to dominate the land masses other than Australasia, to the extent that only the bigger snakes, tortoises, and a few lizards occupy land niches for large animals. In Australasia and, to a lesser extent the Americas, marsupials do well, but eutherian mammals, through migration or through importation for farming, are now common. Eutherian marine mammals have also been successful in marine environments, where whales and dolphins hold their own against large fishes, but less so in the air, where birds dominate. A considerable amount is now known about the return of the mammals to the sea (~50 Mya) but very little on how mammals evolved to fly, as the earliest fossilized bats (52 Mya) so far discovered have a morphology similar to present ones (Teeling et al., 2005).

The evolution of some key mammalian features

Modern mammals are very different from reptiles: they are homeothermic, hairy, and have good hearing, vision, and touch; they also have brains that are disproportionately large compared with other vertebrates. Such properties are consistent with their last common ancestor having been a subterranean animal; for such a life, it would have needed dark-efficient senses such as acute hearing, touch sensitivity, and vision, together with an enhanced brain for processing these inputs. It would also have needed to be endothermic and well insulated to ensure that, as very small creatures they could respond quickly to predators and eat rapidly moving insects without losing warmth too quickly. Such a perspective is sensible, but it is based on post hoc justifications; we do not know what other soft tissue changes might once have evolved but failed later selection tests.

It seems to have taken most of the Triassic Period (250–200 Mya) for the various mammalian skeletal and other apomorphies to have evolved from the more basal properties of the synapsid proto-mammalian reptiles, which is no longer that it took for the amniote egg to develop. It is not immediately clear whether the length of this period reflects the difficulty of making these changes or the relatively low selection pressures on the synapsids. Nevertheless, the fact that the clade was rapidly reduced to small shrew-like animals does suggest that there was considerable pressure from very active predators for the synapsids to adapt to a subterranean existence. If so, the number of generations required (~50 million for animals that reproduced annually) probably reflects the sophistication of the changes. The molecular mechanisms by which genomic variation leads to phenotypic change will be examined in Chapter 11; here, only the skeletal and soft tissue changes that occurred during the synapsid-to-mammal transition are considered.

Skeletal features

Middle-ear bones and jaw reorganization

Mammals, unlike other vertebrates, have three middle-ear bones, the malleus, incus, and stapes. They also have two upper tooth-carrying jaw bones, the maxilla and premaxilla, and a single lower jaw bone, the mandible (or dentary), which carries teeth and whose proximal ends meet a fossa (or hollow) in the temporal bone to form the temporomandibular joint. Synapsid reptiles were different: they had a single auditory ossicle, the columna (homologous with the stapes), and more complex jawbones. The lower jaw comprised the dentary, the articular, the angular, and several other bones; the upper jaw contained the maxilla, premaxilla, and quadrate. The joint between the lower jaw and the cranium was formed by articulation between the quadrate and articular bones.

The early development of the lower jaw is common to all synapsids: the long thin Meckel's cartilage is laid down first. In late synapsid reptiles, Meckel's cartilage first acted as a template for the various bones of the lower jaw but then ossified, remaining as a distinct bone, (Anthwal et al., 2013). Over time, however, the articular and angular detached, reduced in size, and combined to become the malleus, while the remaining lower bones merged to form the mandible. In the upper jawbone, the quadrate detached and reduced in size, eventually becoming the incus. The jaw joint shifted to an articulation between the dentary (now called the mandible) and the squamosal, which later became part of the temporal bone. The way in which this happened over about 50 My is illuminated by a series of fossilized synapsids (see Figure 7.2) and the following three demonstrate the anatomical trajectory of the evolutionary steps:

- *Thrinaxodon* (~245 Mya) had a single jaw joint where the articular and quadrate bones met.
- *Diarthrognathus* (~200 Mya) possessed a double jaw hinge. It had a synapsid jaw joint between the quadrate and articular bones, and a future mammalian jaw joint between the squamosal and dentary bones.
- *Hadrocodium* (~195 Mya) had a jaw hinge formed by the dentary and squamosal alone, with the malleus and incus now being detached from both the jaw and Meckel's cartilage.

On the basis of its middle-ear structure and also its brain structure, *Hadrocodium* is probably the first well-defined mammal (Rowe et al., 2011). Such synapomorphies are seen in all extant mammals, suggesting that their last common ancestor had such a jaw and ear skeleton, together with enhanced aural sensitivity.

In modern mammals, Meckel's cartilage mostly disappears apart from its most posterior region that ossifies and becomes the malleus and the immediately adjacent region that becomes the sphenomandibular ligament (a cable between the mandible and the sphenoid process of the cranium). It is worth noting that an ossified Meckel's cartilage was maintained in many species, such as *Liaoconodon*, the gobiconodontids, and *Yanoconodon* until at least the Early Cretaceous (Anthwal et al., 2013). Curiously, this feature has very occasionally been seen in modern humans (Shattock, 1880; Keith, 1910).

Changes in the temporal fenestrae and the evolution of the zygomatic arch

Some 275 Mya, the therapsid proto-mammals, like earlier synapsids, had a single temporal fenestra on each side of the cranium just behind the eye orbit whose role was probably to provide a space through which the external mandibular adductor muscle linked the cranium attached to the lower jaw, facilitating biting (**Figure 7.3**).

In time, the postorbital bone between the two fenestrae was lost and the jugal bone originally at the base of the temporal fenestra extended forward and outwards to produce the zygomatic arch, so extending the space available for muscles. By ~250 Mya (Late Permian), this arch had acquired two roles: as a conduit for the temporalis muscle, the equivalent of the external mandibular adductor muscle and as the superior

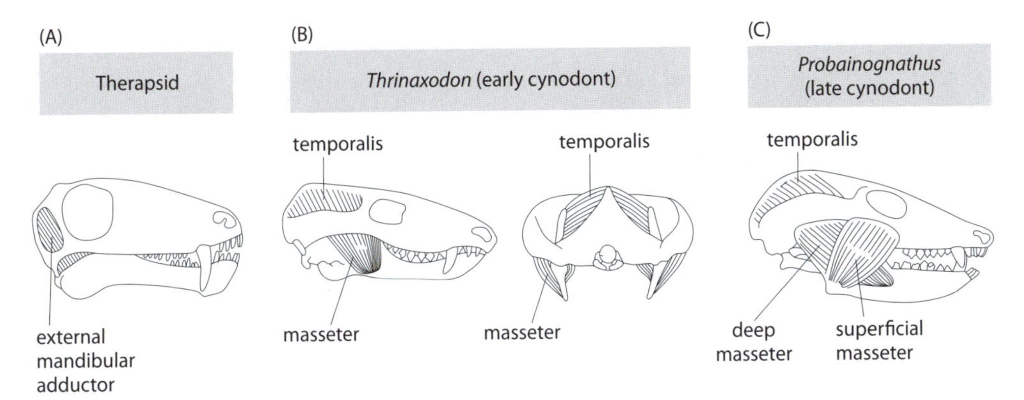

Figure 7.3 The evolution of the zygomatic arch. An early therapsid with a single temporal fossa through which extends the external mandibular adductor muscle. The skull of *Thrinaxodon* (250 Mya). The left view shows that the temporal fenestra has been lost and the (jugal) bone at its base now extends forward increasing the space available for the temporalis muscle. The right view shows that the temporalis. This muscle attaches to the medial surface of the mandible, while the masseter muscle links the lateral surface of the mandible to the jugal bone. The skull of *Probainognathus* (~220 Mya): further mastication muscles have now evolved in the vicinity of the arch.

attachment of the masseter muscle (for mastication), which also extends to the lower jaw and today facilitates chewing.

Secondary palate

A secondary palate appeared at an early stage of mammalian evolution (~230 Mya; Groenewald et al., 2001). The changes involved in making this tissue, which separates the oral and nasal cavities, are now well understood at both the tissue and molecular level. Two shelves of tissue extend from either side of the embryonic oronasal cavity that are stiffened by extracellular matrix. They meet at the midline, where they seal with each other and to the small bony primary palate immediately posterior to the premaxilla, the bone that carries the incisor teeth. In due course, the anterior part of the secondary palate becomes ossified (see Gritli-Linde (2007) for further details on the molecular mechanisms underpinning these events).

Changes to teeth

Dentition has been important in working out when the early mammals separated into the current three taxa, the key synapomorphies being the structure of the molars and their number (in each half jaw bone). Molars are rear teeth characterized by raised enamel cusps, and the organization of these cusps together with the signs of local wear reveal their function. Originally, the molars seem to have been used for grinding food, but the organization of the three cusps on matching upper and lower molars across a wide range of early mammaliaforms such as *Morganucodon* suggests that they were beginning to acquire a shearing function that facilitated chewing (Schultz & Martin, 2014; **Figure 7.4**). The exact details of inheritance are unclear, but both australosphenid and tribosphenid molars evolved: each has three cusps, with differences in their precise arrangement giving the latter a weaker grinding ability than the former (for review, see Davis, 2011). Close analysis of the fossils shows that australosphenid molars are a characteristic of the monotremes, while tribosphenid molars evolved to give those of the

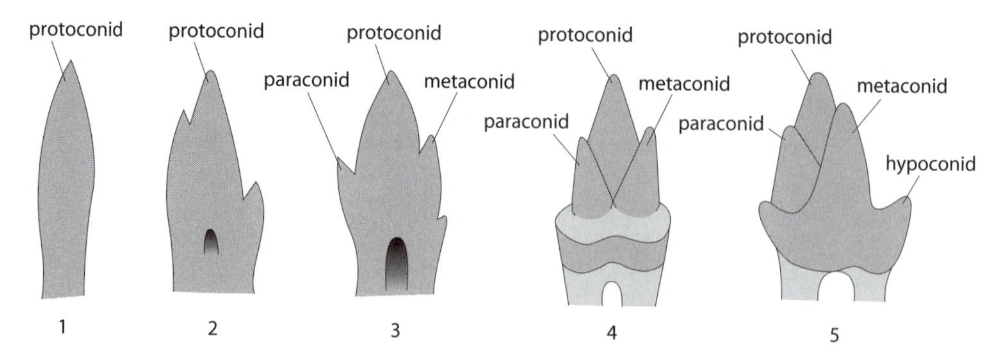

Figure 7.4 Drawings illustrating the evolution of cusped (or conid) molar teeth in a series of synapsids. 1, Reptile; 2, 3, Triassic; 4, Late Triassic; 5, Middle Jurassic mammaliaforms.

theria, with Eutheria having three such molars and Metatheria four. The molecular basis of teeth development is being actively investigated, and it is now known that cusp patterning is controlled by the downstream effects of fibroblast growth factor signaling (Tapaltsyan et al., 2016).

Other tissues

Mammary glands

The 100–200 glands of the monotreme lactatory system produce milk, hair, and sebum, with pups licking the milk that collects on these hairs. It is therefore likely that these glands were modified from those that originally made hair (Oftedal, 2002), with the sebum and hair-making abilities eventually being lost as the glands coalesced into the branching ducts and terminal nipples of metatherian and eutherian mammary glands. Oftedal (2012) has investigated the likely origins of milk: genomic analysis of milk protein sequences has suggested that they date back to early synapsids ~310 Mya (Late Carboniferous Period) with some of the constituents evolving even earlier. By the time of the early and small mammaliaforms, its likely role was to provide fluid and nutrition to the tiny eggs that probably had the porous and parchment-like shells of today's reptiles and monotremes. By the Late Triassic (~210 Mya), rich milk was supplementing the very small amounts of yolk in the tiny embryos of small mammaliaforms.

Hair and homeothermy

Homeothermy or the maintenance of a constant body temperature (usually through endothermy or metabolic activity) requires more food-supplied energy than ectothermy in which body temperature mainly derives from sun heat. It does, however, have the key advantage that the body's metabolic enzymes always work at their optimal temperature. An animal can thus maintain a high level of activity that is independent of the external temperature, which is low underground, and it is noteworthy that mitochondrial activity in muscle, heart, and lungs is much higher in mammals than reptiles (Hulbert & Else, 1989).

Homeothermy is likely to have evolved before hair, as there is little reason for a cold-blooded animal to develop a pelt: its insulation would slow warming as much as it would help maintain heat. The fossil evidence for hair dates back to at least the Middle Jurassic (~165 Mya; Ji et al., 2006), or a little before the earliest evidence of feathers in dinosaurs (~155 Mya, Late Jurassic; Rauhut et al., 2012). Both structures are insulators and have keratin as their primary molecular constituent, albeit that hair uses α-keratin and feathers β-keratin (Alibardi, 2006; Dhouailly, 2009); the former is a normal constituent of the integument (skin), first seen in amphibia, while the latter is not (Alibardi, 2003). Hair evolved to have a further property; it can also have touch sensitivity through its associated mechanoreceptors, an ability particularly enhanced in whiskers.

Producing hair required the evolution of two mechanisms: the first for producing hair follicles, the second for regulating interfollicular distance (hairs and whiskers have different separations). Alibardi (2012) has suggested that the former evolved through co-opting groups of cells with keratin-production machinery to produce long strands of protein rather than the matrix seen in skin. As to the latter, Dhouailly and Sengel (1973) showed that the patterning mechanisms for generating hairs and feathers are essentially interchangeable with those for producing scales in reptiles. This mechanism involves still-unknown interactions between the surface epidermal epithelium and the underlying dermis of the skin: the former lays down a rough grid of locations in the latter at each of which a hair-making papilla forms. An interesting example of a mammalian plesiomorphy that extends back to their reptilian ancestors is the scales that are found on pangolins (scaly anteaters).

Changes to the brain and sensory systems

An early and key feature of the mammalian line was an increased brain size, as assayed by the internal volume of the cranium and compared with nonmammalian vertebrates of a similar body size (Kemp, 2005). Rowe et al. (2011) used X-ray tomography to study the morphology of brain cases of *Morganucodon* and *Hadrocodium*. The former had a brain some 50% larger than that of a cynodont of the same size, with a particularly large olfactory bulb, pyriform cortex, and neocortex. *Hadrocodium* had an even larger brain that was within the expected range for a modern mammal, with further enlarged olfactory bulbs, neocortex, and cerebellum. Rowe et al. (2011) viewed these increases as being driven by the demands of improved neuromuscular coordination, an enhanced olfactory system that included ossified turbinates and an enlarged olfactory epithelium, and the tactile receptors associated with body hair, particularly whiskers. Here, it is of note that rodents and squirrels have a morphologically defined area of the neocortex, the barrel cortex, to which whisker sensors map (Erzurumlu & Gaspar, 2012).

The evidence suggests that the key difference between mammaliaform and earlier synapsid brains was the development of the neocortex from the earlier and simpler dorsal cortex, a feature most pronounced in eutherian mammals, particularly primates (Medina & Abellán, 2009; Kaas, 2011). The mechanisms by which this happened are not yet known, but studies on comparative transcriptomics in the context of vertebrate brain development are beginning to identify some of the genomic and other changes that may have been responsible for the major improvements that led to the modern mammalian brain (Molnár et al., 2014). Perhaps the most important of these in modern Eutheria is the corpus callosum, a large region below the cortex whose fibers link the two halves of the brain. This supplements the more limited communication provided by the anterior commissure of monotremes and marsupials (Suárez et al., 2014). The complex signaling that instructs neurons to cross between the hemispheres is now beginning to be worked out (Fothergill et al., 2014).

Changes to early embryogenesis

Although internal development is normally thought of as a mammalian adaptation, many reptiles incubate their eggs internally, which has enabled them to colonize environments hostile to development within a laid egg, for example where the temperature is too low, or in desert areas where there is a risk of desiccation. Such ovoviviparous reptiles retain the egg within the oviduct and the only adaptations are loss of the hard shell and vascularization of the lower part of the oviduct, so allowing gas exchange between the allantoic and uterine vessels. In truly viviparous reptiles such as the European skink *Chalcides tridactylus,* whose eggs measure only 3.0 mm, a complex placenta forms through interdigitated apposition of the vascularized chorioallantoic membrane and the highly vascularized uterine epithelium (Blackburn & Flemming, 2012). Such reptilian placentas provide for the exchange of nutrients and nitrogenous waste, as well as gases, and are the most likely evolutionary basis for the various mammalian placentas (Morriss, 1975).

Monotremes resemble viviparous reptiles in having a small egg (the Echidna egg is about 1.4 mm) with very little yolk, in which the early stages of development take place (Selwood & Johnson, 2006). Such eggs have a parchment-like shell so that the embryonic membranes do not make direct contact with the uterine wall. After the small amounts of yolk have been consumed, the vascularized chorioallantois mediates gas exchange, while the yolk sac acts as a placenta, absorbing nourishment from maternal secretions into the uterine environment, some of which is stored for postlaying consumption. This rudimentary placental arrangement is active for only a few days, up to about the 19-somite stage in the case of the platypus, and the egg is then laid directly into a temporary pouch on the mother's abdomen (Echidna) or into a well-insulated nest (platypus), where it is maintained at a temperature of ~32°C until the fetus-like hatchling emerges about 10 days later. Milk is secreted onto the abdominal skin surface by two patches of modified sebaceous glands, and lapped by the young (Schneider, 2011). Thus, although monotremes have an essentially reptilian reproductive system, they are less fully viviparous than some reptiles, while having the mammalian characteristics of maternal care through homeothermy and lactation.

Marsupial development is different from monotremes in egg size, placental support, and developmental detail (see Table 7.1; Krause & Cutts, 1986), as well as in anatomy. Marsupial eggs are small, having a thin membrane coat and little yolk. The early cleavage stages give a spherical morula, with nourishment in the early stages being by uterine secretions that pass across the membrane. Following rupture of this shell coat, the embryo is sustained by a vascularized yolk sac placenta, although secretions from the closely apposed uterine epithelium continue to provide some nutrition until birth. In late pregnancy the yolk-sac placenta is supplemented by a vascularized chorion (choriovitelline placenta; Krause & Cutts, 1985) and there is an umbilical cord (Freyer et al., 2003).

After 4–5 weeks the marsupial is born, at a stage similar to that of hatching monotremes, and roughly equivalent to a 14-day mouse embryo. The two unusual features of the tiny neonate marsupial are its precocious clawed forelimbs, which enable it to climb to the pouch, and its advanced mouth and gut, which enable it to latch on to the teat for lactation. The Tammar Wallaby (*Macropus eugenii*), for instance, is born at 28 days of gestation weighing 350–400 mg; it remains permanently attached to the teat for 14 weeks, by which time it weighs 100 g. It remains in the pouch until about 27 weeks after birth, accessing the teat as required. It continues to suckle for several months after it leaves the pouch. One possible driver of the evolution of such development was the benefit of holding the fetus close to the mother and so denying predators access to her offspring.

Eutherian eggs have very little, if any yolk (typical egg diameter is ~70 μm) and thus require an early forming placenta. This starts when the trophoblast cells that surround the morula-stage embryo become embedded in the uterine wall and elicit a strong immunological response that results in locally increased vascularization. These early stages of embryogenesis are primarily concerned with the production of the extraembryonic membranes; the chorion, allantois, and umbilical vasculature soon combine to initiate a chorioallantoic placenta that forms in close association with the domain of enhanced uterine vasculature. The exact details, including the extent of trophoblast invasion, depend on the clade. The result is an enhanced supply of nourishment that allows the female to give birth to what can be a fully functioning neonate. A particular advantage of the chorioallantoic placenta is that it enables internal development to proceed for a long time; one benefit of this is the opportunity for advanced brain development.

Changes to the reproductive and urinary systems

Although the urinary and reproductive systems of monotremes differ very little from those of reptiles, major changes evolved in the urogenital systems of marsupial and eutherian mammals (see Table 7.1 and **Figure 7.5**), In female monotremes, there are paired internal ovaries and oviducts that terminate in the cloacal opening, while in males the vas deferens tubules from each internal testis meet in the future prostatic area to produce a common duct that penetrates the penis, whose terminal regions may have up to four exits. This develops from the genital tubercle at the distal end of the cloacal membrane situated between the hindlimb buds. In females, the tubercle becomes the clitoris. In both, the cloaca also serves as the defecatory exit from the gut. The urinary system is also simple: the ureter from each kidney exits inside the bladder, which voids into the cloaca.

As the monotreme line evolved, the major change was to the female reproductive system. First, the urorectal septum, which separates the bladder from the gut, descended to the perineum, so producing a separate anus for the gut and a urogenital sinus for the reproductive and urinary systems. In marsupials, these both exit into a common vestibular area that is the remaining trace of a cloaca. Second, each oviduct evolved to become a uterus and a lateral vagina that was the entry point for sperm. In marsupials, a third central connecting tube evolved to become the birth canal (Figure 7.5). As the two lateral vaginas terminated at the proximal part of the sinus, the marsupial penis became bifid in its distal region and so was able to enter both vaginas. It is worth noting that such a bifid penis is a very rare human congenital anomaly, one that argues for the eutherian and mammalian lines separating after the bifid penis had evolved. Meanwhile, the testes in marsupial males descended into an external sac that was ventral to the penis, unlike the scrotum of eutherian mammals, which is dorsal to the penis (see Figure 7.5). There has been interesting but unresolved discussion (Kleisner et al., 2010) about the reasons why marsupial and eutherian mammals need an external scrotal sac while all other vertebrates do not, even birds, whose body temperature can be as high as 43°C (Prinziger et al., 1991).

The major innovations seen in placental mammals were to the external anatomy of the urinary and reproductive systems. In Eutherian females, the urogenital sinus separated into vaginal and urethral components, so giving three separate exits, with the anus and the vagina separated by the perineum. Internally, the ventral parts of the two reproductive ducts fused to give a single uterus and vagina (so obviating the need for a bifid penis). In the male, the distal part of the urethra changed so that, instead of voiding into the cloaca, it had its own exit that was initially into the cloacal membrane. Thus, the male had an anal and a urethral opening. The other major change was the development of two sets of folds on either side of the cloacal membrane. In the female, these became the inner labia minora and outer labia majora, which protect the urethral and vaginal exits, together with

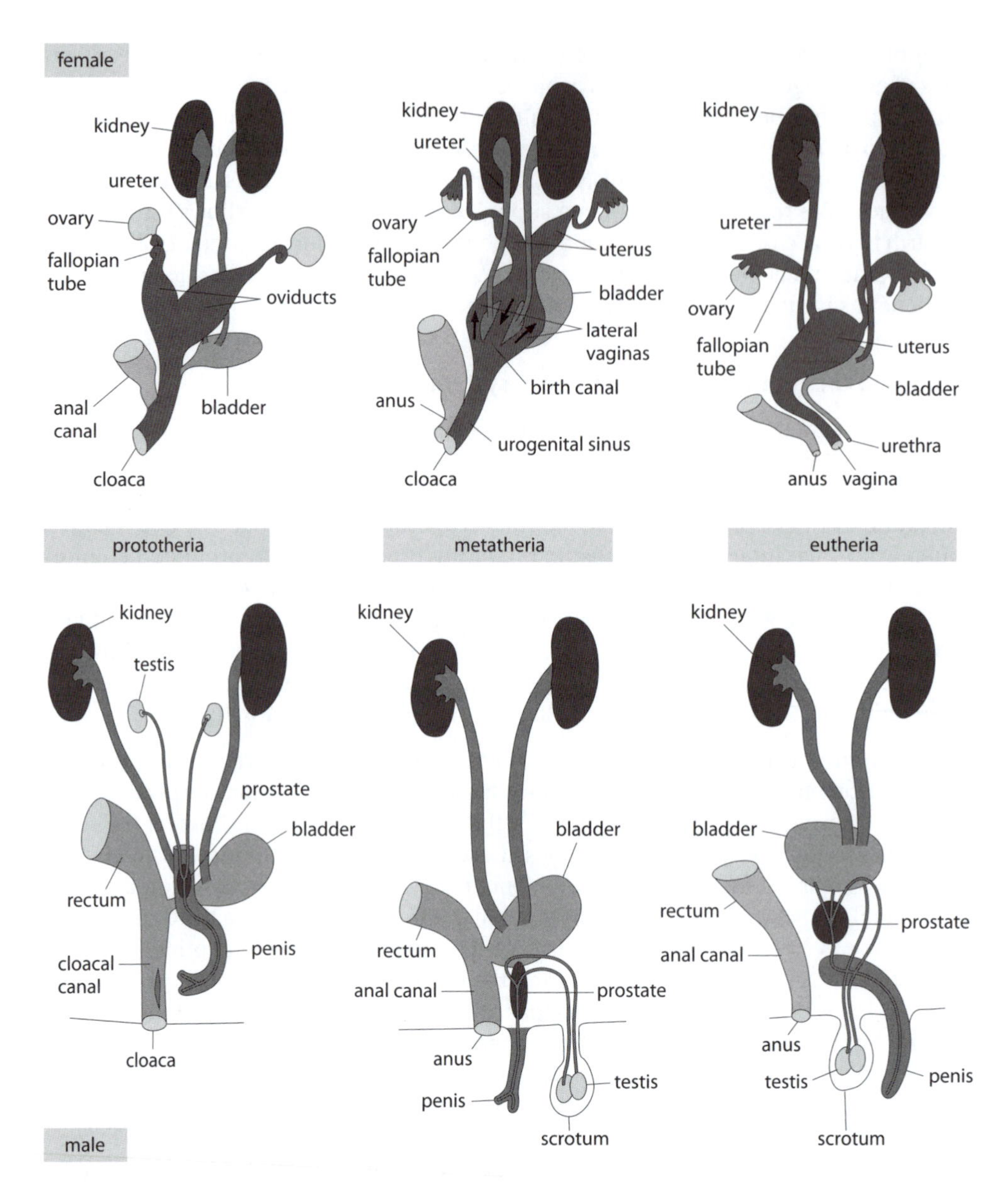

Figure 7.5 The core features of the urinary and reproductive systems of female (upper) and male (lower) prototherian, metatherian, and eutherian mammals. These diagrams are intended to be typical of the different taxa rather than to represent particular animals.

the tubercle, which enlarges to become the clitoris (for developmental and molecular details of these events in the mouse, see Greenfield & Bard, 2015).

In the male, the two sets of folds have roles that are very different from those in in females. The outer pair extend posteriorly and seal over to form the scrotum into which the testes descend, with the distal ends of their vas deferens tubules meeting in the region of the cloacal membrane that will form the prostate. The inner folds seal and form a tube, one end of which encloses the urethral exit so extending the urethra to the prostate region. There, this tubule meets the two vasa deferentia and coalesces with them to form a single tube that will carry both sperm and urine and that extends some two-thirds of the way up the penis, eventually coalescing with it. The terminal part of the penile duct derives from a tube that forms within the penis.

The net result was that, unlike all noneutherian vertebrate males that use their penis for reproductive purposes and urinate through their cloaca, eutherians developed a penis intended for both reproduction and urination. It is still hard to see the selective advantage of a penis with a dual reproductive and urinary function and a very complex mode of morphogenesis, given that it is unlikely to matter to a four-legged eutherian male vertebrate whether it urinates through a penis or a cloaca.

For all of the complexity of the eutherian reproductive system, it is clear that the advantages of an extensive internal period of development have allowed eutherian mammals to become the largest, most successful, and most sophisticated of all modern land animals. They have had limited success in the air, and the contemporary taxa of gliding squirrels and flying bats are few compared with those of the birds. Mammals have, however, had considerable success in the sea, to which they returned several times (Uhen, 2007), with perhaps the most dramatic example of evolutionary opportunism being the evolution of the Cetacea ~50 Mya, or about 320 My after the first amphibian colonized dry land.

Back to the sea and the evolution of whales

Although Aristotle noted some 2000 years ago that dolphins were not fish, it seems to have been John Ray, an important seventeenth-century biologist, who was the first to realize that whales and dolphins were mammals. The structural adaptations that enabled the Cetacea to cope with marine life involved major changes to the tail, the body shape, and the limbs, all of which required growth modifications to pre-existing anatomical features rather than the evolution of novel ones. Equally important were the anatomical and physiological modifications to the respiratory, auditory, and renal systems that enabled land mammals to adapt to a marine environment (Reidenberg, 2007). Little is known about how these latter changes occurred.

Over the last 50 or so years, however, the evolutionary origins of the Cetacea have become clear: the discovery and analysis of a series of fossils have shown the series of steps by which an unknown species of ungulate (hooved mammal) from the Artiodactylae (the even-toed ungulates) gave rise to two clades some 50 Mya. One eventually led to the anthracotheres, whose earliest fossilized skeleton is relatively recent (~15 Mya) and whose current descendant is likely to be the hippopotamus; the other gave rise to the cetaceans (McKenna, 1975). The fossil evidence shows that a clade of four-legged carnivore evolved to become a marine mammal (Bajpai et al., 2009): the fossils show a slow extension of the tail, the nasal opening moving cranially and unifying, and the external hindlimbs being lost, leaving only a rudimentary pelvis and small internal femur.

These fossil data are not however rich enough to construct a cladogram back to the last common ancestor of whales and land tetrapods because the anatomical differences among the fossils are too substantial (the branches of the cladograms are long) to generate an unequivocal line of descent. For similar reasons, it turned out not to be possible to work out the lines of descent using phylogenetics based on analyzing DNA sequence data: there are too few contemporary Artiodactylae for an unambiguous phylogenetic tree to be produced on the basis of interspecies gene or protein sequences (Chapter 8). Fortunately, there was another class of sequence data that turned out to be helpful here; these were retrotransposons or mobile genetic elements that are found in all eukaryotes but are usually not transcribed. These interspersed elements can be short (such short interspersed nuclear elements [SINES] have ~500 base pairs [bp]; Kramerov & Vassetzky, 2011) or long (long interspersed nuclear elements [LINES] have ~6000 bp), but the reason why they are useful here is that once they have integrated into the genome, they tend to remain there and so can be treated as just another heritable feature.

Nikaido et al. (1999) set out to identify SINES and LINES that were characteristic of the Artiodactyla and that were therefore potential synapomorphies for the clade. Although these are usually thought of as reflecting anatomical features, this is not necessary, and any characteristic that is shared and derived satisfies the synapomorphy criterion. Nikaido et al. (1999) identified 20 such sequences that were shared by various current members of the Artiodactyla and also seen in whales. They were therefore able to use a standard phylogenetics package to reconstruct a unique tree with 20 insertion events on the basis of which organisms carried particular SINES and LINES. The results (**Figure 7.6**) show clearly that whales are, indeed, a member of the Artiodactyla, with the hippopotamus being their closest and the camel their most distant ungulate land relative. Such a phylogeny gives no direct clue as to when the lines leading to the hippopotamus and the whales separated, but the fossil evidence suggests that it was at least 55 Mya.

Further, more dramatic evidence showing that cetaceans evolved from tetrapod mammals comes from embryology. Unlike all other mammals, cetaceans lack hindlimbs, although they have tiny, rudimentary pelvises and bilateral internal femurs with no connecting joints in a similar location to the hindlimbs of a normal mammal. Bejder and Hall (2002) have analyzed limb formation in dolphin embryos and found that the 24-day embryo has a hindlimb bud very similar to its forelimb bud and indistinguishable from the early limb buds of other mammals such as the mouse (**Figure 7.7**). By 48 days, however, things are very different: the dolphin

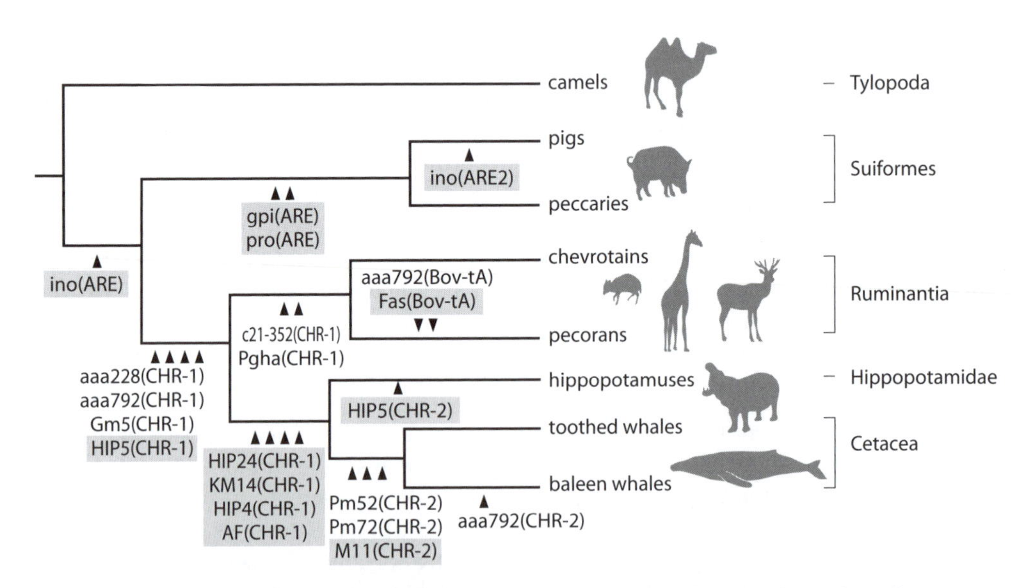

Figure 7.6 A cladogram showing the phylogeny of whales and other even-toed ungulates from an early artiodactylan ancestor using short interspersed nuclear elements (SINES) and long interspersed nuclear elements (LINES) as characters to determine branching points.

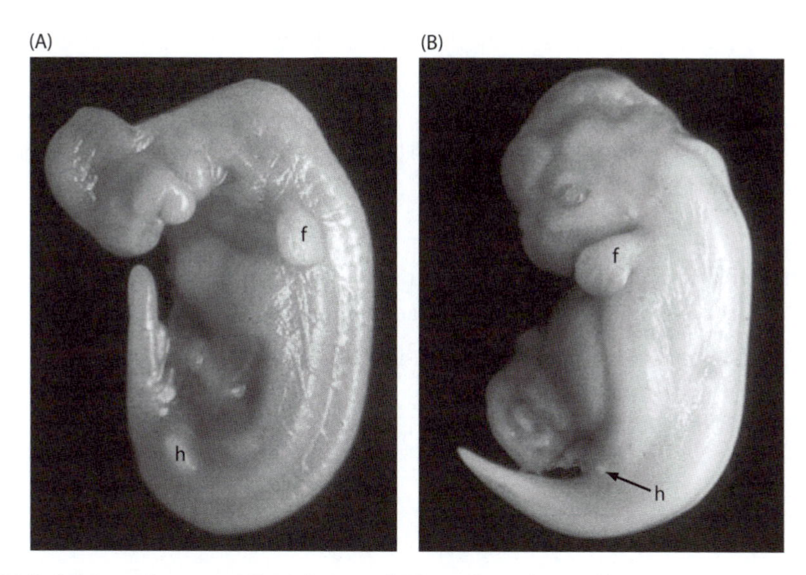

Figure 7.7 Dolphin embryos at (A) 24 days and (B) 48 days. The 24 day embryo (~7.5 mm) has both a forelimb (f) and a hindlimb bud (h). The 48 day embryo (~17.5 mm) has a well developed forelimb bud with digits, but the hindlimb bud has almost disappeared.

forelimb bud has developed as expected, forming digits in the normal way; its hindlimb bud has, however, almost vanished.

The embryological explanation for this is straightforward: patterning in the early limb bud is controlled by its distal region, the apical epidermal ridge (AER). This ensures that the bones are laid down in proximal-to-distal order: in the hindlimb bud, the hip bones are laid down first, then the femur, and so on (Gilbert, 2014), and this is followed by a sustained period of growth; similarly, in the forelimb AER, the scapula is laid down first, then the humerus, and the digits appear last. The presence of a rudimentary pelvis and femur in whales suggests that patterning activity in the early hindlimb ceased immediately after the femur had formed, with such small amounts of tissue as are present in these animals reflecting the minimal amount of subsequent differentiation and growth. The reason why the femur is internalized is not clear as later stages of dolphin development have not been studied. It is, however, likely that the diminutive hindlimb as a whole simply became internalized as the fetus enlarged. The whale carries its evolutionary history as a plesiomorphy (the ability to make a hindlimb) but has its own apopomorphy (early cessation of hindlimb development). This story is the exact embodiment of Darwin's *descent with modification*.

Key points

- Today's mammals originated from synapsid reptiles some 300 Mya. Then, they were the dominant clade but almost all taxa were lost in the Permian extinction, with mammals descending from cynodonts, one of three surviving synapsid taxa.
- During the Early Mesozoic Period, the synapsids survived only as very small animals that lived in burrows (the nocturnal bottleneck), showing evolving properties that are today associated with living in the dark.
- The three extant clades of mammals separated well before the K-T extinction (~66 Mya).

- Information as to how the specific anatomical features for each clade evolved are beginning to become apparent.
- Whales and dolphins are descended from hooved mammals and share a last common ancestor with the hippopotamus; this has been established on the basis of anatomical, embryological, and genomic evidence.

Further reading

Kardong KV (2014) Vertebrates: Comparative Anatomy, Function, Evolution, 7th ed. McGraw-Hill.

Kemp TS (2005) The Origin and Evolution of the Mammals. Oxford University Press.

CHAPTER 8 THE GENOMIC EVIDENCE FOR EVOLUTION

A set of homologous DNA sequences from a group of related species provides important evolutionary insights because it reflects not only current genetic information about these species, but also, through their mutation-derived differences, the history of how they diverged. Phylogenetic analysis aims to reconstruct that history as a phylogram whose root is the most likely sequence from which the set evolved. The various ways of constructing such trees differ in the nature of the information that they use, the way in which they analyze it, and the amount of information that the analysis provides. This chapter summarizes the various approaches, the language of phylogenetics, the problems that can be encountered in making phylograms, and the quality of the resulting evolutionary information, which can include population histories. It thus provides the background for the following chapter on investigating the early stages of evolution, a period for which there are minimal amounts of fossil evidence.

All cellular organisms have DNA-based genomes that use essentially the same coding, and this fact alone reflects the fact that they all descend from a common ancestor (albeit that there are a few very minor coding variations, particularly in mitochondrial DNA) (Bernt et al., 2013; Elzanowski & Ostell, 2013). The genomic perspective on evolution starts with the hypothesis that contemporary sequences that are similar almost certainly reflect a common ancestor sequence from which both evolved. The slight doubt here allows for two possibilities: first, that one organism donated a gene-containing DNA sequence to another through horizontal gene transfer (Keeling & Palmer, 2008); second, that the similarities in short sequences could have arisen by chance. In practice, a sequence of about 17 bases is more than enough to identify a unique sequence in a human genome of 2×10^9 bases; the odds on two nonrelated such sequences being the same are 4^{17}:1 or about $>10^{10}$:1. Nevertheless, and to reduce even further the chances of coincidence, phylogenetic analyses are normally carried out on sequences of well over 100 nucleotides.

There is now a very large amount of evidence to support the view that contemporary differences in related sequences derive from mutations that have occurred since they shared an ancestral sequence. There are immediate predictions to be made from this evidence: the closer two species are on the basis of anatomy-based cladograms, the fewer should be the numbers of mutations between homologous gene sequences. The corollary of this is that evolutionary relationships among living organisms based on phylogenetic analysis of sets of their gene sequences should mesh with relationships derived from cladograms constructed on the basis of the fossil record. Both predictions are found to be correct to the extent that sequence-based phylogenies can be used to probe groups where the fossil evidence is incomplete or absent (for example, see Schneider & Sampaio, 2013).

Gene phylogenetic trees (phylograms) can really only be constructed computationally, but, once the algorithms are in place, analyses can be extended to as many sets of homologous sequences as are available. In such cases, the result for each will be a tree with the most likely original common ancestor sequence at its root and the current sequences as terminal leaves. It is worth noting that one of the computational origins of such sequence trees was work done on reconstructing the history of sets of medieval manuscripts through analyzing copying errors (Buneman, 1971), a problem that is formally equivalent to that of constructing phylogenies on the basis of mutation.

Because DNA sequencing is so cheap and fast today, there are now full sequences for the genomes of at least 80 plants, invertebrates, and vertebrates and almost 30,000 prokaryotes, together with protein-coding data for many genes in other organisms, and details of many sequence variants (polymorphisms). As this information is stored in massive public databases such as ENSEMBL (http://ensemblgenomes.org/info/genomes), there is an essentially infinite amount of sequence data publicly available online, together with a host of sophisticated informatics tools for exploring their relationships. This chapter considers how to analyze these data, while the next shows how such informatics approaches can be used to explore the early stages of evolution, long before any fossils were laid down.

Phylograms and cladograms are subtly different

Although phylograms and cladograms are both types of phylogenetic tree and look superficially similar, they are not quite the same. Consider two contemporary organisms, **B** and **C**, that share an unknown last common ancestor, **A**, with homologous sequences **Q** in species **B** and **R** in species **C**, with both deriving from sequence **P** possessed by an unknown common ancestor **D**, which may or may not be the same as **A**.

In formal terms, cladograms are composed of triplets of the general type

<species **B**> <*descends with modification from*> <species **A**>

<species **C**> <*descends with modification from*> <species **A**>

while a phylogram is composed of triples of the general form

<sequence **Q** of species **B**> <*descends through mutation from*> <sequence **P**>

<sequence **R** of species **C**> <*descends through mutation from*> <sequence **P**>

For such phylograms based on sequences from many organisms, the first higher-level nodes represent the last common ancestor sequences of each pair of current sequences but give no indication of what those ancestor organisms were. Thus, the last common ancestor **D** that carried sequence **P** may have lived very much earlier than species **A** if the sequence that led to **P** underwent more random mutation than expected; alternatively, sequence **P** might have survived for some time on the lines leading to **B** and **C** if it received less than its fair share. Where, however, the sequences come from species that are in different phyla (for example, mouse and *Drosophila*) and are clear homologs, we can be sure that the sequence possessed by the last common ancestor of the two species lived in the early Cambrian, at the very latest.

There are three other major differences between phylograms and cladograms. First, anatomical changes can be traced back in time through the fossil record, whereas sequence changes cannot, other than through occasional samples from ancient organisms whose DNA has been fortuitously preserved. It is occasionally possible to study specific sequences in such ancient DNA and compare them with sequences in extant organisms, but sufficient homologs of a specific sequence to construct a phylogram have yet to be found. That said, the study of ancient DNA is becoming ever more interesting and gives insight into protein stability and change over the Pleistocene Period. The recent work on sequencing bear mitochondrial DNA from some 300 Kya (Dabney et al., 2013) and on horse DNA perhaps twice as old (Orlando et al., 2013) suggests that there may be more to come here from other sources where there are traces of DNA. It is also possible that robust cross-linked proteins such as collagen from ancient organisms have maintained their primary sequence well enough for parts, at least, to be sequenced (Buckley et al., 2008).

Second, DNA sequences are not samples of the phenotype of the organism but of the genotype and thus a far weaker assay of change than an anatomical feature. One cannot tell whether a mutation changed the phenotype of a past organism or not, although mutations that changed amino acids are more evolutionarily significant than those that did not (neutral mutations) because their effects are subject to selection. What this means in practice is that one cannot tell whether the last common ancestor of two contemporary DNA sequences was the same as the last common ancestor of two organisms with slightly different anatomy, be they old or new. That said, phylogenetic trees do establish the relationships among a group of current organisms extremely well, and the more genes that are sampled in the study, the better the estimate is. Although ancestor species cannot be identified, if enough sequence trees are generated for a group of species and integrated, one would expect that the resulting tree would be equivalent to the species tree.

Third, the relationships of phylogram triplets are quantitative, whereas those of cladogram triplets are qualitative: the latter are essentially truth statements with the length of a cladogram edge carrying no quantitative meaning and hence is unscaled. In phylograms, however, the length of an edge or branch has a very specific meaning: it is a measure of the minimum number of mutations required to change a parent sequence into daughter ones so that the fewer the number the shorter is the branch length. It is difficult to be precise here because, in assigning this length, one has to decide how to weigh the relative likelihoods of a single nucleotide change, a back mutation, a deletion, or even an insertion. In most cases, it is not hard to do this and so obtain a quantitative estimate of the length of branch in a phylogram. Even where this is difficult, one can assume that the longer the branch length generated in a phylogenetic tree, the longer the time period that it represents.

Although cladograms explicitly exclude time in their construction, they can be linked to the fossil record through the age of the earliest organisms to include particular synapomorphies; such data give a minimal estimate of when a feature evolved. Time can also be included in phylograms provided that there is a measure of how long it takes, on average, for a mutation to become an established feature of the genome. In this case, the length of that branch becomes a fairly precise measure of time between two contemporary sequences and their last common ancestor. The addition of time to a phylogenetic tree enables it to be used as a history of the evolutionary relationship of the organisms being analyzed that is, to some extent, independent of cladograms that incorporate the fossil record. The hedge here is because estimates of mutation rates often link back to one or another aspect of the fossil record.

In short, a phylogram indicates the closeness of the relationship amongst homologous sequences, and implicitly among the species possessing these sequences. If the problem of horizontal gene transfer is ignored (next section), a phylogram for a set of homologous sequences for a particular protein will provide an estimate of the closeness of the parent sequences to daughter sequences. The more genes that are included in the analysis, the nearer will be the estimate to the correct relationships and hence to the true cladogram. Indeed, if the quality of the cladogram is poor owing to the weakness of the fossil record, then the phylogenetic tree may be a more accurate and more useful measure of evolutionary history.

Choosing sequences for phylogenetic analysis

For the purpose of making a phylogenetic tree, the essential data are a set of homologous sequences from a group of species (orthologous sequences such as the *hox* genes in different species) or even within a single one (paralogous sequences such as the many *hox* genes in a single mammalian species, such as the mouse). This sequence may code for a protein, an RNA molecule, or may even be untranslated - function is irrelevant here. In practice, the choice of sequence is less simple than it might seem because there are a series of potential pitfalls.

First, one needs to be clear about what homologous means in this context. The standard implication is that two sequences are homologs if they derive from a common ancestor sequence. If a sequence was acquired through horizontal gene transfer, that is where one organism donates to another a piece of DNA (or RNA that can be copied into DNA) that becomes incorporated into the recipient's genome, then the donor and recipient will not share a common ancestor in any immediate sense. Such xenologs certainly occur among bacteria (see Chapter 9), but turn out to be far less common in multicellular organisms because the incorporation is unlikely to be carried through to the germline. The best-known exception here is *Wolbachia*, a parasite bacterium that is commonly found in insects; the two organisms lead a symbiotic existence, with the insect genome often including many *Wolbachia* sequences (Le et al., 2014). Any phylogenetic analysis done using sequences acquired through horizontal gene transfer will be flawed because the common descent assumption will have been violated.

Second, there is little point is conducting phylogenetic analyses among a group of closely conserved sequences as the number of mutations required to generate a last common ancestor are so few that it is hard to decipher relationships and one is merely left with the conclusion that the sequence probably evolved very early. Examples where this sort of problem occurs are in the various DNA-binding motifs, such as homeodomains, zinc-finger domains, and paired-box domains that characterize the different classes of transcription factors. Such motifs alone show very minor differences across the phyla and are of little use in making phylogenetic trees, although, of course, complete protein sequences that include such motifs are routinely used for phylogenetic analyses.

Third, where the homologous sequences are very different, alignment becomes difficult because of the number of ways that the analysis can be done. It then becomes hard for the sorting routines to calculate a unique phylogenetic tree because there are several possible options, with none having a much stronger likelihood than another. This is the long-branch problem that made analyzing the origin of whales so hard (Chapter 7). There is also the interesting question of whether, for protein-coding sequences, one should analyze the genomic or the equivalent amino acid sequences. DNA-based analyses are more sensitive because of the redundancy in codons and hence more useful for comparisons among closely related species. Amino acid-based analyses are a little coarser and hence more useful in analyzing sequences that are from distantly related species; such sequences have two further advantages: first, because they code for functioning proteins, they less noisy; and, second, they are directly subject to selection.

Fourth, where there is a group of similar sequences within the genome that clearly represent duplications, it may not initially be clear when and how many times the duplications took place. The homeotic family of transcription factors, each of which carries a homeodomain DNA-binding motif, highlights the problem. *Drosophila* has eight such homeotic genes that form a group of ordered paralagous sequences (the *HOM-C* cluster), all on chromosome 3. The mouse has 38 such genes, separated into four clusters on different chromosomes (2, 6, 11, 15). The genes on each cluster are orthologous to those in *Drosophila,* with each group of paralogs being in the same genomic order (**Figure 8.1**; Gehring et al., 2009).

Working out the relationships among all these homeotic genes was not simple and it took a considerable amount of detailed phylogenetic analysis to decide which sequences were closest to one another (Holland & Takahashi, 2005). The results of the analysis suggested that the last common ancestor of mouse and *Drosophila* had a cluster of eight homeotic genes, and these probably reflect three duplications of a single ancestor gene coding for a transcription factor. The original homeodomain-containing protein probably dates back to ancestral Cnidaria (for example, jelly fish) in the Ediacaran Period and probably before the evolution of the Bilateria (Ryan & Baxevanis, 2007).

A very early step in bilaterian evolution was the split into protostomes (the first opening in the embryo becomes the mouth) and deuterostomes (the second opening in the embryo becomes the mouth - the first becomes the anus), with the former giving rise to almost all of the invertebrates (for example, *Drosophila*) and the latter to chordates (for example, the mouse), hemichordates, and sea urchins. The phylogenetic analysis shows that the

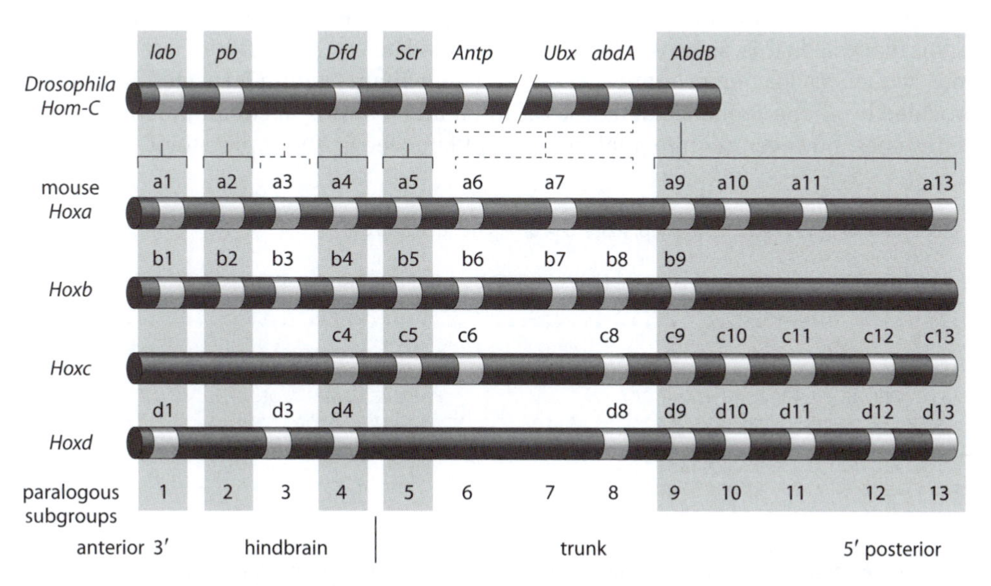

Figure 8.1 Comparison of paralagous and homologous *Hox* genes in *Drosophila* and the mouse. The genes are expressed in a rostral-to-caudal sequence, so patterning the long axis of both animals, although the downstream effects of this patterning are, of course, very different.

Drosophila line maintained each of the homeotic genes, but things became far more complicated in the line leading to the mouse. There was first a single duplication of the second gene (the ortholog of *Pb* in *Drosophila*) and then multiple duplications of the terminal gene (*Abdb* in *Drosophila*), eventually giving five copies. Later, this group of 13 genes underwent two duplications with three of the groups being translocated to other chromosomes and with some of the possible 52 sequences being lost (Holland & Takahashi, 2005; Pascual-Anaya et al., 2013).

The fifth and final major complication in making phylogenetic trees derives from the assumption that the sequences that are used computationally represent the consensus sequence of that gene in each organism. The correct choice of sequences is not necessarily obvious, given that even the two alleles on the paired chromosome of a diploid organism would not be expected to be the same, but will differ through at least a few single nucleotides. Indeed, one might even differ from the other through deletions or additions, and both may differ from the eventual choice of consensus sequence. Identifying the most likely nucleotide for each position within a population may require so much sequencing that compromises may have to be struck on the basis that variation within a species is usually very much less than that between species.

In practice, the types of sequences that easily lend themselves to phylogenetic analysis are, as has long been realized, those that share similar regions, such as the many homologs of cytochrome C and Pax6 (**Figure 8.2**), and those where there is a single gene in each of the organisms under consideration. It thus pays to take some care about the choice of sequences before moving to computational phylogenetic methodologies. In general, it is sensible to analyze rapidly changing sequences when examining short periods of evolution and slowly changing sequences, ideally of amino acids, when analyzing long periods.

Constructing sequence-based phylogenetic trees

Phylogenetic trees are always constructed computationally, and there is a host of programs that will do this, some of which are mentioned below. All start with a set of homologous sequences for one or more species, or even variants within a species, information provided by all the sequence databases. A full understanding of computational phylogenetics does, however, require a fair amount of knowledge about probability theory and

93.9% identity in 132 residues overlap; score: 644.0; gap frequency; 0.0%

```
PAX6_MOUSE     5 HSGVNQLGGVFVNGRPLPDSTRQKIVELAHSGARPCDISRILQVSNGCVSKILGRYYETG
PAX6_DROME    57 HSGVNQLGGVFVGGRPLPDSTRQKIVELAHSGARPCDISRILQVSNGCVSKILGRYYETG
                 ************ ***********************************************

PAX6_MOUSE    65 SIRPRAIGGSKPRVATPEVVSKIAQYKRECPSIFAWEIRDRLLSEGVCTNDNIPSVSSIN
PAX6_DROME   117 SIRPRAIGGSKPRVATAEVVSKISQYKRECPSIFAWEIRDRLLQENVCTNDNIPSVSSIN
                 **************** ****** ******************* * **************

PAX6_MOUSE   125 RVLRNLASEKQQ
PAX6_DROME   177 RVLRNLAAQKEQ
                 *******   * *
```

Figure 8.2 A comparison of parts of the Pax6 amino acid sequences from mouse and *Drosophila melanogaster* (DROME) made using the ExPASy comparison tool. Although the former has 422 and the latter 857 amino acids, there are regions such as that shown where amino acid matching (*) is almost perfect – there has been virtually no evolutionary change here in more than 500 My.

population genetics, together with some informatics and statistical skills (see "Further reading" at the end of the chapter). The treatment here is just intended to provide a qualitative explanation of the key principles behind the various types of analysis, together with some important results.

Phylogenetic trees can be constructed using four different approaches that use increasing amounts of sequence information.

- The most basic is to just analyze the presence or absence of sequences (such as the example of whale evolution given in Chapter 7).
- For analyzing sequence data, the simplest approach is to choose a set of taxa and investigate the differences between them. To do this, one first constructs a distance matrix whose terms are numbers that reflect the genetic distance between each pair of sequences. A computer program then uses neighborhood-joining algorithms to group taxa that are genetically close, constructing a tree whose branch lengths are as short as possible and so represent the most likely paths of sequence descent.
- Rather richer and more commonly used are the more sophisticated techniques that analyze details of the nucleotide differences across the sequences. These methods not only generate trees, but can also construct ancestral sequences on the basis that mutations in each will lead to modern sequences in the most parsimonious way.
- These methods all use consensus sequences and exclude information on sequence variation or polymorphisms in a population. The final approach uses this information too and is known as coalescent theory: this analyses sequence variation within a species on the basis of a model of population genetics and a defined mutation rate. The theory is used to run evolution backwards to the most recent common ancestors (MRCA) that had the original parent sequence. While it can be used to reconstruct sequence history, its major importance is that it sheds light on the past population dynamics of the species carrying the sequence. Coalescent theory is discussed at the end of the chapter.

In practice, several approaches are used for any analysis and the slightly different trees that they generate are integrated to construct a consensus tree on the basis of their relative likelihoods.

Phylogenies based on shared sequences

Where a set of sequences is distributed across a group of species, it is straightforward to work out a phylogeny for the group on the basis of that distribution. Such a tree is actually more a cladogram than a phylogram, as inherited features are being analyzed rather than sequences. The insertion of a new sequence in the genome is really an apomorphy and its inheritance by a set of daughter species represents a shared, derived characteristic or a synapomorphy. Constructing the tree is computationally simple. The best-known example of this sort of analysis is the use of retrotransposons to unpick the evolutionary line of whales and so confirm that their closest living relatives are hippopotamuses (Nikaido et al., 1999; Chapter 7).

Phylogenies based on distance matrices

The first step in constructing a phylogenetic tree for a set of sequences is to make a distance matrix. For this, each sequence (S_i) is aligned against each the others ($S_{1\to n}$) as closely as possible to obtain a measure of the differences (D_{ij}). This would be straightforward if all gene sequences were the same length and mutations restricted to simple base changes. Unfortunately, possible mutational changes include deletions, insertions, secondary mutations (both backwards and forwards), and even duplications; the net result is that homologous sequences vary both in detail and in length. The program has to cope with all possibilities and usually starts with a progressive sequence alignment that initially considers sequence elements where the differences are small and then adds those that are more and more distinct (this often means including gaps, which are usually of different sizes and may be in different locations).

It is unusual for there to be a unique way of producing this sequence alignment; indeed, one would not expect there to be because the effects of random evolution and selection over many millions of years are being compared and there is no logic linking the genomes of different organisms that have mutated independently. Because of the complexity of the task, such alignments are always made computationally, and the simplest way of analyzing a sequence alignment is to assign a number to each pair of sequences (i, j) based on the number of mismatches between them (a gap counts as a mismatch). The most straightforward way of constructing this genetic difference is to use the Hamming distance for a pair of aligned sequences:

$$D_H = (\text{number of sites that differ in a sequence})/(\text{sequence length})$$

The difficulties here are that no account is taken of secondary mutation and that all differences are given equal weight so that, for example, gaps are equated with simple mutations; the net result is that D_H is always an underestimate of the correct distance. Two sorts of corrections are usually applied to D_H: the first makes allowance for the possibility that successive mutations can occur at a site, the simplest correction here being the Jukes–Cantor model whereby:

$$D_{JC} = -3/4 \ln (1 - 4/3\, D_H)$$

The second correction gives more weight to related changes where the nature of the mutation is clear than to complex changes where it is not. Here, the aligned sequences are first broken up into clusters for each of which a D_{JC} is calculated and this is the material used for the Fitch–Margoliash method. Because assigning a distance to closely related sequences is more accurate than to those that are very different, this method uses a least-squares optimization algorithm that assigns more weight to the former than the latter in computing D.

The net result of all distance methods for analyzing N sequences is an $N \times N$ **distance matrix** where the components are the corrected $\mathbf{D}_{ij}$ and that is symmetric (the distance measure of sequence **X** and sequence **Y** is the same as that for sequence **Y** and sequence **X**). From this matrix, it is possible to compute a tree linking sequences on the basis that the lower the $\mathbf{D}_{ij}$ for a pair of sequences the more closely related they are. There are a host of programs (for example, Clustal and Mega6) that will construct such phylogenetic trees from distance matrices, using procedures such as neighbor-joining. In practice, and because different methods produce slightly different alignments, it is normal to analyze the whole dataset with several techniques, one of which is usually maximum likelihood.

One question that immediately arises from such an analysis is the degree of confidence that one can have in it, and one way of answering this is to use the technique known as bootstrap resampling (this is more common than the alternative jackknife resampling technique, which systematically removes a small part of the data and so can reset codon triplets). Here, the tree-generating programs are repeated many times, with each reflecting the replacement of a different part of the data. Nodes that are robust are only rarely affected if a small amount of the data is altered, while nodes for which the support is weak immediately show up with a low bootstrap number. The general view is that a bootstrap figure of 70% is the absolute minimum for a reasonable hypothesis, while a figure of 95% or above is very unlikely to be wrong. The tree shown in **Figure 8.3** was the result of 1000 such simulations and only one link has a bootstrap value of less than 85%. However, putting exact confidence limits on bootstrap figures is difficult (Buckley & Cunningham, 2002). In practice, bootstrapping can be used with both distance-matrix and tree-searching approaches.

It is also important to note that such a tree is calculated only on the basis of $\mathbf{D}_{ij}$ values and so has no explicit root or origin; it cannot therefore localize the position of the last common ancestor sequence. There are two ways of rooting such phylogenetic trees. The first is to include in the analysis a sequence only distantly related to those being studied (see Figure 8.3). The analysis will show where this outlier meshes onto the hierarchy and the node will link directly back to the root of the tree. Were, for example, one to construct a phylogenetic tree of the vertebrate *Pax6* genes, the *Drosophila Pax6* gene would be an obvious outlier. The other way of rooting a phylogenetic tree is to use implicit timing on the basis that the length of an edge is a measure of the time taken for divergence. If this is

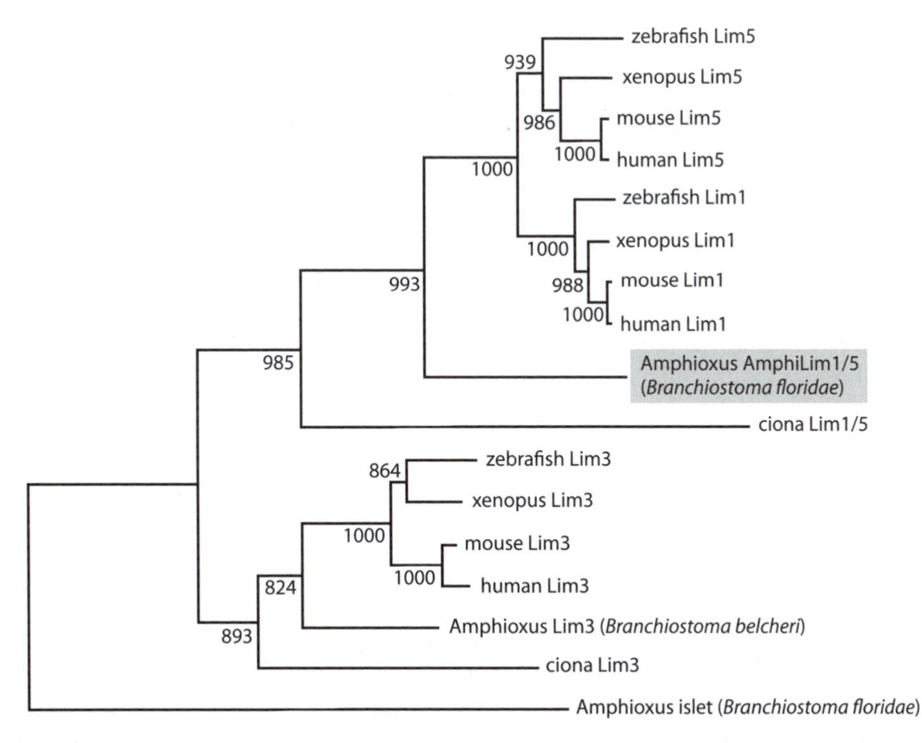

Figure 8.3 A distance matrix phylogenetic tree designed to show the evolutionary status of the amphioxus *AmphiLim1/5 gene*, which contains a homeodomain. The outlier gene used for rooting the tree was for the amphioxus islet, which also contains a homeodomain. The program used for constructing the tree was ClustalX and the numbers are bootstrap values based on 1000 trials (see text).

so, the root should be at the midpoint of the longest distance between two terminal nodes (such as those sequences for two very different contemporary species).

Phylogenies based on tree-searching methods

Because distance-matrix methods analyze the presence rather than the kinds of differences between sequences, their use is restricted to the production of phylogenetic trees. Methods that include actual nucleotide differences in their analysis are not only more accurate because they are based on more data but also provide additional information as they can suggest likely ancestor sequences for each node in the hierarchy. Making such phylogenies is known as tree-searching because the procedure involves searching for the tree that generates all the current sequences on the basis of mutation from an ancestral root sequence. As there are many possible such sequences, the procedure incorporates the test criterion that the tree should require the minimum number of mutations. A tree-searching program, in principle at least, thus looks at every possible tree and identifies the most likely node and associated sequences. As there can be a very great number of trees, this is an insoluble problem for very large numbers of long sequences, but there are a series of heuristics for excluding unlikely options and so shortening the computing time. In practice, such programs start by constructing a starter tree from sequences that are not too different, and then add increasingly different ones to the analysis. There are three approaches to the computational analysis: maximum parsimony (MP), maximum likelihood (ML), and Bayesian statistics.

Maximum parsimony

This approach sets out to find the phylogenetic tree on the basis of identifying those ancestral sequences that generate contemporary ones with the minimum number of mutations. The MP approach is thus to search every possible scenario and see which is the most efficient. As the number of possibilities increases exponentially with the number of taxa, so does the amount of computational time required to identify the most parsimonious tree. In practice, computational algorithms have been devised that explore more likely solutions and reject less likely ones. While the confidence that one has in the final tree can be improved by bootstrapping (see above), MP methods are not much used today because they do not include an evolutionary model and so exclude any information about the relative likelihoods of mutation possibilities.

Maximum likelihood

These methods also search through all possible trees but incorporate a model of evolutionary change based, for example, on the details of mutation. Thus, it is known that (1) the rates of base mutations from purine to purine (A → G or G → A) or pyrimidine to pyrimidine (T → C or C → T) are higher than those for transversions (for example, A → T) and (2) that mutations in the final base of a codon that do not change an amino acid are more likely to be inherited than those that do. This is because the latter change may alter the final phenotype and so be lost as a result of natural selection. Any analysis that includes such information about mutation will be more accurate than one that does not. ML methods use programs such as MetaPIGA or PAUP to examine all plausible trees. The programs assign likelihoods to each and, as it were, recommend the tree with the highest likelihood of predicting the data, together with a statistical analysis of the confidence that one can have in it.

Bayesian methods

This approach is now also widely used; it aims to generate the set of hypothetical phylogenies that are supported by the data, and assign probabilities to each, implicitly recommending that with the greatest. They are thus subtly different from ML methods that essentially look for the phylogeny that best generates the data. As the methodology is based on Bayesian statistics, which, in essence, use existing information to refine expectations, it can also incorporate a model of evolution whose rules help generate that set (these rules are known as the prior probabilities). Neither the theory nor the methodology are simple as they are based on sophisticated statistical theory, but the practice is straightforward. The user submits to a program such as MrBayes or BEAST a set of sequences that perhaps includes a likely outgroup sequence to help root the phylogeny together with a model of mutation (which, if it includes rate constants or other clock models, can generate timings).

The aim of such a program is to explore the full space of trees that is suggested by the data and incorporate the prior probabilities (this is the evolution model) and find that which is the most likely. The program uses Monte Carlo and Markov chain methods; it starts by constructing a possible tree from the data, and then uses a random number (the Monte Carlo component) to mutate the sequences and construct another tree solely on the basis of the new sequences alone; this is the Markov criterion, which essentially says that history does not influence the future. If the new tree is better on some criterion, it becomes the starting tree for the next simulation (or link in the Markov chain); if worse, it only has some assigned probability of replacing the original tree for the next simulation. The program runs for as many as 1,000,000 Monte Carlo simulations (ignoring, perhaps, the first 25%, as they might be dependent on the start tree), building up a frequency table for the various possible trees. One way of knowing when to stop is to run two simulations in parallel and, when their distributions are the same to within some predefined criterion, one can be reasonably sure that the simulation is complete.

What emerges from the simulation is a set of trees, each of which could generate the data and with each having a frequency based on the number of times that the simulation "visited" that tree as it explored tree space. The branch lengths within the tree reflect the number of mutations required to make the link, and this is a measure of how timing events were included in the original model. Integrating the set of simulation outputs gives a set of posterior probabilities that each tree is correct and is based on mathematically rigorous methodology. It should, however, be said that there are some minor concerns about Bayesian trees. First, the distribution depends on both the prior probabilities and the inheritance model used, so one needs to test the quality of this and the robustness of the tree by using other priors and models. Second, Bayesian phylogenies turn out to be more sensitive to the quality of the algorithms for measuring genetic differences between sequences than are other methods. Third, some doubt has been cast as to whether it is entirely appropriate here to use the Markov criterion, that it is only the immediate state that determines the subsequent one (Cartwright et al., 2011).

In practice, it is normal to use at least two methods to construct a phylogram for a group of species. In general, the different methods each give very similar results (**Figure 8.4**).

How accurate are phylogenies?

For all the very large amounts of computation needed to construct phylogenies for large datasets, it is not easy to decide which mode of analysis is the most appropriate for

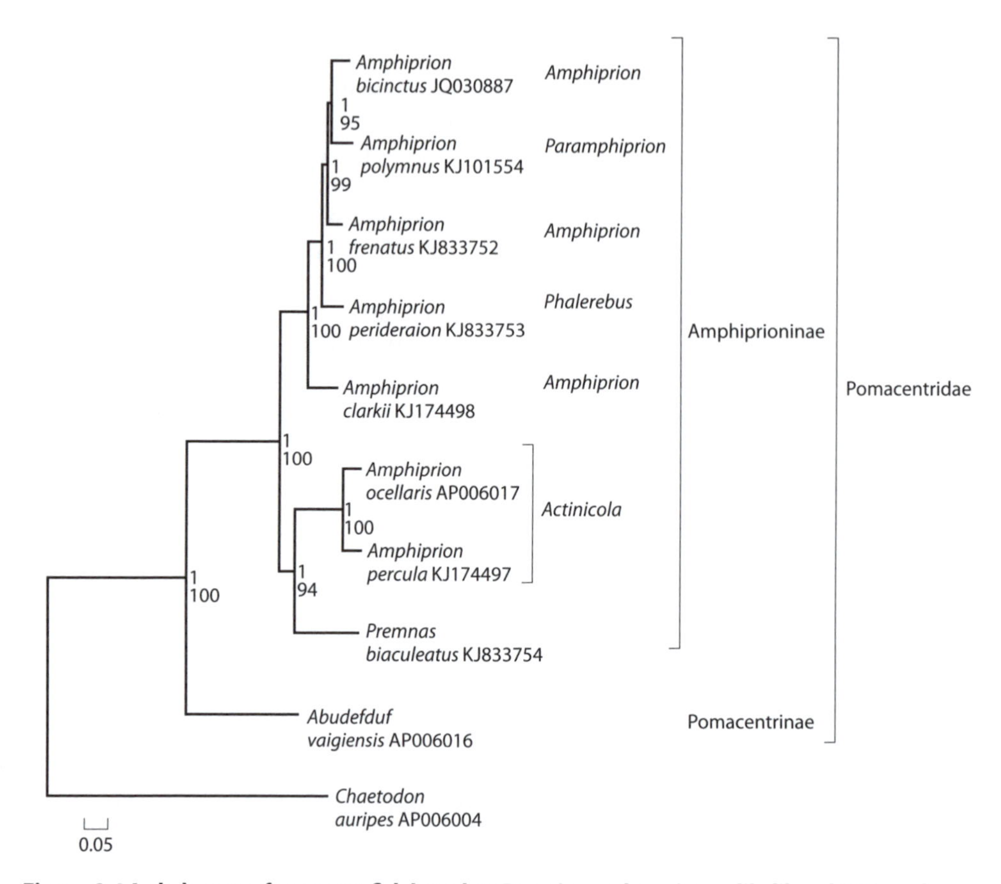

Figure 8.4 A phylogeny of anemone fish based on Bayesian and maximum likelihood (ML) analyses using the butterfly fish *Chaetodon auripes* as an outgroup to root the tree. The analysis used ~40 protein- and nonprotein-coding genes, and the Bayesian probabilities for the tree are the upper figures at each node (all equal one) and the bootstrap numbers from the ML analysis are the lower figures. The scale bar represents nucleotide substitutions per site.

determining a particular phylogeny. Hall (2004) therefore constructed an interesting experiment. He took an *Escherichia coli* protein-coding DNA sequence and generated successive generations of that sequence using a mutation-generating program. The net result was a tree with seven generations and more than 100 nodes, with a precise sequence associated for each, and with successive generations having increasing amounts of randomly generated deletions, mutations, and substitutions. He then used the full range of phylogenetic techniques on the final-generation DNA sequences to reconstruct estimates of the original sequences, the intermediate sequences, and the complete tree. To improve the quality of the analysis, he produced the sequences in three formats: first the nucleotide sequences, then the corresponding amino acid sequences, and, finally, the DNA sequences organized so that common codons were aligned (DNA-CA - it turns out that this format facilitates the handling of gaps).

His initial results illustrated the relative speeds of the various methods: neighbor-joining on the basis of distance matrices took less than a second for all formats; tree-searching methods using the same computer took much longer with analyses on DNA-CA sequences giving more accurate results than the equivalent DNA or amino acid sequences. Here, MP,

ML, and Bayesian methods took 5 minutes (without bootstrapping and about 1 hour with), 6 hours 42 minutes, and 7 hours 43 minutes, respectively (these figures would be much shorter today because computing speeds are faster). As to accuracy, tree methods were, as expected, better than those based only on distance matrices with, on average, ML methods giving the most accurate results. While Bayesian techniques could give good results, they turned out to be too sensitive to alignment choices to make them as reliable as ML methods, for this analysis at least. Hall therefore recommended standard distance matrix methods for a quick analysis and ML methods for reliable phylogenetic trees.

The added advantage of tree-searching methods is, of course, that they generate estimates of ancestral sequences, and these can be used as the basis for the synthesis of ancient proteins (Thornton, 2004). If the analysis, for example, uses sequences from both protostome and deuterostome animals and they can be shown to derive from a common root, then the generated last common ancestor sequence is an estimate of one that came from a very early bilaterian ancestor, one that probably lived more than 550 Mya. Thus, Mirabeau and Joly (2013) were able to show that peptide hormones and receptors co-evolved before the protostome/deuterostome split. At a more physiological level, Shi and Yokoyama (2003) synthesized ancient short-wavelength-sensitive receptors on the basis of sequences generated from phylogenetic analysis of modern avian, fish, reptile, and mammalian sequences. They then analyzed the sensitivity of these proteins to ultraviolet and blue light, and were able to follow how the sensitivity changed as clades evolved. As such physiological probes can now be used to study how ancient organisms lived, this important use of phylogenetic analysis is now opening up an intriguing area of research.

Gene trees and species trees

It is important to remember that sequence phylogenies are not the same as species phylogenies: the variation that takes place in a single sequence over time is a very limited assay of the extent of the changes in a whole genome, which should, of course, be equivalent to the species. The techniques for constructing species trees from phylogenetic trees are based on finding a tree that represents a minimization of the differences between the individual gene trees (for example, see Chaudhary et al., 2013) and, as the number of gene trees increases, that tree will converge onto the species tree (Maddison, 1997). The use of large numbers of gene sequences to construct such trees is a branch of a new subject known as phylogenomics (Schraiber & Akey, 2015).

Such multi-gene phylogenetic analyses have been useful in, for example, understanding the taxonomy of single-celled eukaryotes (or protists), which are discussed in more detail in Chapter 9. A large number of these can be distinguished, for example, by such morphological characters as number of flagellae (one, two, or many) and the presence or absence of mitochondria and chloroplasts. More than a decade ago, Cavalier-Smith (review - 2014) proposed a classification whose major clades were unikonts (a single flagellum and mitochondria - basal organisms for animals and fungi), bikonts (two flagellae, mitochondria and chloroplasts - root organisms for plants), and excavates (two or more flagellae and diminished mitochondria - this group included slime molds); there were also some minor clades. A phylogenomic analysis using 135 gene sequences showed that this high-level classification was likely to be correct (Burki et al., 2008).

Because the quality of extant genetic data is generally far better and more precise than that of anatomical differences taken from the fossil record, considerable effort is being put into estimating cladograms on the basis of multi-gene phylograms. A recent innovation here is the use of coalescence methods and other sophisticated programming approaches (Szöllősi et al., 2015) in generating species trees on the basis of genetic information. The difficulty, of course, with such methods is that it is hard to link the high-level nodes of multi-gene phylograms with the ancient species that they represent.

Adding timings – the ticking of the molecular clock

In the 1960s, when amino acid sequences were becoming known, it became clear that the extent of the differences between protein sequences depended strongly on the time that had passed since their host organisms had shared a last common ancestor, as based on the fossil record. The discovery of the genetic code confirmed that some DNA mutations changed the amino acids, whereas others did not - these are the neutral mutations. This, in turn, led to the idea of a molecular clock (Kumar, 2005) whose regular ticking represented neutral mutation change becoming established within a sequence, spreading throughout the population as a result of genetic drift at a rate assumed to be constant.

The importance of such a clock was that, because it only counted mutations that did not affect protein function, its speed was essentially independent of selection, unlike the fossil record. Such clocks cannot, of course, be linked directly to a cladogram as its branches do not measure time but represent truth values (for example, a taxon does or does not carry a synapomorphy). It can, however, be included within a phylogenetic tree as the branch lengths here represent a measure of the quantitative difference between sequences, albeit that in using the clock for comparisons, distances are not linear: as discussed earlier, one has to take account of backward and repeated mutations - long branch lengths turn out to represent underestimates of time.

The idea that there was a single, uniform clock rate for all species at all times turned out to be oversimplistic, and for two reasons. First, analyzing sequence differences is complicated: not only are there concerns about balancing the relevant weights of the different types of mutations (replacements, deletions, and insertions), but there are decisions to be made as to which mutations are neutral and which are active, and so subject to selection, particularly when nonprotein-coding sequences are being discussed. Second, the rate at which sequence differences accrue between two taxa depends on the generation times, the degree of selection that may vary over time, and the influence of genetic drift that depends on population size (Ayala, 1999).

In short, the speed of the clock cannot be assumed constant even within a clade, at least over long periods. In practice, therefore, the clock always needs to be calibrated for particular applications against the general fossil record or for the species under consideration (van Tuinene & Blair Hedges, 2001; Ho & Duchêne, 2014). An interesting exception here is the work of Shih and Matzke (2013; Chapter 9), who studied the events of early eukaryote evolution. They were able to add evidence from horizontal gene transfer events to calibrate their clocks on the basis that the same events represented in different phylogenetic trees had to have occurred at the same time. They considered that using such additional information from maintained duplications of ancient genes in their Bayesian analyses improved estimates of the times of evolutionary change by ~20% and were able to estimate that mitochondria and chloroplasts had been endosymbiosed ~1200 and ~900 Mya, respectively. Unfortunately, such additional data are rarely available.

Where the clock idea turns out to be of more general use is over limited periods of time and for groups of related taxa with the same generation time. Weir and Schluter (2008), for example, carefully analyzed the sequences for cytochrome b, a mitochondrial gene, for 12 orders of birds. They used detailed fossil data for absolute timings and a set of corrections to allow for known computational difficulties. The clock rates ranged from 1.03 to 3.63 mutations per million years, and the combined mean rate was 2.21 ± 0.68 (assuming a generation time of 1 year). A reasonable working figure for adding timing to vertebrate phylogenies in the absence of any better data is probably about 2, but this figure should be taken as no more than indicative.

Historical population analysis using coalescent theory

Both distance-matrix and tree-searching analyses ignore two important aspects of the available data: first, they use sequences that are assumed to represent the consensus (or most common) sequence for a species, rather than allowing for the spread of sequences within a population due to mutation; second, they ignore the reproductive behavior of the population that carries the sequences, be it haploid or diploid. It is possible but not simple to overcome these limitations through coalescent theory, a term derived from an original insight by Fisher. In modern words, this was that, if the variation in a gene sequence within a population is traced backwards in time, this variation coalesces back onto a single sequence possessed by the most recent common ancestor (MRCA; the term used for ancestors within a species), taking what is known as the coalescent time (measured in generations) to do this.

Coalescent theory (Kingman, 1982, 2000) provides a means of modeling the reversal of mutations in a sequence as a population goes back in time to a MRCA that possessed the original sequence from which later ones diverged - in essence, it reverses mutation and genetic drift. Application of coalescent theory not only gives estimates of phylogenies and original sequences, but also of the number of generations between nodes and how population size changed over time (Li & Durbin, 2011; Wall & Slatkin, 2012). The basics of coalescent analysis are covered here and its use in analyzing human evolution is discussed further in Chapter 16.

Before looking at the theory, it is helpful to briefly consider population genetics (Chapter 14). This investigates the future behavior of a population carrying several alleles of a gene, predicting how the distribution of the alleles within the population will vary over time as a result of genetic drift and selection in accordance with the breeding and other aspects of the behavior of that population. The simplest population model is that of Fisher and Wright (for example, see Fisher, 1930); this assumes that mating is entirely random and that generations do not overlap (the Moran and other models do not make such assumptions). Under this simple scheme, one can analytically model future drift for a population with **N** sexually reproducing members and **J** alleles (or sequences). The Fisher model thus gives formulae for predicting the future distribution of the alleles; in more sophisticated population models, such distributions can only be obtained computationally.

Coalescent theory starts with a set of **J** genes, each with spread of sequences across the population, an assumed mutation rate, and a model of population genetics; it then runs genetic drift backwards in time for as many generations as needed with the aim of deriving a function called the coalescent, a quantitative model of the genealogy of **n** individuals (a sample of the population) on the basis of sequence variation. Two points should immediately be clear. First, two closely related sequences will coalesce at some point in the past that we recognize as the node representing the sequence of their MRCA. If the simulation is then continued for the whole set of sequences, we will eventually reach the basal MRCA for them all, on the way generating a phylogenetic tree or coalescent for the set of **J** sequences (Liu et al., 2009; Wakeley, 2010). Second, the larger the population of breeding individuals between MRCAs, the more generations, and hence the longer it will take for two sequences to coalesce (**Figure 8.5**). While coalescent theory can be used for constructing phylogenetic trees over relatively short periods (typically ~10 My), its more important use is in exploring the history of the size of the population back to that most recent common ancestor whose sequence gave rise to the range that is now present.

It should be said that the mathematical basis of coalescent theory is complex, and analytical solutions for obtaining the coalescent are only possible for the most simple of conditions (Hein et al., 2005). In practice, although the modeling and computational methodology methods used to generate coalescents are quite sophisticated, the ideas behind the process are relatively straightforward. One takes the sequences and the population model and then, using a random number to assign mutation to the sequences, sees how many generations it takes for genetic drift to be reversed back to an MRCA for each gene; eventually, a genealogy tree is produced. The simulation is repeated many, many times, each with different randomly generated parameters for localizing the sites for mutation. At the end of the set of simulations, one can analyze the resultant spread of genealogies statistically and obtain a most likely genealogy linking the MRCAs, together with estimates and confidence limits for the coalescent time for each, the population history and the most likely sequence at each node.

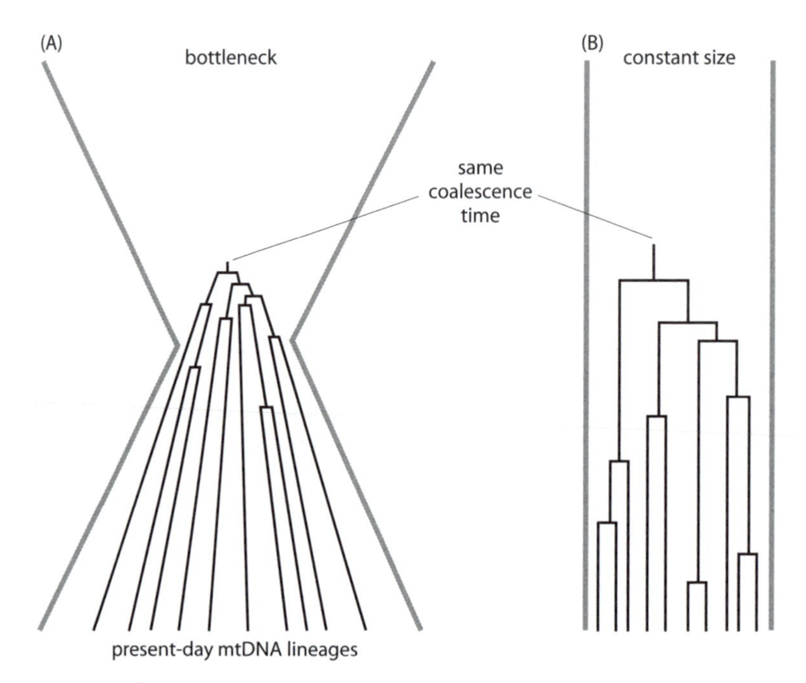

Figure 8.5 A drawing illustrating two scenarios for the coalescence of a set of 10 mitochondrial DNA (mtDNA) sequences back to an ancestral sequence with both taking the same amount of time. (A) There is an early bottleneck, detected by the short lengths of the node branches, followed by a period of exponential growth (long branch arms) and diverging population (broad outer lines). (B) The population remains constant and there is no obvious pattern to the lengths of the branches.

One the most interesting sets of numbers that emerges from the analysis gives the estimates for branch lengths between nodes: these are measures of the number of generations (and hence time) between MRCA nodes. As mentioned above, the larger this number, the larger the reproducing population is at the beginning of the coalescence for that MRCA branch. The analysis thus provides a means of tracking population sizes over time. Moreover, it turns out that the accuracy of these estimates, seen in the breadth of the confidence limits, depends far more on the numbers and lengths of genes than on the number of individuals analyzed. In practice, significant results can be obtained with large amounts of sequence data from relatively few individuals. Li and Durbin (2011) were thus able to reconstruct the history of African, Chinese, and European human histories over the past 200 Ky on the basis of the complete diploid sequences of just seven individuals. In particular, they inferred that the African population went through a bottleneck (population decline) from around 16,100 breeding individuals around 150–100 Kya to about 5700 at around 50 Kya, while the European and Chinese populations each experienced a severe bottleneck some 40–20 Kya when the population was reduced from about 13,500 to about 1200 (Chapter 16 has more on this subject).

Coalescent approaches can be expanded to incorporate other aspects of population dynamics such as migration or links between two groups through the choice of model of population genetics used in the analysis. Today, coalescent theory is a standard technique used by computational biologists who wish to explore population genetics over evolutionary history and many other aspects of the history of genetically diverse populations. It is thus a particularly exciting area of evolutionary biology.

The major limitation on the use of coalescent theory is the width of the confidence limits on estimates of the various parameters that emerge from the statistical analysis; these can be very wide to the extent that the information can only be viewed as indicative rather than quantitative, unless very large amounts of information are included in the analysis. Here, it is worth noting that phylogenetic analyses also have some minor limitations. Apart from the previously mentioned use of consensus sequences, all methods assume that parent sequences generate daughter ones with minimal amounts of mutation and this may not always be so. Perhaps of greater importance is the weakness of the assumption that only the genetic drift of neutral mutations is being analyzed. In practice, it is hard to distinguish neutral mutations from those that turn out to have an advantageous role; the latter will not, for example, be lost as a result of back mutation.

Nevertheless, even allowing for these very minor limitations, the evidence that has emerged from analyzing sequence data computationally has illuminated great swathes of the record of life and the complex evolution of gene sequences and the species that include them in their genomes. Such insights have the added quantitative advantage that they are independent of the fossil record other than for time measurements; qualitative comparisons, particularly within trees, clearly show which steps were fast and which slow. Particularly interesting in this context because there is so little fossil evidence are the insights that computational analysis has generated about the early evolution of bacteria, the origins of eukaryotic cells, and the early stages of the evolution of the major groups of multicellular organisms. This area is discussed in the next chapter.

Key points

- Comparisons across groups of similar DNA sequences in very different contemporary organisms show that they descend from a common ancestor and so are homologous.

- Phylogenetic approaches enable diverse species to be grouped into trees on the basis of genomic sequence comparisons using standard computer programs.
- These groupings are independent of those from cladograms as they are based on completely different information. They do, however, give the same or higher resolution groupings of extant species that are related.
- Tree-searching algorithms can generate likely ancestor sequences that, if using sequences from organisms in different phyla, represent those of very ancient organisms.
- The branch lengths of phylogenetic trees are a measure of mutation differences. If neutral mutations accumulate at a constant and known rate per generation (this is the molecular clock hypothesis, which often applies over short periods of evolution), then these branch lengths are also a measure of time.
- Coalescent approaches, which run genetic drift and mutation backwards in time for sequence variants within a species, can provide estimates of historical population numbers.

Further reading

For the basics of phylogenetic analysis, see:

Desalle R & Rosenfeld J (2012) Phylogenomics. Garland Press.

Felsenstein J (2004) Inferring Phylogenies. Sinauer Press.

For more mathematical treatments, see:

Caetano-Anollés G (ed.) (2010) Evolutionary Genomics and Systems Biology. John Wiley.

Koonin EV & Galperin MY (2003) Sequence - Evolution - Function: Computational Approaches in Comparative Genomics. Kluwer Academic.

Lemey P, Salemi M, Vandamme A-M (eds) (2009) The Phylogenetic Handbook: A Practical Approach to Phylogenetic Analysis and Hypothesis Testing. Cambridge University Press.

Saitou N (2013) Introduction to Evolutionary Genetics. Springer.

CHAPTER 9
THE FIRST THREE BILLION YEARS OF LIFE: FROM THE FIRST UNIVERSAL COMMON ANCESTOR TO THE LAST EUKARYOTE COMMON ANCESTOR AND BEYOND

Life evolved almost 4 billion years ago with the formation of the first cell, which is known as the first universal common ancestor (FUCA) of all subsequent cells, which probably had an RNA-based genome. This very primitive cell evolved, increasing its abilities and acquiring a DNA-based genome, and so became the last universal common ancestor (LUCA), whose progeny diverged to become the future eubacteria and the archaebacteria. A little later, cyanobacteria evolved; their ability to photosynthesize and so produce oxygen led to the Great Oxygenation Event ~2.45 Bya and a change in the earth's atmosphere. Around 2 Bya, eukaryotes started to evolve. It is now clear that the first eukaryotic common ancestor (FECA) acquired the genes for its nucleus mainly through endosymbiosis of an archaebacterium and, much later, its mitochondria from endosymbiosis of a further eubacterium. Once organelles had been acquired, this cell evolved to become the last eukaryotic common ancestor (LUCA) of all future eukaryotes. The plant and algal clades diverged from the others through endosymbiosis of a cyanobacterium to give a chloroplast and the acquisition of two flagella, while the fungal and animal worlds had a single flagellum. Plant, algal, and fungal cells also became distinguished from animal cells through acquiring various polysaccharide cells walls, with all being in place soon after 900 Mya. This chapter looks at the phylogenetic, timing, and other evidence on which this early history of life is built.

The use of computational phylogenetic techniques to explore the diversity of homologous DNA sequences has transformed our understanding of early evolution and facilitated the investigation of areas such the early evolution of prokaryotes and unicellular eukaryotes (the grade of protists) for which there are only traces of a fossil record. Without such analysis, very little would be known about the first three billion years of life. The essential methodology is to analyze homologous sequences across contemporary prokaryotes and protists and so identify their phylogenetic relationships and likely roots. Such approaches have, of course, little to say on the actual origins of life as represented by the existence of the first universal common ancestor (FUCA), the most primitive of prokaryotes. They are, however, informative in probing the history of all organisms back to the last universal common ancestor (LUCA), the prokaryote parent of all subsequent species, including eukaryotes (**Figure 9.1**).

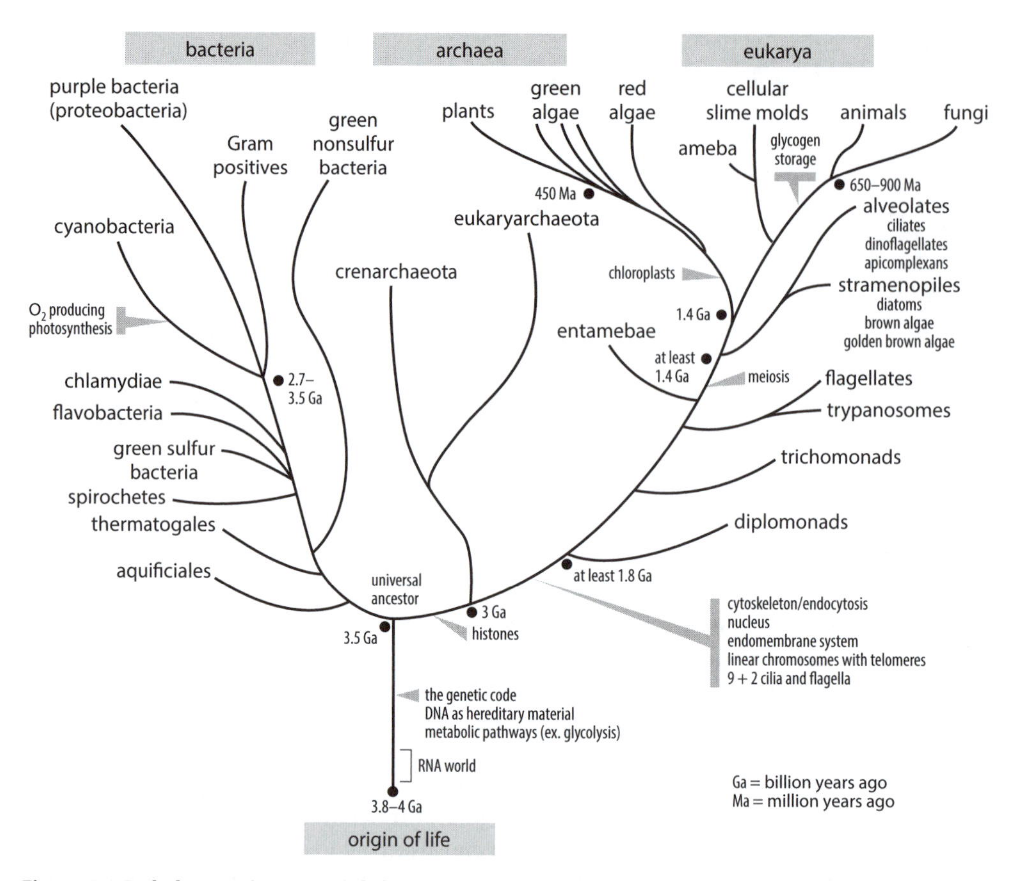

Figure 9.1 A phylogenetic tree mainly based on the phylogenetics of the small subunit (SSU) of ribosomal RNA showing the separation of eubacteria, archaebacteria, and eukaryotes, and including some timings and apomorphies. Note that Aquificiales, apparently the most basal member of the eubacterial clade, is a Gram-negative eubacterial taxon (with a double-bilayered membrane). Other protein-based phylogenies give slightly different cladograms. This picture is misleading because the eukaryotes form from the bacteria and archaea through endosymbiosis and horizontal gene transfer.

It should not be thought that these key events of unicellular life were simple or quick: the earliest evidence of prokaryotic life dates back to more than 3.5 Bya, of eukaryotic life to about 1.8 Bya, and of a chloroplast-associated organism to around 900 Mya (Shih & Matzke, 2013). The immediate question is why these times were so extraordinarily long in comparison with, say, the period of less than 30 My that it took very primitive life to evolve from the early worms of the Ediacaran biota, to the Cambrian explosion, the basis for the many extinctions and diversifications that led to the richness of contemporary life. The answers almost certainly have to do with just how difficult it was, given the randomness of mutation, for the reciprocal protein networks and genomes to generate a baseline of complexity that was required for multicellular organisms to develop.

The first aim of this chapter is to use data from contemporary organisms to try to explain and date the details of early prokaryotic and eukaryotic life. The second is to summarize the evidence on how primitive eukaryotes evolved to give the first stages of the major phyla of multicellular organisms that are seen today: the animals, fungi, plants, and algae. The data for doing this come from genetics, biochemistry, and the fossil record.

The chapter ends with a brief discussion about the many times single-celled organisms grouped themselves together to form multicellular organisms.

Prokaryotes

Originally, all living prokaryotes were clustered together, but it became clear in the late 1970s that they fell into two very different taxa, now known as the eubacteria and the archaebacteria (this terminology will be used here rather than the alternative bacteria and archaea because it is less ambiguous), differing in some key properties (**Table 9.1**). The eubacteria are by far the more common and fall into two major groups, those with a single-layered bounding membrane (monoderm) and those with a double layer (diderm), with the major diagnostic difference being that the cell walls of the former can usually take up Gram stain (and are Gram-positive) while those of the latter cannot (and are Gram-negative).

Archaebacteria were originally found in areas such as hot springs and salt lakes where little else could survive and were therefore known as extremophiles and thought to be relatively uncommon. More recently, however, they have been found in most habitats and it is now accepted, following Woese et al. (1990), that there are three broad domains of life: the eubacteria, the archaebacteria, and the eukaryotes. The fossil data on the relationships among members of these taxa are very limited (Chapter 4), and most of what is known comes from phylogenetic and other sequence analyses.

Table 9.1: Some properties of cells from the three domains

Property	Eubacteria (E)	Archaebacteria (A)	Eukaryotes (A or E denotes origin)
Genome	Circular DNA	Circular DNA	Linear DNA
Nucleus	No	No	Yes (A)
Introns	No processing of type I introns	Processing of introns in RNA-coding sequences	Processing of introns (A)
Histones	No	Yes	Yes (A)
DNA replication and repair	Simpler	Complex	Complex (A)
Transcription machinery	One RNA polymerase	Three RNA polymerases, complex machinery (for example, TATA box)	Three RNA polymerases, complex machinery (A)
Transcription initiation code	Formylmethionine (fMet)	None	None in nucleus. fMet in chloroplasts and mitochondria (E)
Ribosome	30S + 50S	30S + 50S	40S + 60S (nearer A)
Photosynthesis	Yes – cyanobacteria	No	Yes – chloroplast (E)
Citric acid cycle	Some of it	Yes	Yes – mitochondria (E)
Membrane	Gram-positive: single bilayer Gram-negative: double bilayer, mainly R-glycerol ester lipids	Single bilayer L-glycerol ether lipids	Single bilayer (E) Mainly R-glycerol ester lipids (E)
Cell wall polysaccharide	Murein	Pseudomurein	Today: variable (see Table 9.2)

A great deal is now known about the morphology, structure, biochemistry, and genetics of the members of the three domains (see Table 9.1). The archaebacteria and eubacteria both have circular chromosomes, use the same genetic code, and have similar transcription machinery and metabolic pathways, together with an outer membrane based on lipid bilayers. The similarities are sufficiently profound to demonstrate that they both derived from the LUCA, a very simple prokaryote for which the core molecular apparatuses for reproduction, metabolism, and genetic information storage were in place.

Molecular analysis shows, however, that archaebacteria have more complex genomes than eubacteria, possessing introns and a sophisticated genetic and protein machinery for transcription and replication, as well as histones. This indicates that the simpler eubacteria are closer to the LUCA than are archaebacteria, which, on the basis of phylogenetic analysis, clearly evolved later (Cavalier-Smith, 2006). More complicated to interpret are their respective membranes. Monoderm eubacteria have a single bilayer membrane mainly composed of R-glycerol phospholipids held together with ester bonds covered by a coat or wall of murein polysaccharide, while diderm eubacteria have a double bilayer membrane with a thin intervening layer of murein. Archaebacterial membranes predominantly have a single bilayer of L-glycerol isoprene linked together with ether bonds, which are more heat-stable than ester ones (perhaps helping to explain their extremophile behavior), and are surrounded by a protein coat of pseudomurein. Particularly curious is the fact that eubacteria use the R-isomer of glycerol for their bilayer, whereas archaebacteria use the L-isomer. These membrane differences are considerable and hard to interpret in evolutionary terms; it is, however, possible that there were ancient parent taxa with complex bridging membranes that have been lost.

The origin of life

Because all contemporary organisms share a basic structure, genetic code, and molecular machinery, it is a fair assumption that life only evolved once. Prokaryotic life thus has at its root a cell known as the FUCA about whose evolution there has been a great deal of discussion, but whose origins and details remain hard to access. One fanciful line of speculation has suggested that the FUCA evolved not on earth but from elsewhere and landed in the sea, an environment in which it flourished. Two suggested sources for such an event were civilizations outside the solar system who sent the FUCA in a rocket, or random chance whereby a FUCA was trapped in a meteor that eventually reached earth; indeed, organic molecules have been isolated from meteors. This idea, known as directed panspermia, was put forward in the middle of the last century by a group whose most famous member was Francis Crick. It has the advantage to its advocates of not being disprovable, but it is obviously unsatisfactory because it just transfers the problem of how life evolved to some other star or planet rather than trying to solve it.

There was actually no need to invoke such a hypothesis because there was a better, testable theory, put forward in the 1950s. This was that the early seas were a primordial soup containing complex chemicals, the building blocks of life whose self-assembly eventually led to the FUCA. The first real evidence to support this view came from Stanley Miller (1953), who, assuming that the earth's early atmosphere included ammonia, methane, hydrogen, and water vapor, but lacked oxygen, ran an electric discharge through these gases (simulating the effect of lightening) and found that the dissolved products included amino acids. Later, such experiments with other reducing gases produced nucleic acids such as adenine, while today it is known that a major contribution to the organic mix in the early seas came from comets and extraterrestrial dust (Bernstein, 2006).

The advantage of the "soup" approach is that it provides a good explanation of how large amounts of essential organic chemicals could have been present in the early sea, sufficient to build up concentrations adequate for the number of chemical interactions needed for the formation of cell components to take place essentially simultaneously. The model as it stands is, however, unclear on the origins of the energy needed to drive the synthesis of complex molecules. There are two obvious options: electrical energy from lightning bolts near the surface, and thermal energy from the sea, particularly in the vicinity of the thermal vents present in deeper water, the current habitat of thermophile archaebacteria. It is also reasonable to assume that life evolved at the interface between the sea and the solid surface of the shore or bottom, so that rock or clay could act as a catalyst for polymerization and other reactions of adherent molecules by lowering the free energy required for them (Ferris & Ertem, 1992).

As to location, Mat et al. (2008) have summarized, giving their advantages and disadvantages, five possible temperature zones where the FUCA might have formed. They refer to the most likely model as the *hot cross origin*; in this, the FUCA formed at a tepid temperature and slowly spread, but it was in a far hotter environment that it acquired the more sophisticated chemical reactions that led to its acquiring its major functions and so went on to become the LUCA that then diversified.

While the Miller experiments give an indication as to how organic molecules could form, they do not suggest how L-amino acids could link to form proteins or how nucleotides based on D-ribose could link to form chains of nucleic acids. This problem immediately raises the chirality question of why either L- or R-isomers were specifically included rather than a mix, and the answer is still not known. Of equal importance was the ability of molecules to self-assemble and so make larger structures held together by hydrophilic and hydrophobic bonds. Such biophysical approaches are beginning to illuminate possible ways in which the larger-scale properties of the FUCA emerged, and they hold considerable promise (Budin & Szostak, 2010).

FUCA, the first universal common ancestor

The FUCA clearly had a genome and some associated protein machinery, metabolizing and synthesizing proteins within a containing sac. It is now widely accepted that information in the FUCA was stored in RNA rather than DNA chains, partly because RNA chains can act as ribozymes with protein-synthesizing and other enzymic properties, and partly because of the wide range of functions that RNA molecules undertake today compared with the very limited, albeit key, role taken by DNA chains in all extant organisms except RNA viruses (Moore & Steiz, 2005; Higgs & Lehman, 2015; Kun et al., 2015).

As to the minimum number of proteins required for reproduction, construction, and metabolism, the general view is that this is around 300 (Mat et al., 2008). Today, symbiotic bacteria such as *Nasuia deltocephalinicola* can have as little as 137 protein-coding genes, while the very simple nonsymbiotic eubacterial mycoplasma can have fewer than 500 protein-coding genes. Insight into the likely protein set of the LUCA comes from a phylogenetic analysis by Kim and Caetano-Anollés (2011) of conserved structural domains of proteins (fold superfamilies) in 420 diverse organisms that included eubacteria, archaebacteria, and eukaryotes. This analysis was designed to give upper and lower bounds on the relative numbers of these domains in last common ancestors. The results indicate the small numbers of such protein domains in the LUCA compared with contemporary organisms (**Figure 9.2**).

A significant step towards answering the question of how the FUCA's bilayer membrane evolved comes from the work of Mansy et al. (2008), who showed that fatty acid esters can self-assemble to form vesicles with many of the expected properties of an early bilayer, particularly in their ability to allow unaided molecular transport across them and to extend easily. A deeper problem is how such vesicles came to contain the machinery of life and how the various reactions within them became compartmentalized (Dziecio & Mann, 2012). There is, however, no consensus view on this yet.

It is important to note that the FUCA protein systems could not have worked independently: cell division would clearly have failed if, for example, mitosis had been independent of genome reproduction, protein synthesis, or membrane expansion. Some degree of molecular communication between the various protein systems within the FUCA would have been required for its success. The difficulty, of course, is that development of such feedback really requires a living, reproducing cell so that its evolution could have been encouraged by the blind hand of selection. It will be interesting to see if a set of steps can be suggested and modeled that start with a membrane containing the sorts of complex organic molecules likely to have been in a primeval soup and end with a cell with the abilities of the FUCA. This may be the only way to obtain insight into how the FUCA cell evolved, but there is little reason to doubt the conclusion of Glansdorff et al. (2008) that life was born complex.

Putting dates on when life evolved is not easy, but analysis of the radioactive profile of zircon fragments from Australia that date back to about 4.4 Bya argues that oceans were present about 4.2 Bya or earlier and so available to contain a molecular soup of complex

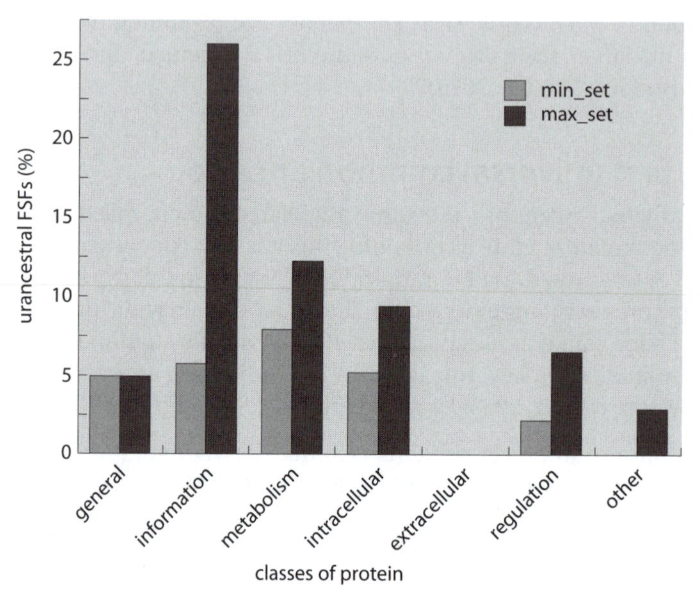

Figure 9.2 The classes of protein possessed by the first universal common ancestor on the basis of phylogenetic analysis of fold superfamilies (FSFs) of protein domains. The pairs show upper and lower bounds, while the percentages are comparisons with today's organisms.

organic molecules (see Appendix 3 for details of dating methods). The first traces of life date back to around 3.7 Bya: structured carbon deposits lacking the ^{13}C isotope of carbon found in Western Greenland were compatible with their originally having been organic compounds, and so likely to represent traces of very early life (Ohtomo et al., 2014). The first direct evidence of life is the presence of rocks in Western Australia dated to 3.48 Bya with a morphology that is characteristic of contemporary stromatolite microbial mats. These, however, contain a wide range of prokaryotes and cyanobacteria rather than the much more primitive prokaryotes that would have formed the original stromatolites (Noffke et al., 2013). If these dates are even roughly correct, several hundred million years of chemical jostling were required to produce the FUCA.

LUCA, the last universal common ancestor

In due course and probably after a fair amount of genetic and molecular strengthening, the LUCA evolved in the early seas. If the FUCA did have an RNA-based genome, the evolution of the LUCA would have required a transition to a DNA-based genome. Two nonexclusive drivers for this have been suggested: either enhanced stability against attack from RNA viruses, or the improved thermal stability that DNA chains have over RNA ones (Mat et al., 2008). Today, as all cells share a DNA genome with essentially the same genetic code and basic transcriptional machinery and metabolism, it is highly likely that this transition only occurred once and led to a LUCA with a DNA genome, basic protein toolsets for replication, transcription, and protein synthesis, and some core biochemical pathways, all enclosed within a lipid bilayer membrane.

There has been considerable discussion as to whether the LUCA was a simple monoderm eubacterium with a single bilayer membrane, a diderm eubacterium with a double bilayer (and a morphology similar to *Escherichia coli*), an archaebacterium, or a very primitive eukaryotic forerunner, or something with a mix of such properties such as having an external membrane with mixed esters with internal extensions. The general feeling is that the most unlikely candidate was a diderm eubacterium. This is partly because of its double membrane and partly because monoderm bacteria make antibiotics to which other monoderm, but not diderm, eubacteria are susceptible: it is thought likely that this antibiotic resistance was one of the drivers for the later evolution of diderm eubacteria (Gupta, 2011).

There is still no accepted view as to the taxonomic status of the LUCA as the three following papers illustrate. Cavalier-Smith (2014) argues for the traditional view that the LUCA was more like a very simple monoderm eubacterium, but goes on to suggest that archaebacteria and single-celled eukaryotes are more sister taxa than parent and child. Xue et al. (2003) have argued on the basis of an analysis of various contemporary tRNA populations that the LUCA was far closer to archaebacteria than to eubacteria. Glansdorff et al. (2008) view the LUCA as being a community of varied cells with eukaryotic features so that eubacteria and archaebacteria both represent derivation. The choice depends on which aspects of the data seem primary and which secondary.

All approaches have to explain the differences between the components of the three cell types, and probably the simplest suggestion is that the LUCA had a mix of esters in its external cell membrane and properties common to both eubacteria and archaebacteria. One argument in favor of this view is the existence of the contemporary *Aquifex* genus of eubacteria (the Aquificiales in Figure 9.1). Contemporary *Aquifex* have a short and limited genome that is only ~30% the size of *E. coli*'s; they contain many more genes commonly associated with archaebacteria than do other eubacteria, and are thermophiles that thrive at ~85°C and survive at 95°C. For all that they have properties associated with archaebacteria, they are clearly Gram-negative diderm eubacteria because they have the characteristic double cell membrane of this taxon rather than the very different membrane expected of archaebacteria. Their mix of properties suggests that their ancestors were evolutionally upstream of both archaebacteria and eubacteria. This is, of course, the parsimonious view, and contrasts with the alternative hypothesis that today's *Aquifex* represent the downstream simplification of a far more complicated ancestor.

Readers of the primary literature will soon realize that the current discussion about the features of the LUCA is wide ranging and complicated, with various authors appealing to the parsimony, elegance, and relative simplicity of their particular view. The difficulty is that prokaryote (some authors prefer the term *akaryote*, which just means *without a nucleus*) evolution may not have been very efficient and the route from FUCA to LUCA may have explored many blind endings with indirect, even slow, pathways turning out to be more advantageous in the longer term than apparent highways whose descendants were lost.

Whatever the nature of the LUCA, it is clear that some time later its descendants gave rise to the early eubacteria and archaebacteria, which went their very different evolutionary ways with, in due course, the diderm clade separating from the monoderm eubacterial line, perhaps developing its double membrane as a response to antibiotic production by Gram-positive cells. In the early sea, where there were no predators or, indeed, any obvious brakes on evolution other than the physical and chemical environment, there was clearly a great deal of genetic experimentation with the most important and successful of these leading to photosynthesis. This ability to capture light quanta and to use their energy for organic synthesis led to the production of simple sugars that could be used as a future energy source and polysaccharides that could be used structurally for cell walls and coats. One diderm taxon that particularly succeeded here became the cyanobacteria (or blue-green algae) that are now, on the basis of the detailed structure of later stromatolites in the Tumbiana Formation in Western Australia, thought to date to about 2.72 Bya (Flannery & Walter, 2012) or perhaps a little earlier (Nisbet & Nisbet, 2008). Their metabolism, which produced free oxygen, led to the Great Oxygenation Event of ~2.45 Bya and a complete change in the earth's atmosphere.

FECA, the first eukaryote common ancestor

Over a very long period, the descendants of the early archaebacteria and eubacteria diversified until there was a rich prokaryotic population in those ancient seas, with sufficient numbers of unicellular organisms for the likelihood of contact to become significant. It was from such encounters that the first eukaryote common ancestor (FECA) probably arose; this was a cell with a nucleus. As to when this happened, there is both fossil and molecular evidence. The earliest known microfossils that may be of eukaryotic origin are found in both Canada and China, and date to some 1.8 Bya (Knoll et al., 2006). This figure is much the same as that obtained with molecular dating analyses (Parfrey et al., 2011). If these dates are anything near correct, it took the best part of two billion years to produce a nucleated cell from its earliest prokaryotic ancestors.

Much of the evolution of the FECA remains obscure, but one fact is clear: standard cladograms such as that in Figure 9.1, which give three domains independently spurring off an original root organism (LUCA), are not only wrong, but also misleading. The evidence summarized in the last column of Table 9.1, together with a great deal of phylogenetic analysis, unequivocally point to the FECA having components from both eubacteria and archaebacteria - there is no clear and single line of descent with modification. The most likely way in which this happened was through endosymbiosis; here, one cell engulfs another and, instead of the former breaking down the latter, the two either come to develop a symbiotic relationship, or the cellular structure of the engulfed cell is lost, but some, or all, of its genome becomes incorporated in the host's genome through horizontal gene transfer. The idea that cell organelles were acquired through endosymbiosis has a long history: it was originally put forward by Mereschkowski very early in the last century and taken up and developed by Margolis in the 1960s (Katz, 2012), but its proof required DNA sequence analysis.

Phylogenetic analysis has suggested that the likely source for the majority of the FECA's genomic information was an archaebacterium, while the engulfing cell was, on the basis of its membrane structure, more likely to have been a eubacterium. In this way, horizontal gene transfer between the two (Dunning-Hotopp, 2011) would have led to the eubacterial genome acquiring the necessary archaebacterial genes and molecular machinery needed for a primitive eukaryote (Williams et al., 2012). This view was buttressed by the discovery of Lokiarchaeota, a novel archaebacterium phylum from deep marine sediments in the Arctic Mid-Ocean (Gakkel) Ridge (Spang et al., 2015). Phylogenetic analysis supports a common genetic ancestry of this organism and the eukaryotes, an argument reinforced by the archaebacterium carrying genes for actin proteins, vesicle trafficking, ubiquitin modification, and small GTPase enzymes belonging to the Ras superfamily. All of these genes had previously been considered as being unique to eukaryotes.

Although this discovery highlights the importance of the archaeal provenance of the eukaryotic genome, it does not explain the origins of eukaryotic membranes: the single cell membrane of an archaebacterial cell is, for example, very different from that of the double bilayer that encloses the eukaryotic nucleus. Therefore, this approach has to suggest that the FECA would have to have replaced the archaebacterial single-layered membrane that includes L-glycerol and isoprene with a double-layered D-glycerol phospholipid membrane. Williams et al. (2013) argue that this may have been simpler than once thought; this is not only because many of the proteins for both synthetic pathways are common in the eubacteria and archaebacteria, but also because membranes containing both classes of lipid are stable (Shimada & Yamagishi, 2011).

The main complication with this scenario is that it would have required considerable further evolutionary change or perhaps a further endosymbiotic event for the FECA to acquire its internal membranes such as the nuclear membrane, the endoplasmic reticulum, and the Golgi apparatus. This problem would, however, have been circumvented were the original host cell to have been an early member of the Planctomycetes-Verrucomicrobia-Chlamydiae (PVC) eubacterial superphylum, such as *Gemmata obscuriglobus*. These cells are unusual in having a doubled external membrane, with the inner membrane extending into the cytoplasm and folding to give internal, double membrane-separated compartments (Sagulenko et al., 2014), one of which is a nucleoid containing the genome (**Figure 9.3**). Not only is the double-layered morphology of PVC membranes related to the current eukaryote endomembrane components, but so is the protein composition, a fact discovered from an informatics search of the databases (Santarella-Mellwig et al., 2010). This implies that an early PVC eubacterium was a good candidate for the original host cell, needing relatively small changes to its cell membranes to achieve eukaryote membrane morphology.

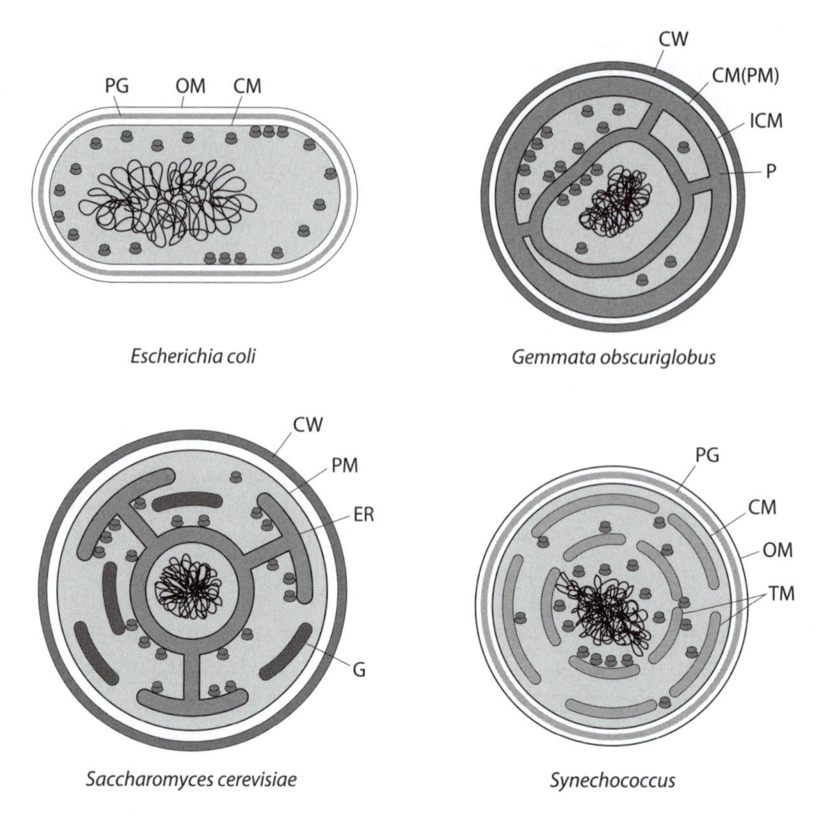

Figure 9.3 Drawings of four contemporary cells with important evolutionary features. *Escherichia coli* is a simple diderm prokaryote showing an absence of nucleus and a diderm outer membrane. *Gemmata obscuriglobus* is a planctomycetes diderm prokaryote with internal double membranes derived, in part, from the inner cell membrane layer and a nuclear region enclosed by a double membrane. *Saccharomyces cerevisiae* is a simple yeast eukaryote with a nucleus and limited amounts of endoplasmic reticulum. *Synechococcus* is a diderm prokaryote with thylakoid internal membranes and a candidate endosymbiont for a chloroplast (CM: cytoplasmic membrane; CW: cell wall; ER: endoplasmic reticulum; G: Golgi apparatus; ICM: intracytoplasmic membrane; OM: outer membrane; P: paryphoplasm; PM: plasma membrane; PG: peptidoglycan; TM: thylakoid membrane).

If this were the case, the FECA would have formed through endosymbiosis of an archaebacterial cell that only donated its genome, with perhaps the remainder of the cell being lost through breakdown. This possibility has added strength because the planctomycetes, unlike almost all other prokaryotes but in common with eukaryotes, have membranes that endocytose external vesicles containing proteins and other constituents (**Figure 9.4**) (Fuerst & Sagulenko, 2012). This scenario for the origin of the nucleus of the eukaryotic cell is attractive because it is parsimonious, but it should be mentioned that other means of deriving the FECA have been suggested (Cavalier-Smith, 2010; Koumandu et al., 2013).

LECA, the last eukaryote common ancestor

Even assuming that the FECA came from a planctomycetes eubacterium with a nuclear membrane and an endoplasmic reticulum, several further changes were needed before the FECA evolved to become the last eukaryotic common ancestor (LECA). These included, at the least, a linear, diploid - rather than a circular, haploid - chromosome, a

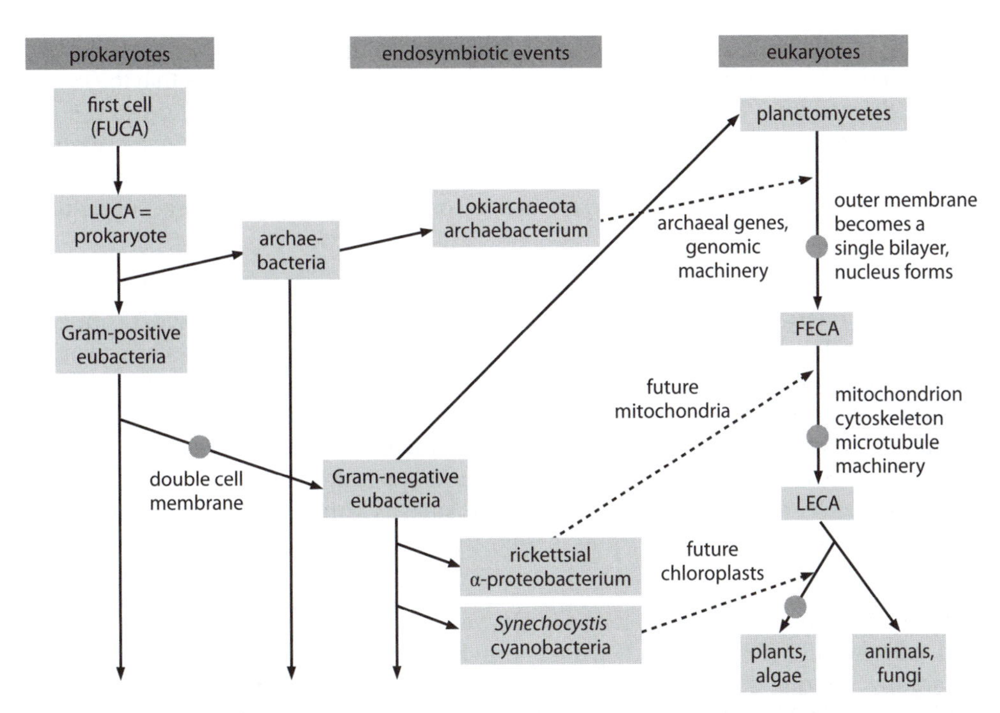

Figure 9.4 One scenario for the line of evolution from the first universal common ancestor (FUCA) to the last eukaryotic common ancestor (LECA) based on the parent cell of the first eukaryote common ancestor (FECA) being an ancestor of contemporary planctomycetes eubacteria.

cytoskeleton, and a means of organizing microtubules for cilial or flagellar movement, and for organizing chromosome movements during the anaphase stage of mitosis. The most obvious and largest change, however, was the acquisition of mitochondria, intracellular organelles containing a basic circular genome. These are present in all extant eukaryotes and their key role is to generate adenosine triphosphate (ATP) for their parent cell through the citric acid cycle and the electron transport chain.

Mitochondria have a double membrane, with the internal bilayer forming the cristae, or internal membranes. Although this double membrane could, in principle, have derived from that of the FECA nucleus, the view generally held today is that the mitochondrion descended from a small diderm eubacterium that was engulfed by a parent eubacterium that already had a nucleus and that survived as a symbiont. We know nothing of the processes by which horizontal gene transfer led to the great majority of its genes becoming integrated with the host genome, leaving only a few dozen behind in the future mitochondrion, or how they came to be present in several copies.

Based on sequence homologies and phylogenetic analyses, the most likely eubacterial source for mitochondria were early relatives of the rickettsial group of the α-proteobacteria, which are all diderms. Perhaps the most obvious reason for this is that all eukaryotic mitochondrial genes reflect homologs of those found in *Rickettsia* (for review, see Gray et al., 2001). Apart from being the major aerobic energy source of the cell, mitochondria have other biochemical roles and contain many hundreds of proteins; however, only a few of these are coded for now by their own DNA (Friedman & Nunnari, 2014). It thus seems clear that the original engulfed cell kept some but transferred part of its protein-coding genome to its host via horizontal gene transfer, with much of the original diderm structure surviving in a symbiotic relationship with its host cell. Not only are all extant mitochondria morphologically similar, but phylogenetic analysis suggests that they seem to have derived from a single gene set in an organism that lived around 1.2 Bya (Shih & Matzke, 2013). It thus seems likely that the events that led to an early FECA successfully engulfing a potential mitochondrion seem only to have occurred once, and the differences among extant mitochondrial genomes across the phyla simply reflect subsequent mutation.

Very little is known about the mechanisms by which horizontal gene transfer enabled the archaebacterial and eubacterial genes to be integrated and rationalized, or, indeed, how the circular genome of the parent organism became linear. If, however, the original host cell was, indeed, a member of the Planctomycetes, the first steps towards the evolution of an endomembranous system (the endoplasmic reticulum, the nuclear membrane, and the Golgi apparatus) had already been taken. As to acquisition of the cytoskeleton and the microtubule-based mitotic apparatus, the origins of actin and tubulin were clearly eubacterial, since proteins such as MreB and FtsZ, their homologs in contemporary prokaryotes, play a role in organizing the eubacterial cell wall (Margolin, 2014).

The other important innovation in the LECA was the flagellum or cilium with its associated ciliary body (or centriole), whose structures remain fairly uniform across the eukaryotes. Among the very few taxa lacking them are the red algae and the heliozoan protozoa (**Table 9.2**): while both have microtubules, the role of centrosomes is taken by the very much simpler polar rings and centroblasts, respectively (Dave & Godward, 1982; Cavalier-Smith & Chao, 2003). Although eubacteria and archaebacteria both possess one or more of these whip-like extensions that they primarily use for propulsion, eukaryote flagella differ from these in their molecular constituents, propulsive mechanism, and morphological organization: eukaryote flagella include nine microtubule doublets surrounding two central microtubule doublets. There is no obvious way that a eu- or archaebacterium flagellum could easily change to become the eukaryote equivalent, and Carvalhos-Santos et al. (2011) have argued that they were apomorphies that evolved as the FECA developed into the LECA, and suggested possible ways in which this could have happened.

A further important question that still remains unanswered was whether the LECA, by now much like a simple contemporary ameba, was haploid or diploid; in other words, did it mate? The origins of diploidy (and sex) remain obscure: the evolutionary advantages of two sets of genes are obvious, and a degree of mating can occur today between members of some prokaryote species. Because all extant eukaryotic cells are either fully diploid or go through a diploid phase, it seems likely that the LECA possessed at least the basics of diploidy, which it acquired after the evolution of the FECA.

Today, mitochondria and chloroplasts have an importance in evolutionary analysis that complements their functional role. Mitochondria are almost always inherited solely from the female gamete; chloroplasts are similarly inherited from female plant gametes, other than in gymnosperms, where they come from the pollen. This contrasts with the male sex chromosome whose genome is, of course, carried by sperm alone. Phylogenetic analysis of mitochondrial, chloroplast, and Y chromosome genomes thus gives direct, unbroken maternal or paternal lines of inheritance back to last common ancestors. In the case of humans, these ancestors, respectively known as mitochondrial Eve and Y chromosome Adam, are considered in Chapter 16.

Table 9.2: Properties of the major groups of organisms

	Nucleus	Mitochondrion	Chloroplast	Flagella or cilia[a]	Main cell wall polysaccharides[b]	Centriole	Type
Eubacteria	N	N	N	1–many	Murein	N	Single cell
Archaebacteria	N	N	N	1	Pseudomurein	N	Single cell
Amebae	Y	Y	N	0 or 1	None	Y	Single cell
Animals	Y	Y	N	1	None (but chitin is the main constituent of exoskeletons)	Y	Multicellular
Fungi	Y	Y	N	1	Chitin	Y	Both
Red algae	Y	Y	Rhodoplast	0	Cellulose, agarose, carrageenan	Centroplast	Both
Green algae	Y	Y	Y	2 (heterokont – the flagella are different)	Polysaccharides (cellulose) and glycoproteins	Y	Both
Plants	Y	Y	Y	2 (dikont – the two flagella are the same)	Cellulose + pectin	Y	Multicellular
Excavates (for example, slime mold, euglena)	Y	Y or partial	Some	2–4	Polysaccharides (for example, cellulose), glucans	Y	Single cell + aggregate

N: no; Y: yes.

[a]The three different whip-like appendages of eukaryotic cilia and eubacteria and archaebacteria flagella are defined by function not structure. Eukaryotes with one or two cilia are referred to as unikonts or dikonts (or heterokonts if the flagella are different). Note that the figures here exclude cells such as those of the respiratory tract in animals that have mats of cilia.

[b]This list is not comprehensive but is intended to note the most important constituents.

Early eukaryote evolution and the roots of the major groups

A unicellular organism with a nucleus, a diploid genome, mitochondria with their own mitochondrial DNA, an endomembranous system (all the internal and external membranes), and a cytoskeleton satisfies the basic criteria for the LECA. For further eukaryotic evolution and the emergence of the various major groups of organisms, further diversification was required. This included possible changes in the number of eukaryotic flagella, endosymbiosis of a cyanobacterium that became a chloroplast, and the development of cell walls with a range of polysaccharides. The distribution of these properties across groups of today's organisms (see Table 9.2) is sufficient to construct a cladogram showing the likely lines of descent of the major groups of organisms (**Figure 9.5**).

The most obvious taxon-distinguishing property here is the possession of chloroplasts and the phylogenetic evidence based on analyzing chloroplast genes from a diverse set of organisms allows two important conclusions to be drawn (McFadden, 2001). First, the acquisition of a chloroplast was through the endosymbiosis of a photosynthesizing cyanobacterium by an early eukaryote; this probably occurred some 900 Mya (Shih & Matzke, 2013). Second, all plants and algae descend from this early symbiont, and hence such an event only occurred once.

The reasons for accepting that a cyanobacterium was the primary source of chloroplasts is partly because their thylakoid bimembrane architecture and molecular structure are very

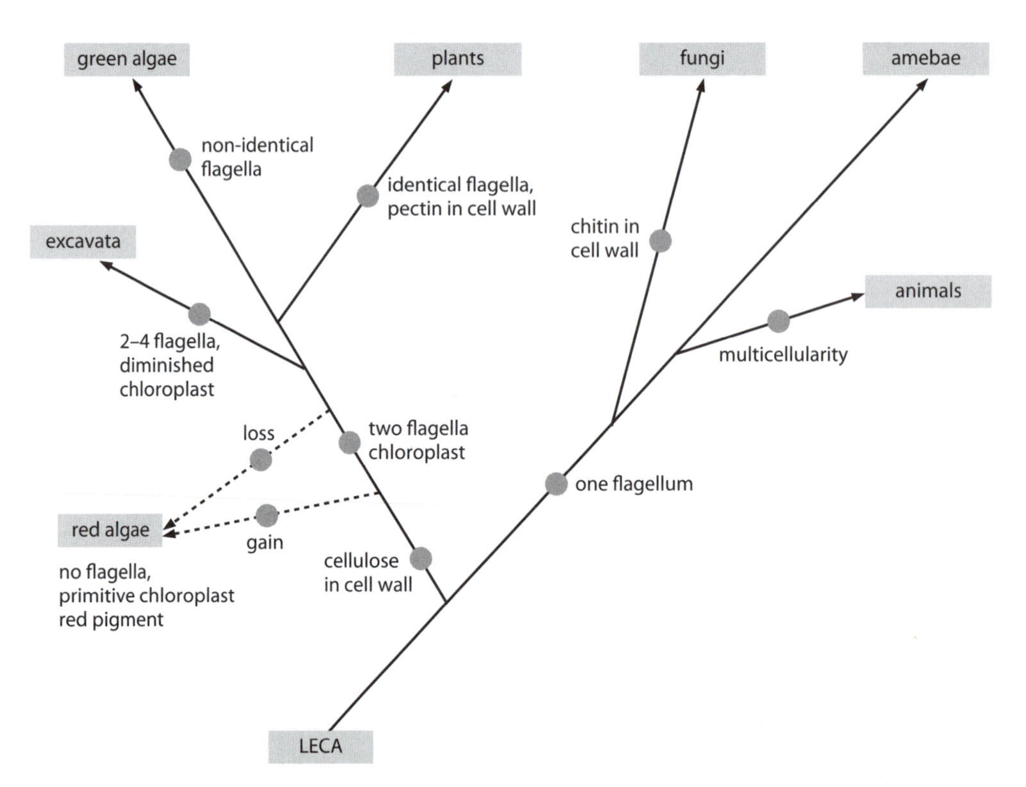

Figure 9.5 A cladogram based on the data in Table 9.2 showing how the last eukaryotic common ancestor (LECA) gave rise to the major groups of organisms, particularly the multicellular ones. The two possibilities for the evolution of the red algae are shown by broken lines. Omitted from the cladogram are the many taxa of unicellular organisms.

similar (Petroutsos et al., 2014), and partly because the core genes show a strong phylogenetic relationship. One gene in particular has been useful in identifying a possible taxon whose ancestor was the source of the chloroplast: Nowitzki et al. (1998) found that the spinach glucose-6-phosphate isomerase was far closer to its homolog in the cyanobacterium *Synechocystis* PCC6803 than to any other cyanobacteria. It is worth noting that the original genomic content of the endosymbiont has been drastically reduced, with some genes being transferred to the host genome and some being lost; today, only 120–130 genes remain in the chloroplast genome (Jensen & Leister, 2014).

Over time, the descendants of that early eukaryotic protist carrying a chloroplast evolved to give plants, red and green algae, the simple microscopic algae known as glaucophytes, and some of the collection of very simple organisms grouped within the excavates (see Figure 9.5 and Table 9.2). It became apparent some 30 years ago, however, that there were some simple organisms that did not fit into this scheme because they had diminished photosynthetic abilities and shared with plants and algae various sequences not seen in cyanobacteria. It now seems that the explanation is that there have been a few examples of secondary endosymbiosis where simple cells endosymbiosed chloroplast-like structures (Keeling, 2013). One such case was the endosymbiosis by an ameba ~60 Mya of a photosynthetic chromatophore (a much simpler organelle than a chloroplast) from a cyanobacterium probably related to a contemporary *Synechococcus* (Nowack et al., 2008).

The other important distinguishing feature that characterizes the various major groups of multicellular organisms is the nature of their cell coats. These are seen in almost all prokaryotes: monoderm eubacteria have an outer coat of murein (or peptidoglycan), diderm eubacteria have a thin murein layer between their two membranes, while archaebacteria have an outer coat of pseudomurein. Most eukaryotes, other than amebae and animals, have a cell wall made of one or another polysaccharide, with a high-level clade-defining example being cellulose, a characteristic of all taxa that have, or at one stage had, a chloroplast (see Table 9.2). This clade is subdivided into red algae that also have carrageenan, plants that also have pectin, and the excavates that also have glucans. The fungi, recognized as the sister clade of the animals through having a single flagellum and lacking chloroplasts, have a cell wall that includes chitin, a protein that is also present in invertebrate exoskeletons (vertebrates have keratin, the main constituent of horns, hooves, nails, and hair). What this means, of course, is that the synthesis of a range of cell-wall polysaccharides occurred at an early stage in eukaryotic evolution, with chitin and cellulose being two particularly early and important examples.

The one group of photosynthetic organisms that is hard to place precisely in the cladogram shown in Figure 9.5 is the red algae (see Table 9.2): they lack flagella and have rhodoplasts rather than chloroplasts; these organelles contain unstacked thylakoids and use the red pigment phycoerythrin, as well as chlorophyll, to facilitate light capture. It is not at first sight clear whether such properties reflect normal chloroplast endosymbiosis and flagella acquisition followed by a substantial loss of structure, or whether they evolved before plants and green algae as a result of an earlier endosymbiotic event. Nozaki et al. (2003) undertook a detailed phylogenetic analysis of a range of sequences carried by red algae and a diverse group of other organisms. Their results showed that rhodoplast-carrying red algae and chloroplast-carrying organisms shared a common root, with the red algae separating early off the main line. It thus seems clear that red algae have their current photosynthetic apparatus partly as a result of chloroplast diminution and partly through the evolution of phycoerythrin.

Table 9.3: Core dates in the origins and evolution of early life

Date	Event	Evidence
~4.2 Bya	Earliest oceans	Zircon radioactive data
~3.7 Bya	Organic molecules compatible with life	Carbon isotope balance
~3.5 Bya	Stromatolite microbial mats	Radioactive dating of fossils
~2.9 Bya	Early evidence of photosynthesis	Radioactive dating of oxides
~2.45 Bya	Significant amounts of atmospheric oxygen due to cyanobacteria	Radioactive dating of oxides
~2.1 Bya	Eukaryotic cell with nucleus	Radioactive dating of fossils, phylogenetic dating
~1.7 Bya	Small multicellular organisms	Radioactive dating of fossils
~1.2 Bya	Eukaryotic cells with mitochondria	Phylogenetic dating
~0.9 Bya	Eukaryotic cells with chloroplasts	Phylogenetic dating

replace the methane and carbon dioxide atmosphere of the early earth can be found by aging the rocks in which the oxides of iron and other metals were first deposited. This oxidation was the result of photosynthetic activity by cyanobacteria, with a key component of this process being rubisco I, a primary enzyme involved in fixing carbon dioxide and releasing oxygen. Radioactive dating suggests that these oxides were first deposited some 2.9 Bya (Nisbet & Nisbet, 2008), and this gives an approximate date for the likely appearance of cyanobacteria.

If it is accepted that the earliest multicellular organisms lived around 2.1 Bya, then it took a very long time before anything resembling an animal or a plant left any identifiable trace that has so far been found. Although there are trace fossils dating to about 1 Bya that could be due to worm burrows (Seilacher et al., 1998), the first direct evidence of large-scale organisms comes from biota that date to around 600 Mya. It seems that, with very few exceptions (Chapter 4), any earlier such organisms were too soft, too small, or too rare to be fossilized or found, and that the evolutionary processes that led to, say, the Metazoa were very slow, only acquiring momentum at the beginning of the Cambrian period some 540 Mya.

Finally, it should be emphasized that, although the essential features of the progress from the LUCA to the LECA are understood, there is more work to be done. In the steps towards the production of the LUCA, for example, the phylogenetic analysis is not yet detailed enough to sort out the relative contributions made by the archaebacterial and eukaryotic parents (Gribaldo et al., 2010), nor is it clear how endosymbiosis occurred or the mechanisms by which genes were horizontally transferred and integrated. And there will always be more to do on the origins of the FUCA.

Key points

- Little is known about the origins of the FUCA, but it probably formed almost 4 Bya with an RNA-based genome.
- The LUCA of all subsequent life was a very primitive bacterium with a DNA-based genome that probably used today's genetic code, and had a bilayer membrane and sufficient molecular machinery for reproduction and maintenance.

- Bacteria diversified widely, with the stromatolite fossil record showing that mats of prokaryotes were present some 3.5 Bya.
- The FECA formed by endosymbiosis ~2.1–1.6 Bya, with a eubacterial cell probably engulfing an archaebacterial cell and incorporating much of its genome through horizontal gene transfer.
- The LECA occurred when an early eukaryote further engulfed a small eubacterium (perhaps an ancestor of today's α-proteobacteria) that maintained its morphology and became the mitochondrion with its own circular genome.
- Single-celled eukaryotes (protists) diversified and one key event was the endosymbiosis by an early protist of a cyanobacterium, which evolved into a chloroplast, so forming the last common ancestor of plants, red and green algae, and probably the excavates.
- Other diversifications were in cell-wall polysaccharides and in the number and types of flagella. Such differences are synapomorphies for the major groups of organisms.
- The step to multicellularity probably occurred many times in each major group; this step was probably reversible as seen in extant organisms such as *D. discoideum*.

Further reading

Fenchel T (2003) The Origin and Early Evolution of Life. Oxford University Press.

Knoll AH (2004) Life on a Young Planet: The First Three Billion Years of Evolution on Earth. Princeton University Press.

CHAPTER 10
EVO-DEVO 1: EMBRYOS

This chapter provides the core information on embryogenesis that is needed to understand the third main area of evidence about evolution, that from molecular function, mainly during development. The first part considers the early evolutionary history of embryos, through to the protostome–deuterostome divide. Most of the chapter, however, discusses the development of the two most important model systems for studying evolutionary genetics, these are the fruit fly (*Drosophila melanogaster*), a protostome, and the mouse, an amniotic deuterostome. The approach, which focuses on the basic principles of development, highlights the role of the complex protein networks responsible for signaling, growth, differentiation, and morphogenesis that drive change in all embryos. This chapter provides the context for the next three, which, discuss the evolutionary evidence on how homologous proteins and networks control development in very different embryos (evo-devo) and the various ways in which mutation leads to anatomical variation, mainly through its effects on development.

The evidence for evolution comes from comparisons in three independent areas of study: anatomical similarities across both contemporary and fossilized organisms (cladistics); sequence similarities across likely homologous proteins (phylogenetics); and similarities in the way that very different organisms use homologous genes for similar functions, particularly during development. The last of these is the area known as evo-devo, and the main aim of this chapter is to provide the developmental context for discussing this area of evolution (Chapter 11). Current work on evo-devo started in the 1980s, but the subject has deep roots: it seems to have been Thomas Huxley who first made the obvious point in discussions with Charles Darwin that the anatomical differences among adult organisms reflect the different ways in which their embryos develop.

This chapter starts by discussing the evolution of form in protostome and deuterostome embryos and goes on to consider how their development depends on the activity of complex protein networks. The corollary of this is, of course, that adult anatomical variants derive from the effects of changes to the proteins in these networks, which themselves result from mutation (Chapters 11 and 12). It should be emphasized that the treatment here is very light, concentrating mainly on animals rather than plants and principles rather than detail: our knowledge of development is far too broad to be properly summarized in a single chapter, and the reader who needs to know more about any aspect of the subject is referred to Gilbert (2014) or Wolpert et al. (2015).

An advantage of focusing on protein networks, and thus taking a systems biology perspective, is that it provides a helpful context in which to discuss evo-devo comparisons across the phyla. There are three other reasons for taking this approach. First, it is parsimonious - there are relatively few networks compared with the vast number of proteins, and many are homologous; second, it naturally links genomic and phenotypic activity. Third, and of particular importance, it provides a natural framework for analyzing the functional way in which anatomical differences between related organisms, that is those that share a last common ancestor, might have arisen through variation. This is because networks occupy a mid-way step in the mechanisms by which the genotype affects the phenotype.

Diploblast and triploblast embryos

The primary feature that categorizes the development of an organism is the number of germ layers in its very early embryo. The simplest multicellular organisms, the sponges, have a single epithelial germ layer (basic epithelia are two-dimensional single-layered sheets of cells), which produces a range of derived cell types; sponges have little, if any, symmetry. Organisms with radial symmetry, such as the cnidarians, which includes jellyfish, sea anemones, and ctenophores (the comb jellies), are diploblasts; they have two epithelial germ layers, an ectoderm and an endoderm. In these animals, the endoderm forms the gut and the ectoderm forms the external layer, with some cells soon detaching and forming nerve cells, while others may detach later and form muscles.

The very great majority of animal species, however, are triploblasts: these have an ectoderm that forms the outer cell layer and the nervous system, an endoderm that forms the gut and a middle layer, and an intervening mesoderm, which produces muscles, fibroblasts, tendons, and, in vertebrates, the skeleton and the circulatory system. These species all show bilateral rather than radial symmetry, at least in their embryos or larvae, and are

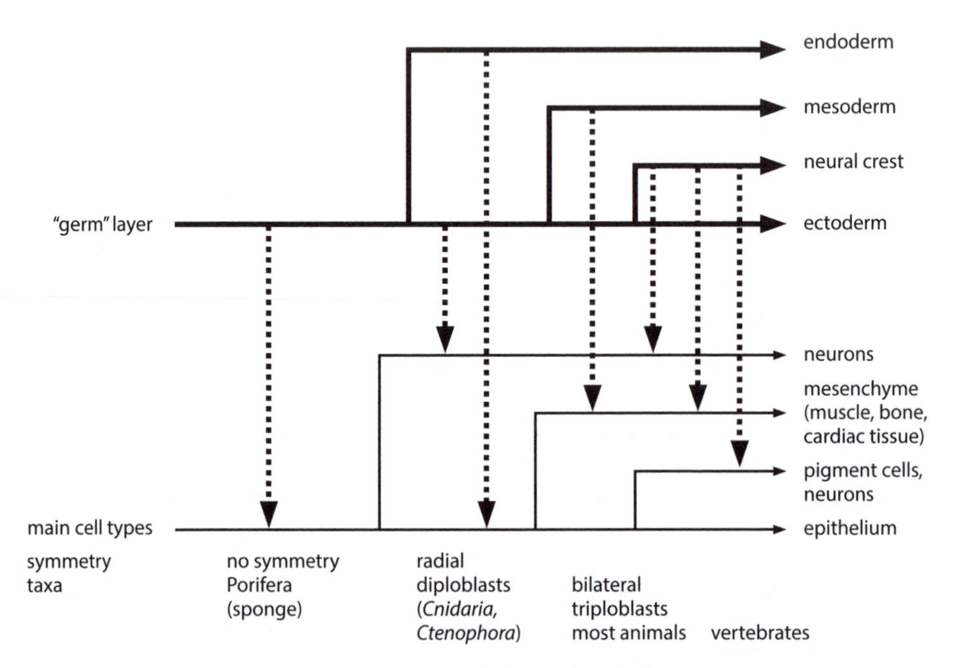

Figure 10.1 Diagram showing the evolution of the germ layers in animals together with the classes of organism, their associated symmetry, and some basic cell types associated with each germ layer.

hence collectively known as the bilateria. In all vertebrates, there is what can be considered as a fourth germ layer, the neural crest: this forms within the ectoderm from the edges (crests) of the region that will form the neural tube. Neural crest cells migrate away from the tube and colonize much of the embryo; they produce a wide range of cell types that include the peripheral and enteric (gut) nervous system, pigment cells, and much of the skull and face (**Figure 10.1**).

Early development and the protostome/deuterostome divide

In most animals other than large-yolked amniotes, development starts with the egg going through a series of divisions to produce a ball of epithelial cells, the morula, which then epithelializes to give the blastula (or blastocyst in mammals), within which is a fluid-filled center, the blastocoel. An indication of the nature of the embryo comes from the morphology at this early stage: if the cleavage pattern is radial, the embryo is probably a deuterostome; if it is spiral, it is more likely to be a protostome, and this choice has important implications for gut formation and nerve cord location.

A little later, various groups of cells become committed to their appropriate germ layer. It is around this stage that anteroposterior patterning is established. Gastrulation, the set of events that forms the gut and organizes the mesoderm, then starts: the basal cells (the endoderm region) form an invagination that extends across the blastocoel, meeting the other side where a second cavity opens - this internal tube is the future gut. The next categorization of embryos depends on the fate of those two openings: if the first becomes the early mouth (stomatodeum) and the second the anus (proctodeum), the organism is a protostome (most animals); if the second becomes the mouth and the first the anus, the animal is classified as a deuterostome (the chordates and echinoderms). Gut formation in deuterostomes is usually hard to follow under the microscope, but an exception is the sea urchin, which is both simple and transparent (**Figure 10.2**).

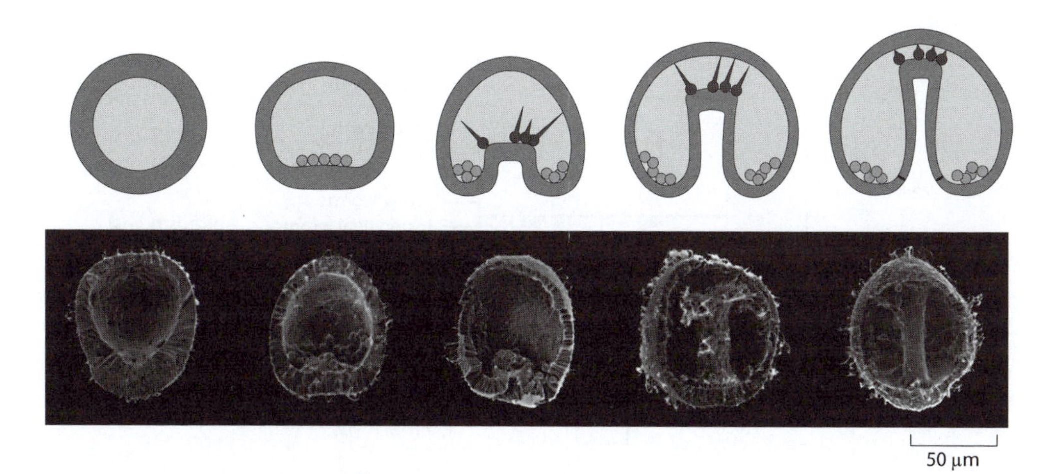

Figure 10.2 Sea urchin gastrulation. The diagrams of the upper picture shows how the primary mesenchyme of this deuterostome, the future internal skeleton, detaches from the basal area. This is followed by the extension of the endoderm to form the gut. The mouth develops at the dorsal surface as a result of interactions between the endodermal tip and the ectoderm. The lower picture is a set of scanning electron micrographs.

This simple picture reflects an ancestral state, and breaks down in many extant organisms. The guts of invertebrates can, for example, form in several ways, while the manner in which the vertebrate gut forms depends on whether the egg has a large yolk or develops a placenta. It is, however, possible to link all embryos back to a protostome or deuterostome clade on the basis of other morphological homologies. There was, nevertheless, serious discussion for a long time as to the evolutionary validity of the protostome/deuterostome divide. Phylogenetic analysis based on 18S ribosomal DNA has clarified the picture and not only substantiated the reality of these two classes, but also highlighted evolutionary relationships across the phyla (Giribet et al., 2000) (**Figure 10.3**). The analysis affirms a key fact: any sequences and associated functional homologies that cross the protostome/deuterostome divide reflect, after allowing for the rare possibility of homoplasy, activity in the last common ancestor of protostomes and deuterostomes. This early organism, known as Urbilateria, was almost certainly a protostome and lived before about 530 Mya (Lowe et al., 2015), the beginnings of the Cambrian Period, and before the time that protostome and deuterostome animals were both present (Chapter 4).

There are several other core events that determine the basic geometry of the triploblast embryo. First is the anteroposterior patterning of the embryo that determines head-tail polarity. Second is the other major event of gastrulation, the organization of the mesenchymal tissues that mainly derive from the mesoderm, events that are associated with the morphogenetic movements of cells and tissues as the embryo begins to assume polarity and shape (many time-lapse movies of these beautiful events are downloadable from the internet). Third is neurulation, the determination of the future nervous system, and fourth is the patterning of the future organization of the head and body; this often leads to segmentation, particularly in arthropods, worms, and chordates.

Neurulation, the development of the main nerve cord, is particularly interesting in an evolutionary context because it forms very differently in protostomes and deuterostomes. Protostomes produce their nervous system from ectodermal cells on the ventral (lower)

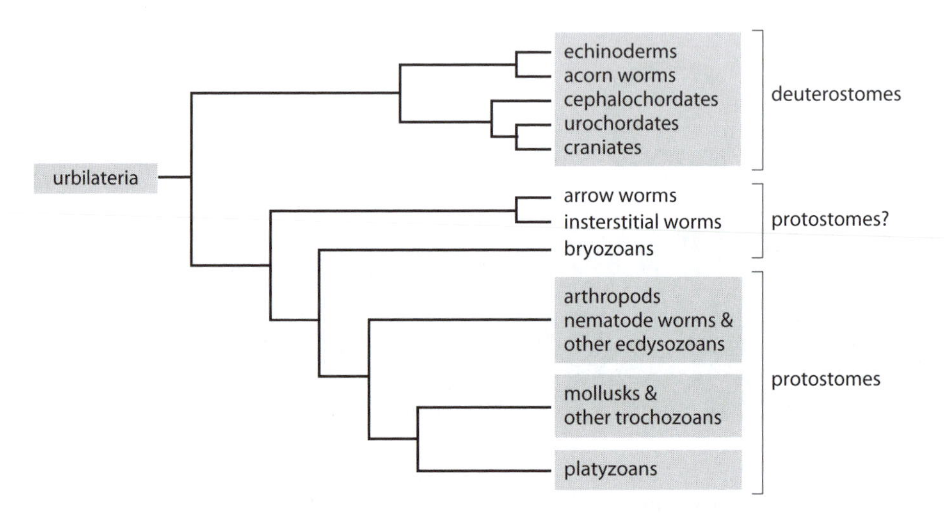

Figure 10.3 A phylogenetic tree linking modern phyla back to urbilateria, the last common ancestor of deuterostome and protostome triploblasts. Note that the evidence for classifying arrow worms, interstitial worms, and Bryozoa here is only molecular as their developmental status is not completely clear. Tree compiled using data on 18S ribosomal rDNA.

side of the embryo. In *Drosophila*, for example, the nervous system derives from two lateral domains of ventral ectoderm, from which future neuroblasts detach and migrate medially to form a nerve cord that is ventral to the gut. There, the cells differentiate into nerves and put out axons that extend over the embryo. In contrast, the neural tube of deuterostomes, which becomes the brain and spinal cord, forms in the dorsal (upper) region of the embryo and the gut lies ventral to it. The peripheral nervous system develops from neuroblast axons that originate within the neural tube and from differentiated neural crest cells after they have migrated to their final locations.

The realization that protostomes have a ventral nerve cord and deuterostomes a dorsal one was immediately problematic to early evolutionary biologists. The functional similarity of the two cords implied that the patterning of future neurons in the ectoderm had shown some fluidity in the deep past, but they did not indicate the original location of the nerve cord in urbilateria, or even prove that the two are homologs. The molecular data (Chapter 11) make it clear that, as the transcription factors that initiate nerve cord formation in both are homologous, the tissues are also homologous, with the ventral morphology of protostomes being likely to represent the original location, while that of deuterostomes was derived.

Although many of the early events in the development of protostomes and deuterostomes are different, there is one important similarity. Once anteroposterior patterning starts, it is manifested by the differential expression of sets of *Hox* genes (Chapter 8); this is most clearly seen in the different fates of the segments in arthropods and vertebrates. Such anteroposterior Hox patterning is also seen in many organisms that do not have ectodermal segments (Biscotti et al., 2014). Other patterning events organize the dorsoventral and the mediolateral axes, with these three sets together setting the fates of most of the early tissues. Once all this basic patterning is in place, much of the rest of development involves filling in the enormous amount of species-specific anatomical detail.

The great degree of diversity that exists in adult organisms across the phyla is, at first sight, bewildering. It is thus to the great credit of Ernst von Baer, an early nineteenth-century, pre-evolutionary embryologist from Estonia, that he was able to see the broad picture of the wood through the detail of the trees and in 1828 produce his laws of development, which made some sense of adult diversity. These are that:

- During development, more general features appear earlier in the embryo than the more special characters
- Less general forms develop from more general ones
- Embryos do not pass through the forms of lower embryos, but instead become separated from them
- The embryo of a higher form thus never resembles any other form.

These laws have stood the test of time to the extent that more than a century of evolutionary work has done no more than confirm their validity as a description of the relationships between very different modes of embryonic development and adult morphology.

Model systems

The study of development has always focused on a few model organisms chosen both for their ease of use in one or another specific roles and for their cheapness (**Table 10.1**).

Table 10.1: Common model systems for developmental biology

Organism	Class	Use
Dictyostelium discoideum	Slime mold	Origins of multicellularity, cell movement
Arabidopsis thaliana	Angiosperm	Genetics and mutation analysis
Caenorhabditis elegans	Nematode	Genetics and mutation analysis
Drosophila melanogaster	Arthropod	Molecular genetics, developmental mechanisms
Sea urchin	Echinoderm	Morphogenesis, molecular genetics
Danio rerio (zebrafish)	Fish	Genetics, mutation analysis, and morphogenesis
Xenopus laevis	Amphibian	Experimental manipulation
Chick and quail	Birds	Experimental manipulation
Mus musculus (mouse)	Mammal	Molecular genetics, model for disease
Homo sapiens	Mammal	Identifying genetic abnormalities

Since the early 1980s, two of these model organisms have become of particular importance in linking genes to evolution: the fruit fly, *Drosophila melanogaster*, is particularly used in genetic studies as its breeding time is about 2 weeks, while the mouse is a model system for mammalian development in general and for human genetic disorders in particular; this is because there are many molecular and other homologies between the two. Unexpectedly, and because of the power of molecular genetics technology, the human species itself has also become a model system for genetics and for developmental anatomy: medical research has led to the identification of very large numbers of genetic abnormalities whose molecular basis can now be readily identified.

Mouse homologs of important genes identified in humans can now easily be identified and their developmental and other roles studied in mouse embryos, using gene technology if necessary. Such parallel studies of humans and mice have not only established how development can go wrong if particular genes are mutated, but has also established the roles of the normal genes in normal development. As it is now also easy to identify homologs of these genes in *Drosophila*, it is straightforward to study their roles in this very different organism, and these comparisons have provided a great deal of evo-devo knowledge. In the next chapter, the role of the *Pax6* gene in eye development across the phyla is discussed; this is a particularly well-known exemplar of cooperative effort among biologists working on very different organisms.

In order to illustrate the diverse ways in which embryos can develop, the next two sections describe how *Drosophila* and mice go through early development, producing in their very different ways core body plans typical of an arthropod protostome and a mammalian deuterostome. These anatomical descriptions provide a framework for discussing the molecular mechanisms by which new tissues are generally produced in embryos, and how sometimes unexpected homologous mechanisms can underpin the development of very different organisms.

Drosophila development

The fertilized *Drosophila* egg is a small ovoid structure, ~0.5 mm long, that is laid by the mother, who takes no further part in its development. This egg consists of a developing oocyte together with nurse cells that produce maternal RNA that will pattern the cytoplasm and follicle cells that will provide yolk and the outer vitelline layer. The fertilized egg initially contains a single central nucleus; after about seven divisions, the 128 nuclei migrate to lie under the external cell membrane. Those that arrive in the posterior region will become the pole or future germ cells, and are the first to cellularize (**Figure 10.4**). After about 12 divisions and some 3 h 30 m in at 25°C after fertilization, the blastoderm cell layer forms as the outer membrane of the cell extends inwards to surround each

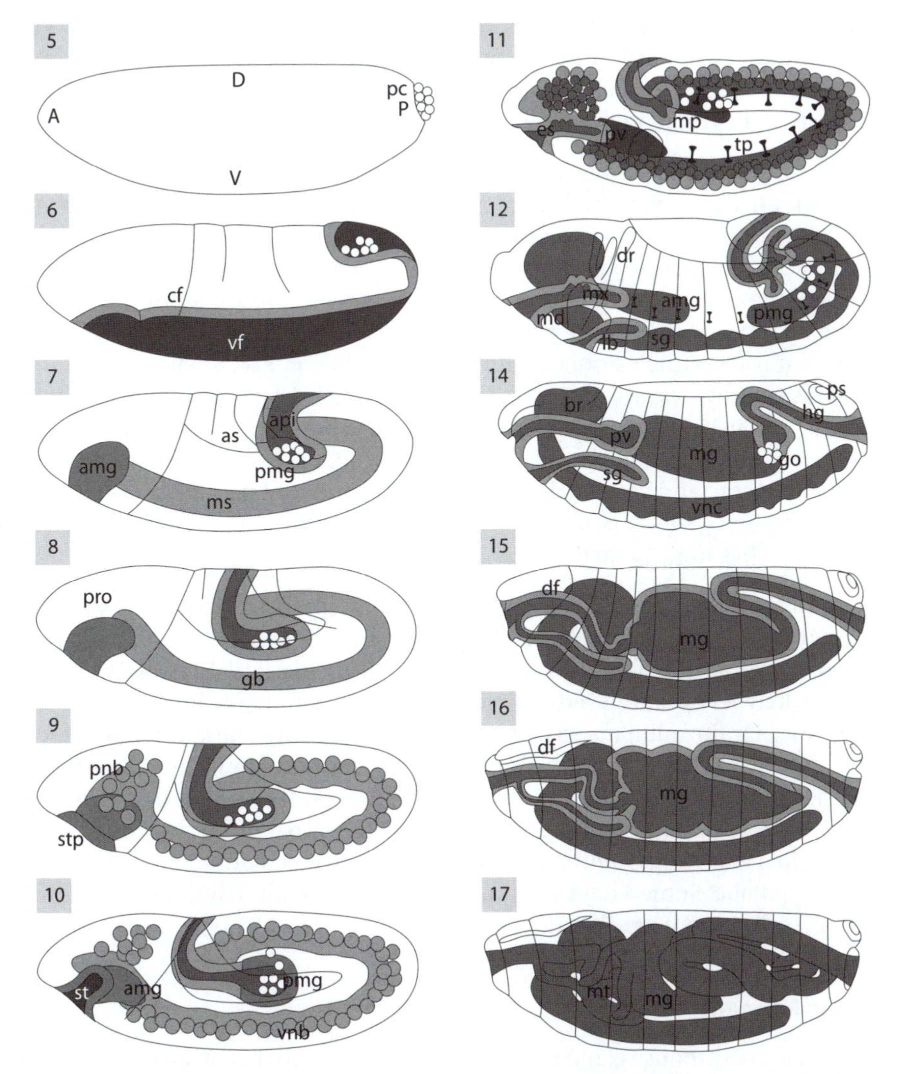

Figure 10.4 Some important stages in the development of the *Drosophila* egg from just before gastrulation (stage 5) to hatching (stage 17). (A, D, V, and P: anterior, posterior, dorsal, and ventral, respectively; amg: anterior midgut rudiment; br: brain; cf: cephalic furrow; df: dorsal fold; dr: dorsal ridge; es: esophagus; gb: germ band; go: gonads; hg: hindgut; lb: labial bud; md: mandibular bud; mg: midgut; Mt: Malpighian (kidney) tubules; mx: maxillary bud; pc: pole cells; pmg: posterior midgut rudiment; pnb: procephalic neuroblasts; pro: procephalon; ps: posterior spiracle; sg: salivary gland; st: stomodeum; stp: stomodeal plate; tp: tracheal pits; vf: ventral furrow; vnb: ventral neuroblasts; vnc: ventral nerve).

nucleus with its immediate cytoplasm, so forming an ovoid epithelial sheet underlying an acellular vitelline membrane.

After about 12 h, cell division ceases until the start of the larval stages: the cell cycle networks are turned off and most of these cells never divide again. The several thousand cells that have formed now start to produce embryonic tissues. The pole cells first segregate at the posterior end of the embryo, lying between the epithelium and the external membrane, and eventually migrate to the gonads where they form germ cells. The remainder of the epithelium soon segregates into the three germ layers: the ectoderm that will form the outer ectodermal layer and the nervous system, the endoderm that will form the gut, and the mesoderm that will form muscles. Tissues that will develop later, particularly during pupation, are made from small groups of imaginal discs of ectodermal cells and small internal nests of mesenchymal cells; once in place, they are quiescent until needed.

Gastrulation in *Drosophila* involves two separate events. First, regions of anterior and posterior epithelium, the endoderm, invaginate, extend, and eventually meet to form the gut (in this highly evolved embryo, the original way of making the gut in protostomes has changed). Second, a furrow develops along the already-patterned ventral region of the embryo, with the cells inside the furrow breaking away to form mesoderm that extends around the body. Soon after, as the culmination of a series of molecular events that started soon after fertilization, the ectoderm and mesoderm become segmented, and this is the most obvious external feature of the embryo. Small, paired domains of cells segregate as internalized buds in most ectodermal segments, and these will become the imaginal discs and histoblast nests. Simultaneously, the mesenchymal tissue differentiates to give muscle, somatic, visceral, and cardiac tissues and a set of internal nest cells. About 24 h after fertilization, embryogenesis is over and a functioning and independent larva hatches from an egg that has not changed in size since fertilization.

Over the next four or so days, the organism goes through three larval stages that mainly involve growth, with successive layers of external cuticle being sloughed off. Pupation then starts: over the next few days of metamorphosis, the larval ectoderm is lost and replaced by a new set of external adult structures that form from the imaginal discs (imago is another word for the adult insect) and histoblast nests, while new internal tissues form from the other nests. The development of each disk depends on its initial segment location: anterior ones will form the antennae, the eyes, and the mouth of the head, while those from body segments form the wings, the halteres (balancing organs that evolved from the original rear wings), and the limbs, with the most posterior ones developing into the external genitalia. Some 4 days after pupation starts, the adult fly emerges; this is about 9 days after fertilization if development is at 25 °C.

Mouse development

Embryogenesis in the mouse could not be more different from that in the fly. The very small and almost yolkless mouse egg (~75 μm in diameter) is internally fertilized and is never laid; it spends its first 4 days floating in the uterine tube and dividing; by then, it has become the blastocyst, consisting of an external trophoblast layer with an inner cell mass of stem cells and blastocoel space. At around embryonic day (E) 4.5, the blastocyst implants itself in the wall of the uterus where the trophoblast cells cause a local vascular and proliferative response, the first stage of placentation. Around then, the inner cell mass segregates into the epiblast and the primitive endoderm layer and, over the next 2 days, the former will form the early embryo, while the latter, together with some of the trophoblast cells, will form the extraembryonic membranes, the placenta, and the umbilical cord. At this stage, the extra-embryonic part of the conceptus is very much larger than the embryo (**Figure 10.5**).

At E6.5, this embryo is just a small, cup-shaped, and apparently featureless structure made of a few hundred cells that derive from the epiblast, but which has now acquired an anteroposterior polarity. Embryonic development now gathers momentum as gastrulation starts: a small part of the anterior ventral region, known as the node, moves posteriorly, taking about 48 h to reach the tail region. As it moves, it signals to local cells that respond by reorganizing themselves to form an overlying ectoderm, internal mesoderm, and ventral endoderm, with anterior regions continuing to develop while posterior regions await the gastrulating signal. Accompanying all this activity is considerable cell proliferation and growth (at its fastest, embryo cell numbers double in less than 10 h). The main lines of development and some early features in the mouse are laid out in Figure 10.5, while the lineage of the primary germ layers and their descendants is given in

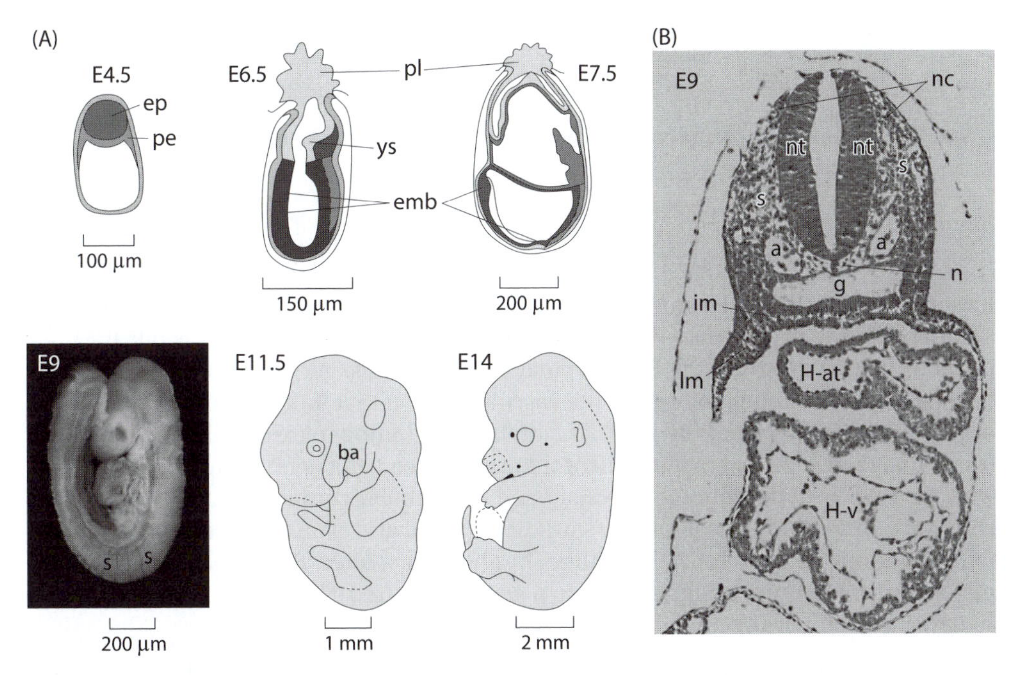

Figure 10.5 (A) Some stages of early mouse development. Embryonic day (E) 4.5: just before implantation, the blastocyst has an epiblast (ep) and primitive endoderm (pe). E6.5: the embryo (emb) is a small cup of cells within the egg cylinder linked to the yolk sac (ys), extraembryonic membranes, and the early placenta (pl). E.7.5: gastrulation is taking place, the embryo is lengthening, and the extraembryonic membranes and early placenta are developing. E9: a three-dimensional reconstruction of the embryo based on sections; this shows the early brain, heart, eye placode, and the segmented somites (s). E11.5: the limb buds are beginning to develop and the jaws are forming from the maxillary and mandibular branches of the first pharyngeal arch. Internally, the metanephric kidney and reproductive system are beginning to develop. E14: the embryo is developing autopods, all the basic organ systems are in place and the embryo is growing rapidly. (B) Mid-thoracic transverse section of an E9 mouse embryo showing some of the major features (ba: branchial arches; nt: neural tube; nc: neural crest cells; s: somites; g: gut; a: dorsal aorta; im: intermediate mesenchyme; lm: lateral mesenchyme; n: notochord; H-at: heart common atrial chamber; H-v: heart common ventricular chamber. Distance bars show microns).

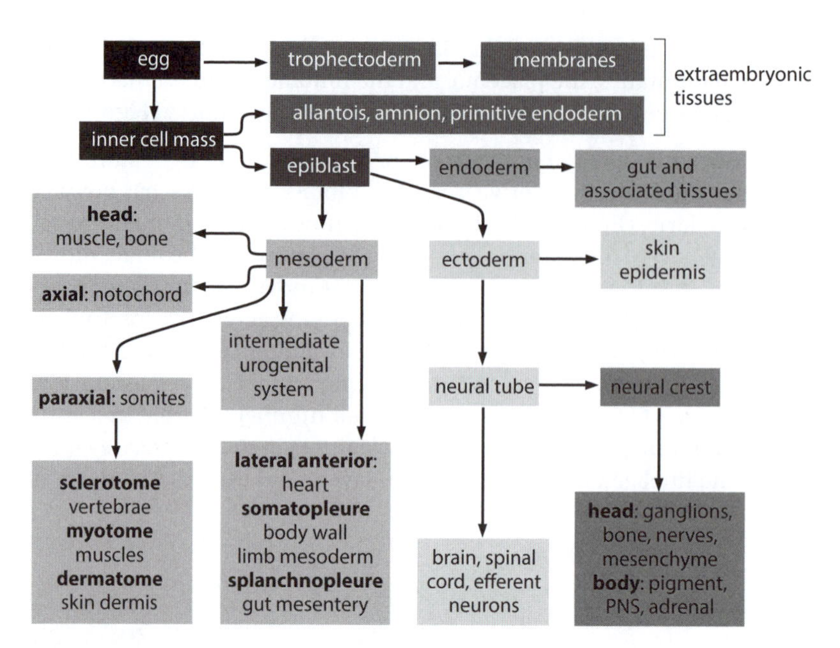

Figure 10.6 Lineage in the mouse embryo: some major paths of tissue development.

Figure 10.6. Particularly interesting here are the derivatives of the mesoderm: it separates into four different regions, each of which has specific fates.

By about E9, the essential features of the embryo are in place: the head region includes the neuroepithelium of the future brain, together with eye and ear rudiments, while the body includes the neural tube (the future spinal cord), somites, gut, and a heart that is now beating, so circulating blood round the vascular system that links to the placenta via the developing umbilical cord. By E11.5, the brain is beginning to form, while the branchial (or pharyngeal) arch system that will form the lower head and neck tissues is now present, with the jaws forming from the maxillary and mandibular components of the first branchial arch. In the body, the limbs are starting to extend and even the musculoskeletal system is beginning to develop - the embryo has become a fetus. By E14.5, almost all of the major tissues and organ rudiments are in place (see Figure 10.5) and the remaining five or so days until birth are occupied by filling in anatomical detail and growth.

There are some basic similarities between *Drosophila* and the mouse. Both have bilateral symmetry, together with very similar neurons, muscles, tendon mesenchyme, epithelia, and other cell types. At the anatomical level, they each have an outer ectodermal layer, musculature, a gut, a nerve cord, and brain, although the locations of nervous system are different. The anatomical differences between them are, however, far more obvious, as examination of the head shows. Most of the *Drosophila* head derives from ectoderm, but the various tissues and sense organs of the mouse are each built from several of the derivatives of the mesoderm, neural crest, ectoderm, endoderm, and the anterior neural tube (the future brain). The camera eye of the mouse, for example, includes a lens, cornea, retina, pigmented epithelium, choroid, and sclera; these structures include contributions from all of these early tissues except endoderm. This is in sharp contrast to the compound eye of *Drosophila*, which forms its set of about 800 ommatidia, each of which has its own optical components, exclusively from the epithelium of the appropriate imaginal disc.

There are also substantial differences between the limbs of the mouse and *Drosophila*. Those of the former derive from ectoderm (epidermis), neural crest cells (pigment-producing cells), and four different mesoderm derivatives that separately give muscles, bones and tendons, blood vessels, and dermis. The limbs of *Drosophila*, however, mainly derive from ectoderm and a little mesenchyme that produces the musculature. Even many of their functional requirements are carried out by different organs and in different ways: respiration in *Drosophila*, for example, is handled by tracheoles (oxygen-carrying epithelial tubules that extend internally from the ectoderm) rather than lungs, which derive from mesoderm and endoderm, while the urinary system in *Drosophila* is based on endoderm-derived Malpighian tubules as opposed to the mesoderm-derived kidney system in the mouse.

As there are such profound differences in the ways that the cells of these two organisms are organized into tissues, their respective internal and external skeletons, and the way that basic physiological functions are handled, it had been hard, until relatively recently, to envisage that they could ever have had a common ancestor, other than a very primitive organism with just a few cell types. Some 30 years of evo-devo work has shown that this pessimism was misplaced: comparative molecular anatomy has identified a range of homologous proteins in these two organisms that mediate homologous functions in their very different modes of development. This triumph of evo-devo research is discussed in the next chapter.

The processes that drive development

So much happens so rapidly during embryogenesis that one needs to stand back to get a general perspective of how developmental change in embryos is achieved. Over a hundred years of experimental manipulation has shown that, in every embryo, the developmental activity of the cells in a particular tissue or region at a particular stage depends partly on their lineage (parental history) and partly on molecular signaling from their neighboring tissues and cells. The balance of importance of these influences varies between tissues, and may be species-dependent. It is really only since the 1980s, however, that we have had a clear idea of the molecular nature of those signals and how their reception causes the cells in that tissue to change their properties on the basis of their lineage.

It is now known that, at any given time, the main effects of lineage in a cell at a particular development stage are to ensure that it expresses first the appropriate receptors to receive signals from its neighboring tissues, and second the specific set of transcription factors that are required to initiate transcription of the proteins needed for its future development. A cell in this state is said to be competent to receive the appropriate signals and act on them. Sometimes, a cell can have two or more sets of receptors and transcription factors so that it has developmental options; in the mouse, for instance, this happens in the early somite, a key mesenchyme precursor of the segmental organization of the vertebrate body (see Figure 10.5 and below). Receptors are usually located in the membrane and have an external signal-binding region and an internal activation region. Signal-binding activates the internal region to initiate a signal-transduction network (for example, see Appendix Figure A1.1) as a result of which, a transcription activator enters the nucleus, binds to a transcription complex located at a DNA promoter site that becomes activated. This leads to the synthesis of new proteins, some of which are involved in activating new developmental processes and so driving subsequent events in the embryo (**Figure 10.7**).

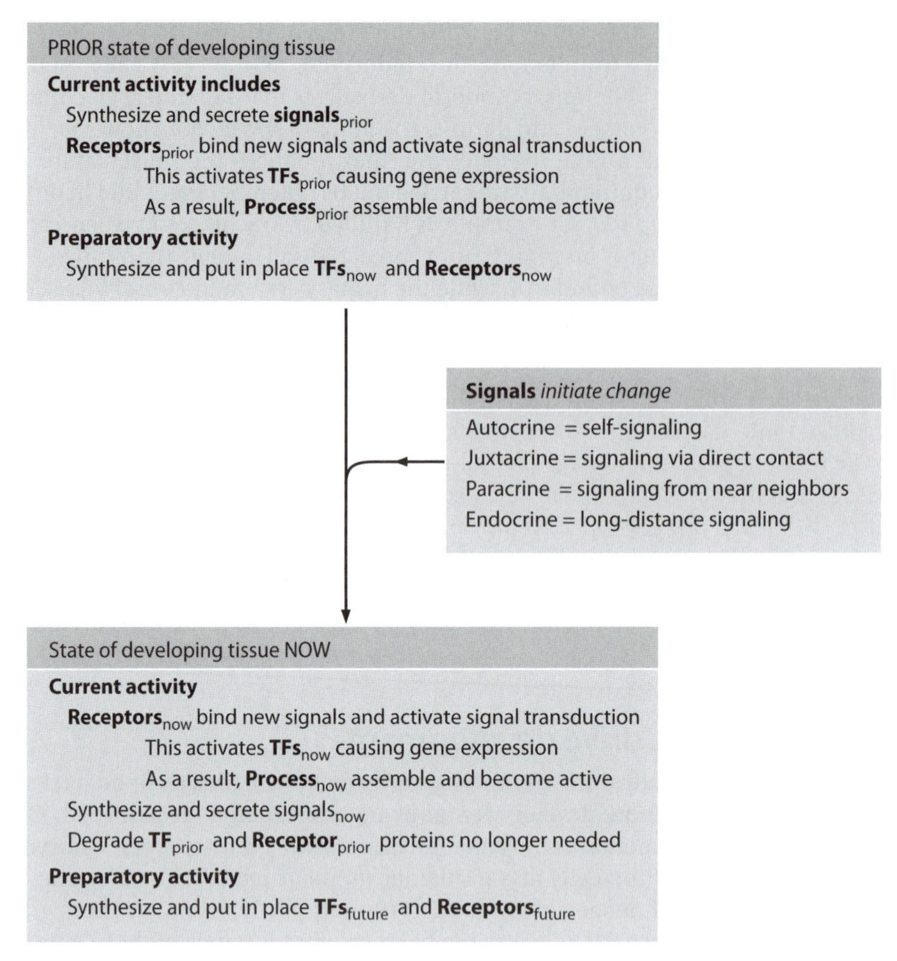

Figure 10.7 Diagram illustrating the effects of lineage and signaling in driving change. The lower box shows the current (NOW) activity of the cells in a developing tissue: these are partly to effect its current role and partly to prepare for the future. The top box shows the earlier (PRIOR) state of that cell, and its preparation for the NOW state. The central box shows the various types of signaling that can cause the state of the tissue to change (TFs: transcription factors).

Signaling

The main way in which signals from a nearby neighbor reach developing cells is through diffusion; they then normally bind to cell-surface receptors. This is known as paracrine signaling and well-known examples are members of the Hedgehog, Wnt, and the various growth-factor families; homologs of these signals are found in many organisms Retinoic acid, an important paracrine signal is an exception as it binds directly to receptors in the cell nucleus. There are also two classes of signal–receptor pairings that work through direct contact (juxtacrine signaling) with both members located in the membrane of a particular cell type: these are the Notch–Delta and Eph–ephrin families; the former is particularly involved in neuronal differentiation, while the latter facilitates the establishment of boundaries between tissues. The best-known form of signaling in adults, endocrine (hormonal) signaling via the bloodstream, is rare in early embryonic development because most endocrine tissues form relatively late in development.

The relative importance of signaling and lineage is highlighted by the behavior of the sonic hedgehog signal in early mouse development, which is secreted by the notochord; this is a long, thin cylinder of mesenchyme tissue that lies immediately below the neural tube. This signal diffuses to the overlying ectoderm-derived neural tube, an epithelium at this stage, and to the adjacent mesoderm-derived somites (see Figure 10.5B). The result of this signaling is that the nearby neural tube cells become motor neurons, while the adjacent somite cells enter the pathway that leads to them forming vertebrae (other somite cells become muscles or dermis). The effect of the signal thus depends entirely on the lineage of the cells to which it binds.

Sometimes, as in these cases, development is initiated by a single signal; sometimes, however, signaling is better seen as a complex conversation between two tissues with signaling going both ways, and with the signaling molecules changing over time (a good example occurs in mouse kidney development; Bard, 2002). Of particular interest, however, are the cases where the strength of the signal varies spatially and there is a concentration-dependent reaction from the responding cells. Such an effect underpins, for example, the development of somites (Aulehla & Pourquié, 2010) and, almost certainly, the pigmentation in the patterns seen in butterfly wings and vertebrate skins (see Chapters 11 and 12). The generic name for such phenomena is, not surprisingly, pattern formation.

Proteins and networks

Sometimes, the proteins synthesized as a result of signaling activity can be seen as directly functional. Good examples are actin and myosin (key muscle components), collagen (a major constituent of bones and tendons), and enzymes. In many cases, however, the new proteins either activate or participate in a protein network; here they work cooperatively and the result is the initiation of some novel process. Networks that are particularly important in development include those for the signal-transduction pathways that drive gene expression and those for the processes that drive developmental change (**Figure 10.8**). These include the networks for cell proliferation, programmed cell death (or apoptosis; Baehrecke, 2000), differentiation to a new cell type, and morphogenesis (for example movement, folding, and aggregate formation; **Figure 10.9**), each of which is usually activated by a particular signal. These processes are considered in more detail in Chapter 12 and Appendix 1).

In signal transduction, the network responds to a signal by translocating a transcription activator from the cytoplasm to the nucleus where it interacts with an appropriate transcription complex that includes a specific transcription factor. A few such transduction pathways, such as Notch, are relatively simple and involve only a few proteins, but most are more complex. A typical and important example is the network activated when the epidermal growth factor (EGF) signal binds to the external domain of its ErbB1 cell-surface receptor (a receptor tyrosine kinase or RTK). This activates a complex network of ~60 proteins with a range of functions that includes enzymes, activators, transferases, and kinases (see Figure A1.1). This and other signaling pathways have homologs in a wide variety of organisms (see www.genome.jp). The processes that the EGF-activated pathway initiates depend on the competence of the activated cells in a particular embryo. The most common process is proliferation, but others include protein degradation, angiogenesis, cell migration, implantation in mammals, and neural cell patterning, particularly in *Drosophila*.

Many other processes can be activated by protein networks including those that maintain the organism's physiological abilities such as the heartbeat cycle (which starts

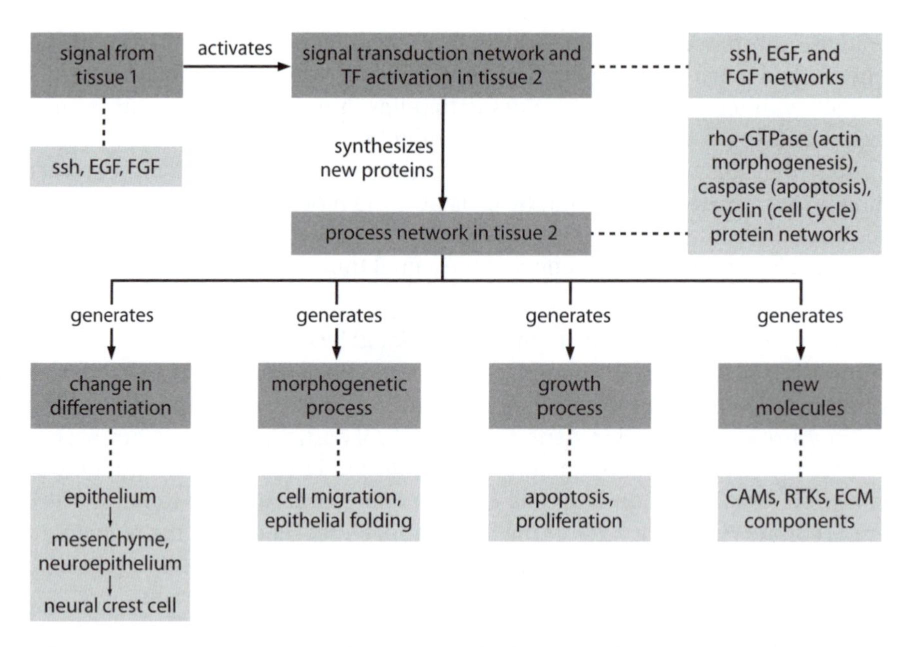

Figure 10.8 A diagram (based on triplet facts, Appendix 1) summarizing the various processes that drive developmental change. The blocks linked by dashed lines represent examples (TF: transcription factor; ssh: sonic hedgehog; EGF: epidermal growth factor; FGF: fibroblast growth factor; CAMs: cell adhesion molecules; RTKs: receptor tyrosine kinases; ECM: extracellular matrix).

at E8–E8.5), the circadian clock cycle, neuronal activity, and muscle contraction, together with biochemical pathways such as the Krebs cycle, which is present in the mitochondria of all eukaryotic cells. Very often, more than one of these new processes operate simultaneously as a tissue develops: **Table 10.2** gives some indication of the wide range of processes involved in making a mouse forelimb, a model system considered briefly below and in more detail in Chapter 13. Many of these networks, together with their roles, are detailed at www.sabiosciences.com/pathwaycentral.php. In almost every case, however, knowledge about their internal workings is very thin, to the extent that even the functional role of many of their constituent proteins in the network remains unknown.

Investigating the details of how such complex networks operate is part of systems biology, but one does not need to appreciate all the details of a network to realize that the key role of each is to produce a specific functional output, be it transcription, differentiation or apoptosis (**Table 10.3**). It is worth emphasizing that the same signal and signal transduction network can be used in many different contexts because the downstream response always depends on which transcription factors are in place, and these are lineage dependent (as the sonic hedgehog example above shows). Some of the signals and networks involved in the development of *Drosophila* and the mouse will be examined in the next few chapters.

How tissues form

The modern view of development sees a new tissue forming as a result of some external signal or patterning system stimulating a small rudiment that is competent by virtue of its lineage (it expresses the appropriate receptors and transcription factors). These signals

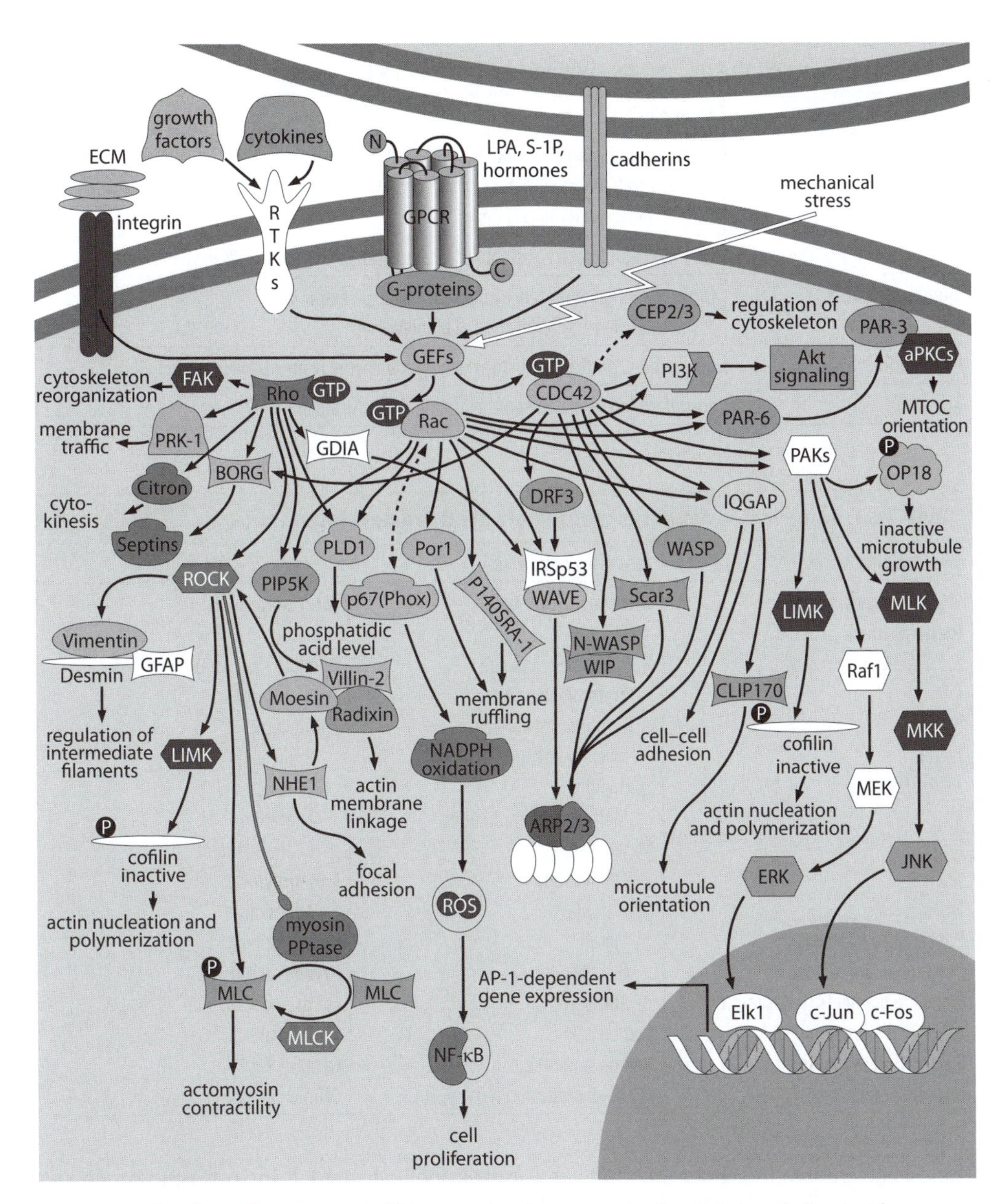

Figure 10.9 The rho-GTPase network. This very complex network of proteins mediates a wide variety of changes in the cytoskeletal system of the cell. This network has several external receptors, including those for growth factors and cytokines (these are receptor tyrosine kinases (RTKs)) and the various ligands that activate G-protein-coupled receptors; it can also be activated by physical stress. The effect of activation is to change the internal dynamics of the cytoskeletal activity, but the ways in which it does this are not known.

result in the activation of process networks within the rudiment that cause the tissue to enlarge, produce new cell types (differentiate), change its shape, and perhaps interact with other tissues. In short, process networks drive anatomical development under the direction of signaling networks; they thus represent the executive arm of the genome (Bard, 2013b; Davidson, 2010; **Figure 10.10**).

Table 10.2: Some of the processes involved in mouse forelimb development

Processes involved	Examples
Patterning	Proximodistal: long bone and muscle specification Dorsoventral: the positioning of claws on paws Lateromedial: patterning of digits
Differentiation	Includes chondroblasts, osteoblasts, osteoclasts, and muscle cells from mesoderm
Morphogenesis	Distal migration of mesenchymal cells and axon extension Distal extension and bifurcation of blood vessels
Proliferation	Everywhere, particularly in the distal mesenchyme (zone of polarizing activity)
Apoptosis	Loss of mesenchyme and epithelium between future digits

Table 10.3: Some major networks whose outputs drive development

Signaling networks	Processes networks	
Signaling pathways (gene names)	**Differentiation to:**	**Morphogenesis**
Egf (Erk/Mapk pathway)	Hematopoiesis lineage	Boundary formation
fgf	Erythroid lineage	(Eph–ephrin)
Hippo	Lymphocyte lineage	Epithelium:
Jak/Stat	Myeloid lineage	Branching
NOTCH-DELTA	Epithelium	Folding
hh	Multilayered epithelium	Migration
Tgfb	Mesenchyme	Rearrangement
vegf	Chondrocyte	Mesenchyme:
Wnt	Fibroblast	Adhesion
Patterning networks	Muscle	Migration
Notch oscillator system	Osteoblast	**Proliferation**
Signaling gradients (for example,	Adipocyte	Cyclin + cell division
FGF, Wnt & RA in somites)	Neuron	**Apoptosis**
Patterning systems	Neuron support cell	Caspase, Fas
Hox set of genes	Pigment-producing cell	Cellular apoptosis

There is an interesting difference between morphogenesis, which includes the processes that organize the early embryo and shape later tissues, and the other activities downstream of signaling and patterning: apoptosis, cell division, and differentiation are all cell autonomous. These properties reflect the internal responses of individual cells. Morphogenetic responses, other than individual cell migration, involve groups of cooperating cells (such as in boundary formation, the folding of epithelial sheets, and the formation of condensations of mesenchymal cells). Such morphogenetic activity is often constrained by the features in the local environment, such as the boundaries provided by the external basal laminae of adjacent tissues (Figure 10.10 has an the *adjacent tissue* box to describe this).

A typical example of vertebrate organogenesis is the development of the metanephric, or future adult kidney in the mouse (**Figure 10.11**): this forms as a result of reciprocal signaling between metanephric mesenchyme and the ureteric bud, an epithelial spur off the

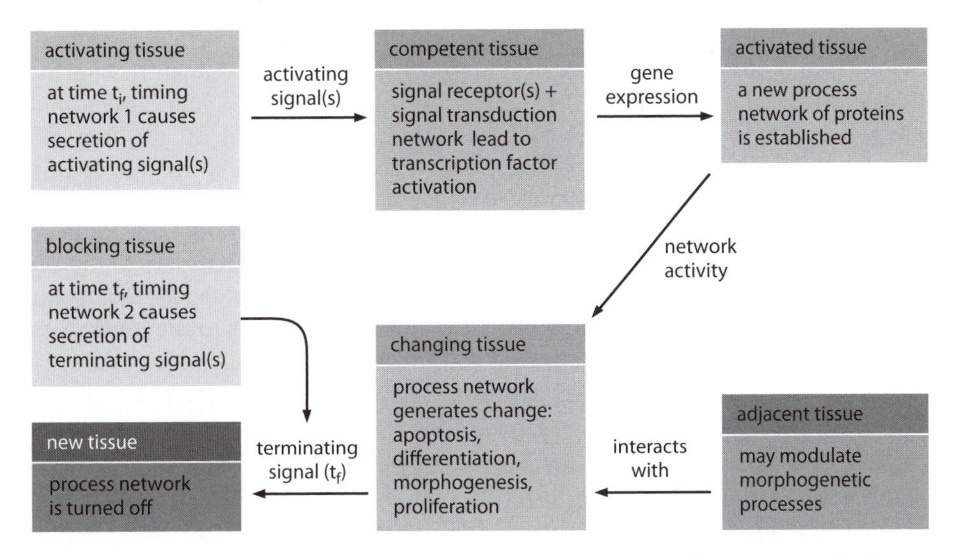

Figure 10.10 A systems view of how anatomical change takes place in embryos: as a result of signaling, process networks are activated in a competent tissue and these processes generate change. The "Adjacent tissue" box represents the effect of the local environment.

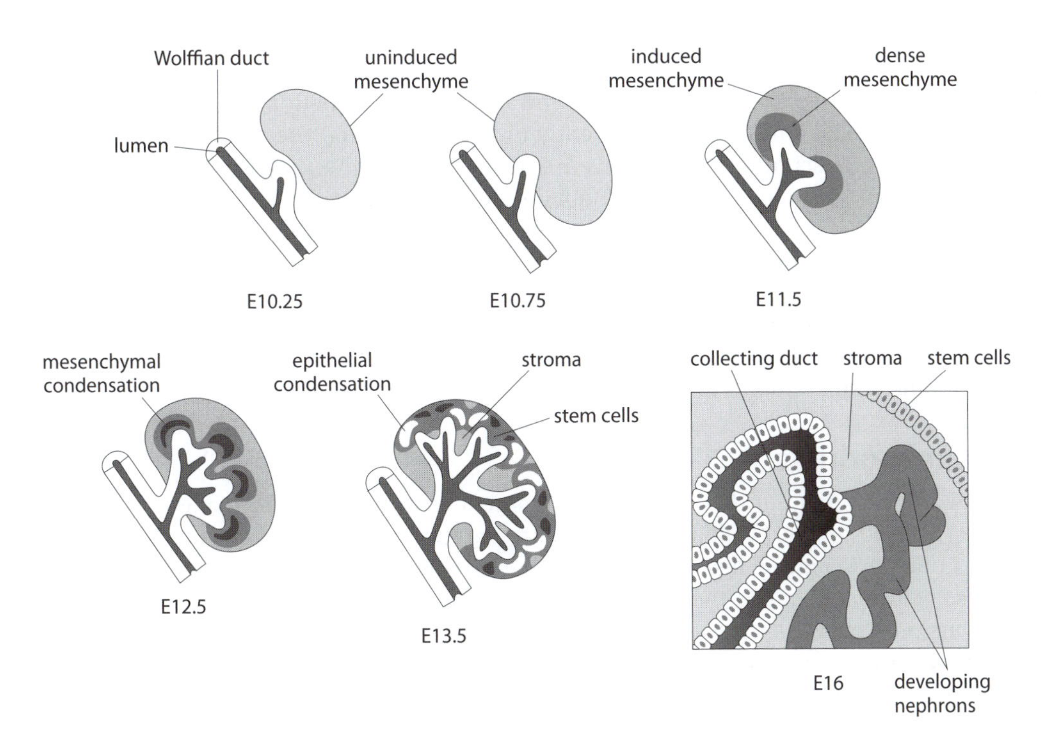

Figure 10.11 Diagram showing some key events in the development of the mouse metanephric kidney over the period embryonic day (E) 10.25–16. Not shown are the many diffusing signals by which one tissue instructs another to undergo change.

main nephric or Wolffian duct, both of which originally derive from the intermediate mesenchyme (Figure 10.6). The metanephric mesenchyme (MM) is a small mesenchymal condensation that forms at about E11 and soon starts to secrete glial-derived neural growth factor, (GDNF), a signal that diffuses to the nearby nephric duct, which, at this stage, extends down the body of the embryo. There, it stimulates the formation of the ureteric bud and its extension towards the MM which it soon invades. As a result of further reciprocal signaling between the MM and the bud, the former differentiates to become nephron stem cells and undergoes proliferation, while the proximal part of the ureter starts to bifurcate, producing branches throughout the MM. Meanwhile, the distal parts of the ureteric bud and the nephric duct form the ureter, which links to the newly formed bladder.

Small groups of these stem cells form the nephrons, the basic functional unit of the kidney. Each group forms a condensation near the tip of a ureteric branch that first epithelializes (an example of a mesenchyme-to-epithelium transition) and then extends to make the early nephric tubule. One end of the tubules connects to nearby blood capillaries, thus forming the glomerulus, while the other end connects to the branch tip. As the branches continue to form and extend, further nephrons form, while earlier ones start to become functional; these soon produce urine, which drains down the ducting system to the ureter and into the bladder from which it is excreted into the amniotic sac. In the mouse, the kidney starts to secrete urine at about E15.5, while the last nephrons of the many hundreds that are made do not form until well after birth.

Every step in these processes involves signaling (Bard, 2002), with the responses depending on location and timing. The cell-autonomous process networks activated in the ureteric bud as a result of this signaling include that for proliferation, while those activated within the metanephric mesenchyme include those for proliferation, for the mesenchyme-to-epithelium transition, and for the differentiation of the various specialized epithelial types that subsequently form in nephrons. The networks that direct the morphogenesis of groups of cells are also activated: the ureteric bud derivatives require the process that leads to epithelial bifurcation, while the MM first requires the interactions that initiate the formation of mesenchymal condensations and then, once the mesenchyme-to-epithelium transition is complete, the processes that lead to growth and the convolutions of each epithelial tubule. Although there is now a large amount of molecular information about the instructions and processes that direct kidney development in the mouse (Davidson, 2009), much of the detail on the networks that drive development still remains to be elucidated.

Tissue modules

It is worth pointing out that, during development, some transcription factors are more powerful than others. While most cause a limited number of proteins to be synthesized and so perhaps activate a single network, there is a higher level of transcription factors whose activation leads to the production of a tissue module. Such transcription factors activate proteins whose roles are, amongst others, the activation of further downstream networks. There can be several tiers of transcription that initiate a whole series of downstream signaling, patterning, and process networks whose activity can lead to the formation of whole organs, a process that may be repeated several or many times.

There are numerous examples of such repeated modules across the world of multicellular organisms, some of which are initiated many times. In vertebrates, there are, in addition to the nephrons discussed above, the skeleton that contains more than 100 long bones and synovial joints, epithelial tubule bifurcations, digits, and teeth. In many invertebrates, there are segments, compound eyes composed of hundreds of ommatidia, and often many limbs (for example, in millipedes). Plants, in particular, provide dramatic examples of repeated structure: they can produce thousands of leaves, flowers, fruits, and seeds. Further evidence to support this view comes from the study of developmental mutants that affect structure numbers: ectopic insect eyes (Chapter 11) have all their cell types, and extra digits in mammalian limbs normally have joints and nails (but note, in Chapter 13, the interesting exception of sesamoid-derived digits).

It should be pointed out that these modules are essentially templates in the sense that what they produce may be essentially identical replicates (for example, hairs, leaves, or fruits) or may be variants whose details depend on their locale (for example, the different-sized digits on a forelimb are, at least partially, dependent on the level of a sonic hedgehog gradient; Sanz-Ezquerro & Tickle, 2003). In the invertebrate world, the hundreds of legs on a millipede have roughly the same morphology and size, but the limbs on a lobster vary greatly in size, even though their essential morphology, other than that of the claws, is much the same.

Although the idea of an all-encompassing program of development that is coded within the genome is usually inappropriate because so much of tissue formation is contingent on interactions, it seems sensible to view these tissue modules as genomic subroutines or encoded modules that are activated as a whole, but whose phenotypic details can be modulated. While the metaphor seems sensible, it is still not known how such a module might be coded or implemented. The challenge therefore is to identify a particular module that is both experimentally and genomically accessible; one possible tissue that is relatively simple in this context is the synovial joint (Decker et al., 2014).

The origins of anatomical differences

If most organisms carry developmental networks with very similar functional outputs, two immediate questions arise; these concern the origins of the minor differences among individuals within a species and the wider differences between different but related species that reflect more than scale (for instance, the right mouse lung has five lobes, whereas the right human lung has three lobes). The differences among the individuals within a single species are mainly quantitative and reflect size and gender. That all animals within a species are proportionate and roughly the same in size almost certainly reflects the systemic effects of endocrine signals such as growth hormone. Superimposed on any variation in amount of hormone and systemic intensity of response, however, are secondary effects that reflect the activity of local growth networks. An example here is face shape all normal human faces have the same form and what marks one from another are minor differences such as those in the relative size of the nose, the breadth of the forehead, and the length of the chin. These differences reflect the effects of alleles that have minor effects on the local growth pathways that operate as the face develops; they just reflect normal variation and are discussed further in Chapters 12 and 13.

The origins of the larger and qualitative differences across species within a clade almost always lie primarily in the activities of the signaling pathways that underpin patterning and growth. These are, for instance, responsible for determining that the mouse left lung has a single lobe, while the right one has five, or why all mammals have seven cervical vertebrae, while the number in the birds and reptiles is highly species dependent. The differences in the details of such pathways, which go beyond the effects of the minor mutations seen in variants, reflect either different proteins in the networks or that a particular key protein has one rate constant in species A, while its homolog has different constant in species B. One interesting possible variable here is that a protein with alternative splice forms may express one or the other in a network. Livi and Davidson (2006) have found examples of proteins where the choice of which alternative splice form should be employed within a network depends on the cell's lineage.

Changes in the behavior of a network can also be due to the genetic context in which they operate. Here, the detail of the differentiation of the vertebrate ectoderm is a good example: it can be thick or thin in different parts of the body, and the same basic patterning system can support hairs, feathers, or scales, depending both on the species and the body region (Dhouailly & Sengel, 1973). The effects of patterning can also be species-dependent and even individual-specific, as a glance the patterning of teeth across the vertebrates or the external pigmentation patterns across the butterflies, giraffes, and zebras, makes clear. Many of these subtle differences derive from differences in the fine details of the behavior of individual proteins, and reflect opportunities for mutation-derived variation. Such behavior is discussed in Chapter 13.

The genome

Finally, it is worth considering the relationship between the genome and the early organism. Once the nature of DNA coding had been elucidated, it was initially thought that the genome drove embryogenesis, acting much like a computer program. Occasionally, this metaphor is helpful: the *Caenorhabditis elegans* worm develops from an egg to an adult with about 1000 cells, and, after some very early cell interactions (Han, 1997), the fate of almost every cell is determined by its parent cell on the basis of lineage – there is little flexibility in this highly determined system. However, things are very different in complex invertebrates, vertebrates, and plants, as a century of experimental work has shown. If, for example, a group of cells from a mid-stage vertebrate embryo is transplanted to a new location, the subsequent behavior of those cells can be highly influenced by their new environment.

This is clearly shown in the classic experiment in which a fragment of mesenchyme that would normally form a femur in a developing quail hindlimb was transplanted to the apical bud (the tip) of an early developing chick forelimb (**Figure 10.12**). The transplanted tissue was mainly found to form toes: it had "remembered" that it was fated to be leg but responded to the spatial patterning signals in its new location that instructed the tissue to display a distal phenotype (Krabbenhoft & Fallon, 1989). This emphasizes the point made earlier, that the behavior of a tissue depends partly on its lineage (transcription factor expression) and partly on its environment (signaling).

Such experiments show that the future state of a cell within a tissue is determined by the demands that the current cell and its environment make on the genome rather than being dictated to by the state of the genome. The main line of instruction is thus from the phenotype back to the genotype via the transcription factors in place, rather than the other way round. Perhaps the clearest way of seeing this comes from work on stem cells: it is now possible to take a differentiated cell expected to go down one pathway of development, modify its set of transcription factors, and totally change its developmental fate (Pereira et al., 2012).

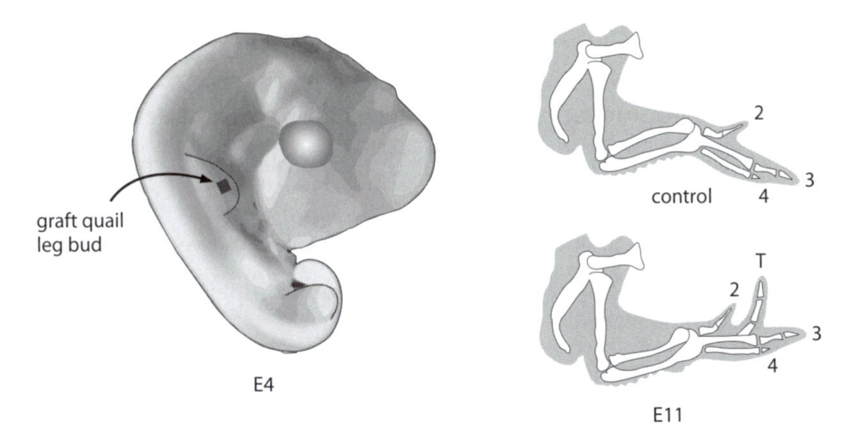

Figure 10.12 The relative effects of lineage and environment. A piece of quail limb-bud mesenchyme likely to become the femur is grafted into the apical tip of a chick wing bud from a 3 day embryo (E3). Because quail cells carry a nucleolar marker, the fate of the graft can be checked against the numbered digits; histology on E11 shows that the limb tissue changes its fate and becomes a toe (T). The rudiment thus carries its hindlimb history but responds according to its new position. The formation of leg- or wing-specific structures by leg bud cells grafted to the wing bud is influenced by proximity to the apical ridge.

The idea that the genome acts as a program really arose because of the apparent inevitability of embryonic development, but it was one of those clever ideas that are disproved by simple experiments. In the context of development, it is now much more helpful to view the genome not as a program, but as a database of resources that is available on demand to a cell. Under normal circumstances, of course, the options that are available to a developing cell, once the early embryo is past neurulation, are quite restricted (see Figure 10.6). Most changes to the state of differentiation are minor, and mesenchymal cells rarely, for example, become epithelia or vice versa, although it can happen (for example, in the development of the mammalian nephron from metanephric mesenchyme and in the formation of the mesothelial covering from the peripheral cells of mesenchymal tissues in the thoracic cavity). There are, however, two areas where the computing metaphor is helpful. The first is the many networks that are used repeatedly in development; these can be viewed as subroutines with variable parameters. Second are the tissue modules discussed above; these represent a higher category of encoding.

Key points

- The general features of an organism develop early in development and provide the foundations for later, more specific characters (von Baer's laws).
- In development, the genome does not so much act as a central controller but as a resource, available on demand, as required by the state of a cell within a particular tissue at a particular time. This resource includes the genes and regulatory sequences for proteins, many of which participate in networks, and tissue modules.
- Developmental change usually starts when signaling molecules from one tissue bind to surface receptors on another. This binding activates signal transduction to the nucleus that, in turn, initiates transcription from genomic regions carrying the appropriate complexes.

- Such signaling may be constant across a tissue, so causing uniform change, or it may vary in strength, so leading to a pattern of responses.
- The proteins that are produced as a result of signaling have many developmental roles:
    - They may have functional importance (for example, as enzymic and structural proteins)
    - They may contribute to or activate one or more process networks
    - They may prepare the cell for future change (for example, new signals, receptors, and transcription factors).
- Developmental change in animals and plants is driven by four main classes of process: differentiation, proliferation and apoptosis (which together modulate growth), and morphogenesis. The first three are cell autonomous, while the last may involve groups of cells.
- A developing cell's current state of differentiation also includes the signals it can send to the cells of other tissues and the proteins needed for possible future change.
- A cell's lineage (or parentage) determines these possible future states through the expression of cell-surface receptors and primed transcriptional complexes on promoter regions of those genes that it will need. If there are alternative possibilities of change, there will be two sets of these proteins and the choice will be made by signals that bind to one or another receptor.
- The results of morphogenetic changes such as movement, adhesion, and folding depend on the local geometry of the embryo, as well as on the properties of the participating cells.
- The control of development is thus distributed, with a cell's lineage and all aspects of its environment playing roles in effecting change.

Further reading

Gilbert SF (2014) Developmental Biology, 10th ed. Sinauer Press.

Wilkins AS (2002) The Evolution of Developmental Pathways. Sinauer Press.

Wolpert L et al. (2015) Principles of Development. Oxford University Press.

CHAPTER 11 EVO-DEVO 2: THE EVIDENCE FROM FUNCTIONAL HOMOLOGIES

This chapter discusses the third main area of evidence about evolution, that from homology of molecular function, or how proteins that are homologous often have very similar roles in very different organisms, particularly during development (evo-devo). Evolution predicts that each species reflects its broad history through ancestral innovations (plesiomorphies), while displaying its own novelties (autapomorphies). This chapter discusses examples of such plesiomorphies that occur across the phyla, particularly at the levels of proteins and protein networks. The chapter ends with a discussion about what these network and gene homologies across the phyla imply about the last common ancestor of the protostome and deuterostome phyla, the primitive bilaterian worm known as *Urbilateria*. This chapter, which focuses on how the activities of homologous proteins generate different phenotypes during animal development, thus complements the next two, which consider how mutation and the subsequent changes in protein function lead to phenotypic change.

Evo-devo, the study of the evidence from development that illuminates evolution is essentially the study of how homologous proteins, functioning either alone or in networks, have similar roles in sometimes very different embryos. This was not an obvious expectation; even the idea seemed inconceivable before the molecular basis of regulatory control mechanisms in eukaryotes began to be understood, late in the last century. There was thus no reason to suppose that there should be any similarity in the means of producing functionally similar, but morphologically different, organs, such as, for example, the eyes of vertebrates, mollusks, and arthropods. It was in this context that Salvini-Plawen and Mayr (1977) suggested that eyes had separately evolved some 40 times across the phyla; they would certainly never have predicted that a single gene, *Pax6*, had a key role in the development of each.

It had actually been known since the early 1970s that there were sequence and functional homologies in proteins such as cytochrome C that were involved in biochemical pathways across the phyla, and it was then straightforward to identify structural proteins and enzymes that were sequence and functional homologies. It was not, however, until the mid-1980s, when the molecular regulation of developmental change was first being investigated, that it started to become clear that homologous proteins also had very similar roles in regulating the development of similar tissues in very different organisms.

It should, of course, be emphasized that the fact that distantly related organisms express homologous proteins does not of itself imply that these proteins fulfill the same role in each: the mutations that each will have acquired since their host organisms shared a last common ancestor could well have led to changes in function. An enzyme, for example, may have changed its substrate, while the descendants of the original homeodomain-containing transcription factor have diversified to fulfill a wide range of roles. The possibility that homologous proteins do share a common function is thus only a hypothesis that has to be tested against the evidence on a case-by-case basis.

Such tests have been done for a wide range of proteins and it is now known that there are many functional homologies in the underlying signals, transcription factors, and protein network systems that regulate the building of tissues in embryos across the phyla (Held, 2014). This is so despite the obvious and extensive differences in the external appearances of individual species, their internal tissue anatomy and their modes of development. If one just considers mouse and *Drosophila*, whose last common ancestor lived before the Lower Cambrian, the profound differences between their phenotypes today represent more than 540 My of evolutionary divergence. The many homologies in the regulatory genes that underpin their very different tissues and that are discussed below reflect the functional stability of a set of ancestral genes that were present in an organism that lived at the beginning of bilaterian evolution; for this reason, they are known as deep homologies (Shubin et al., 2009).

It is, of course, obvious that the more recently the last common ancestor of a group of species lived, the closer are both the sequences and the homologies. The differences in the molecular underpinnings of the development of mouse and man, whose last common ancestor was ~30 Mya, are relatively minor, as one would expect from two organisms whose major phenotypic differences in the adult are in size, hair coverage, and relative cranial capacity. Experimentation has shown that human proteins with particular developmental functions usually have similar effects when expressed in mouse embryos at the equivalent time and in the appropriate place. It is this observation that makes the mouse such a good model system for studying mutations that lead to human genetic disease.

It may seem surprising that, as this chapter discusses, deep homologies still exist between organisms such as *Drosophila* and mouse, given that their genomes, and particularly their phenotypes, have diverged so dramatically from that of their last common ancestor. These are particularly seen in those regulatory networks that drive developmental change, and the question immediately arises as to why these are so similar, given that both their genomes and their phenotypes are so very different. If other genes have mutated beyond recognition, why are their regulatory mechanisms still relatively similar?

The answer is not known but there are two possibilities. First, the functional inter-relatedness of the proteins in a network, such as a signal transduction pathway, conferred a degree of stability against mutation on such systems so that change was faster to achieve through modulation and the addition of new components than through the production of new systems. Second, selection acts on the phenotype and only indirectly on underlying control systems: networks are thus buffered against selection through the effects of minor mutation effects that create variants in their output, which, in turn, lead to novel phenotypes, at least some of which will survive. The expected result of these two effects would be that these protein networks remain relatively stable at the interface of genotype and phenotype, maintaining an ability to use whichever genomic resources survived mutation to produce phenotypes that would survive selection.

Evo-devo work is of particular importance in two contexts. First, the functional evidence is completely independent of and so complements the evidence on evolution from analyses of anatomical comparisons and of genomic sequences. Second, the common developmental plesiomorphies shown by embryos as different as those of *Drosophila* and the mouse illuminate the ways in which tissues must have formed in the early embryos of ancient organisms, for which there are, of course, no fossil data (see Chapter 4). Such work, in particular, suggests that such homologous genes and mechanisms were present in their last common ancestor, known as Urbilateria, that lived before the Cambrian explosion.

The first part of this chapter thus discusses some of the molecular and network homologies, while the last part considers what this and other information tells us about Urbilateria. The information integrates much of our understanding of how early bilaterian life evolved, and almost all of the work is recent. Those biologists who started to explore the molecular basis of development in the 1980s opened a door to what may have been the most exhilarating period in the history of evolutionary science since the age of Darwin.

Functional homologies

This section discusses some well-known examples of homologous signals and receptors, transcription factors, and networks with, following the last chapter, a particular focus on *Drosophila* and the mouse.

Signal and receptor homologies

The development of the *Drosophila* embryo involves a set of signaling proteins, many of which were identified through analyzing mutant flies that were often named on the basis of their odd appearances. Thus, the *hedgehog* gene acquired its name because the mutant embryo looked prickly as it had double the usual number of denticles and was also dumpy in shape, while the *disheveled* embryo had denticles that were abnormally oriented. The genes for *Drosophila* signals (receptors in brackets) include *Decapentaplegic* (*Thickveins* and *Punt*), *Hedgehog* (*Patched*), *wingless* (*Frizzled*), and *Delta* (*Notch*).

The mouse embryo uses ~30 signaling molecules and many were identified through searching for homologs of *Drosophila* signals using either molecular genetics or bioinformatics approaches. The mouse turned out to have three homologs of the *Drosophila hedgehog* gene, which reflects early genetic duplications of the original gene; these are named *Sonic, Indian,* and *Desert Hedgehog* (Pereira et al., 2014). Indeed, informatics analysis of all the sequences of the early signals seen in *Drosophila,* but not its later ones such as juvenile hormone, led to the identification of sequence homologs in the mouse. The key point, however, is that the mouse homologs of *Drosophila* signals were also signals; sequence homologs turned out to be functional homologs. Closer analysis of many of these mouse signal proteins showed that at least one of their effects was on the growth pathways that are so active as the mouse embryo develops. Some are therefore known as growth factors, with well-known examples being the epidermal, fibroblast, and vascular-endothelial growth factors.

Because the mouse signals are so similar to those of *Drosophila*, it was not surprising to discover that there are similarities in the binding regions of mouse homologs of the *Drosophila* signal receptors, and also in their transmembrane and intracellular regions. Again, sequence homologs were also functional homologs. These homologies extend to the rest of the receptor and on to many of the internal members of the transduction pathways (below). Note, however, that the homologies in these signaling pathways carry no implications as to the nuclear transcription factors that they eventually activate. In these two very different embryos, their respective signals are used in many different contexts, with their effects depending on the lineage and hence on the transcription factors expressed in the receptor cells.

Transcription factor homologies

DNA-binding transcription factors are the key proteins in driving developmental change because it is their activity that is responsible for producing the new proteins that lead to changes in phenotype. Many *Drosophila* transcription-factor families are known that have homologs in the mouse. Here, four important examples are considered; more will be mentioned at the end of the chapter when the anatomy of Urbilateria is discussed. These pairs of *Drosophila* and mouse homologs should not be viewed as special, such homologs will exist across all organisms and, the evolutionally closer these organisms are, the more will be found and the closer will be the genomic sequences.

Pax6 *and eyes*

The *Pax6* gene was discovered through the analysis of the *smalleye* mouse mutant: the heterozygote (with only one mutant gene) has small eyes with no iris, a disorder known as aniridia; a similar phenotype is seen in humans (Hanson et al., 1994). The mouse homozygote phenotype (both gene copies carry mutations) is more profound: the animal has only very rudimentary eyes, together with other facial and brain abnormalities (Grindley et al., 1997). *Drosophila* records were immediately examined to see if there were known mutants that lacked eyes, and it took little time to identify the *eyeless* mutant, named on the basis of its dramatic phenotype, which had first been identified many years earlier. Molecular analysis showed that the underlying mutant fly gene was a homolog of *Pax6*, with both genes encoding a transcription factor containing two specialized DNA-binding domains, a homeobox and a paired box, with the two proteins having regions of similarity and regions of difference (see Figure 8.1).

This discovery, made in the 1990s, was completely unexpected because the two organisms make their eyes in such different ways. The compound eye of *Drosophila* forms as a wave of hedgehog-dependent activity headed by a morphogenetic furrow passes over the future eye domain of the appropriate head ectodermal imaginal disk (Chapter 10). This leaves in its wake a developing array of about 800 small clusters of cells, with the central one initiating signaling to its neighbors. When the eye disks evert during pupation, each group becomes an ommatidium with a nerve, a lens, photosensitive cells, pigment, and accessory cells (**Figure 11.1**A). The compound eye is essentially a hexagonal array of separate optic fibers, derived only from ectoderm, each of which projects back to the optic lobe that undertakes the first steps of integrating the 800 separate images into a single one (Otsuna et al., 2014).

The mouse eye is completely different (**Figure 11.1**B): this is partly because it uses a single cornea and lens to focus a single image on its retina, working much like a camera, and partly because so many different tissues contribute to the organ. The retina (with its pigmented and light-sensitive cell layers) forms as an outgrowth of brain neuroepithelium, while the lens and cornea form from surface ectoderm. The sclera that protects the retina forms from mesoderm, as do some of the muscles and the posterior layer of the cornea, while the neural crest contributes further mesenchyme, iris musculature, and pericytes, the small muscle cells in the walls of arteries (Gage et al., 2005). In the camera eye, the

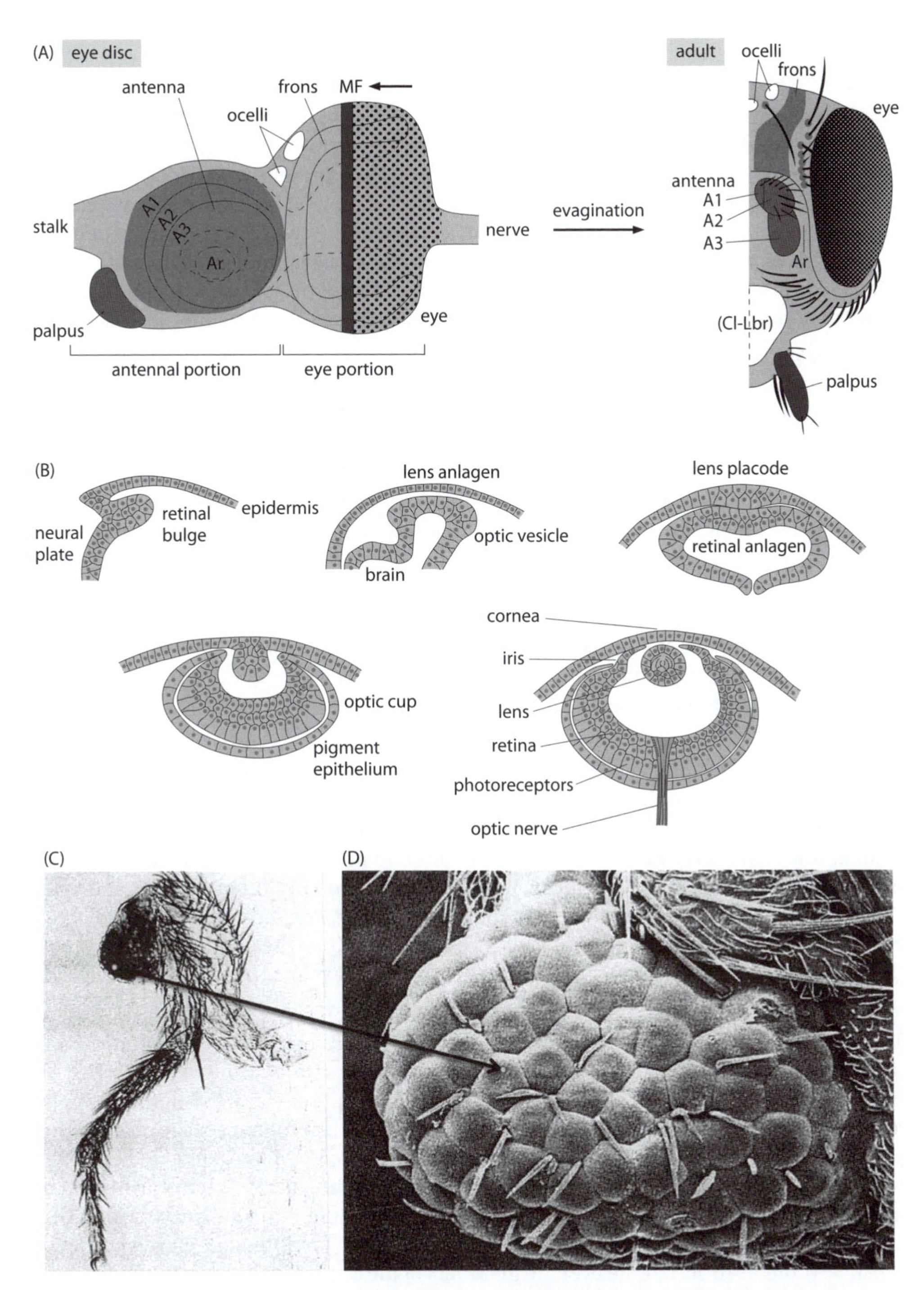

Figure 11.1 (A) Diagram of the development of the compound eye in *Drosophila* as a wave of activation headed by a morphogenetic furrow (MF) moves across the epithelium of the antenna-eye imaginal disk. The final eye forms during pupation when the disk everts. (B) Development of the camera eye of the mouse Pax-6: where to be conserved is not conservative. (C) Ectopic eyes resulting from the expression of *Pax 6* in the *Drosophila* leg. (D) An enlargement of the ectopic eye shown in (C).

retinal fibers project back to the optic lobe which undertakes the first steps of integrating a single image essentially made of pixels, one from each neuron.

It still seems remarkable that the early development of these two profoundly different eyes is each activated by orthologs of the same transcription factor, Pax6. It is, however, important to emphasize that the mouse and *Drosophila* Pax6 protein sequences are not the same, with some regions of the sequences being similar, while others are very different (see Figure 8.1). Moreover, the role of Pax6 in *Drosophila* is much more restricted than it is in vertebrates, where mutation causes a wide range of facial and brain abnormalities (Hanson et al., 1994). These differences reflect the effects of more than 500 years of divergence and mutation since the two genes shared a sequence in a last common ancestor.

Sequence homology and protein similarity do not, of course, predict functional homology and, indeed, it was originally thought that, even though the mouse and *Drosophila Pax6* genes had homologous sequences, each was most unlikely to work in the "wrong" organism. A classic experiment showed that this intuition was wrong. Using some clever genetic tricks, Halder et al. (1995) were able to express the mouse Pax6 protein in *Drosophila* limbs and other external ectodermal tissue, and achieved a remarkable result: those regions of ectoderm in which the mouse *Pax6* was expressed developed into compound eyes (**Figure 11.1**C, D). Perhaps even more surprising, given that the vertebrate camera eye is so complex, is that ectopic expression of the *Xenopus Pax6* homolog alone is enough to generate a small but essentially normal eye in the toad (Chow et al., 1999). Given the many cell lineages that contribute to its eye, this result clearly means that the descendants of the host cell containing Pax6 can mobilize nearby cells to participate in eye development and differentiate accordingly.

Of course, the production of eyes in *Drosophila* and the mouse requires the expression of many proteins other than Pax6 that are under the control of further transcription factors. Analysis has shown that there are both sequence and functional homologies among several other transcription factors regulating the development of eyes in the two organisms (Kumar & Moses, 2001). These include sine oculis and optix, eyes-absent, dachsund, and teashirt in *Drosophila*, which are the respective homologs of the Six family, Eya1 and Eya4, Dach1 and Dach2, Mtsh1, and Mtsh2 in the mouse. In the ectopic toad eye, homologs of all these genes are expressed (Chow et al., 1999). This confirms the point made earlier, that the genes for the regulatory systems are better buffered against the ravages of mutation than are those that give the final phenotype.

There are two further interesting similarities in the very different eyes of *Drosophila* and the mouse. First, the wave of ommatidia induction in *Drosophila* is regulated by Hedgehog signaling and the effects of Atonal, another transcription factor. Something similar happens with the induction of ganglion cells in the zebrafish retina: a wave of induction passes across the retina propagated by the zebrafish homolog of hedgehog and activated by ath, the zebrafish homolog of the Atonal protein (Jarman, 2000). Patterning here may, however, not be a homology but a homoplasy, based on the availability of transcription factors not otherwise being used. Second, and at a more functional level, the key light-sensitive proteins in vertebrate eyes are members of the opsin family. Such proteins are, in fact, ubiquitously used: Opsin homologs, such as Rhodopsin, are present in the visual systems of all diploblast and tripoblast organisms (Arendt et al., 2004; Shichida & Matsuyama, 2009).

It is, however, Pax6 that seems to be the key transcription factor here because detailed cross-species work across the phyla has demonstrated that, wherever an animal has an eye, upstream of its early development is the expression of a *Pax6* homolog. This has been confirmed in mammals; arthropods; mollusks; box jellyfish, whose eyes have a cornea, retina, and lens; and even Cnidaria, the simplest organism with eyes (Kozmik et al., 2003). What these results almost certainly mean is that, very early in the evolution of multicellular organisms, a protein with paired-box and homeodomain regions was associated with the development of light-sensitive cells. As the fossil evidence (Chapter 4) shows that Lower Cambrian organisms had eyes, while worms and jellyfish may have had light-sensitive cells even earlier, the ancestor of Pax6 may first have been involved in light sensitivity almost 600 Mya. That so many organisms use a *Pax6* homolog for eye development and an Opsin homolog for light detection provides spectacular evidence for animals with light-sensitivity sharing a common ancestor that may extend back to the Ediacaran Period.

Hox *genes and body patterning*

Another of the remarkable homologies is the close and completely unexpected similarity in the molecular mechanisms by which invertebrates such as *Drosophila* and vertebrates such as the mouse pattern their very different anteroposterior tissues during embryogenesis (see Chapter 10 and Gilbert, 2014). Consider *Drosophila*: its larvae each have 14 ectodermal segments when they pupate, most of which contain imaginal disks that will then produce the external tissues in the adult appropriate to their position, from antennae at the front to the external reproductive apparatus at the back. The basis of the initial anteroposterior molecular gradient that will eventually generate this pattern is laid down by nurse cells and is present in the unfertilized egg. Once the embryo starts to develop, a complex series of molecular interactions takes place, which involves many transcription factors together with a great deal of signaling through, among others, hedgehog and wnt. The net result of all this molecular activity is that each segment has its own molecular identity and future fate (Gilbert, 2014).

This identity is conferred on the segments late in their development by a group of genes that encode transcription factors that are first expressed when these segments are becoming specified. The associated mutants were first identified more than a century ago and their names reflect the abnormal, identity-changing (homeotic) phenotypes that alter the spatial identity of one or more segments. The *Antennapedia* mutant, for example, has a leg on its head instead of an antenna. What is important here, however, is that these homeotic patterning genes were shown in the 1980s to encode a set of eight transcription factors, known as *Hox* (or homeotic) genes because each includes a 180 base-pair sequence known as a homeobox that encodes a 60 amino acid DNA-binding domain (or motif), the homeodomain (Chapter 8). *In situ* hybridization work showed that, as would have been predicted, segments had distinct patterns of *Hox* gene expression. Much more surprising was the observation that the sequence of *Hox* genes along *Drosophila* chromosome 3 was the same as the sequence of their expression in the embryonic segments along the anterioposterior axis (a relationship known as colinearity; **Figure 11.2**).

The discovery of Hox patterning in *Drosophila* led to an examination of whether anything similar might be happening during the development of the mouse, whose paraxial mesoderm (that adjacent to the neural tube) also undergoes segmentation to give somites. Genetic analysis soon showed that each somite had its pattern specified by a unique set of mouse *Hox* genes. The number of mouse *Hox* genes was, however, very much greater than that of *Drosophila* (Chapter 8). Instead of a single set of eight genes, the mouse had four sets of paralagous genes, each of which was on a separate chromosome and had anything up to 13 homologs arranged in four groups (a, b, c, d). A further surprise was that the sequence of

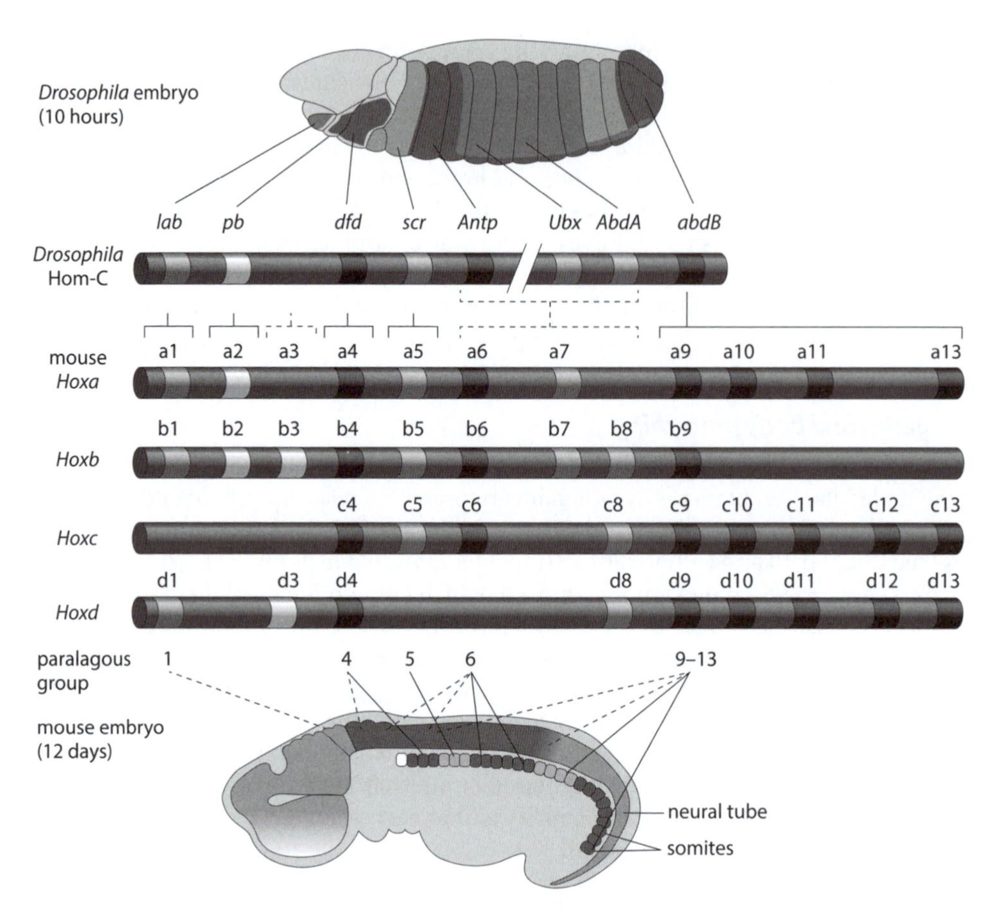

Figure 11.2 Hox patterning in *Drosophila* and mouse embryos (showing the most anterior genes expressed in a segment) illustrates homologous colinearity in the two embryos. For the mouse, the dotted lines show Hox patterning in the neural tube and the solid lines show the Hox patterning in the somites.

genes for each of the four mouse homeotic clusters down its particular chromosome was the same as the homologous sequence in *Drosophila*. The greatest surprise, however, was the observation that the expression sequence of the *Hox* gene set along the neural tube and somites of the mouse embryo was the same as in *Drosophila* (they, too, were colinear). Gene knockout and other mouse work has now shown that the fate of each group of cells depends on the exact subset of *Hox* genes that it expresses (the complete group of subsets is known as the Hox code).

Given these similarities, what added to the sense of surprise was that the respective mechanisms for setting up the patterns of homeotic gene expression are so different in the two organisms. In *Drosophila*, Hox patterning is the end result of a complex and many-stage sequence of molecular patterning, signaling interactions, and transcription that is initiated in the unfertilized egg. In the mouse, Hox patterning is set up rapidly during or soon after gastrulation, a little before segmentation occurs, through signal gradients (fibroblast growth factor 8, bone morphogenetic proteins (BMPs), and Wnts) interacting through the CDX transcription factors. The *Hox* gene expression pattern in the mouse is much more complicated than in *Drosophila*, and its establishment and interpretation are not well understood.

The actual process of segmentation in *Drosophila*, where they all form at about the same time, is also different from that in the mouse where a newly epithelialized posterior somite forms every 2 hours from paraxial mesoderm as the posterior region of the embryo extends, eventually forming the tail. The mechanism for this involves growth, a clock, and a mesenchyme-to-epithelial transition (Hubaud & Pourquié, 2014). Apart from both *Drosophila* and mouse embryos using their *Hox* genes for the anteroposterior patterning of the adult organism, all other aspects of their segmentation and subsequent development are thus different.

Since the original discovery of these sets of *Hox* genes, a great deal of work has been done on the wide family of genes that includes a homeobox domain; it now seems that members of this gene family are present in all the Bilateria, and phylogenetic analysis shows that they share a common ancestor. There was clearly an original *protohox* gene that appeared very early in multicellular evolution, going through many rounds of mutation, duplication, and diversification over time and across the phyla as they evolved (Holland, 2013). It is not that the *Hox* genes are the same in each organism or have exactly the same function: the evidence shows that some species-specific sets of *Hox* genes control anteroposterior patterning in a very wide range of organisms across the Bilateria, while the rest have other roles. Many other genes, such as those of the Pax family also include homeodomain regions. It is thus clear that hox patterning of the anteroposterior axis was an important feature of urbilateria, the last common ancestor of mouse and *Drosophila*.

The role of homeobox genes in patterning may actually extend even further back in time than this. The Wox family of plant-specific homeodomain-containing transcription factors plays an important role in seed and flower development (Costanzo et al., 2014). If the role of homeodomain-containing transcription factors in both plants and animals really does represent a homology, this takes the time of the last common ancestor sequence for this homeodomain-containing transcription factor almost back to the time of their last common ancestor, a simple eukaryote that lived more than 900 Mya (Chapter 9).

Hox *genes and patterning the vertebrate limb*

If there is any theme that runs across the whole of evolution, it is that of opportunism, the idea that if a genetic option can generate some phenotype it will eventually do so through a mix of mutation, genetic drift, and associated random breeding. Most options will fail, but occasionally that phenotypic variant may thrive in a novel environment - this is normal selection. An interesting aspect of opportunism is when that novel environment is within the organism itself, and occurs when a new tissue forms. The development of this tissue needs regulatory systems, and there is little point in, as it were, reinventing the wheel. Once such signal transduction and process networks were in place for one context, they were available for use in others, provided that their expression could be made tissue specific. The molecular device for this is the enhancer/repressor system: these are small cis-acting regions of the genome that may be up- or downstream of the gene in question and to which proteins associated with a transcription complex can bind: activators enhance transcription, while repressors silence it (Nord, 2015). Opportunistic use of such networks requires either the modification of existing or the evolution of new enhancers and repressors.

An example of such molecular opportunism is the secondary role adopted by the set of *Hox* patterning genes. Their initial role was in anteroposterior patterning of the embryo, but, once in place, they were available to be used in new tissues that arose later. Soon after fish evolved, a subset of the *Hox* genes was used to pattern their pectoral and pelvic fins, particularly in the sarcopterygian fish, which have homologs of the proximal bones seen in tetrapods. With the evolution of digits in these fins, the Hox system was further re-used to pattern amphibian and subsequent autopods (**Figure 11.3**). Close study of limb development showed that its patterning was handled by Hoxa and Hoxd proteins, particularly under the influence of Wnt, FGFs, and Sonic Hedgehog signaling (Duboule et al., 2007).

The details of tetrapod limb organization today are determined by two rounds of Hox patterning in the embryo. The first is proximodistal (scapula to hand in the forelimb) with specific tissue identity being determined by Hoxa,d 9–13; this clearly derived from the original patterning of the sarcopterygian fin. The second is the patterning of the early autopod (for example, the footplate) that gives rise to the digits: this is the result of antero-posterior patterning of Hoxa13 and Hoxd10–13 (Tickle, 2006). This system has turned out to be important in a wider context of understanding variation because the effects of the various mutations that alter patterning are easy to identify. Some of this work will be considered in Chapter 13.

It is worth noting that many of the genes whose proteins are involved in patterning vertebrate limbs have homologs that are involved in patterning *Drosophila* limbs. Of unclear significance is the fact that homologs have similar functions in the two systems (Tabin et al., 1999). Included among them are *dlx* and *Dll*; *Shh* and *hh*; *Bmp* and *dpp*; *Rfng* and *fng*; *Ser-2* and *ser* (all in mouse-*Drosophila* order). There are two possible explanations for this coincidence, which come down to the question of whether Urbilateria had appendages. If it did, then the behavior of these genes in limb formation represents a plesiomorphy and their behavior in contemporary vertebrates and invertebrates reflect homology. The alternative view is that Urbilateria did not have appendages and that the similar roles of these genes reflect independent evolution or homoplasy. There are, however, too many similarities to suppose that each gene had an independent role and, if Urbilateria did not have limbs, its descendants probably only had a limited number of potential networks available for employment. The fossil record from the Lower Cambrian is not particularly helpful here: neither *Haikouella*, a very early chordate, nor *Maotianshania cylindrica*, a priapulid worm, had limbs, while *Aysheaia*, which looks like a velvet worm, did. As it is not known whether the former lost them or the latter inherited them from their ancestral Urbilateria, it has seemed sensible to take the conservative view, which is that, because Urbilateria was a very early organism, it did not have limbs.

Such opportunistic reuse of existing proteins and networks has been a parsimonious way of facilitating the evolution of new tissues or of making existing ones more

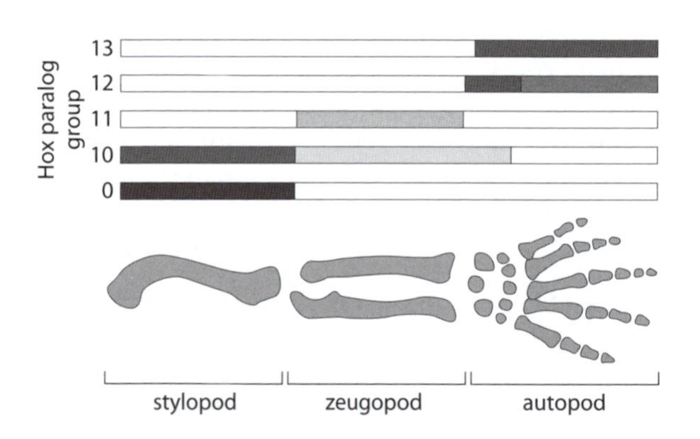

Figure 11.3 The adult mouse forelimb skeleton showing the domains patterned by the *Hoxa* and *Hoxd* paralogs.

complex; indeed, the gene-expression databases include many examples of the reuse of transcription factors, signals, and process networks, often but not always under tissue-specific control. This opportunism does, however, carry a medical caveat in the context of building on such information to cure human disease. When considering possible manipulation of the regulatory systems in the context of treating disease, it is important to check the broad effects of such manipulation in mice. Although rodents are not humans, such work is a first step in avoiding the risk of unpleasant side effects occurring in humans.

Nerve cords in protostomes and deuterostomes

This third example examines a classic problem: the origins of the different locations of the nerve cord in protostomes, where it is ventral (below the gut), and in deuterostomes, where it is dorsal (above the gut). This very basic difference has intrigued evolutionary biologists since the nineteenth century, as it seemed to deny the possibility of a common ancestor and so reflect independent origins for the two systems. Comparative molecular genetic analysis of brain determination in mouse and *Drosophila* has greatly clarified the situation and it is now clear that the homologs of the same signals and transcription factors pattern both embryos, which are themselves representative of their wider clades in this context (Figure 11.3).

The dorsoventral patterning of the neural region has now been shown to be set up in both embryos by a pair of counteracting protein gradients: one is for a signal and the other is for its antagonist; these are Dpp and Sog in *Drosophila*, and BMP4 and Chordin in the mouse. The relative concentrations of these gradients at any position in the early embryo activate one of three transcription factors; these are Vnd, Ind, and Msh in *Drosophila*, and NKX2, GSX, and MSX in the mouse (**Figure 11.4**). There are two points of particular interest in these systems. First, the signals and transcription factors are each homologous pairs, so the patterning systems as a whole are also homologous and therefore reflect common descent (Lichtneckert & Reichert, 2005). Second, the polarity of each system is reversed with respect to the other.

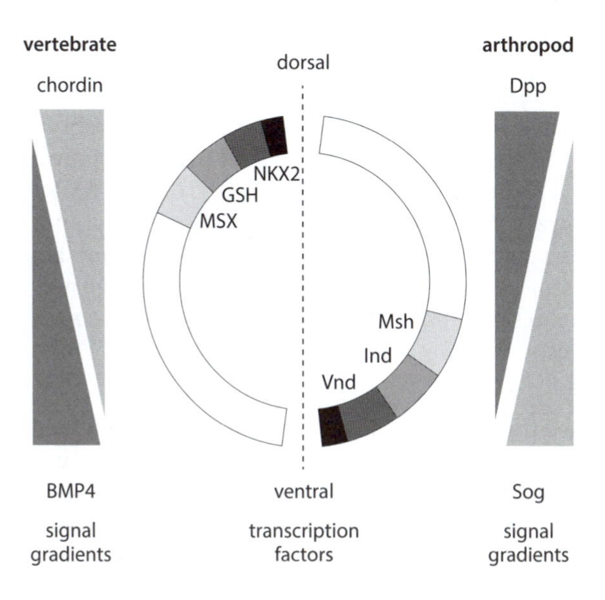

Figure 11.4 Diagram showing patterns of signaling gradients and transcription factor expression in vertebrates and arthropods in the anterior parts of typical vertebrate and arthropod embryos. Chordin/Sog, BMP4/Dpp, NKX2/Vnd, GSH/Ind, and MSX/Msh are all homologous pairs.

The direct implication is that, very early in the evolution of the bilateria, there was a polarity reversal in the neuronal patterning system that led to the neural cord becoming relocated with respect to the gut. The experimental evidence does not imply which polarity was basal, but the fact that the very great majority of animal phyla have a ventral cord, with only the chordates having a dorsal chord, argues that the protostome organization represented the original morphology (Denes et al., 2007; Lowe et al., 2015).

Protein network homologies

While signals and transcription factors initiate and direct change during embryogenesis, they do not mediate it; this is the task of protein networks, and these fall into two categories. First, there are the signal-transduction pathways triggered by signals binding to their receptors and whose downstream result in the nucleus is transcription-factor-activated gene expression. Second are the process networks that result from this new gene expression and these include, as discussed earlier, those that drive cell differentiation, morphogenesis, proliferation, and apoptosis. In Chapter 13, the effect of mutation on these latter networks will be examined; here the emphasis is on the signal transduction and growth pathways because it is on these that the most work has been done.

Phylogenetic analysis has shown that most of the signal-transduction pathways seen in mouse and *Drosophila* are closely related, with each pair sharing many homologous proteins that are closely related in sequence, function, and role within the pathway. Demonstrating this using the standard literature is quite difficult because the names of homologous proteins are organism dependent, as this chapter has shown, often because they were named on the basis of the phenotype of the mutant that was then abbreviated when the mutation was discovered. Such comparisons are far more effectively done computationally.

Information on signal transduction and other pathways and networks is now easily accessible in the various bioinformatics databases. The key one for making such interspecies comparisons is KEGG (Kyoto Encyclopedia of Genes and Genomes, www.genome.jp/kegg; Kanehisa et al., 2012); this includes interspecies comparisons among many networks for both embryos and adults, albeit that its focus is more on vertebrates than invertebrates. **Figure 11.5** shows the pathways generated by KEGG for the Notch–Delta signal transduction pathway in mouse and *Drosophila*, and one does not need to understand the details of how these pathways operate to see just how closely related they are. Similar homologies are seen when comparing the cell proliferation networks in mouse embryos and *Drosophila* larvae: proliferation can be activated through several paracrine growth factors that are themselves homologous (for example, WNT1/Wg, SHH/Hh, and BMP4/Dpp) and that excite homologous pathways. *Drosophila* does, however, have two important endocrine factors associated with pupation and moulting, juvenile hormone and ecdysone, that do not have homologs in vertebrates. In all pathways associated with growth, a key output of the network is the signal that activates the cyclin/cyclin-dependent kinase pathway that leads to mitosis.

Growth reflects the balance between proliferation and apoptosis (programmed cell death), with the latter being implemented through the caspase pathway. This system is not only present across the Bilateria, but even has components in sponges and other very primitive organisms (Sakamaki et al., 2015). There are two ways in which the caspase pathway can be activated, the extrinsic and the intrinsic pathways, with the former being used to model tissues and to remove unneeded cells during development. An example is

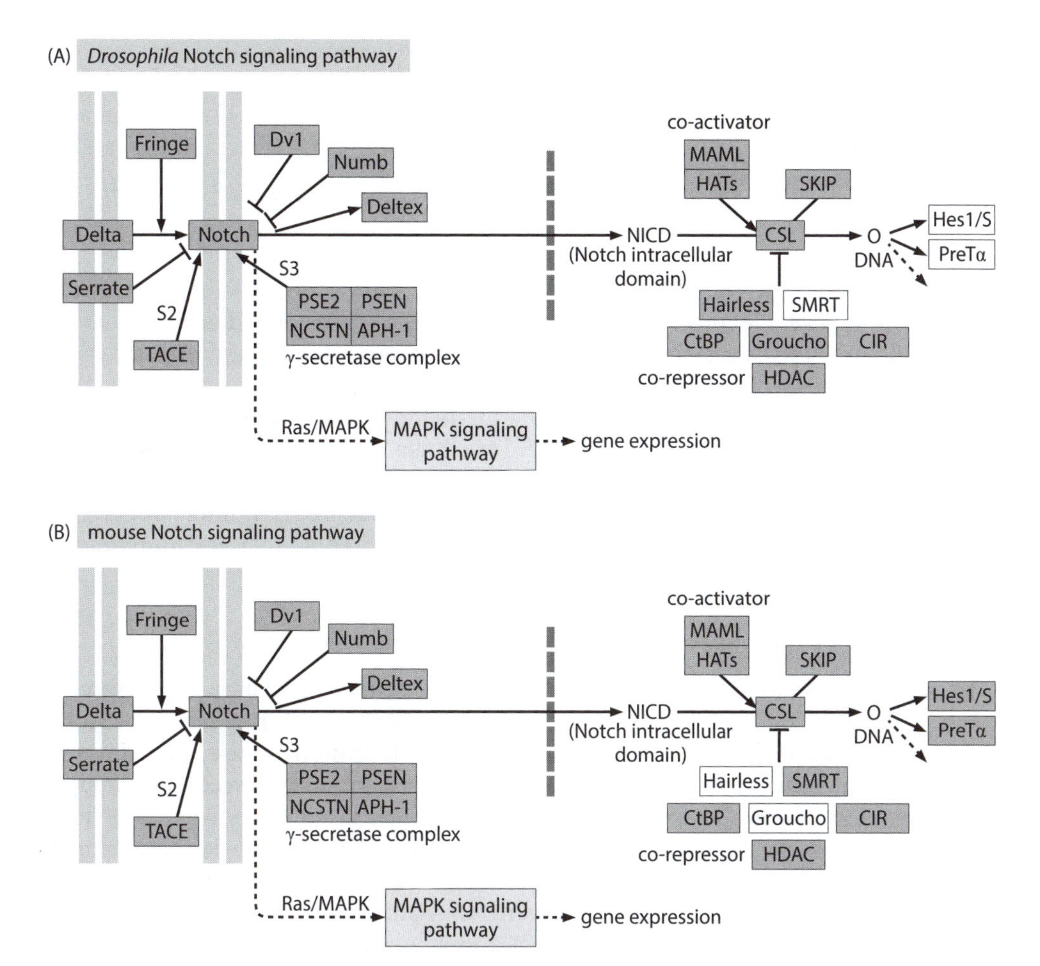

Figure 11.5 The Notch–Delta pathways in *Drosophila* and mouse. Proteins in gray boxes are present, while those in white boxes are absent. The two networks are surprisingly similar considering that they are separated by more than 500 My of evolutionary history. Note that proteins in the two diagrams with the same names are not identical, but are organism-specific orthologs, named using the KEGG convention.

in early autopods where, for example, the footplate initially includes unseparated digits: a patterned signal such as tumor necrosis factor or BMP implements the caspase pathway in the ectoderm and mesenchyme between the metatarsals, so causing their death and separating the digits (Sanz-Ezquerro & Tickle, 2003). Apoptosis can also be activated through the intrinsic pathway, which is used when a cell is damaged. Here, the caspase pathway is triggered through the release of mitochondrial cytochrome c into the cytoplasm, often through the action of P53, BCL2, and some associated proteins. *Drosophila* has much of this apparatus, although there are, of course, some differences between it and the mouse (Denton et al., 2013).

Of particular interest is the fact that, in both mouse and *Drosophila,* the regulation of apoptosis and proliferation are both interlinked through homologs of the Hippo signaling pathway (Zhao et al., 2011). The networks for many other systems across the phyla also show remarkable degrees of homology with respect to both their proteins and their structure, and a summary webpage of such networks is maintained by the Society for Developmental Biology (www.sdbonline.org/sites/fly/aimain/aadevinx.htm). While the molecular biology of these interlocking networks whose homologs are present in a host of animals is extremely complicated, the core conclusion is that many represent contemporary versions of networks with an extremely ancient evolutionary provenance.

Urbilateria

For all of their obvious differences, there are some basic morphological similarities between the mouse, a deuterostome, and *Drosophila*, a protostome: both are bilateral, triploblastic, and have similar cell types. Although the morphological resemblances break down at the tissue level, both have similar physiological requirements, even though they are based on different anatomical structures. Thus, the mouse has a four-chambered heart to pump blood to oxygenate it in the lungs and to circulate it through the vascular system, while *Drosophila* just has a simple heart tube to move hemolymph round the body cavity, acquiring oxygen via diffusion of oxygen from its simple tracheal respiratory system. The mouse has a liver for many biochemical functions, many of whose roles in *Drosophila* are fulfilled by the fat body, while excretion is handled by kidneys in the mouse and by the Malpighian tubules in *Drosophila*. Both, of course, have reproductive systems and associated gender differences, but their respective morphologies are quite different.

It has always been assumed that deuterostomes and protostomes share a last common ancestor, a primitive organism known as Urbilateria ("ur" means original or primitive). What needs to be established is whether pairs of different organs in mouse and *Drosophila* that have common functional abilities reflect homologies (common descent) or homoplasies (convergence from different ancestral features). If the former, they were likely to be present in Urbilateria. The only way to approach this problem is through computational analysis to identify possible proteins whose homologs are expressed in such pairs of tissues and then to confirm their function through genetic engineering, particularly of knockouts. This approach to analyzing the last common ancestor of the deuterostomes and protostomes was initiated by De Robertis and Sasai (1996). Analyzing mouse and *Drosophila* genes alone will not, however, give a complete picture of this ancient organism. This is because individual genes can, over time, be lost from one or another species and their functions replaced by others so that neither the mouse nor *Drosophila* would be expected to include all of the key genes of Urbilateria (de Robertis, 2008; Matsui et al., 2009).

A better approach to identifying at least some of the Urbilateria proteome is identify the overlap of the genes in a wider range of organisms, a task undertaken by Wiles et al. (2010); they looked for homologs in genomes of mouse, man, *Drosophila*, *Caenorhabditis elegans* (a nematode worm), and yeast (a unicellular fungus). Their results, best illustrated in a Venn diagram (**Figure 11.6**), identified over 2000 genes whose orthologs are common to vertebrates, the nematode worm, and *Drosophila*, a number that they viewed as conservative (Figure 11.6). They also found that many of these genes are also present in yeast, whose last common ancestor with animal phyla was probably more than 1 Bya. These homologs are likely descendants of those present very early on in the history of eukaryotic life and are candidate genes for those in Urbilateria.

Analysis of such genes showed that they primarily reflect basic cell biology, such as the proteins that mediate mitosis, meiosis, membrane function, cytoskeletal activity, and the biochemical machinery of the cell. In the context of exploring the anatomy of Urbilateria, the key genes are those that deal with tissue generation in multicellular organisms, and

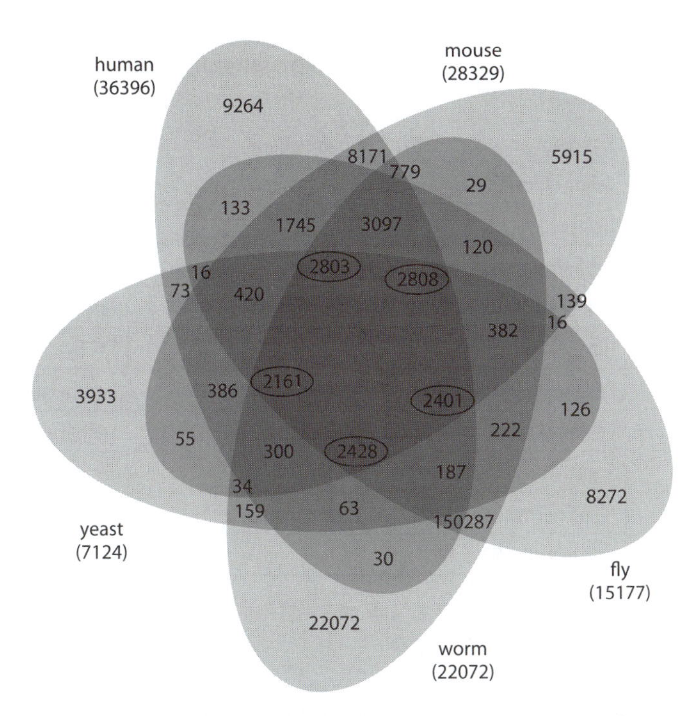

Figure 11.6 Five-way Venn diagram of orthologous genes in humans, mouse, fly, worm, and yeast. It should be noted that many genes in each species are paralogs. The numbers of genes overlapping between one species and all the others are circled. All other intersections reflect the genes found in the organism with the largest number in that group.

these particularly reflect signaling and signal-transduction pathways. As discussed earlier, there is a considerable set of *Drosophila* signals and receptors with homologs in the mouse. At the level of systems and networks, there are partial homologies in the signal transduction, cell cycle, apoptosis, and proliferation pathways. One other system worth mentioning is the β-catenin mechanotransduction pathway that links mechanical stresses to changes in gene expression (Brunet et al., 2013): closely related versions are found in both *Drosophila* and the zebrafish, and a much earlier version was thus probably present in Urbilateria. While it is natural to assume that these networks and pathways were probably much simpler than those in today's organisms, there is evidence that this may not have been so and that network complexity arose very early (Zmasek et al., 2007).

Because signaling and process networks are widely used, however, they give little insight into the actual anatomical features of Urbilateria; specificity here is determined by transcription factors. The best test for homologous function is the prediction that knocking out a pair of homologous proteins in two very different organisms leads to similar defects in each. In such cases, it is a fair assumption that the last common ancestor of the two species had an ancestral gene whose protein had a similar function (not necessarily an identical function because the ancestor was anatomically different from both of its descendants). If the two species are from either side of the protostome/deuterostome divide (as are *Drosophila* and mouse), then it is likely that today's homologous genes were present in Urbilateria, doing at least some of what they do today.

Some of this knockout work has been done, and some important transcription factors whose homologs are expressed in functionally equivalent organ systems of mouse and

Table 11.1: Some homologous transcription-factor genes in mouse and *Drosophila* (adapted from Held, 2014)[a]

Drosophila gene	Knockout phenotype affects	Mouse gene	Knockout phenotype affects
Hox system	Segment identity	*Hox* system	Segment identity
tinman	Heart tube	*Nkx2-5*	Heart muscle
eyeless	Eye formation	*Pax6*	Eye formation
bagpipe	Gut mesoderm	*Bapx1*	Gut mesoderm
byn	Hindgut formation	*Brachyury*	Tail formation
ems	Peripheral olfactory neurons	*Emx1/Emx2*	Peripheral olfactory neurons
otd	Central olfactory neurons	*Otx/Otx2*	Central olfactory neurons
dsx	Sexual development	*Dmrt1*	Testis and sperm development
Gata factor (*pannier*)	Blood cell maturation	*Gata1, Gata2, Gata3*	Erythroid maturation
Gata-ABF (*serpent*)	Gut, hematopoeisis, fat body[b]	*Gata4, Gata5, Gata6*	Gut, liver

[a]Some knockout phenotypes have wider abnormalities than those mentioned here.
[b]The fat body is the *Drosophila* equivalent of the liver.

Drosophila are given in **Table 11.1**, together with the phenotypes of the knockouts. Thus, for example, the very different hearts in the two organisms require the activity of homologous transcription factors, while the identities of the very different segmented structures in both species are determined by Hox coding, albeit that the mechanisms by which this coding is established and is translated into morphology remain unclear.

Combining such molecular data with current morphological similarities in mouse and *Drosophila* gives us a broad picture of Urbilateria. It seems likely that the fertilized egg of this organism produced a bilaterally symmetric triploblastic embryo whose anteroposterior patterning enabled it to develop into a worm-like organism having a gut, a tail, segmented muscles, a nervous system with perhaps a small ganglionic brain that was probably ventral to the gut, an olfactory system, small areas of ectoderm sensitive to light, and a reproductive system. It also had a wide range of differentiated cell types that included nerves, muscles, epithelium, and blood cells. Of particular interest is the likely presence of a heart, as this suggests that the adult worm, unlike, say, a current nematode worm, was large enough to require a circulation.

Many of these features were clearly present in the early deuterostome chordate, *Haikouella*, that lived 420–415 Mya (Figure 4.5), which is thus likely to have evolved fairly soon after the nerve cord had reversed from ventral to dorsal, if that is what happened. It is interesting here that this organism has many more anatomical features than *Pikaia*, a more primitive chordate apparently lacking light-sensitive organs (Figure 4.3A), whose fossils date to some 15 My later than those of *Haikouella*. This difference has always seemed strange and has been generally assumed to reflect the fact the simpler organism evolved first and, although early examples were lost, it found a niche in which its morphology was stable over time. The phylogenetic analysis on which the anatomy of Urbilateria is based suggests an alternative explanation: the anatomy of *Pikaia* reflects a loss of some original tissues.

Urbilateria clearly flourished in its marine environment so that, by the beginning of the Cambrian, its descendants had produced a wide range of species with very different body forms. It is probable that Urbilateria had few, if any, predators and the selection pressures on its early variant descendants were initially very low. The net result was that a host of different species rapidly evolved with the diverse body forms that are now seen in the fossil record (Chapter 4). Perhaps it was only towards the end of the Cambrian that populations were sufficiently large and crowded that selection by the environment rather than by functional failure started to become important, but by then phylum diversity was on its way (de Robertis, 2008).

The obvious question is how such diversity occurred and the simplistic answer is through mutation and variation. This cannot, however, be the full answer because there are very few extant systems where there is a direct Mendelian link between a set of simple mutations and a spread of anatomical variants. The most that can be observed are the alternatives of a normal phenotype, a minor dosage effect in the heterozygote, and a major abnormality in the homozygote mutation, as in the mouse *Pax6* homozygous mutant. The key problem is to explain how mutation, or what is known as genetic tinkering, can give rise to a spread of phenotypic variants. This is the topic of the next chapter.

Key points

- The development of each species reflects its evolutionary history (plesiomorphies) and its own novelties (autapomorphies).
- The area of evo-devo is mainly concerned with identifying the functional roles of protein families that have homologs in very different phyla and represent early plesiomorphies or heritable innovations.
- Classic examples are the transcription factor Pax6, the activity of whose homologs underpins the development of every known eye, and the Hox transcription factors, homologs of which determine anteroposterior patterning in most animals. Other examples include whole signaling networks.
- Animals fall into two classes based on the fate of the first morphogenetic invagination of the blastula: in protostomes (most phyla), it forms the mouth; in deuterostomes (chordates and echinoderms), it forms the anus and the second invagination forms the mouth (although these early invaginations were lost when the amniote embryo evolved).
- Plesiomorphies that link proteins with tissues and that are shared across protostomes and deuterostomes have identified some of the likely features of their last common ancestor, this was probably a primitive invertebrate worm known as Urbilateria that lived at or a little before the beginnings of the Cambrian Period.

Further reading

Bard J (2013) Driving developmental and evolutionary change: a systems biology view. *Prog Biophys Mol Biol* 111:83–91.

Gilbert SF (2014) Developmental Biology, 10th ed. Sinauer Press.

Held LI (2014) How the Snake Lost its Legs: Curious Tales from the Frontier of Evo-Devo. Cambridge University Press.

Shubin N (2007) Your Inner Fish. Random House.

Todd Streelman J (ed.) (2013) Advances in Evolutionary Developmental Biology. John Wiley.

SECTION 3
THE MECHANISMS OF EVOLUTION

This section considers the question of how evolution happens, or, more specifically, what are the mechanisms by which existing species give rise to new ones. The answer is not obvious as the core and defining difference between parent and daughter species is that the latter cannot breed with the former. It was, however, clear by the 1950s that speciation starts with a subpopulation becoming reproductively isolated from its parent population in some novel habitat. Variants within the subpopulation that flourish in their new habitat will leave more offspring (are fitter) and will soon come to dominate that population. If they carry morphological differences from the parent population, they will be recognized as a subspecies, albeit one that can still interbreed with the parents to produce hybrids. Over long periods of time, further mutations in the two populations will render them genetically distinct and, in due course, unable to interbreed. They will then have become separate species.

During the second half of the twentieth century, the major focus was on the genetic basis of how one population gradually becomes distinct from another and this area of work is covered in Chapter 14. Evolutionary population genetics clearly showed that a small subpopulation that separated from a parent population would immediately carry only a fraction of the genetic variance of the parental total because of genetic drift and would thus have a slightly different phenotype distribution. Whether variants within that subpopulation thrived or failed in its new environment depended on whether the blind hand of selection, be it natural, sexual, or kin, favored them. If it did, they would clearly be fitter and so leave more offspring, eventually becoming the predominant and then the only form. At this point a new subspecies would have formed characterized by adaptations suiting it to the new environment. The latter part of the chapter discusses the quantitative genetic theory of how this happens. The key step in speciation is the achievement of total reproductive isolation and this is considered in Chapter 15. It covers the ways in which isolation can occur and in which populations can come to be genetically separate. Both chapters include experimental work that has speeded up elements of these processes and so investigated aspects of speciation directly.

The processes of selection and speciation are both dependent on phenotypic variation within a population. This, of course, originally derives from mutation that gives variant alleles while genetic drift that ensures that such alleles are distributed throughout the population, although the step from mutation to anatomical and physiological variant is generally very complicated. Chapter 12 discusses the ways in which mutant proteins lead to anatomical variants, and it is mainly through their effect on protein networks. Chapter 13 considers the ways in which mutation affects the dynamics of such networks through looking at how such changes may have happened. These and subsequent chapters end with a more theoretical, systems-based discussion of the generally distant relationship between genes defined by DNA sequences and genes defined by their phenotype (Mendelian or trait-defined genes), and how mutation in the former eventually leads to changes in the latter and so generate novel phenotypes that are able to spread across a population.

The section ends with a chapter on human evolution, the taxon about which evolutionary knowledge is the broadest and deepest. It starts with a brief description of the mainly African fossil record back to a last common hominin ancestor shared with the Panini (chimpanzees and bonobos) some 7 Mya. Such fossil data, together with the genetic evidence, have shown that there were several migrations out of Africa, with different populations colonizing various parts of the world at different times but still able to interbreed when they eventually encountered one another. The chapter ends with a brief discussion of the evolution of the mental abilities of humans and of the number of hominid species that there have been.

CHAPTER 12
VARIATION 1: POPULATIONS AND GENES

Because of historical random mutation and genetic drift, any trait within a population includes a distribution of phenotypic variants. Further recombination and mutation in the genes underpinning that trait can occasionally lead to a phenotypic variant outside the normal distribution. Provided that this change is heritable, there is then the possibility of evolutionary modification. This chapter discusses how such changes can occur, with the particular focus being on how genomic mutation leads to novel anatomical features that are beyond normal variation. The first part discusses the ways in which an offspring's phenotype is observed to differ from that of its parents. The second part examines the possible ways in which the phenotype can be changed by mutation and epigenetic mechanisms. The focus of this chapter is on the effect of mutation on individual genes and it introduces the next chapter, which examines the effect on the phenotype of mutations that alter networks.

In 1859, Darwin started the last paragraph of Chapter 5 of *On the Origin of Species* with the phrase: "Whatever the cause may be of each slight difference in the offspring from their parents – and a cause for each must exist ...". One can feel Darwin's frustration at not knowing anything about the origins of variation, which he had realized was basis for all subsequent evolutionary change; it is no coincidence that the ease of selective breeding in domestic pigeons on the basis of identifying variants occupied a major part of his first chapter. Indeed, variation is ubiquitous in the living world, and can be observed in all species at all levels from the genome to the adult anatomy, although it is most obvious in external features such as pigmentation and size (**Figure 12.1**).

The basic explanation of the origin of variation came some 35 years after *On the Origin of Species* with the discovery of mutation by Bateson and de Vries in the context of the rediscovery of Mendel's work on genetics. Little of normal variation across a population is generated by new or even recent mutations in DNA sequences as the likelihood of this happening at a particular site is very low; the estimated rate for humans is $\sim 2.5 \times 10^{-8}$ per base pair per generation. Most variation in sexually reproducing organisms comes from the random shuffling of genes through mitotic recombination during zygote formation and from random mating, although this latter factor may be modified through sexual selection (Chapter 14). This random sampling of genes is known as genetic drift and the downstream effects of the resulting mix of genes comprise the major reason why the offspring can have phenotypes that differ from their parents (Masel, 2011).

The question of how changes in DNA protein-coding and control sequences can lead to a change in an organism's anatomical and physiological phenotypes is not one that, even

Figure 12.1 Pigeon strains. (A) Ice pigeon. (B) Frillback pigeon. (C) English Trumpeter pigeon. (D) Pigmy Pouter pigeon. (E) Oriental Frill pigeon. (F) Capuchin red pigeon.

a century later, has been well answered. It is not hard to see how small phenotypic variants in proteins can be generated by mutation and propagated across a population over time – the theory of population genetics describes how this happens and is discussed in Chapter 14. There are, however, three harder questions: first, how do such minor changes in protein sequences produce substantial anatomical changes; second, how do big changes that are not the obvious sum of small changes happen; and, third, how are such big changes propagated? In the nineteenth century, the prevailing view was that large changes occasionally happened and could be inherited – evolutionary change by such large steps was called saltation.

A more sophisticated approach, put forward by the evolutionary geneticist Richard Goldschmidt in 1940, was that occasional hopeful monsters appeared as a result of mutation that mainly affected events in early embryos. Organisms carrying such major mutations could, in principle, be fertile, and the novel phenotype thus stood a low but finite chance of becoming part of the normal population; they could thus help found new species. This view was ridiculed at the time partly because there was no clear way in which such a phenotype could occur or be propagated across a population, but, as will be discussed in the next chapter, this approach is compatible, in part at least, with modern thinking on the effects of mutation on networks that control patterning.

The prevailing opinion in the 1940s and 1950s was that of the neo-Darwinist Modern Synthesis: this was based on analyzing phenotypic traits that were assumed to be underpinned by one or a very few genes, although there was then no clear idea of what a gene was or how it actually worked. The inheritance of complex traits and their variants could, in principle at least, be worked out analytically using evolutionary population genetics, but the analysis is impractical for more than three or four such genes. The Modern

Synthesis is, in fact, unable to put forward a satisfactory explanation for large-scale change other than as the sum of a series of small changes that derived from mutation. Perhaps curiously, there seems to have been relatively little discussion then about the origins of variation. One reason was that most evolutionary geneticists knew very little about embryology and so did not give adequate weight to the fact that evolutionary change starts as embryological change. One of the very few biologists of the premolecular age to integrate evolutionary change with embryology was C.H. Waddington, and some of his work is discussed in Chapter 14.

Phenotypic variation

Variation within a single species is ubiquitous in the living world. No two humans, other than identical twins, look the same, and even they slowly diverge in appearance as environmental effects and somatic mutations affect the maturing phenotype. Each animal has its own, unique genome and, if one looks carefully enough, its own unique phenotype. There is even variation in bacteria, which can change their phenotype if selection is high enough, as the rapid acquisition of antibiotic immunity shows. This is a particularly interesting area as it involves both hypermutation for its origin and conjugation, the prokaryote mechanism of genetic transfer, for its spread (Culyba et al., 2015).

If one just considers populations of humans and dogs, two species in which strains can easily interbreed, the obvious differences are in size, facial detail, pigmentation, and behavior. The first three of these are relatively straightforward to discuss in the context of development, but the last is not, and for two reasons. First, we have little idea of how behavior is generated by neuronal anatomy and physiology, let alone coded in the genome; second, it is hard to separate out how much of a behavioral phenotype is intrinsic to the organism and how much is the result of environmental influences (the nature/nurture argument). Behavior, for all its importance in evolution, will not be considered here because not enough is known about its genetic basis, although it will be touched on in Chapter 16, which discusses human evolution.

Although there is no such thing as the normal phenotype for any species, it is reasonable to assume that any individual member of a particular species can be viewed as normal if it can easily breed with other species members and if each heritable aspect of its phenotype (or trait) falls within an expected range. Outbred populations, where there is free interbreeding, have a considerable degree of genetic variation. This is much less in strains selected from small numbers on the basis of specific features. In Chapter 16, some of these effects are noted in human founder populations, but limited genetic pools are also seen in inbred dog and mouse strains. Thus, black C57 BL/6 and white BALB/c mice strains have been inbred to have a limited spread of within-group genetic variation. This limited spread defines what is known as the genetic background of the strain and facilitates, in particular, immunological experiments as transplants can be done across strain members without rejection.

It is also reasonable to assume that a healthy offspring will have much the same set of protein types as its parent, with just a proportion - because their amino acid sequences are slightly altered through recombination or mutation - being either slightly less or more effective. Overall, the offspring's phenotype will thus be distinct from that of each parent, allowing for sexual differences, but these modifications will appear as natural variants, in the sense that they will not seem remarkable. Knowledge of the extent of such variation comes partly from observing normal breeding and partly from direct sequencing. The most important model system here is the human mainly because of medical work: the great majority of humans have an anatomical phenotype within the expected range, with about 2% having a diagnosable anatomical disorder and a further 1% having a serious deformity with most of these reflecting major chromosomal abnormalities.

The difficulty with such figures is that, because babies with serious disorders are unlikely to reproduce, it is hard to know whether such abnormalities are heritable or reflect embryological accidents, which sometimes happen, or the effects of somatic mutations. More to the point perhaps, is that few, if any, offspring or even adults can be identified as likely to have a heritable feature that, on its own, seems likely to give that individual a selective advantage. This is the difficulty in using normal organisms to explore evolutionary change: the extent of the genetic and phenotypic variation seen in a normal population of any organism is usually unimpressive and it is far easier to identify phenotypes that seem detrimental rather than advantageous.

Abnormal variation: sports

Although most of an offspring's traits lie within the normal spectrum, parents occasionally produce offspring whose phenotype lies outside the normal spectrum. In breeding circles, these are known as sports and their features are often heritable, particularly those that affect size and pigmentation. An example is the spotted zebra (**Figure 12.2**), whose pattern almost certainly derives from a degenerate version of reaction–diffusion kinetics within a complex protein network (Chapter 13). It is not, however, known whether this phenotype bred true. The selective mating of sports is the basis for breeding pigeon and dog variants (see Figure 12.1). Such breeding for anatomical features is usually more successful than attempts to breed in physiological abilities such as speed: the likelihood of two champion horses producing an equally talented offspring is much less than their owners might hope. This is probably because so many complex traits underpin speed that the chances of an offspring inheriting all the appropriately tuned proteins are low.

Rarer are sports whose phenotypes affect embryonic patterning and the best-known examples are the homeotic mutants in *Drosophila* where one organ is replaced by another from a different segment (Chapter 11). Such homeotic mutants can breed and their inheritance was a source of study in the middle of the last century, before the molecular basis of the phenotypes was known. These occasional pattern variants are important as they indicate that mutations in the genes that code for network constituents can underpin major anatomical change and make their organisms candidate sources for evolutionary change. These hox mutants were, perhaps, the key stimulus that drove research into the molecular basis of development during the 1980s.

Figure 12.2 This spotted *Equus burchelli* zebra is an example of a sport. The pattern of spots represents a reversion from a complex to a simple solution of reaction–diffusion kinetics.

Such outliers are occasionally found in humans, as two examples illustrate. First, the skeleton of Charles Byrne (1761–83), now in the Hunterian Museum in London, is over 2.30 m high, and its extraordinary growth was almost certainly due to a large and probably hyperactive pituitary gland, which would have produced excessive amounts of growth hormone (de Herder, 2012) – the abnormality certainly turned out to have no selective effect as Byrne left no children. Second, there was no reason to suppose that just one of the six children of Hermann and Pauline Einstein, who were each no more than on the clever side of normal, would be the most brilliant physicist of the twentieth century. Here, it is noteworthy that neither of Albert Einstein's sons inherited his intellectual abilities. In short, it has proven impossible to identify heritable genes that are likely to give a selective advantage to a species under normal conditions. The situation is different under experimental conditions where selection pressures can be high (Chapter 14).

Abnormal variation: human disease

Abnormal phenotypes that can be linked to mutation may not be advantageous, but they are important partly because they help understand normal gene function, and partly because they indicate the types of change that can be generated if the boundaries of normal variation are stretched. Humans turn out to be a particularly good model organism in this context: many heritable genetic syndromes have been identified and published in the medical literature and so are accessible to investigation. They are cataloged in the Online Mendelian Inheritance in Man database (OMIM; www.omim.org), and are often underpinned by mutations to a single gene. Many disorders can be understood on the basis of inadequate protein activity: these include potential medical conditions such as a predisposition to breast cancer through mutations in the *BRCA1* or *BRCA2* genes, which are involved in repairing DNA breakages, and the various connective tissue diseases, many of which derive from mutations in the various collagen proteins.

There are also diseases where an aspect of normal anatomical development fails, and an example is craniosynostosis. Normally, the developing skull bones enlarge to accommodate the expanding brain, and this growth is achieved through the proliferation of mesenchyme cells within the sutures that join the separate bones. As the skull grows, there is a balance between proliferation in the mid-suture region and osteogenic differentiation at the edges abutting existing bones so that new bone is then added progressively (**Figure 12.3**). In a heterozygous mouse with a constitutive fibroblast growth factor receptor (FGFR), cell proliferation is blocked, the system is capable only of bone differentiation, and the coronal suture between the parietal and frontal bones closes prematurely (Morriss-Kay & Wilkie, 2005; Martínez-Abadías et al., 2013). Because the brain is still expanding, other sutures are forced to expand more than normal and the net result is a misshapen skull. Premature fusion of the coronal suture in humans can also be caused by mutations in a group of very different gene types that includes receptors of the FGFR family, transcription factors such as Twist1 and MSX2, and signals such as EPHA4 (Holmes, 2012). The involvement of this range of regulatory proteins clearly points to the involvement of several networks being responsible for bone growth and differentiation.

There are also congenital abnormalities that affect patterning and whose phenotype is seen as missing or duplicated tissues. Among human examples is the Holt–Oram syndrome: its characteristics include a missing thumb and aberrant heart septation that result from mutations in the TBX5 transcription factor gene (Al-Qattan & Abou Al-Shaar, 2015). The issue is complicated to analyze because this protein, like many others, seems to have

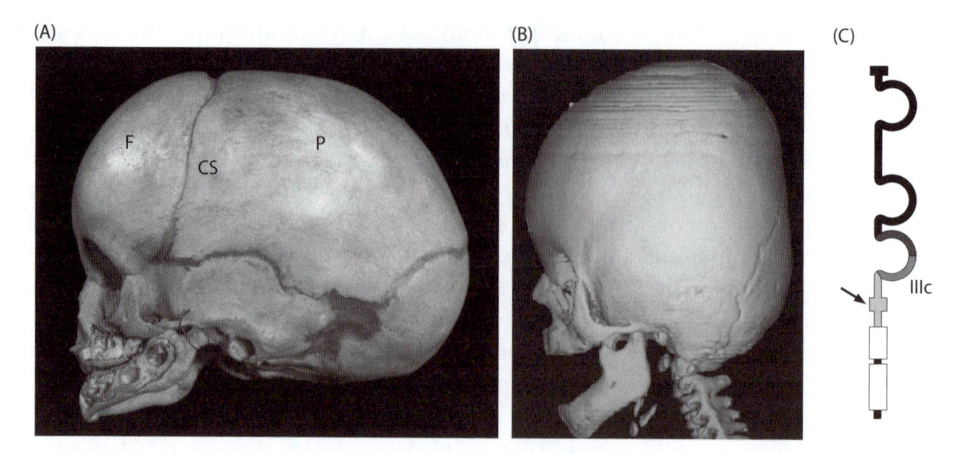

Figure 12.3 (A) The skull of a normal newborn child (CS: coronal suture; F: frontal bone; P: parietal bone). (B) A computed tomography scan of the skull of child with Crouzon craniosynostosis syndrome where proliferation in the coronal suture between the parietal and frontal bones has ceased and the suture has closed prematurely so fusing the adjacent bones. (C) This syndrome is caused by a dominant Cys342Tyr mutation of the extracellular IIIc domain of *FGFR2*, which encodes a fibroblast growth factor receptor (the arrow points to the membrane-binding domain of the receptor).

different roles in different tissues; it is still difficult to analyze events downstream of transcription factors. Other genetic disorders that affect developmental anatomy can often be traced back to genes that code for proteins that are signals, receptors, and kinases (energy-transfer proteins often functioning within networks), as well as transcription factors. A further degree of complexity is seen in heritable disorders that affect behavior; even where a candidate gene can be identified, it is rarely possible to understand how a mutation in the protein can lead to the altered phenotype.

It is, however, important to note that, where a mutant gene has been shown to have a severe impact on some trait, this does not mean that a gene for that trait has been identified, as is sometimes claimed. All that has been identified is a gene whose role in creating that trait is sufficiently important that a mutation in it can have a severe impact on the final phenotype. A car analogy makes this clear: a faulty spark plug will stop a car engine from working, but the plug is not the engine. Thus, a homozygous mutated signal receptor that is unable to bind a signal ligand will fail to activate the downstream process, but the receptor is only the gateway to the process, not the process itself.

The importance of genetic diseases in an evolutionary context is that they show that relatively simple mutations can have two effects. First, they can modulate the function of an individual structural protein: mutations in myosin can lead to muscle disorders (Tajsharghi & Oldfors, 2013); second, they can affect the function of network activators such as receptors, and transcription factors together with the network proteins themselves. The net effect is that such mutations can block or modulate all the changes that normally follow from the activity of such proteins. Such deleterious changes give at least a clue as to what might have been the effect of advantageous mutations in times past, a topic discussed in the next chapter.

Secondary variation

The discussion so far has looked at the genetic origins of variation, but it is important to realize that a novel change to an anatomical features of a developing organism rarely stands alone. If extra digits form as a result of a familial predisposition to polydactyly, for example, the novel bones form in association with the requisite musculature, vascularization, and skin. This is because the inherent plasticity of the developing embryo allows local tissues to adapt to and participate in the formation of the new one. This has been demonstrated by experimentally manipulating the misexpression of a single key gene, as two examples show. First, expression of Pax6 in an abnormal site in the head of *Xenopus* leads to the formation of a small, ectopic eye with all the major tissues expected (Chow et al., 1999). Second, the insertion of FGF-containing beads into the flank of an early chick embryo results in an ectopic limb with all of its normal tissues (Kawakami et al., 2001).

Sometimes, however, the success of an innovation requires secondary features to form that are embryologically independent of that innovation and require completely different anatomical changes to take place that will, for example, affect the physiology of the organism. The example of the giraffe makes the point. The advantage of its long neck is that it allows the animal to forage on trees too high for other ruminants, but this change required major adaptations of the cardiovascular system: the heart has to pump blood some 2 m higher than is required for normal ruminants. The heart has thus co-evolved with neck elongation: while its overall size is much as expected for the size of mammal, its ventricular wall has thickened substantially (Mitchell & Skinner, 2009). The corollary of this is that blood pressure in the limbs is far higher than normal and, to stop edema here, the lower limb arteries have also thickened substantially to withstand the pressure increase (Petersen et al., 2013).

The important point made by this example is that a particular anatomical change that evolved to meet the needs of a new niche often required changes to other anatomical and physiological systems. Such linked evolutionary changes are known as co-evolution. In the case of the giraffe, one can envisage that the extent of initial neck lengthening was limited by the abilities of the cardiovascular system. For the neck to continue extending, the cardiovascular and other systems also had to change, so that there was a slow cooperative, step-by-step change to the animal's phenotype. A similar effect probably occurred in the evolution of long-necked dinosaurs such as *Diplodocus*, but there is, of course, no evidence, as none of the soft tissues have been fossilized. A good example of co-evolution is shown by the set of fossils that mark the series of anatomical features that characterized the successive descendants of an ancient herbivore as they became the clade of sea mammals (Chapter 6; Bajpai et al., 2009).

Genotype variation

Phenotypic variation within some trait is underpinned by variation at every level of bioactivity: there are mutations in the DNA, polymorphisms in proteins (for example, the human blood groups) and hence in networks, and small differences in anatomical structures, such as in the routes that nerves take and in the very minor details of the vascular system, and in physiological function – some people are athletes and others will never be, no matter how hard they train. One immediate difficulty is that, once one gets to these higher levels, it is not always clear whether an aspect of an individual's phenotype derives from its genome, from random events during development (for example, the pattern of blood capillaries or the connections of a nerve) or from the environment. Indeed, in the original work on population genetics by Fisher (1930), the variance parameter, which is a measure of the spread of some trait in a population, was partitioned into an inherited component (due to genes whose composition could not then be specified) and an environmental component, with each being given equivalent weight in the formal theory.

If, in a breeding experiment, however, the phenotype of a particular offspring shows a novel feature, one's immediate response is to suggest that the change reflects the effect of a novel germline mutation, and this may be so. Alternatively, it may reflect the effects of either a somatic mutation or a rare combination of parental gene variants already present within the population. Basic breeding work, if it can be done, will show whether the trait reflects a somatic or germline mutation and, for the latter case, will indicate whether the mutation is recessive, dominant, or sex-linked, and whether it breeds true or whether it only sometimes shows its phenotype (this is imperfect penetrance and reflects a degree of genetic buffering).

Normal variation in the sense of discrete change, as for example the petal colors of flowers, is conceptually straightforward to understand in terms of the activity of a single or a few genes and perhaps their polymorphisms. Thus, the various polymorphisms of trichohyalin, a protein expressed in hair follicles, are associated with whether European hair is straight, waved, or curly (Medland et al., 2009). Continuous changes in a phenotypic character, such as pigment intensity, growth, height, and shape, are much harder to explain on the assumption that they reflect mutations in a single gene sequence. It is much easier to see continuous change as the result of the action of several genes, and the proof that this was compatible with Mendelian genetics was an early breakthrough made by Fisher in 1918 (see Fisher, 1930). Today, we are more likely to view continuous change as a result of the sum of the actions of the mutation-induced protein variants in a network. Jacob (1977) suggested that the appropriate word for such small mutation-induced changes was tinkering.

The effects of mutation on the phenotype

It has been clear since the iconic work of Watson and Crick in the 1950s that mutation in the genome underpins the genetic component of the variation in phenotypes across a population. It is, however, only recently, with the availability of relatively cheap whole-genomic sequencing, that it has become possible to discover the extent of that genomic variation, with most work being done on humans. The publication of the early results of the 1000 Genomes Project (1000 Genomes Project Consortium, 2010) has provided important baseline data on this. If one takes what one can think of as a reference human genome and looks at the likely variants in an individual human's genome, the results have shown that this individual will have, in their 20,000–25,000 protein-coding sequences:

- 10,000–12,000 variants that will not give an amino acid change in a protein
- 10,000–11,000 variants that will lead to an amino acid change in a protein
- About 200 in-frame deletions and about 90 premature stop codons
- 50–100 variants that may well to lead to an inherited disorder.

Fortunately, most of these abnormalities are on one chromosome only and so unlikely to lead to a serious abnormality or a substantial trait outlier, unless the mutation is dominant. What is not yet known is how many further mutations there are in this genome that are likely to affect the noncoding regions, such as the enhancer and repressor sequences that regulate gene expression and are still mainly unidentified. Perhaps the most important implication in the context of evolutionary change is that there is a significant reservoir of mutation across the population, most of which is evolutionally neutral but is available to be amplified by further mutation or by recombination through mating to produce downstream changes in the phenotype (Shapiro, 2011). An example of this will be discussed in Chapter 14, where the work of C.H. Waddington on what he called genetic assimilation is described.

It is also worth noting that the mutation figures given above probably reflect an asymmetry in the contributions from the egg and the sperm. All oocytes are laid down early in ovarian development and, as Haldane noted, mainly include the original mutations in the parents because relatively few cell divisions will have occurred in these germ cells since fertilization. Mature sperm, however, reflect the many mutations that will have occurred in the primary sperm cells (spermatogonia) that keep dividing throughout life. Although there are DNA repair mechanisms within cells, one or two mutations get through each division and these accumulate, progressively reducing sperm quality (Sartorius & Nieschlag, 2010). Similarly, mutations can arise in nongerm cells (somatic mutations) and build up in rapidly dividing cells such as those of the gut, hematopoietic system, and skin, where they can, if the mutations affect the growth and apoptosis pathways, result in cancer.

Although one sometimes thinks of new mutations as the obvious source of variation, it should not be forgotten that most of the novelty in an offspring's genome derives from the fact that each of its chromosomes differs from that of the parent from which it is inherited. This is because each pair of parental chromosomes undergoes crossover and genetic recombination at some random position during meiosis, so that each daughter chromosome includes parts of both parental chromosomes. Each offspring therefore has a gene complement that, of course, reflects that of each parent but is unique. Novel mutations formed during meiosis as opposed to those that are inherited are just another addition to the mix.

The effect of mutation on individual genes

DNA sequences that underpin any phenotype fall into three classes: those that code proteins and are both transcribed and translated; those that are only transcribed such as intron sequences and the various RNA populations (for example, ribosomal RNAs, transfer RNAs, small interfering RNAs (siRNAs) and microRNAs (miRNAs)); and those that are untranscribed, such as, enhancer and repressor sites. Mutations in enhancer and repressor sites can have important effects on gene expression but are particularly hard to identify as they can be a megabase or more distant from the genes whose expression they regulate (Anderson & Hill, 2014). Histones play an important role here as the genomic folding that they mediate can bring distant regulatory sequences into close proximity with the promoter sequences that they affect (Harmston & Lenhard, 2013). Enhancer/repressor analysis, together with that of miRNA and siRNA, which both modulate mRNA translation, is a difficult but important area of current research. As to proteins, while the occasional mutation can introduce a stop codon that truncates them, the more common effect of mutation is to change the protein's shape and/or charge distribution. Each affects its ability to interact with other proteins and with substrates: the former primarily through steric and the latter through binding interactions, which, in the case of enzymes, affect rate constants. This is the effect of Jacob's tinkering or bricolage, the original French word.

Because there are normally two copies of a gene, the phenotypic effect of a mutation on one allele is usually recessive: the normal protein is usually present in sufficient quantities to cover for the mutated one, and the heterozygote shows a weakened phenotype (a dosage effect) or no effect at all. In many such cases, the effect of mutation is thus minor, with a minimal effect on the phenotype; in other cases, however, the mutation leads to an anatomical or physiological disorder, many of which have been discovered and studied in humans. Two classic examples here are cystic fibrosis and sickle cell anemia: the former is due to recessive mutations in the *CFTR* gene, which codes for a cell-surface regulator involved in chloride transport, and the latter to mutations in the β-globin gene which codes for an oxygen-transporting protein. If the homozygote shows a severe phenotype, it is likely to be selected against and the mutation will rapidly be lost; it will only survive if the heterozygote is fertile and it is likely to spread through the population only if the heterozygote phenotype has a selective advantage. The classic example of such a balanced polymorphism is, of course, sickle cell anemia, where the heterozygote affords some protection against malaria (Chapter 14).

Occasionally, a gene mutation is dominant and this is usually because either there is only a single copy of the gene in the genome (for instance, it is on the X or Y chromosome in the male) or the phenotype of the mutant protein negates that of the normal one. An example here is a gene that codes for a receptor that the mutation makes constitutive: such a receptor behaves as if it is always in the ligand-activated state; this results in its associated intracellular network always being active, irrespective of the behavior of the normal receptor. Such mutations have been found in the various FGFRs and result in, for example, dwarfism and craniosynostosis. Although dominant or single-copy genes can, in principle, spread rapidly through a population as the phenotype will be shown by 50% of all offspring of an affected parent, most are deleterious, and their reproductive capabilities are likely to be low. The frequency of many dominant genes that have a deleterious effect on development is that of the background mutation rate. There is certainly little evidence to suggest that this is a normal route for driving evolutionary change.

Sometimes, mutations in individual proteins can lead to large-scale structural abnormalities because the strengths of interprotein bonds are altered. This happens in Ehlers–Danlos syndrome, which is characterized by stretchy skin and hyperextensible joints due to mutations in collagen, which weaken interprotein binding (Malfait & De Paepe, 2014). More common are mutations that effect a protein's interactions with other proteins and with smaller molecules. An example of the former is the binding of ligands to receptors and, here, binding may fail, be too weak or too strong; these changes not only reflect a change in the rate constants associated with the interaction, but also, for example, on whether the downstream response to the signaling is blocked or constitutive. For proteins that interact with small molecules (for example, enzymes and kinases), the kinetics of binding may likewise be increased or decreased.

Perhaps the most interesting class of mutations in an evolutionary context is that which generate duplications, with the diverse *Hox* genes being a classic example (Gehring et al., 2009). Although an original gene and its duplication probably start off as identical sequences, differential mutation over time can lead to paralogs, one of which maintains its original role, with the other becoming available for a different function. The standard way of studying this is to look for gene homologs in species who shared a relatively recent last common ancestor and that are absent in nonmembers of the clade (for review, see Long et al., 2013). There are several mechanisms that lead to duplication (Shapiro, 2011), but more important is the fact that a coding sequence will not be differentially expressed until it acquires the appropriate noncoding regulatory sequences such as tissue-specific promoter and activator sites. The evolution of novel, functional genes is not a simple matter, but is a key evolutionary step (Sassa, 2013).

The role of the environment in generating variation – epigenetic inheritance

It is easy to forget the extent to which the environment affects the phenotype. In humans, pigment intensity may well depend on the degree of exposure to the sun, weight depends on food consumption, and physical endurance depends on exercise, while much of what humans do depends on what they are taught (cultural inheritance). Moreover, and as every gardener knows, the development of plant morphology is highly dependent on the environment (Gilbert & Epel, 2015). A particularly dramatic animal example where the environment plays a deterministic role is in the gender of crocodiles: in mammals, gender is fixed at conception by the presence or absence of a Y chromosome; crocodiles do not have a sex-specific chromosome and their gender is determined by the temperature at which the embryo develops its gonadal system. At 31°C, males and female crocodiles are equally likely to form; below 30°C, females predominate and above 32°C, males predominate (Crews, 2003). It is also known that temperature-dependent sex reversal can take place in more basal adult reptiles such as the Australian bearded dragon (Holley et al., 2015).

An interesting and long-asked question here is whether the changes to an organism that result from environmental stimulation or physiological activity might be inherited by the offspring. Such ideas can be traced back to Lamarck, who was the first serious evolutionary biologist, albeit that he was feeling his way through uncharted territory. Because there were then no other means known for effecting evolutionary change, his views were widely accepted throughout the nineteenth century. Thus, Darwin (1859) explicitly writes about "variability from the indirect and direct action of the external conditions of life, and from use and disuse" because he knew of few alternative sources of variation.

The idea of acquired inheritance was, however, dismissed by Weismann (1889), who discovered the continuity of the germplasm, as he knew of no mechanism by which it could be implemented. The situation is little different today: inheriting an environmentally activated physiological response requires a mode of inheritance that does not depend on random mutation or genetic drift. Such mechanisms are generally referred to as epigenetic, a term originally coined by Waddington (1953) to mean above genetics in the sense of how genes were used but that now usually means regulatory changes to genes that may well be heritable. The most common use of the word is in the context of site-specific CpG methylation in the genome, which is known to regulate gene expression (Varriale, 2014), particularly in vertebrates (Albalat et al., 2012). Such changes happen a great deal over the lifetime of an organism in somatic cells, and are an obvious candidate for heritability with the particular advantage that they may become common faster than mutation effects.

The difficulty in understanding how any novel methylation that might take place in an individual could be passed on to offspring is that, in in early mouse embryos at least, any methylation present in early germ cells is stripped away when they start to differentiate into eggs and sperm (Hajkova et al., 2002). The best-known examples here are the about 100 imprinted genes that are differentially methylated in male and female germ cells; however, this methylation occurs some time after any earlier methylation has been stripped out (Monk, 2015). A further difficulty in assigning a heredity role to methylation is that germ cells are essentially isolated populations that differentiate very early on in development.

It is thus hard to see how any environmentally induced methylation acquired by an adult could be transferred to its eggs or sperm. Iqbal et al. (2015) therefore deduced that the best way to proceed here was to induce methylation in pregnant mice and study the resulting children and grandchildren. They did, indeed, observe novel methylation in the offspring, but found that this effect was lost in the grandchild generation.

Nevertheless, there are many observations of inheritance effects that are hard to explain through standard mutation. A recent example comes from Yehudah et al. (2015). They found a statistically significant difference in the methylation of the *FKBP5* gene between Holocaust survivors and their children: in the former it was high and in the latter it was low (Yehuda et al., 2015). This work did not, however, examine whether the methylation patterns were present in the newborn children or acquired as a result of their development and environment.

An alternative and possibly better candidate for epigenetic inheritance than methylation is one or another classes of the RNA populations known to be present in sperm and eggs (Liebers et al., 2014). These particularly include various noncoding RNA types that are capable of regulating gene expression (Jodar et al., 2016) and hence could play a role in epigenetic inheritance (Yan, 2014). Gilbert and Epel (2015) have reviewed a broad spectrum of phenomena in a wide range of organisms that come under the general rubric of epigenetic inheritance. Some of these can be traced to RNA, and others to methylation combined with chromatin modification, while the basis for the remainder is still unclear. However, such epigenetic inheritance seems not to persist: of the about 30 examples in 13 organisms listed by Gilbert and Epel (2015), only four persisted beyond 10 generations.

Epigenetic inheritance is currently an active area of research and it will be interesting to see whether it provides a convincing mechanism by which environment-induced changes initiated in parents can be maintained in the long term. Indeed, as most of the phenotype derives from events taking place in the embryo and hence insulated from the effects of the environment, one would expect any epigenetic inheritance to be of minor significance. Nevertheless, there are indications that we do not yet have the full story of inheritance, and possible mechanisms of function-directed inheritance are still being explored (Gissis & Jablonka, 2011; Gilbert & Epel, 2015).

The dynamics of variation across a population: the broader view

When discussing a gene, a protein, or a phenotype, it is easiest to consider its change as just a single event occurring against an otherwise static background. This, however, is the wrong perspective: that single mutation exists within the context of constant change across the genome of an individual. Throughout the population, mutation and genetic drift are altering every genomic site in every individual, albeit very, very slowly. If one only considers the protein-coding and protein-regulating genes in this context, the genomes of the whole population can be visualized as an enormous multidimensional object, with each dimension reflecting the probability distribution for all possible sequences of a particular gene across the population. Because mutation never ceases, that distribution varies very slowly over time, pulsing gently in each of its many dimensions. However, as speciation can easily take a million generations, these pulsations are relatively fast on an evolutionary timescale.

Directly above this hypothetical multidimensional dynamic *gene* object is a related *protein* object in which each dimension reflects the probability distribution of properties for each protein as detailed by the possible range of polymorphisms that mutation can generate. The shape of this object changes more slowly; this is because many mutations are silent or do not change the function of the protein. This object is, however, more complex because it now has a space- and time-dependent component: each protein may be expressed in different tissues and at different times as embryos develop and organs begin to function. As the work discussed in the last chapter has emphasized, there is a further multidimensional dynamic object above this protein object that represents the distribution of *protein networks* over time and space in the developing and adult organisms, with there being strong functional links between the two objects. This is, however, a smaller object than the protein one because it is based on groups of proteins that cooperate to generate specific functions.

DNA sequences provide information, while the activity of proteins and protein networks drive development and maintain physiological and biochemical stability. As one moves up the object hierarchy, the closer one becomes to generating the anatomical and physiological phenotype. Above the *network* object is a *trait-generation* object in which networks are grouped in time- and space-dependent ways. Each dimension now represents the local trait options; examples include the range of protein and network options for shaping a human nose, the length of a horn, or the types of pigmentation in a cat hair. The distribution shown in each dimension of the top-level object represents the range of a particular aspect of the actual phenotype in some species subpopulation.

These hypothetical objects of course interact, and the obvious way is through upwards causality: genes define proteins, proteins assemble in networks, and so on. Downwards causality is, however, equally important: the environment can change the phenotype through, for example, food availability and ultraviolet exposure, and changes here affect lower levels of object through feedback to the activities of networks and proteins. During development, time- and space-specific protein signals feed back to the genome to initiate new rounds of protein expression. The degree of complexity is very great and, because so much detail is still not understood, the details are likely to become more complex. Even at the level of a single organism (**Figure 12.4**), the complexity subsumed within these various umbrella terms covers a great deal of that organism's development, physiology, and behavior.

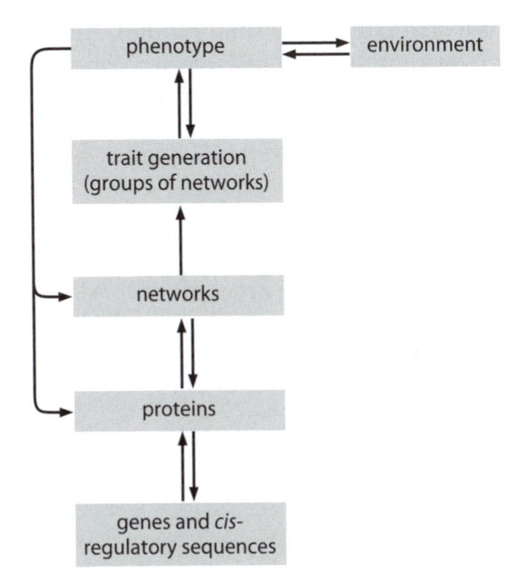

Figure 12.4 Diagram showing the hierarchy of levels and their interactions that link genes to the final phenotype of an organism.

This set of complex objects at successive layers are, of course, meant to be no more than metaphors that broaden one's outlook: they are intended to change the perspective from which change is viewed during development and over evolutionary timescales from one that is gene-centered to one that looks at the complexity of the relationship between the genotype and the phenotype, and how a single mutation can eventually affect a whole population. This viewpoint, which will be developed in the final sections of subsequent chapters, builds on an idea from Fisher (1930), who considered the behavior of the total distribution of genes across a population, and used the idea to analyze their long-term effects in populations.

This chapter has specifically focused on the effect of mutations on individual genes; the next looks more closely at the effects of mutations on protein networks. Changes in the resulting network dynamics can have ramifications on the ways and the speed with which the multidimensional trait object, in particular, can change. The implications of this for the phenotype distribution of the species become particularly important in the subsequent chapter on selection.

Key points

- Phenotypic variation is present in every trait, tissue, and feature across a population. In a particular organism, it mainly derives from a mixture of random mutation in DNA sequences, random chromosome recombination in parental gametes, random breeding, and genetic drift.
- Variation in some traits, such as pigmentation and overall size, can also be affected by the environment, as well as by genes.
- Much of our knowledge of mutation and its effects come from the study of human disease, together with genetic engineering and the use of mutagens in model organisms.
- Organisms whose features, such as pigmentation and size, are outside the expected distribution are often called sports. Those features that are heritable to some extent are the basis for selective breeding of new varieties and strains, and selection may, in the wild, lead to their becoming adopted throughout the population.
- Although it has long been felt that there are mechanisms other than random mutation that will allow the rapid inheritance of adaptations to the environment and changes in behavior, the evidence for the inheritance of such acquired characteristics is still very thin.

Further reading

Gilbert SF (2014) Developmental Biology, 10th ed. Sinauer Press.

Held LI (2014) How the Snake Lost its Legs: Curious Tales from the Frontier of Evo-Devo. Cambridge University Press.

CHAPTER 13
VARIATION 2: CLADES AND NETWORKS

The last chapter considered variation seen in normal populations and the effect that mutations in proteins have on phenotypes. The former included the occasional sport, while the latter were almost all deleterious. Neither gave obvious examples of variants that were likely to be evolutionarily advantageous. This chapter first looks at how mutations whose effect is to alter networks outputs can change the phenotype. It then considers the problem of generating variants from a very different perspective. It asks questions of the form: given a pair of extant species that are known to have evolved from a common ancestor or a single species characterized by an autapomorphy, what variation would have been necessary in that ancestor to achieve the phenotypes seen today? A series of examples is given, with many focusing on limb morphology because change here is easy to recognize. All imply that phenotypic novelties can be linked back to heritable changes in the output of complex protein networks, particularly those for signaling, patterning, and growth. Such phenotypic changes are, of course, readily heritable because they derive from mutations in one or more of the many proteins that participate in such networks. Furthermore, because networks have a single output that can be affected by mutations in many of these proteins, heritable changes in network outputs can occur far more rapidly than changes that depend on mutations to a single protein.

The question of how changes in DNA protein-coding and control sequences can lead to downstream changes in the anatomical phenotype is not one that, more than a century after the discovery of mutation, has been well answered. This is because, as the last chapter demonstrated, single proteins do not themselves create structures and rarely even structural features, although they may activate or play an important role in their development or physiological function. Indeed, the changes that result from mutations in a single protein are not particularly impressive, mainly being deleterious and hence unlikely to give any selective advantage. It is now, however, becoming clear that generating structure and function in the phenotype depends on the activity of complex protein networks, and the effect of mutation in them is to change their outputs, which, in turn, modifies the normal phenotypes. It is from this viewpoint that variation is discussed in this chapter.

After a section discussing the effect of mutation on networks, the approach taken is explicitly evolutionary, attempting to answer questions of the general form: if a pair of organisms differs in some feature that derived from a common ancestor, what would have been the nature of the variation that achieved these different phenotypes? Indeed, if an extant organism differs from the common ancestor in some very specific way, a similar question can be asked. Although this is not a standard way of approaching variation, it has some qualitative similarities to the far more quantitative approach of phylogenetics (Chapter 8), which attempts to identify the most likely common ancestor sequence that could have generated contemporary ones, and the way in which this happened.

This chapter includes a set of examples that test the validity of this approach. Of the many that could have been chosen, the majority are from the vertebrate limb for two reasons; first, because limbs have well-defined bones and associated tissues, change is easy to identify, and, second, because so much molecular and experimental embryological work has been done on limb development, many of the events underpinning change are now known (Gilbert, 2014). The examples, as a whole, clearly suggest that many limb and other variants can be linked back to heritable changes in the output of the complex networks that operate within embryos, particularly those that affect patterning, timing, and growth.

The effect of mutation on protein systems and networks

Process networks are normally activated by signaling networks. Such signaling may be uniform over a region, giving the same result in each responsive cell, or it may vary in concentration so giving rise to patterning. Either way, signaling activates transcription factors and the resulting changes in gene expression during development have the net effect of mobilizing protein networks whose outputs mediate differentiation, morphogenesis, and proliferation. This section considers the effect of mutation on these networks and their activation.

Signals, receptors, and transcription factors

Signaling starts with a signal binding to the external part of a receptor located in the membrane of some tissue. Such ligand binding changes the intracellular domain of the receptor, so activating a signal transduction pathway that, in turn, activates an expression complex already located on the genome by a transcription factor. The resulting gene expression can drive a wide variety of events that particularly include activating networks. Such signaling pathways are repeatedly used and the effects of mutation on them are unrelated to the details of the network that they activate. It is, however, highly dependent on the transcription factor activated and the lineage of the cell in which it is expressed: some transcription factor can have different roles in different locations and at different times, as the example of the *HoxA* and *HoxD* genes has shown (Chapter 11).

Provided that both copies of a particular protein involved in this process are rendered nonfunctional by the mutations, the most common effect will be that the expected downstream network(s) will not be activated, and an abnormal anatomical phenotype will result. In most heterozygotes, however, either the normal protein compensates for the deficiency in the mutated protein and the organism is normal, or there is a dosage effect, where the number of mutated genes determines the phenotype, which is known as semidominant. Thus, homozygotes in the mouse fibroblast growth factor (FGF)18 signal have major bone abnormalities deriving from the failure of this occasionally used ligand to bind to its receptor, while the FGF18$^{-/+}$ heterozygote seems normal (Liu et al., 2007). Similar effects can derive from mutations to a receptor: mice heterozygous for the R-II receptor of the bone morphogenetic protein (BMP) signal are normal, but the homozygotes die at gastrulation (Beppu et al., 2000). Particularly important here are mutations to transcription factors: in the case of Pax6, for example, all homologs of the Pax6$^{-/-}$ homozygote lack eyes; in contrast, the mouse Pax6$^{-/+}$ mouse has small, but normal, eyes.

Heterozygotes can also, of course, show an abnormal phenotype, as two examples show. The first is when the effect of mutation is to initiate ectopic signaling, as in the case of the mutation in the FGF receptor, FGFR2, whose effect is to maintain the activity of the differentiation network at the expense of proliferation and leads to craniosynostosis (Morriss-Kay & Wilkie, 2005; Chapter 12). The second, and perhaps more important in an evolutionary context, occurs when a signaling gradient that underpins spatial patterning is affected. An example occurs in the mouse doublefoot mutant, which is semidominant and can have up to nine digits in the homozygote (**Figure 13.1**). This phenotype results from the ectopic expression of Indian hedgehog, a signal that leads to an increase in breadth of the paw and a subsequent modification to the normal sonic hedgehog patterning gradient (Yang et al., 1998; Babbs et al., 2008). This normally initiates five digits in mammals and three or four in birds, and is regulated by distant enhancer sequences (Anderson & Hill, 2014), although the details of how this happens are not known (de Bakker et al., 2013). One importance of the doublefoot mutant is that, because the extra digit modules have normal joints, nails, and histology (Malik, 2013), it emphasizes that part of the role of one network is to help activate further downstream ones. Another is that it provides an evolutionary clue to the original range of digits in early land tetrapods (see Table 6.2) and perhaps how the basic number came to be five (Chapter 6).

The significance of these experimental results in an evolutionary context is clear. The effect of many mutations that affect signaling pathways and associated transcription factors will be to block the initiation of network activity, or perhaps to maintain the activity of a network that should be turned off. If such a change carries some selective advantage in a particular environment, both copies of the gene will probably be required to carry the mutation, and this means that, over evolutionary timescales, the mutant proteins will become the new wild-type proteins. The effects of mutations that alter spatial patterning are equally intriguing because they can lead to qualitative change. Such patterns can be particularly complex because it is now becoming clear that they can be affected by mutations in cis-acting regulatory elements (Gaunt & Paul, 2012).

The effect of mutation on the output of networks

Mutational tinkering with the proteins that function within networks is different from their effect on the proteins involved in network initiation. In practice, every protein in the network, and there may be as many as a hundred, is subject to such tinkering and there will be a considerable distribution of protein phenotypes for a particular network across a population. There is, however, no reason to doubt that the very great majority of

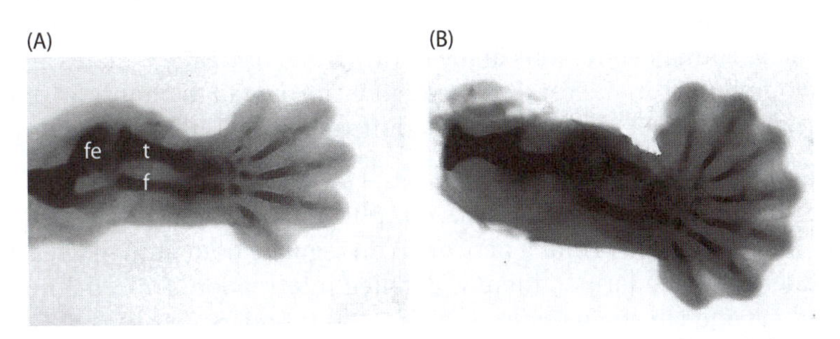

Figure 13.1 Cartilage preparations of the hindlimbs of (A) a 14.5-day embryo control and (B) a heterozygote doublefoot mouse mutant (fe: femur; t: tibia; f: fibula; scale bar: 200 μm).

these networks will produce either the same output or one whose quantitative value is within the fairly narrow limits that define normal variation (such as the extent of the growth in the various features of a human face). In other words, the properties of the network as a whole are stable to even broad amounts of mutation – the network is buffered (Nochomovitz & Li, 2006).

Networks can, however, be amplifiers of variation (Wilkins, 2007), and it is clear that, as minor mutational changes accumulate within a population, it becomes harder for the internal kinetics of the system to buffer their increasing effects. Eventually, a mutated protein, making what might seem an innocuous change to the kinetics of the network, will have a major effect on its output. Should this happen, there are several possible outcomes: first, the network might fail and this would probably be lethal; second, the network output can be altered to give a phenotype that was outside the normal range; and, third, an alternative pathway could be activated within a complex network that would generate a qualitatively abnormal phenotype, perhaps a homeotic effect. The latter two options would appear as sports and, in principle, generate potentially heritable variants in the phenotype.

The best direct evidence on the effects of mutation on networks comes from experimental embryology and the study of human diseases. Mutations in the proteins of standard signaling pathways, particularly in heterozygotes, underpin many disorders, both congenital and those such as cancers, which mainly derive from somatic mutation in the pathways that regulate proliferation. Well-known protein families here are those for ras, raf, and mitogen-activated protein kinase, which are involved in signaling pathways that particularly regulate proliferation: ras mutations are associated with several developmental disorders, particularly neurofibromatosis type 1, whose phenotype includes several abnormalities, with perhaps the most serious being abnormal neural crest cell differentiation, which leads to neuronal tumors (Schubbert et al., 2007).

Further downstream at the level of process networks, it is important to distinguish between those whose output is an on–off state, such as the signaling, differentiation, and apoptosis networks, and those where the response is graded, such those for timing, proliferation, and morphogenetic activity. The differentiation switch seems to be particularly sensitive to major mutation: this is because any mutation that blocked or changed a differentiation network other than those affecting pigmentation would probably be lethal. An example of this is the runx2 transcription factor involved in bone ossification. In the absence of this transcription factor ossification fails and the embryo dies (Takarada et al., 2013). Of interest here is that a minor mutation in the human *RUNX2* gene or its regulation has been suggested as one of the reasons why the morphology of bones such as the clavicle differ between anatomically modern humans and Neanderthals (Green et al., 2010). In this context, apoptosis can be viewed as a state of differentiation, albeit that it is a terminal one for the cells affected. Somatic mutations in the network that normally initiates apoptosis can change development, as in the case of interdigit webs or lead to abnormal development in embryos (Baehrecke, 2000), or cancers in adults, where cells that should be programed to die for one or another reason do not.

Mutations in the timing, proliferation, and morphogenetic networks can lead to qualitative changes in their outputs. Timing networks can regulate both physiological and developmental systems. Of the former, there is detailed information about the networks that run the heartbeat and circadian rhythms (Noble, 2011; Anderson et al., 2013), but knowledge of the details of developmental clocks is limited, with that underpinning somite formation being best known (Riedel-Kruse et al., 2007). These are the types of mutation that result in heterochrony, where the timings of events, normally during development, are altered, but the molecular details of how such changes are made remain unknown.

Perhaps the most common of all evolutionary changes, however, are those that affect growth, and these can derive from mutations in both the signaling and proliferation pathways. Here, there is direct evidence that the rates of proliferation can be under local quantitative control. This is under both normal and abnormal conditions; thus, Hornbruch and Wolpert (1970) found that mitotic rates within the developing chick limb can vary over a factor of five, while the abnormal growth rates seen in cancers can be due to mutations within the proliferation networks. It thus seems possible that the proliferation networks at least are context-tunable, as well as being under the direct control of signaling networks.

As to the networks that control morphogenesis, it is hard to identify mutants where there are changes to such properties as movement, adhesion, folding, and tube formation. In this context, spina bifida, a disorder in which the neural tube fails to close or closes late, shows strong statistical heritability (Jorde et al., 1983). The molecular basis of this heritable change to morphogenesis is not fully understood, but mutations in planar-cell polarity genes, which are involved in the control of cell movement within epithelial sheets, are associated with failure of neural tube closure in some mouse strains (Tissir & Goffinet, 2013).

Examples of anatomical change due to variation in network outputs

The mutations that alter network outputs are important because they indicate that major anatomical change is not as hard to achieve as one might suppose on the basis of the altered phenotype of proteins acting alone. Nevertheless, it is hard to pin down the origins of past or potential change from just looking at variation within current populations because substantial changes to the normal phenotype are so rare. It is easier to analyze the differences between organisms that share a common ancestor. In principle, this could be done on the basis of genomic differences, but this is impractical, partly because those genes that contribute to the phenotypes are not known and partly because, even if they were, it is not yet possible to interpret how genomic and protein properties generate the details of the anatomical phenotype. Even if the DNA of some ancient ancestral organism could be sequenced, the task would still be impractical now because neither the spectrum of proteins needed nor their specific roles are yet known.

Hence the approach taken here, which is to compare anatomical features in extant species with those in an ancestor and to consider the nature of the variation needed for change to have occurred. What follows is a selection of the many examples that are known (for more, see Held, 2014); they have been chosen mainly because they are clear cut. In addition, some have the advantage that they can be studied experimentally. The interesting and perhaps unexpected conclusion is that all reflect phenotypic changes that reflect the output of networks.

Organism size

The most obvious difference between related species is size. The dwarf gecko and the Solomon Islands skink are two lizards that are, apart from pigmentation differences, very similar in morphology (**Figure 13.2**). They do, however, have one important difference in their phenotypes: the adult gecko is about 16 mm long, while the adult skink is over 800 mm long

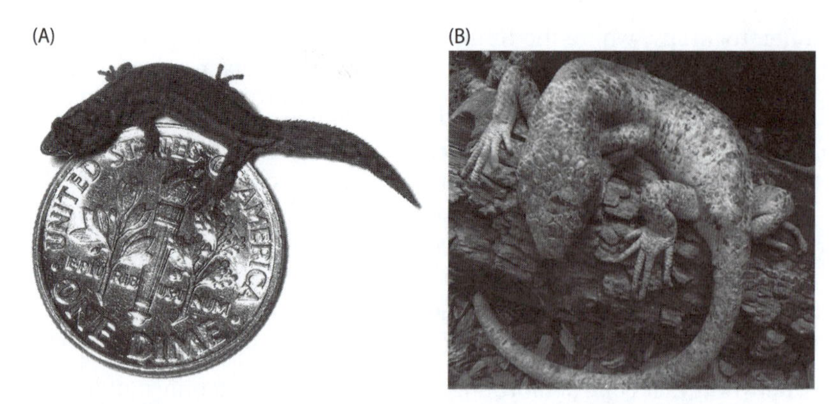

Figure 13.2 ***Sphaerodactylus ariasae*****, the dwarf gecko (A) is 16 mm long, while the Solomon Island skink (B) is 830 mm long.** The only obvious differences are in pigmentation and size.

(Fleming et al., 2009; Hagen et al., 2012), a difference that reflects 5–6 extra cell divisions across the organism. While it is quite possible that species-specific rates of growth may account for some of the size difference, the more important cause for the difference is clearly that the gecko stops growing relatively soon after it has developed, while the skink clearly keeps growing for a long time after development has ceased. One can envisage a last common ancestor whose timing networks controlled when the body-wide growth networks were turned off and that had a rich and varied set of pigment-forming cells. Variation in timing and patterning would be enough over time to lead to the very different sizes of the two reptiles.

Such a change in the time at which a developmental event occurs in a specific organism is known as heterochrony and its experimental study dates back to Rowntree et al. (1935). Important ways in which timing differences modify development and so generate morphological change include predisplacement and postdisplacement, where a developmental event starts sooner or later than expected, and acceleration, where the rate of development is increased. A well-known example here is the rapid growth and development of the forelimbs of marsupial embryos that allows them to migrate to their mother's pouch after birth (below and Chapter 7).

A further type of heterochrony occurs when adults retain juvenile characteristics, a phenomenon known as pedomorphosis. Perhaps the best-known example is the axolotl. Most amphibians develop as tadpoles and at some point in development, their thyroid is activated, thyroxin is secreted, and the larva undergoes metamorphosis. Axolotls have lost this ability and can only undergo metamorphosis if their thyroid is externally stimulated through, for example, ingesting iodine. In its absence, they grow to become what are essentially enormous (up to 300 mm long) larvae. Heterochrony is an important way of producing variants.

Limb differences

An area of work that has been particularly productive in the understanding of variation has been the vertebrate limb. This is partly because abnormal phenotypes are so easy to recognize, partly because the developing limb is accessible to experimental investigation, and partly because a great deal is known about the underlying molecular genetics, as the earlier discussion on the doublefoot mutant and Hox patterning has shown. What follows is a set of species comparisons in which the nature of the variation needed to achieve the differences between them can be examined.

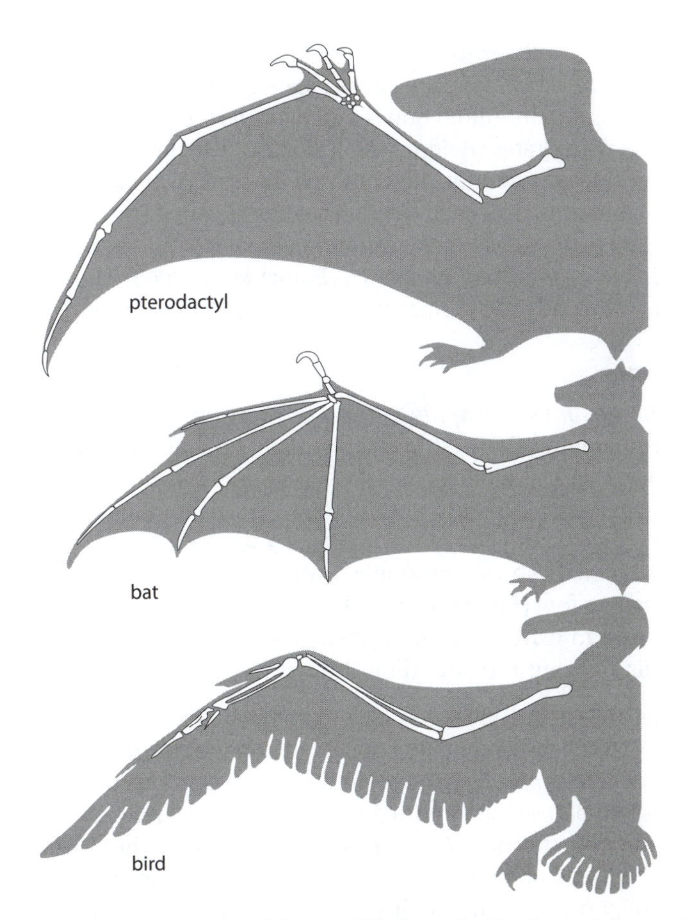

Figure 13.3 The skeletons of a pterodactyl, bat, and bird matched in size and demonstrating how each, in its own way, extends its digits to enable a wing to form.

Growth changes

Digit length: Wings are a characteristic of birds, bats, and pterosaurs, but their digit and phalange proportions are different. In pterosaurs, only the fifth digit extends; in bats, the four fingers extend; in birds, digits 2 and 3 are lengthened (**Figure 13.3**). It is clear that the patterning networks that control the local proliferation networks for phalange growth are operating differentially in both species- and digit-specific ways.

Timing changes

Marsupial forelimbs: In most tetrapods, the hindlimb buds start to form very soon after forelimb bud initiation; development of both is at the same rate and completed by the time of birth so enabling many newborn animals to walk. Marsupials such as wallabies and kangaroos are different: they are usually born in an early fetal state, having the hindlimbs expected for their age but exceptionally well-developed forelimbs with claws that they use to migrate from the exit of the vagina to the pouch (Chapter 7). There is thus a heterochronic difference in the development of marsupial forelimbs, which develop much faster than their hindlimbs (Keyte & Smith, 2010; Chew et al., 2014). It is clear that this change in the forelimb timing networks reflects both a major increase in the activity of the growth networks and premature differentiation in the forelimb tissues.

Hindlimb diminution in cetaceans: Whales and dolphins lack rear limbs and have only internalized femurs and a rudimentary pelvis (see Figure 7.7). At 24 days of development, however, the early dolphin hindlimb bud is much like its forelimb bud and typical of those see in other limbed vertebrates (Bejder & Hall, 2002). At 48 days however, when digits are forming on the developing forelimb, the hindlimb bud has virtually disappeared; development has stopped and growth of such tissues that have formed (the hipbone and femur) is severely diminished so that only their rudiments are left in the adult. Given that whales evolved from normal four-limbed ungulates (Bajpai et al., 2009), the variant that underpinned limb loss was clearly the premature cessation of its growth and development.

Patterning changes

Changes to the lower limbs of leaping tetrapods: Most limbed vertebrates have in their hindlimbs a proximal femur bone, and a more distal tibia–fibula bone pair in their forelimb that articulates with the talus (or astragalus) and calcaneus, which comprise the ankle and heel. The exceptions are amphibians (except for salamanders), the tarsiers (monkeys found only in South-East Asia), and galagos (or bushbabies, small African nocturnal monkeys) which all show a homoplasy: the talus and calcaneus have extended to give a much-elongated foot (**Figure 13.4**; Held, 2014). This extended link in the lower part of their legs gives them extra leverage in their lift-off, with galagos, in particular, being renowned for their jumping and bounding abilities (Aerts, 1998).

The origins of this apparent duplication have only been studied in the *Xenopus* toad and the mechanism by which the second pair of extended bones form derives from a variant in the way that Hox coding normally patterns limbs. In most embryos that have been looked at, the Hox expression patterns are approximately the same in fore- and hindlimbs. Blanco et al. (1998) compared the *hoxa11* expression domains in the developing fore- and hindlimbs of *Xenopus* and noted three differences. The gene is expressed for a longer time, over a larger region, and closer to the distal tip in the hindlimb than in the forelimb. The view of the authors is that the extra time of expression of *hoxa11* in the region that would normally become the ankle is the major reason why the bones that would normally become the calcaneus and talus acquire morphologies similar to those the tibia and fibula. This anatomical homoplasy is thus mainly due to a change in the timing networks that regulate limb patterning.

Webs: While land-living birds have distinct toes, birds that swim often have webbed feet. This latter state actually reflects the way in which autopods (for example, hands and feet)

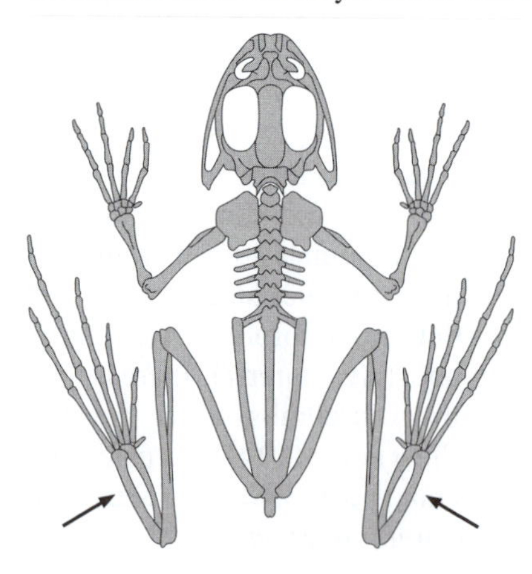

Figure 13.4 A drawing of a frog skeleton showing (arrows) the apparently duplicated tibia–fibula pair that have evolved from two of the ankle bones.

form from a foot- or handplate: the digits develop within mesenchyme, with both enclosed in an ectodermal covering. Webless birds require an additional step for their development: the epithelium and mesenchyme between the still-forming digits of the hindlimb rudiment are lost through activation of the apoptosis process network. It is now known how this happens: in the webless chick, FGF2 signaling in the foot leads to the expression of Msx2, a transcription factor, together with BMP signaling in the interdigital region. These jointly activate apoptosis in the mesenchyme and the ectoderm between the digits (Gañan et al., 1998), which is then lost. In the webbed duck, FGF2 signaling does not take place, Msx2 is not expressed, and apoptosis does not occur (even though BMP is present). All early land vertebrates had distinct digits so apoptosis activation was part of normal limb development. It is clear that the variation required for webbed feet to form was merely the turning off of the FGF signaling network.

Number changes

Variation in digit number: Although most vertebrates have five digits comprising a claw and various small bones, birds have three (forelimb) or four (hindlimb). Experimentation has shown that digit number is primarily controlled by a gradient of the sonic hedgehog signaling protein set up by a small region in the early hand- or footplate known as the zone of polarizing activity. This gradient activates the formation of digit modules (complete digits with nails and joints) in ways only partially understood, although limb bud width is certainly one component (Towers et al., 2008). This is shown by the large number of digits seen in the doublefoot mutant whose plate breadth has been increased by ectopic Indian hedgehog signaling (Figure 13.1, Yang et al., 1998; Babbs et al., 2008).

Ectopic digits: There is another way of forming an extra digit, and this is through the increased growth of a sesamoid bone in the normal autopod. Such bones are usually small and do not link to other bones through joints but are formed within tendons. There are many such sesamoid bones in the vertebrate skeleton and several in the limb, with the best-known being the patella (knee cap), which incidentally evolved separately in birds and mammals - dinosaurs lacked them. Ectopic digits have independently evolved from sesamoid bones several times in specific members of very different families.

Extra sixth digits have separately evolved in pandas who use them to shred bamboo leaves (**Figure 13.5**), in moles to improve digging, and in elephants, where they presumably help spread weight. Although these ectopic digits have evolved in only one taxon within three very different clades, they share three common features. First, they occur in both the fore- and hindlimb autopods, irrespective of whether they have a function in the latter region,

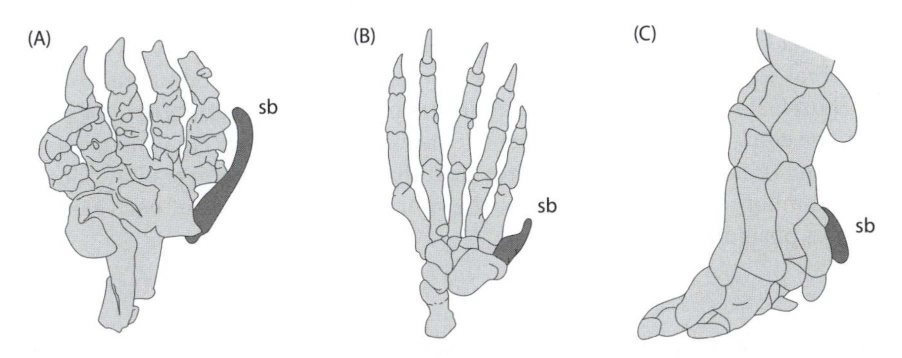

Figure 13.5 Ectopic digits in (A) a mole, (B) a panda, and (C) an elephant forelimb (sb: sesamoid bone).

although the latter digits may be smaller than the former. Second, they all derive from enhanced growth that extends the existing sesamoid bone, which is adjacent to the first digit, and so are not composed of jointed phalanges. Third, they each lack nails or claws – none is a standard digit module, although its external appearance is superficially normal apart from the claw (Mitgutsch et al., 2012; Davis, 1964; Hutchinson et al., 2011). In each case, the ectopic bone clearly derives from the maintenance of the proliferation network that enlarges the early sesamoid bone. This, in turn, may derive from a change in the networks controlling when and where proliferation occurs.

Four further examples of evolutionary change

Stickleback spines

An example where the basic molecular biology and the timescale over which variation has occurred is now understood comes from anatomical changes in the stickleback fish. Marine sticklebacks have pronounced pelvic spines for protection; freshwater sticklebacks lack these spines, reflecting a variant occasionally present in marine sticklebacks (Cresko et al., 2004). This variant seems to have become a defining feature of freshwater sticklebacks several times, some as recently as 12 Kya. It is not, however, known whether this loss is always associated with genetic speciation because it seems that marine males are unwilling to mate with smaller female freshwater sticklebacks, perhaps because of the size difference. Molecular analysis has shown that pelvic spine formation results from local Pitx1 activity; the wide expression of this transcription factor in the fish is controlled by four enhancer regions. In the freshwater sticklebacks, the enhancer region for the pelvic region has been lost and the gene is not expressed there (Chan et al., 2010; Schluter & Conte, 2009). As a result, the downstream module of signaling and process networks responsible for spine production is not activated.

The number of neck vertebrae

The number of cervical vertebrae in extant tetrapods is highly variable: amphibians have very few, while diapsids, including birds, crocodiles, and dinosaurs, have numbers ranging from about 10 to 40. Synapsids, particularly mammals, have, with very few exceptions, only seven cervical vertebrae; in this respect, there is no difference between mice, giraffes, and whales. The last common ancestor of the synapsids and diapsids was probably a mid-Carboniferous (~315 Mya) sauropsid reptile, and examples such as *Hylonomus* had six or seven cervical vertebrae. It thus seems that an early postseparation event was to fix this number in the former and to make it far easier to modify in the latter.

The molecular basis for regulating this number is unclear, but Burke et al. (1995) showed that vertebral specification in both groups is controlled by Hox coding: for the cervical region, *Hoxa4*, *Hoxb4*, *Hoxc4*, and *Hoxc5* in the mouse (seven vertebrae), together with *Hoxd4* in the chick (14 vertebrae), all need to be expressed. The way in which this number is determined remains unknown, but it is clear that the downstream result is the production of a number of vertebral modules, the morphological details of each being determined by its position.

Beak size

The Galapagos finches are the group of bird species whose individual adaptations to particular habitats were a key stimulus for Darwin's thinking about evolution. The birds fall into two groups: the ground finches have broad, strong beaks for tearing at cactus roots and eating insect larvae, while the cactus finches have narrow beaks that enable them to punch holes in cactus leaves and eat the pulp. It is now known that that beak morphology is controlled by two factors (Abzhanov et al., 2006): BMP4 signaling and calmodulin expression, a protein that regulates calcium levels within cells. The strength of BMP4 signaling controls beak breadth and depth, while calmodulin upregulation facilitates beak elongation. It is clear that variation in upstream patterning networks determines the amount of BMP4 and calmodulin expression in the anterior head and these, in turn, control the growth pathways responsible for beak size (Mallarino et al., 2011).

Vertebrate skin patterns

There are obvious patterning differences in the hair pigmentation in strains of cat, zebra, and giraffe, and in the scale pigmentation of fishes. It has long been suspected that reaction–diffusion kinetics underlies these diverse patterns (Bard, 1977, 1981). Such kinetics were invented by the great mathematician, Alan Turing (1952), who, in his only foray into biology, asked the following question: can a group of identical cells with identical biochemical networks linked to allow the diffusion of small molecules (that Turing called morphogens) produce a nonuniform concentration distribution? The intuitive answer is "no" because diffusion would iron out any spatial differences in concentration that might arise, but Turing showed that this intuition was only right if the kinetics describe the behavior of a single morphogen. If, however, there are two morphogens whose synthesis depends on one another as a result of a network of chemical activity, then autocatalytic effects within the network can create spatial patterning because the system becomes unstable under a limited range of parameter values (this is an early example of chaos theory).

Having first proved this by working through the formal mathematics of the kinetics and diffusion operating over a domain of cells, Turing then devised a relatively simple quantitative example of what are now known as reaction–diffusion kinetics for two morphogens operating in a ring of cells, and showed that the solutions were two out-of-phase concentration patterns of peaks and troughs round the ring. Although this seems an odd example, it is mathematically tractable, and Turing had to solve the complex equations by hand because he did not have access to a computer. It is then only a small step to say that these kinetics represent a signaling network and that the peaks and troughs of morphogen concentration represent switch and nonswitch regions for directing pigmentation production.

This paper has excited continuing interest since its publication and it has now been shown that Turing kinetics, operating over two-dimensional cellular arrays, can generate a wide repertoire of vertebrate hair patterns, such as those in zebras, as well those seen in mollusk shells and fish scales (Bard, 1981; Meinhardt, 1984; Barrio et al., 1999). The basic pattern generated by these kinetics is of spots, but varying the kinetics and the boundaries of the patterning domain can change wavelengths and generate stripes that can even bifurcate. One interesting feature of the kinetics is that they depend on an instability in the chemistry of the interactions, and the intrinsic randomness of this instability means that the precise detail of the patterns cannot be predicted. This tallies with the observation that no two zebras have exactly the same striping pattern, and even the two sides of a single animal are different.

Of particular interest is the effect of heterochrony on the patterns that Turing kinetics would generate if they were to be used during embryogenesis. If the kinetics have a fixed wavelength, then the earlier they form in the embryo, the fewer will be the patterning elements that will fit onto the embryo and the larger they will end up because there will be more embryonic time for them to expand; the later they form, the more will be the number of elements and the smaller they will end up because there will be less time for growth after they are laid down.

Turing kinetics thus provide a useful context for examining the origin of the patterning in the three different species of zebra: *Equus burchelli* has about 26 stripes and a complicated pattern, *Equus zebra*, which has about 40 mediolateral stripes, and *Equus grevyi*, which has about 70 such stripes (**Figure 13.6**). Morphological and size details of the early development of horse embryos were investigated by Ewart (1897), and it is straightforward to show that, if 400 μm mediolateral stripes are spread across equine embryos of 21, 25, and 35 days, subsequent growth will generate the body striping patterns seen in the adults of the three different species (**Figure 13.7; Bard, 1977**). Striping variation in zebras is thus likely to derive from heterochronic changes in a basic patterning network.

For all of the theoretical significance of there being a class of potential networks that are able to explain complex two-dimensional biological patterns, there is still no unambiguous experimental evidence that Turing kinetics actually operates in biological systems. Of interest here is some recent theoretical work based on experimental analysis of the system that patterns pigmentation of melanophores and iridophores in zebrafish skin. Bullara and De Decker (2015) have shown that while the mechanism appears to be formally equivalent to a Turing system in its behavior, it is actually slightly different in its underlying assumptions, particularly in the necessary role of growth, from those required to generate Turing kinetics.

In summary, this section has shown that a very wide variety of the types of change as species diverge over evolutionary are most likely to have derived from mutations that change the outputs of the various signaling, patterning, and process networks, the activities of which are responsible for normal tissue formation in embryos. Although little is known about the details of these networks and how they operate, three further points seem clear. First, heritable changes in networks can lead to large-scale anatomical change initiated in the embryo but manifested in the adult. Goldschmidt's (1940) suggestion that major changes were initiated by mutations that produced the large-scale phenotypic changes, albeit short of being monsters, does not seem quite as fanciful now as it did then. Second, changes in networks are due to mutations in their constituent proteins or to genetic drift, and are thus as heritable as any other mutation. Third, because there are so many proteins in networks, similar phenotypes will probably result

Figure 13.6 (A) *Equus burchelli* with about 26 stripes; (B) *Equus zebra* with about 50 stripes; and (C) *Equus grevyi* with about 75 stripes. Each has its own detailed pattern.

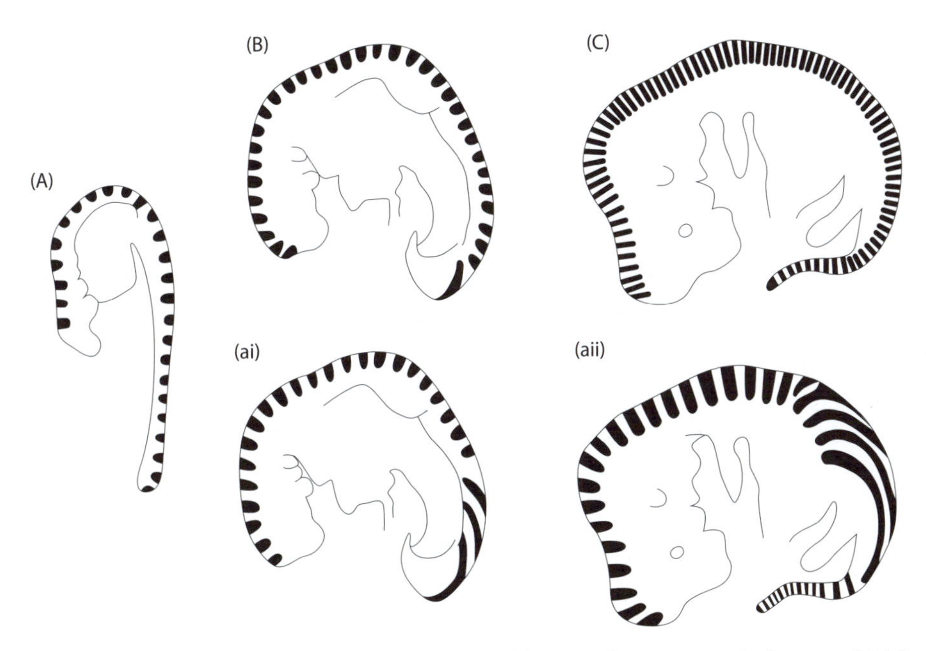

Figure 13.7 (A–C) Three, three-and-a-half-, and five-week horse embryos, respectively, onto which have been drawn stripes with a 200 μm spacing. (ai, aii) Three-and-a-half- and five-week embryos, respectively, showing how stripes laid down at 3 weeks would be expected to look as the embryo grows.

from mutations in more than one protein; heritable change in networks can thus occur more rapidly than in individual proteins.

Developmental constraints on variation

There is a joke exam question from the 1970s, a time when nothing was known of the molecular basis of embryogenesis, which built on the doggerel of Lewis Carroll in *Alice Through the Looking Glass*:

"The time has come," the Walrus said,
"To talk of many things:
Of shoes and ships and sealing-wax
Of cabbages and kings
And why the sea is boiling hot
And whether pigs have wings."

Q: Recode the pig's DNA so that it does have wings.

This joke implicitly poses a further, more serious question: are there limits to the extent to which mutation can produce novel phenotypes? This question is not one that can be easily answered, but it is immediately clear that the possibilities for phenotypic variation are intrinsically limited by various types of constraint. There are physical constraints: is it possible to produce a wing structure capable of generating enough energy to lift the weight of the pig? There are developmental constraints, as variation can only build on what is already part of the phenotype: is it possible to duplicate the basic patterning of the region of the pig's future anterior limb system so that, like a fly's imaginal disc, the lateral mesenchyme can produce both a superior wing like that of a bird and the original inferior limb? Then there are evolutionary constraints: what might be the selective use of a small wing that was physically inadequate to lift the animal?

The most obvious constraints are provided by the laws of physics. Sea invertebrates such as arthropods, and particularly mollusks, can grow almost indefinitely because they are supported by water; squid can thus grow to a length of over 10 m and weigh many kilograms. Nonmarine invertebrates are, however, limited in size by the strength of their chitin exoskeleton and by the respiratory efficiency of tracheal tubes. The largest nonmarine invertebrates were probably the giant *Meganeura* dragonfly-like insects of the Carboniferous (~300 Mya) whose wingspan was up to 70 cm and that weighed several hundred grams, but they lived at a time when there was probably more oxygen in the air than now (Chapter 4). Among extant invertebrates, goliath beetles are ~10 cm long with their larvae weighing ~100 g, while the Chaco golden knee spider has a span of 26 cm and weigh about 150 g. The size limits on land vertebrates are much less because of the strength of their bone skeleton: *Supersaurus*, a Jurassic sauropd, was about 30 m long and weighed 35–40 tons.

The local need for oxygen imposes its own limitations: the largest unvascularized tissues in vertebrates are the lens and cornea; both are sustained through oxygen diffusing from the adjacent ciliary body and this is only likely to be possible because both are essentially passive structures with minimal energy requirements. It is probably for this reason that wheels have not evolved in multicellular organisms: their energy requirements are too great to be met through diffusion, even when aided by a protein pump. The only known examples of a rapid, freely rotating system are the flagella of bacteria that are sustained by hydrogen or sodium ion gradients operating over nanometer distances that drive the necessary adenosine triphosphate production (Atsumi et al., 1992; Oster & Wang, 2003).

There is also the obvious chemical constraint in that it is hard for a novel polymerized structure to evolve that is not only useful, but can also be physiologically manipulated – the chances of both events occurring at the same time in an environment that gives the polymer a selective advantage are very low. However, an example of where this did happen was the evolution of silk by early insects, and it is now a synapomorphy for the spider clade. This example of evolution occurred early in life when the potential for change was far less limited than now, and this raises the most serious contemporary constraint on change, that imposed by evolutionary, particularly developmental, history.

Of particular interest are those constraints imposed on change by the anatomy of the extant organism. Ridley (2004) discusses the interesting example of the route taken by the recurrent laryngeal nerve (RLN), a spur off the vagus nerve that emerges from the early vertebrate hindbrain. In fishes, the RLN is a nerve that innervates a gill slit. In mammals, it takes the very inefficient route of running down the neck, looping round the aorta, and ascending back up to the larynx; in the case of long-necked mammals such as the camel and giraffe, this detour is several meters. This route only makes sense because it is known that the evolution of the gill system involved major changes to the associated blood vessels and, in particular, the posterior locationing of the aorta and its associated nerves. Unlike stem cells, embryonic tissues cannot discard their evolutionary history and novelties have to build on those tissues present when they form.

Within these constraints, there is still considerable freedom for change in early embryos as demonstrated by many of the abnormal anatomical forms documented in OMIM (Online Mendelian Inheritance in Man) that are generated by mutations affecting the activities of patterning and growth networks. Perhaps the major point made in this chapter, and exemplified by the example of the amphibian limb with its remodeled calcaneus and talus (see Figure 13.4), is that it is still possible for anatomical alterations to be generated by activating and repressing network activity and by modulating their activity, and this is likely to happen through relatively modest mutational changes.

Finally, there are the niche constraints. The biosphere is now so occupied that it is hard to envisage a niche that is unoccupied and in which an organism with substantial structural or physiological innovation would flourish. Our flying pig would have to compete with eagles for food, and it does not take an expert to see which would survive that battle! Indeed, the major bursts of anatomical innovation have always come after the major extinctions when whole arrays of niches became unoccupied, and perhaps, given enough time, a flying vertebrate would evolve that had a porcine ancestor, but it would probably change to look more birdlike, as a hippopotamus ancestor that took to the sea became more fishlike.

Taken together, these constraints make it clear that that there are areas of what Maynard Smith et al. (1985) called morphospace that remain unoccupied, and for good reasons. Then, it might have been surprising that so much of this space was unoccupied, but this was because so little was known about molecular genetics and the ways in which tissues were made. The work of the last 20 years, particularly on protein networks and signaling, has now clarified the principles that underlie tissue generation during development, while the work on genetic engineering has shown how major changes can be generated through sequence manipulation that affects network outputs, particularly those that affect patterning and growth. In the limits, however, the potential for anatomical change comes up against the buffers of the laws of physics and the constraints of selection, and the latter are discussed in the next chapter.

The mutational basis of trait change – the broader view

All of the examples in this chapter are compatible with the idea that change derives from mutations in the proteins involved in the various signaling, timing, and process networks, particularly those for growth and patterning. Because there are so many cooperating proteins in a single network, the effects on the phenotype of a mutation in just one of them is thus indirect and so hard to predict. In most cases, however, the abnormality is buffered and the result is within normal variation, but, because the internal properties of complex networks can amplify change (Wilkins, 2007), the mutation will occasionally lead to substantial change. Perhaps the most interesting examples here are those that result from changes to patterning and that can result, for example, in ectopic digits and changes to bone morphology. It is thus not hard to see how random mutation can create the sort of substantive innovations that are typical of interspecies differences.

This conclusion has two important corollaries: first, because the mutations for this are essentially minor in their local effects within the networks, they can be inherited and maintained in the offspring in the normal way; second, because of the complexity of networks and the narrow range of possible outputs, mutations in several genes can lead to the same or a similar variant, as the example of craniosynostosis has shown (Chapter 12). The net result is that the spread of a mutant trait throughout the population can be quite fast if the change in phenotype has a strong selective value. Evidence to support this view comes from the rapid adaptive radiation events in cichlid fish (Danley et al., 2012; Chapter 14), which only started some 12 Kya, and in freshwater sticklebacks (see above), which date back to about 14 Kya (Cresko et al., 2004).

At the end of the last chapter, the dynamic state of the phenotype of a species was represented by a multidimensional object with each dimension containing the population distribution of some phenotypic trait. This top-level object was underpinned by a succession of similar multidimensional objects successively representing groups of networks, networks, proteins, and DNA sequences (see Figure 12.4). At each level, there were influences from below and feedback from above so that causality was complicated. The outer envelope of each multidimensional object as a whole clearly represents the limits that this level can achieve and thus reflects very low probability events.

This chapter adds a further feature to this picture. It shows that, for reasons associated with mutation and genetic drift, the proteins associated with a network can yield an output that is beyond the limits of what can be buffered. In turn, this change affects the properties of groups of proteins and hence some aspect of the final phenotype. In each of these three multidimensional objects, the change leads to a bulge in the envelope of the surface and, in the top-level object, represents a novel, heritable phenotype that is available for selection. In the next chapter, which discusses selection, this multisystems view of the levels of evolutionary change will be used in the context of adaptation and in understanding the theory of population genetics.

Key points

- Most anatomical and physiological characteristics of an organism are underpinned by the activity of systems of protein networks rather than by the activity of a few individual proteins.
- Variation in the phenotype thus results from the summation of the effects of the variation of the individual proteins that participate in these networks. Variation here primarily affects the individual rate constants of protein reactions.
- Variation in networks is buffered so that the output of the network is generally held within the bounds that are seen as normal. Very occasionally, the cumulative effects of protein variation can result in the network output leading to an abnormal phenotype.
- Also very occasionally, major changes in the output of a patterning network may arise that result in an ectopic feature or other change. Because these derive from simple mutations that initially affect single proteins, such novelties are also heritable and, if selectively favored, are a likely source of evolutionary novelties.
- Here it is likely that such a change in phenotype can be independently achieved through mutations in several genes within a network. Heritable change of the resulting phenotype can hence be achieved far more rapidly in a network of many proteins than through a change that depends on mutation to a single gene.

Further reading

Bard J (2013) Driving developmental and evolutionary change: a systems biology view. *Prog Biophys Mol Biol* 11:83–91.

Gilbert SF (2014) Developmental Biology, 10th ed. Sinauer Press.

Held LI (2014) How the Snake Lost its Legs: Curious Tales from the Frontier of Evo-Devo. Cambridge University Press.

Wilkins AS (2007) The Evolution of Developmental Pathways. Sinauer Press.

CHAPTER 14
ADAPTATION, FITNESS, AND SELECTION

The last chapter discussed the mechanisms by which variation in the phenotype is achieved and how every aspect of that phenotype can vary across a population. If a particular organism has a novel feature, it will eventually be lost unless that feature increases fitness in the sense that an organism with that feature will have a greater likelihood of producing fertile offspring in the environment that it inhabits than organisms lacking it. If it does, that feature will be seen as an adaptation that may eventually become a characteristic feature of the population, and perhaps one that defines a new subspecies. This chapter starts by considering the sorts of functional adaptations that are likely to increase fitness and goes on to consider the ways in which such adaptations are selected. These include natural selection (the environment), sexual selection (mate preferences), kin selection (or altruism), genetic drift, and selective breeding, which can lead to quite rapid changes in phenotype. The latter part of the chapter briefly introduces the theory of the modern evolutionary synthesis. This integrates population genetics with evolutionary pressures to examine how trait frequencies in a population can change under these various forms of selection in the context of the reproductive fitness of each resulting phenotype.

The picture of a species given in the last chapter is of a population of similar organisms that that shows a spectrum of phenotypes, with very occasional examples being outliers in some way. The various facets of the overall phenotype of a particular individual in that population mainly derive from that individual's set of germline gene variants, many of which affect protein networks and their outputs. Selection could eventually enable a new distinct population to emerge with some outlying phenotypic characteristics and so lead to the formation of a new subspecies. This chapter considers how this can happen, and so prepares the ground for the next, which discusses the ways in which that subspecies might become a full species.

The first part of this chapter looks at the various ways in which selection can act on variants within a population and so effect adaptation. Selection under normal conditions in the wild is usually, however, far too slow for the process of change be studied there and this part ends with a look at how strong selection under laboratory conditions can lead to rapid and stable changes in phenotype The second part of this chapter discusses the quantitative effect of selection in the context of the theory of evolutionary population genetics (the mathematical formulation of the modern evolutionary synthesis (MES)) that was initiated in the 1920s and 1930s by such pioneers as J.B.S. Haldane, Theodosius Dobzhansky, Ronald Fisher, and Sewall Wright. It is, however, important to note that evolutionary population genetics is predominantly based on phenotypes and anything that the theory says about DNA-defined genes is usually inferred indirectly. Originally, gene alleles were assumed to produce the various phenotypic options directly and were based on Mendelian ideas of genetics and thus were essentially theoretical constructs rather than DNA sequences.

Adaptation

Selection, which acts on variation, results in the appearance of new and stable features in a population that are seen as adaptations, a word that reflects the success of those new features in particular environments. More subtle changes to the phenotype that cannot be linked in any obvious way to pressures from the environment (for example, the human liver has four lobes, whereas the mouse liver has five) are just viewed as minor species differences, perhaps deriving from genetic drift. Adaptation is, however, an umbrella term that covers any stable novelty; examples at different levels include a minor change in a bacterium's biochemistry that allows it to survive on a new food source, the various heritable respiratory adaptions that allow humans to live at high altitude (Beall, 2013), the replacement in dipteran flies of the pair of rear wings seen in early winged insects with haltere-balancing organs that increase flight control (Fox et al., 2010), and the recent changes in pigmentation in *Biston betularia*, the peppered moth, to survive when its tree habitat changed due to carbon deposition (**Figure 14.1**A, B).

One has only to look at any book on animals and plants to see the enormous range of anatomical and behavioral traits that have evolved to ensure that a particular strain or species becomes successful in a particular environment. It can even seem that the environment and the phenotype of the species have shaped each other. Such perfect adaptation of an organism to its environment was, of course, one of the crucial argument for creationism (Appendix 4). The difficulty in countering this argument is that variation in a particular population is unexceptional, selection is generally very weak, and the acquisition of an adaptation so slow that only rarely can it be observed, hence the importance of examples like the peppered moth and the experimental models discussed later, which demonstrate heritable change over shorter times and that are discussed later in the chapter.

It cannot, however, be assumed that an anatomical feature seen today was selected on the basis of its current function. There is now no doubt, for example, that birds evolved from theropod dinosaurs (cladistically, they are still dinosaurs), but it is becoming clear that feathers did not evolve in the context of flight but for insulation and the maintenance of body temperature (Prum & Brush, 2002). They certainly helped birds that had more

Figure 14.1 (A) *Biston betularia*, the normal peppered moth; (B) the black variant; (C) A flatfish camouflaged to look like sand.

extensive feathering than dinosaurs to survive the period of severe cold that followed the meteorite catastrophe responsible for the extinction of all other dinosaur taxa and for ending the Cretaceous Period. Once feathers had evolved, they were able to vary in shape and size and so be subject to selection for their properties in other contexts such as gliding and then flying, as well as for sexual display, as in the peacock. A character that is selected for its advantages in one context and turns out to have another role in a future and different context is known as an exaptation.

Increasing fitness

The range of adaptations that has evolved over half a billion years defines the glorious breadth of species across the living world. Each population with a novel adaptation survived when it first appeared if that adaptation had increased fitness, and so was able to increase the number of successful offspring in a particular environment compared with the parent species. This can be achieved in various ways.

Lowering the chance of death

Many adaptions can be seen as changing the balance between predator and prey: the former may become bigger, stronger, and more vicious, while the latter may become even faster and more nimble, and perhaps develops group protection. The well-known evolutionary trend for taxa to become larger over time is known as Cope's rule and occurs in both vertebrates and invertebrates (Chown & Gaston, 2010). Another important class of adaptations is camouflage, where an organism produces an external morphology and color pattern that mimics its background, rendering it almost invisible to potential predators and so more likely to live longer and breed (**Figure 14.1**C).

The most famous example of camouflage evolution, because the dominant variant changed so rapidly, is the pigmentation of *B. betularia*, the white- and black-bodied peppered moth (see Figure 14.1A, B). Originally, the predominant form of this moth in the north of Britain had a highly mottled, but light wing with this pigmentation pattern camouflaging it when it rested on lichen-rich barks. In the nineteenth century, however, as the industrial revolution proceeded and smoke filled the air of the British midlands, tree lichens died and carbon was deposited on bark. As this happened, what had been an occasional black-winged melanic morph of the moth became predominant and the frequency of the mottled variety dropped. Now, with improved environmental standards leading to lowered pollution levels, the mottled wing variety is becoming more common again, although the detailed reasons for this reversion may be complex (Cook & Saccheri, 2013). The selection pressure driving change is the extent to which the wing pigmentation camouflages the moth against predators. Moths mainly reproduce annually so, over a period of perhaps 150 years and generations, the predominant morph has flipped from variegated to black and back again. This is unusually fast for natural selection, and reflects a strong selection pressure.

Another way of avoiding death is to take the opposite strategy of mimicry: instead of becoming invisible, the organism (the mimic) evolves to look like another organism (the model) that is well worth avoiding, perhaps because it is poisonous (Mokkonen & Lindstedt, 2015). Mimicry through deception is known as Batesian mimicry after Henry Bates, a nineteenth-century English naturalist; there are many examples, some of which exist in the absence of obvious current models (Pfennig & Mullen, 2010). A well-known case is the

Figure 14.2 Batesian and Müllerian mimicry. (A) The nonpoisonous *Dismorphia* butterfly mimics (B) the poisonous Ithomiines butterfly. An example of Müllerian mimicry – both (C) the monarch butterfly and (D) the viceroy butterfly are poisonous.

nonpoisonous *Dismorphia* butterfly that mimics the poisonous Ithomiines butterfly (**Figure 14.2**A, B). There is also a further form of mimicry in which two poisonous organisms have evolved to mimic one another; this is Müllerian mimicry, named after the nineteenth-century German naturalist Fritz Müller, and it gives both species a selective advantage if they have a common predator (**Figure 14.2**C, D).

Increasing offspring numbers

For a species to increase in number over the longer term, each organism in a population needs, on average, to leave more than one fertile offspring; this requires producing enough fertile offspring to survive predation. In sexually reproducing species, males always generate large numbers of spermatozoa; females have, in principle at least, the option of producing any number of female gametes from a single ovum upwards. At one extreme, the female may produce one or very few eggs – this K strategy is common in land vertebrates where there is high parental investment of time, care, and protection of the young hatchlings. At the other extreme, a female may produce many eggs, very few of which would result in adults that reproduced themselves – this r strategy is common in, for example, bony fish and requires little or no parental involvement in the next generation; thus, a mature female cod can produce as many as several million eggs in a single spawning (Secor, 2015). Plants generally follow the r strategy, as demonstrated by the thousands of olives seen on a single tree. The choice of where on the K/r spectrum an organism comes depends mainly on its evolutionary history. The K strategy is associated with a high survival rate for its offspring and the r strategy a very low one; the latter does, however, have the advantage that a small population finding itself in a novel and beneficial environment can increase its numbers rapidly and so capitalise on the associated increase in variation.

An interesting aspect of adaptation is how males and females choose mates (sexual selection, below). Choice is not, however, an option for plants where pollen from the stamens of one individual have to land on the stigma of the pistils of the same or another plant from which they then move to the ovary. Pollen is carried by wind, water, and animals. For the first two of these, the female plant can do no more than have a morphology that facilitates the capture of pollen. Plants for which pollen is carried by insects have evolved strategies to ensure that that a particular insect or group of insects is attracted both to a particular male to pick up pollen and then to the appropriate female plant to deposit it. Examples of successful strategies are odor and color.

These events have been particularly studied in *Ophrys apifera*, an orchid whose flowers look very much like bees. Although these orchids can self-pollinate, their beautifully colored and shaped flowers also attract male bees that try to copulate with the flowers. While this pseudocopulation is being attempted, the male bee's head comes into contact with the stamens and pollen grains stick to its hairs. When that bee next tries to copulate with another such orchid, the pollen gets transferred to its stigma and fertilization is achieved. There are many other such adaptations in plants, and there is as yet no evidence on how such morphologies evolved; indeed, it may be that they arose through fortuitous exaptation (Vereecken et al., 2012).

Mutualism

Another important way of improving fitness is symbiosis, where two organisms come to depend on one another. Well-known pairs of organisms that have evolved to support one another include legumes and the nitrogen-fixing bacteria that live in nodules in their roots, all vertebrates and their gut bacterial fauna, hummingbirds that obtain nectar from bird-pollinated flowers that have, in turn, adapted their flower morphology to fit the bird's bill and so ensure pollination, and *Acacia* trees that provide nourishment to acacia ants, which, in turn, protect the trees from other insects.

Eventually, two species can come to depend on one another to the extent that they require nutritional and other support from one another both for breeding, and for environmental protection - this is symbiosis. The classic example is lichen, where algae or cyanobacteria that provide food and a fungus that provides physical protection and water have become so integrated as symbionts that the resulting structure appears to be a single organism. The net result is a combined life form that can survive in many inhospitable environments. Although such mutual dependencies have long been known, they have a particular resonance in the history of evolution because Pyotr Kropotkin (1902) argued for the importance of mutual aid among plants and animals as the basis for evolutionary change. He held this view, which contrasted to the then prevailing one of evolutionary success depending on the basis of the fight for survival, partly on socialist and anarchist grounds but mainly on the basis of his geographical and other field research in Siberia during the 1860s.

Selection

The corollary of fitness is selectability: the net result of a heritable variant in some trait that increases fitness will be, by definition, to increase the likelihood that organisms carrying that trait will produce more offspring than those lacking it. If a new heritable trait gives a 1% selective advantage, it can be shown using the theory of population genetics and reasonable assumptions of genetic heritability, selection pressures, and population size that it will only take about 1000 generations for the trait to have spread through the great majority of the population (Nilsson & Pelger, 1994). On the scale of evolutionary time, this period is almost instantaneous, but, for practical purposes, this is a very long time indeed: even for *Drosophila*, which has a breeding cycle of about 2 weeks, 1000 generations represents some 40 years.

The implication of this slow rate of uptake of a trait is that it is almost impossible outside of laboratory conditions, where population size and strength of selection can be controlled, to follow evolutionary change. This means that change has to be inferred on the basis of studying morphological and behavioral variants in their different environments and observing the detailed behavior of individual organisms. Very large amounts of such work have been done and it is now known that there are different types of selection.

Natural selection

Natural selection is the umbrella term that covers the influences from the environment in the widest sense that affect an organism's ability to reproduce, and is sometimes called ecological selection to distinguish it from sexual or kin selection. Natural selection occurs when events both outside and within a group of individuals enable one variant in a population to leave more reproductive offspring than another. Note that, if one simply used the word selection, it might carry the implication that the process was active, as in selective breeding. Natural selection may occur because an animal variant has a better chance of surviving against predators or of catching prey, or can survive better in a harsh climate, or can access a novel form of food. A plant variant may be more tolerant of the climate, root better in the soil, or have a leaf display that allows it to access more light and so improves its photosynthetic ability. In other cases, an organism may be immune to a particular disease or produce zygotes that are more robust (for example, seeds may become tolerant of particular soils). The list of possibilities is endless and the situation is made more complex because it is the whole organism rather than a single trait that has the variant and that is subject to selection: a mix of minor traits may prove to be more beneficial in a particular environment than a single, apparently more important one.

For such reasons, the exact explanations for one variant rather than another being adopted can be hard to prove; the reasons are generally inferred on the basis of comparisons among extant and extinct organisms. It thus seems obvious, for example, that because almost all vertebrates have eyes, animals without them, such as moles, descended from animals who lost their eyes when they went underground; there was clearly far less selection pressure on the ancestors of moles that first began to lead subterranean lives to maintain vision than there was for them to improve their touch and digging ability, hence the extra sesamoid digits that they now have (Chapter 13), and perhaps the increase in one sense at the expense of another. A further example, and one that worried people in the past because it was evidence against the goodness of God, is the behavior of ichneumon wasps: these have a modified sting organ that acts as an ovipositor for depositing a fertilized egg in or beside another insect larva. The egg develops and the hatchling feeds on the unfortunate larva.

One feature that is an obvious target of selection is coat coloration, as the butterfly examples mentioned above have shown (see Figure 14.2). Similar arguments apply to vertebrates; thus, polar bears have a much lighter pelt than do brown bears, presumably because the former offers camouflage against ice, whereas the latter are less visible in woods. Such factors, of course, vary from environment to environment, and the differences between them may be major, or may simply reflect the difference between daytime and dusk or night in a particular locale. As an example, it is worth noting that the black and white stripes on a zebra, so obvious during the day, break up their outline at dusk, rendering them almost invisible at a time when they drink and so are particularly vulnerable to predators - this is an example of disruptive camouflage, albeit one that is hard to prove experimentally.

One can often identify plausible reasons why an adaptation increased the fitness of a species in the context of a likely past selection, but it usually impossible to test the ideas. An improvement on this "because it would have worked" argument is to pose the question of evolutionary stability: if the current adaptation was much weaker, would there be any selection pressure for the intensity of the trait to increase now? Sometimes, the only explanation seems to be one based on genetic drift: it is not easy to explain the different selection pressures that encouraged different species of zebra to display different stripe numbers and morphologies in different parts of Africa (see Figure 13.6), or the reasons why the right and left lungs of humans should have three and two lobes, while those of mice should have four and one lobes, respectively. Nevertheless, there seems no reason to doubt that it is the blind power of natural selection that normally decides whether one or, indeed, any of the available variants will actually take over as the norm in a specific subpopulation in a particular environment.

Sexual selection

An important aspect of selection is ensuring that organisms breed successfully: if a variant cannot pass on its genes, the variation will be lost. Considerable effort has therefore gone into understanding sexual selection and the evolution of traits that encourage individuals in a population to increase their fitness by out-producing one another. Darwin was the first person to explore these ideas, and they are an important part of his second major book on evolution, *The Descent of Man, and Selection in Relation to Sex* (1871). The reader interested in more depth than can be given here is referred to Ridley (2004) and Barton et al. (2007), while a fuller treatment is provided by Hardy (2002).

The advantages of sexual reproduction

Because sexual dimorphism is almost ubiquitous in the multicellular world, it is easy to think of it as inevitable, but the situation is not as simple as it seems. There are many animal organisms that are hermaphrodites; these can produce both male and female gametes, and are frequently able to self-fertilize. Similarly, many angiosperms have both male and female flowers on the same plant. There are also many animal organisms that reproduce parthenogenetically (asexually); in this case, however, their offspring form from unfertilized eggs and so are normally female – drone bees and aphids being exceptions. Many phyla have species that always or sometimes reproduce parthenogenetically; common examples include nematodes, scorpions, aphids, and some bees and wasps. Honey bees are interesting here: only the queen bee lays eggs and those that are not fertilized become males (or drones) and these are thus produced parthenogenetically. The only role of drones is to mate with queens, and those fertilized eggs that are female become workers and, occasionally, queens. There are also many more hermaphrodite species than is usually thought: Jarne and Auld (2006) estimated that perhaps 30% of noninsect animals are hermaphroditic, with a small proportion being able to self-fertilize. Although a few fish such as the clown fish, sea bass, and groupers are hermaphroditic, most hermaphrodite animals are invertebrates and particularly include snails, worms, and echinoderms. As is better known, many plants have both types of sex organs, and many of these can self-fertilize.

These examples illustrate the fact that, although sexual reproduction is an option seen in all phyla, it is not obligatory. Given that sexual reproduction is intrinsically inefficient, this is not surprising: for a sexually reproducing organism to maintain its population, females not only have to find or to be found by a mate, but the female also has to produce at least two offspring. For asexual reproduction, each organism merely has to leave a single offspring, without needing to find a mate; more than one, on average, and the population expands. Females in sexually reproducing species need to leave more than two fertile offspring, on average, over the long term, for the population to increase. If all other things are equal, populations that produce asexually will usually expand more rapidly than those that produce sexually.

What, then, is the advantage of sexual reproduction? The obvious answer is that the genetic complements of offspring produced sexually, based as they are on recombination and dual parental contributions, will show far greater variability than those deriving from asexual reproduction or examples where egg and sperm come from the same adult, although recombination during meiosis will provide some variation here. As the previous chapters have made clear, this genetic variability is the key both to buffering against environmental change and to generating the phenotypic variability that confers evolutionary adaptability. The evidence that most animals, and certainly all the larger ones, generally reproduce sexually confirms that the advantages of sexual reproduction outweigh its cost in inefficiency and in the demands that it makes on the phenotype.

The sexual phenotype

Sexual reproduction requires traits that encourage sexual attraction and mating, and a considerable amount of adaptation and selection has underpinned the evolution of secondary sexual characteristics (that is, those only indirectly related to reproduction) that attract the opposite sex. The most dramatic examples are those where there is male–male competition, and this can sometimes lead to one male treating all the females in a group as his personal harem, a situation that encourages the evolution of alpha males. There are many, many examples where males compete for sexual dominance in a group, usually on the basis of strength, and this is one of the drivers of male characteristics. Good examples here are antler size in male deer and physical strength in elephants. Both forms of sexual selection lead to sexual dimorphism. It is clear that sexual selection favors phenotypic properties that enhance an individual's sexual attraction and activity and so increase the frequencies of that individual's gene variants within the next population.

In other cases, the male animal chases and the female chooses; to make themselves eye-catching here, males have evolved display characters that make them attractive to females. Two well-known examples of such sexually dimorphic traits are the lion's mane and the peacock's feather train. The latter is particularly interesting as its advantage in attracting females and so improving the fitness of an individual, in the sense of its leaving more offspring, overcomes its obvious disadvantage in rendering the bird less mobile against predators. Rather less common across the animal world are cases where it is the females who display phenotypic characters that make them attractive to visually undistinguished males. In some cases, both sexes develop secondary sexual characteristics, and an example here is *Homo sapiens*: males develop beards and prominent musculature under the hormonal influence of testosterone while women, unlike monkeys, develop prominent breasts under the hormonal influence of estrogen. Human mating behavior has thus evolved on the basis of mutual attraction, although, in practice, there also are strong and varied cultural constraints across society that regulate sexual behavior.

It is important to note that the possession of secondary sexual characteristics is not always enough to ensure an interest in copulation. Indeed, mating is not the usual primary interest of either sex but requires physiological stimulation. Both sexes can secrete behavior-changing odors known as pheromones, and some of these act as sexual attractants. While female secretion of pheromones to attract males seems more common, the reverse also happens in taxa as diverse as wasps, lizards, and birds. There are also internal biochemical events such as monthly or annual hormonal production, whose effect is to increase sexual desire. Examples here are the monthly change in the color of female genitalia in some simians that attract males, and the roughly annual season of musth in male elephants when their high testosterone levels enhance their interest in mating.

The sex ratio

In almost every species where sexual reproduction takes place, there is a roughly 1:1 ratio of males to females and this primarily derives from the fact that sperm have an equal chance of carrying a male or female sex chromosome. Humans are interesting here because the implications of the Y chromosome being slightly smaller and lighter than the X chromosome have been investigated. This difference enables Y-chromosome sperm to swim slightly faster than X-chromosome sperm and this alters the fertilization balance slightly in favor of males (the primary sex ratio). The result is that slightly fewer females than males are born (the secondary sex ratio). As females have a slightly lower likelihood of early childhood mortality than males (Verbrugge, 1982), the ratio of the two sexes at reproductive age (the tertiary sex ratio) is restored to 1:1. The secondary and tertiary ratios are, however, altered in some societies because, for example, female fetuses are selectively aborted for cultural reasons.

Fisher (1930) provided an explanation as to why the tertiary sex ratio in sexually mature organisms should be 1:1. His argument was that, if producing offspring required an equal contribution from each parent, then that figure provided a stable equilibrium value, alternatively known as an evolutionarily stable state, or ESS (note: ESS also stands for "evolutionary stable strategy," which refers to a behavior strategy that is more stable than alternatives, at least near equilibrium). Deviations from this equilibrium would self-correct in time because, were males to be less common at birth, they would have better breeding prospects than females and so produce more offspring. Thus, parents predisposed to produce males would produce more grandchildren than other parents and male births would become more common (Hamilton, 1967).

This argument fails occasionally in cases where the standard conditions for an ESS are not met. An entertaining example is provided by pyemotic mites (*Pyemotes ventricosus*) that are only ~0.2 mm long: these tiny organisms develop internally, with males inseminating their sisters before they are born. Of the ~250 mites born to a mother, some 92% are female and only about 8% are male (Bruce & Wrensch, 1990). The reason why this figure is stable seems to be either that the mother has some way of reducing the sex ratio, or that there is dramatic and lethal competition among the very young males. The phenomenon is referred to as local mate competition.

Kin selection and altruism

The discussion in this chapter so far has focused on selection associated with anatomical or physiological variation, essentially at the level of what might be advantageous to the individual. Kin selection is very different as it analyses heritable behavior in the context of a group, particularly considering why an individual might do something that benefits its relatives through their reproductive activity at the expense of its own – this is altruism.

How this might happen is not an easy question to answer, and the ideas, the theory, and the modeling of how this might occur have intrigued some of the greatest evolutionary theorists, including Darwin, Fisher, Haldane, Hamilton, Maynard Smith, and Price.

There are numerous examples of behavior that can be viewed as deriving from kin selection (Bourke, 2014). While some come from insect societies (for example, see Wilson, 2000), two involve squirrels. First, red squirrels are much more likely to adopt orphan pups that are related to them than those that are not (Gorrell et al., 2010). Second, a ground squirrel sensing predators will give an alarm signal that allows its group to escape, even though this behavior risks drawing the predator's attention to itself. Of particular interest here is that this behavior is particularly pronounced if the group includes close family members (Sherman, 1977; Mateo, 2002).

Kin selection would be easy to understand if natural selection was only concerned with increasing the frequencies of genes that are both widely distributed within and beneficial to a population in the sense that possession of the gene increases fitness (this is the final cause or purpose). From a population viewpoint, it does not matter which individuals have these genes, and this is most easily seen through Haldane's joke, "I would lay down my life for two brothers or eight cousins," which he later corrected to, "I would truly die to save more than a single identical twin or more than two full siblings." From the point of view of the animal displaying altruistic behavior, however, it can be hard to identify the direct mechanism (the immediate or proximate cause) by which such an increase in fitness is achieved. This is because the individual who suffers needs to have some physiological trait that makes it behave in a way that is detrimental to itself and may even be lethal, a trait that is not obviously heritable, and is certainly unlikely to be dominant. In the case of the squirrels, the recognition of family is likely to be by smell (Mateo, 2002), and it seems that the altruistic behavior is thus innate rather than reflecting an active choice. In the particular case of "viscous" populations, where there is little dispersal of family members, an organism can safely behave as if all or most nearby species members are kin.

The second problem is the notion that there could be a gene for altruism (Dawkins, 1976): complex behaviors depend on sophisticated neurological circuitry that drives muscular behavior. Thus, while the meaning of a heritable behavioral trait may be clear, the meaning of a behavioral gene is not, other than the sense that it represents the sum of the effects of complex neuronal networks, themselves underpinned by sets of protein networks. The idea that all of this can be reduced to a few genes in the normal sense of the word makes little sense and it is clear that the term "gene" in this context has a much higher-level meaning. This point will be considered in more detail towards the end of the chapter.

It is therefore unexpected that the theory of evolutionary genetics, which uses general terms like fitness, the reproductive cost of altruism, and selection pressure together with a macroscopic concept of a gene, provides a good model for explaining altruistic behavior and kin selection. That it does so implies that organisms do behave as if they aim to optimize their gene complement in future generations, albeit that such genes reflect aspects of trait behaviour rather than DNA sequences. The theoretical analysis for this behavior that is now accepted was originally done by Hamilton (1964): he investigated the criterion for a gene increasing in frequency within a population when an *actor* performs some act for the benefit of a *recipient* rather than itself, and showed that this will happen when

$$\mathbf{rB > C}$$

where **r** represents the genetic relatedness of the recipient to the actor, **B** is the additional reproductive benefit gained by the recipient, and **C** is the reproductive cost to the actor. Of the parameters in Hamilton's rule, only **r** can be predicted, whereas the others have to be obtained empirically, and these figures can only be estimates. This criterion has, however, been shown to work for the case of squirrel adoption (Gorrell et al., 2010).

The general theory of kin selection provides the basis for genetic studies of the evolution of social behavior of groups of organisms. A key figure in developing this theory was George Price, who strengthened Hamilton's rule to produce what is known as the Price equation, which describes how gene frequencies change over time as a result of selection (Frank, 2012). It is not discussed quantitatively here partly because the derivation and formulation of the equation is mathematically complex, and partly because it is not easy to use. The essential idea is to separate the evolutionary change in some property from one generation to the next in a population into a part that derives from selection and a part that derives from other factors such as genetic drift. This formalism can be used to describe how a trait or gene frequency changes over time, but not easily.

A second important aspect of Price's work deals with the development of evolutionarily stable strategies for the behavior of groups through the use of game theory, an area further strengthened by Maynard Smith (see Frank, 1995). There is now a considerable body of work, both theoretical and experimental, in this area, which is often known as sociobiology, with much of the experimental basis coming from genetic studies of behavior in insect societies (for example, see Wilson, 2000). Unfortunately, the mathematics required to understand this theory is not simple, while the methodology for partitioning the cause of behavior between heritable and environmental factors (sometime called the balance between nature and nurture) is still contentious. Readers interested in this important but difficult area are referred to Wilson (2000) and Hedrick (2010).

Selective breeding

From time immemorial, humans have undertaken selective breeding to produce new variants of domestic animals and to improve plants both for harvesting and for visual pleasure. Indeed, the whole horseracing industry is predicated on horses being fast because they were bred from parents chosen for their race-winning abilities. In Chapter 1 of *On the Origin of Species*, Darwin focused on pigeons and the ease with which exotic varieties can be selected on the basis of serial interbreeding from minor variants (see Figure 12.1). Today, some of the chance element in breeding can be eliminated because, where a specific gene that confers some trait is known (such as resistance to a selective weed killer in a plant), that gene can be genetically engineered into a host genome and a population bred with the required gene and hence trait.

All selective breeding depends on breeders recognizing favorable features or traits possessed by organisms and doing their best to ensure that such phenotypes are inherited by their offspring. Ideally, this is done through mating males and females that both show the trait on the basis that this will maximize the chances of the offspring also showing the trait. However, as discussed in the last chapter, many anatomical features depend on the inheritance of complex networks of proteins and, because of the random nature of meiotic crossover, it is not always easy to predict the extent of that trait in offspring (horseracing success in parents is not as reliable a predictor of speed in the offspring as owners hope).

Perhaps the most interesting experiment to test the limits of the possibilities of artificial breeding was conducted by C.H. Waddington in the 1950s. Waddington was a major developmental and evolutionary geneticist who predicted the essential features of molecular genetics in the 1940s on the basis of developmental mutants in *Drosophila* whose phenotypes showed abnormal embryogenesis. He was also the first geneticist to realize that the study of mutations responsible for abnormal embryos held the key to finding out about the normal behavior of genes during embryogenesis - work that he did in the late 1930s, long before anyone had any idea that a gene was represented by a DNA sequence.

Waddington's approach to exploring the power of selection in evolution was to identify a rare, abnormal anatomical variant in *Drosophila* that could be induced experimentally, and then see if he could make it heritable through selective breeding. He had little choice other than to use flies as the experiments required around 250,000 organisms; any other choice would have been financially impractical. *Drosophila* flies normally have two front wings in their second thoracic segment (T2), with the hindwings of ancestral insects in their third thoracic segment (T3) being replaced by small balancer organs called halteres. If early embryos of normal *Drosophila* are exposed to ether vapor, however, a very small proportion will replace the T3 region of thorax carrying halteres with a duplicate of the T2 region that carries the forewings (**Figure 14.3**). This ether-initiated anatomical abnormality is a phenocopy of the *bithorax* phenotype; this mutant is seen occasionally in populations of flies that have been exposed to large amounts of X-rays and so carry a heavy mutation load.

Waddington investigated whether selective interbreeding of successive generations of ether-induced bithorax flies would lead to a stable population of four-winged flies that would form in the absence of ether induction - and it did. It took only 20 rounds of ether treatment and subsequent inbreeding for Waddington to produce a population of four-winged flies that mainly bred true without further ether treatment (**Figure 14.4**). Waddington called the phenomenon by which "acquired" characters might be converted into inherited ones genetic assimilation, and showed that there were several other such characteristics for which this phenomenon could occur (Waddington, 1961). There was, however, one prior condition that had to be met for the experiments to succeed: the flies had to come from an outbred rather than an inbred population, so ensuring that there was plenty of genetic variation for selection to act on.

There has been considerable discussion over the years as to the mechanisms by which genetic assimilation occurs, because at first sight, the phenomenon looks like Lamarckian

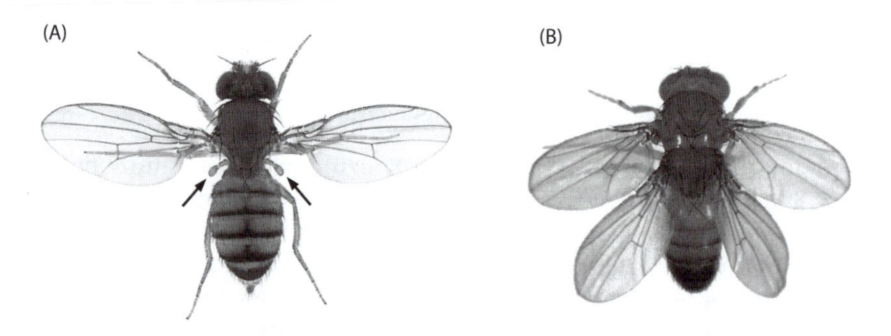

Figure 14.3 (A) A normal *Drosophila* with two halteres behind the wings (arrows). (B) A *Drosophila* bithorax mutant whose third thoracic region that contains the halteres has been replaced by a duplication of the second region that carries wings.

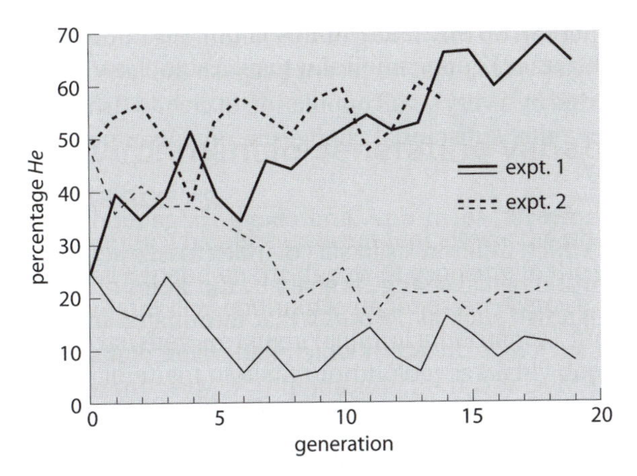

Figure 14.4 The progress of selection for (upper data lines) or against (lower data lines) an initial bithorax-like response to ether treatment in two wild-type populations that reacted with rather different frequencies.

inheritance, although it is not. The implications are now fairly clear: the *bithorax* phenotype normally results from a complex set of mutations in the *ultrabithorax* gene group and its regulatory sequences that lead to abnormal expression of its downstream genes (Pavlopoulos & Akam, 2011). In natural wild-type populations, however, there are neutral mutations within that complex that normally yield a normal phenotype. Ether selection identifies combinations of such mutations that contribute to the bithorax phenotype, although the reasons for this are not clear. Further successive rounds of selective breeding capture further mutations already within the genotype of the selected population that strengthen the likelihood of a bithorax phenotype to the extent that eventually the novel phenotype is self-sustaining. Waddington's experiments, with their extreme selection process, had the effect of concentrating far more *bithorax* phenotype-related mutations in a fly population than could ever happen under normal circumstances.

The important implication of this work is that it demonstrates that, provided selection pressures are strong enough, unexpected novel structural phenotypes can be generated surprisingly rapidly from the genetic variants that exist within a normal, outbred population. In normal evolution, of course, breeding is essentially random and selection pressures are therefore much weaker. Under conditions where there is no limit on the number of generations available for adaptation to take place and selection is relatively low, events that take hundreds or even thousands of generations can seem fast. Here, it is worth remembering that the evolution of the amniote egg with its various extraembryonic membranes took something like 50 million generations (Chapter 6).

The speed of change under natural selection

A few examples are known where heritable change can be rapid under conditions of strong selection, and an example is the alternation of predominant morphs of the peppered moth. Vertebrates can also evolve relatively rapidly, too, particularly when predation rates are low and there are few, if any, constraints on variants. This can happen when a founder population invades a novel environment in which there are few predators. Perhaps the best-known example of this is the current flock of the many hundreds of species of cichlid fish seen in Lake Victoria (Chapter 15; Hajkova et al., 2002). Although there is usually limited information on the timing of speciation, the differences among these various annually breeding species must be recent because the lake formed less than 15 Kya, apparently being seeded by a very small population of cichlid fish (Seehausen, 2002). If so, this is further evidence that substantial changes in organism morphology can be generated relatively rapidly.

Such examples do, however, seem rare, and change is generally very slow, as demonstrated by the rate at which differences in sets of index fossils accumulate. A good indication of the rate of morphological change is given by the series of fossils that bridge the fin of a late sarcopterygian fish such as *Tiktaalik* that terminated in rays and the limb of an early labyrinthodont such as *Icthyostega* which had seven digits (see Table 6.2). The former lived about 380 Mya and the latter about 374 Mya. These changes seem to have resulted from fish being able to look for food in shallows with dense seaweed, with there being a selection pressure to push through the vegetation (Chapter 6). Although exact lines of descent are not known and the fossil record is not complete, it clearly took a few million years and perhaps a similar number of generations for autopods to evolve and rays to be lost. Such information emphasizes the importance of those cases in which traits can be heritably altered under laboratory conditions as it is only here that it may at least be possible to obtain molecular insights into the processes by which this happens.

The evolution of the camera eye

While the easiest means of assaying change is through those few hard tissues that fossilize, equally interesting are changes in soft tissues because they often reflect function directly. Perhaps the most interesting example of this is the evolution of the camera eye. This has a special place in the history of evolution for several reasons, the most important of which is that it is an extremely complex organ that only seems to work if all of its components are already in place. Even into the twentieth century, there was some doubt as to whether it was actually possible for a camera eye to form through a series of small changes, with each change being advantageous to the organism. Darwin thus saw that providing an explanation for the evolution of the eye was a necessary test of any theory of evolution, as he wrote in *On the Origin of Species*:

> *To suppose that the eye, with all its inimitable contrivances for adjusting the focus to different distances, for admitting different amounts of light, and for the correction of spherical and chromatic aberration, could have been formed by natural selection, seems, I freely confess, absurd in the highest possible degree. Yet reason tells me, that if numerous gradations from a perfect and complex eye to one very imperfect and simple, each grade being useful to its possessor, can be shown to exist; if further, the eye does vary ever so slightly, and the variations be inherited, which is certainly the case; and if any variation or modification in the organ be ever useful to an animal under changing conditions of life, then the difficulty of believing that a perfect and complex eye could be formed by natural selection, though insuperable by our imagination, can hardly be considered real.*

The selection pressures driving any improvement in vision are obvious: any organism that can see food, identify a potential mate, or avoid a predator has an immediate advantage over one that does not, and the greater the visual acuity, or ability to resolve detail, the stronger this advantage will be. Light sensitivity is found in many unicellular eukaryotes so this basic facility is very ancient and lies at the root of animal vision. The origins of the

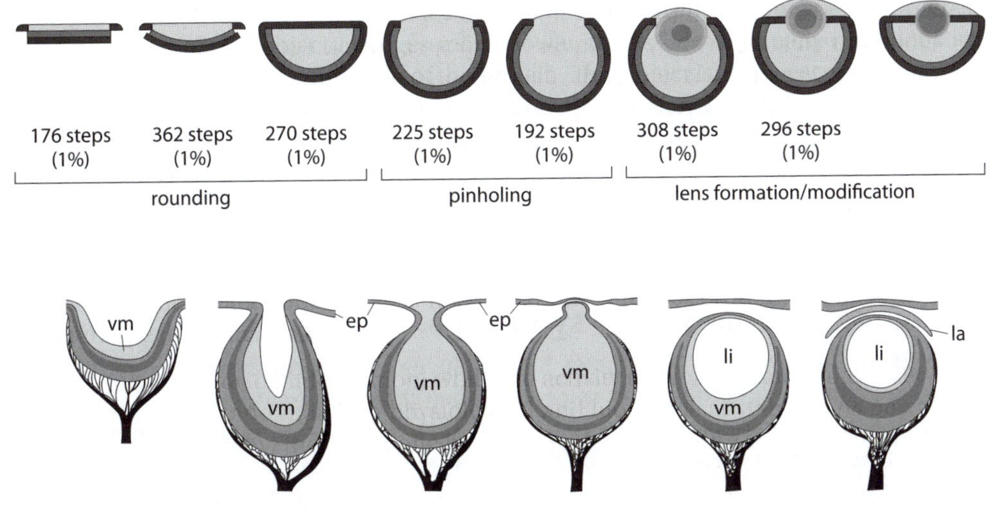

Figure 14.5 Top row: A representation of how the camera eye could develop from a flat placode through rounding, pinholing, and lens formation. The number under each eye is the number of modification steps, each increasing acuity by 1%, required to achieve the next eye in the set. **Bottom row**: Although this set of eyes looks like a developmental sequence, it is actually drawings of six adult mollusk eyes with increasing visual acuity (ep: epithelium; la: lacuna; li: lens; vm: vitreous matrix).

camera eye, which is seen in both vertebrates and mollusks, although the geometries are slightly different (Chapter 5), clearly derive from a patch of light-sensitive cells on the head of an early member of the bilateria that developed neurological connections to the brain, but the route from this primitive light sensitivity to the sophistication of the camera eye is not immediately obvious. Fortunately, there is a set of extant mollusk eyes (**Figure 14.5**; Salvini-Plawen & Mayr, 1977) that indicate one way by which this might have happened.

Nilsson and Pelger (1994) took an analytic approach to explore how many generations it took for the camera eye to evolve from a small, flat plate of light-sensitive cells backed by a pigmented epithelial layer. They worked on the assumption that the driving force for change was an increase in visual acuity, with a 1% heritable increase giving an organism a selective advantage that would be further propagated. They then explored how visual acuity could be increased by simple changes to the geometry of the eye, under the constraint that its diameter remained constant. They showed that the necessary geometric changes to produce the camera eye involved three steps: rounding, pinholing, and then cornea and lens production. These anatomical geometries are particularly interesting because each stage can be seen in the eye of an extant mollusk.

The first step in their study was to analyze the optical properties of these intermediate eyes and calculate the number of steps required to achieve them under the constraint that each increased acuity by 1%. Nilsson and Pelger then used the framework of population genetics theory (discussed below) to work through the numbers of generations that it would take for these small improvements to eventually become established (see Figure 14.5). Using these figures in the appropriate equations, they showed that, under the most conservative of assumptions, the evolution of the complete camera eye formation would take about 360,000 generations or the same number of years for an organism that reproduces annually. On the scale of the Cambrian Period (541–485.4 Mya) when eye evolution occurred, this is a very short time.

This analysis, together with the range of morphologies seen in extant mollusk eyes, suggests that the camera eye evolved sequentially from a simple light-sensitive placode during a relatively brief period during the Cambrian Period or a little earlier, as even some jellyfish have lensed eyes (Garm & Ekström, 2010). It is, however, likely that the set of changes that eventually led to the full camera eye was independently associated with the evolution of several species. The significance of Nilsson and Pelger's work is that it shows that selection pressures can effect significant complex morphological change relatively fast on an evolutionary timescale.

The modern evolutionary synthesis

Much of the first part of this chapter has implicitly considered how trait frequencies change within a population under selection, albeit that the discussion has mainly been qualitative. This discussion has been based on the modern evolutionary synthesis (MES) that brought together ideas from evolution and population genetics (Appendix 2). A key aspect of the synthesis was the mathematical integration of population genetics with ideas about the genetic basis of evolutionary change, and the net result was a new field known as evolutionary population genetics (EPG), which was initiated in the second quarter of the twentieth century by Fisher, Haldane, Sewall Wright, and others. Over the next 30 or so years, the theory was strengthened by geneticists such as Hamilton, Price, and Kimura, who introduced concepts such as kin selection, and strengthened the theory to include genetic drift. The field is still highly active and a current important area, as already mentioned (Chapter 8), is coalescent theory.

EPG provides a formal model that combines quantitative population genetics with evolutionary change based on the interaction of the environment with a population carrying a spread of gene alleles that derived from drift and mutation. The theory introduced concepts such as fitness and selection coefficients to handle changes in the population over time, but formulated the mathematics in terms of underlying allele frequencies, any of which could and, indeed, would spread through the population. Such genes were, of course, hypothetical entities defined on the basis of phenotypes, much as were Mendel's original genes for pea color and shape, rather than DNA sequences. Although the molecular genotypes of such genes were, of course, unknown, it is genetically correct to formulate the theory in terms of hypothesized allele frequencies rather than phenotype frequencies, as the key evolutionary process is how one allele rather than another preferentially spreads through a population over time. One advantage of doing this is that, irrespective of the nature of the gene, the mathematics is tractable, if not simple.

At the time when the basics of the theory were being constructed, it was thought that most aspects of the phenotype reflected the activity of one or a few gene alleles (the latter were complex traits), Although such a model of gene action is now known to be a gross oversimplification of molecular reality, the qualitative predictions of the theory are often correct for reasons that are discussed at the end of the chapter. Perhaps more importantly, the details of the nature of a gene are irrelevant in many theoretical contexts and the theory has successfully explored and explained many aspects of evolutionary change and the movement of genes through a population. Only some of the simpler results are considered here, partly because of the complexity of the mathematical probability theory required to derive them and partly because, while the theory can explain the beginnings of trait asymmetries, it says nothing about when these asymmetries are sufficient for speciation to take place, or even when the early stages of population separation are achieved. This is because there is still no proper formal analysis of the genetic basis of phenotypes. Readers who are interested in a full treatment of the synthesis are referred to the reading list at the end of the chapter.

Population genetics

The core theorem describing how gene alternatives are distributed within a population can be illustrated in the case where there are two alleles of a gene named **A** (frequency **p**) and **a** (frequency $\mathbf{q = 1 - p}$). It is obvious that, in the absence of selection, the probabilities of organisms having as their allele complements **AA**, **Aa**, and **aa** are $\mathbf{p^2}$, **2pq**, and $\mathbf{q^2}$, which, of course, sum to 1. It is then straightforward to show that, if mating is random, the population is large, and there is neither selection nor mutation, then the gene frequencies in the next generation are, again, **p** and **q**. This is what is known as **Hardy–Weinberg equilibrium** and it explains how gene frequencies remain stable over time in the absence of selection. The conditions of random mating across a large population, together with an absence of mutation, describe a very basic model of reproductive behavior in a population. There are two well-known variants here: in the Wright–Fisher model, populations have a constant large size, mate freely, and do not have overlapping generations; in the Moran model, populations do overlap, but when one organisms is born, another is assumed to die. Such idealized populations have the advantage that they are mathematically tractable.

Hardy–Weinberg equilibrium can be altered in two sorts of ways that affect gene distributions. First, the gene proportions and effects may be changed through novel mutation or genetic drift. Second, the population dynamics can be altered through factors such as nonrandom mating and selection acting on one or another phenotype. Evolutionary population genetics explores the mathematical implications of these factors on Hardy-Weinberg equilibrium.

Mutation and genetic drift

Natural variation can occur in a normal population for two genetic reasons: drift and mutation. Both effects are important in leading to phenotypic change, but it is not usually clear which is more significant. The former occurs through the short-term vagaries of reproduction: the smaller the breeding population, the less random is the distribution of genes and the resulting phenotypes will shift in unpredictable ways. The latter occurs through copying errors that effect recombination events, particularly during meiosis, and the resulting polymorphisms are, of course, the core long-term driver of change that leads to phenotypic variation within a population.

One effect of genetic drift in small populations is that gene frequency asymmetries create more homozygotes than heterozygotes. This can easily be demonstrated: suppose, as a result of genetic drift in two small subpopulations, two genes which each had a frequency 0.5 in the original population now have frequencies x and 1 - x. It is easy to show that the proportion of homozygotes to heterozygotes in each population is

$$\mathbf{x^2 + (x - 1)^2 : 2x(1 - x)}$$

and that the ratio has a minimum value of 1:1 when $x = 0.5$, as in the original population. If, however, the gene distributions change so that $x = 0.9$, the ratio of homozygotes to heterozygotes increases to 0.82:0.18, or about 4.5:1. Hence, there are more homozygotes in small, genetically unbalanced populations than in large, genetically balanced ones. The importance of this Wahlund effect is that the phenotypes due to recessive genes become more accessible to selection in small populations.

Selection: gene frequency changes as a result of nongenetic changes

For Hardy–Weinberg equilibrium to break down and trait distributions to change, one or more of the underlying assumptions has to fail: mating must be nonrandom, a subpopulation has to be sufficiently small that genetic drift effects are important, or that one or another allele acquires a selective advantage. When comparing the relative importance of the various alleles of a gene, the model requires a parameter giving each its fitness as reflected in the likelihood of the phenotype of that allele appearing in the next generation of organisms.

The fitness of a gene allele, known as **w**, and its selection coefficient **s** are connected by the simple formula $\mathbf{w} = \mathbf{1} - \mathbf{s}$, where **s** represents the relative disadvantage of the gene for that trait. Hence, a value of $\mathbf{s} = \mathbf{1}$ is lethal, while a value of 0.2 means that 80% of the offspring carry that allele. The fitness coefficient here is yet another umbrella term because it reflects every aspect of breeding from success in mating and in finding food to the ability to survive in a particular environment; it is thus highly dependent on the local ecosystem. As the coefficient reflects the fitness of that trait in such a wide context, it cannot be predicted, and can only be estimated on the basis of population distributions. In breeding experiments, however, selection parameters can be precisely controlled.

An important aspect of evolutionary population genetics is its focus on small populations: this is because their gene distributions reflect sampling from the full population and so usually differ from it through genetic drift. Here, the population size is important in calculating, for example, how fast genes move through the population or whether they are likely to be lost. For such calculations, the important number of individuals is not that of the full population, **N**, but reflects the number of reproducing adults, which is known as $\mathbf{N_e}$, the effective population size. This number is always less than **N** as it excludes sexually immature, unfertile, and postfertile organisms. The number also allows for the breeding habits of the population: if, as in herds of elephants and lions for instance, a few males maintain a harem of females, most males have no opportunity to pass on their genes to the next generation and $\mathbf{N_e}$ is diminished. In practice, the calculation of $\mathbf{N_e}$ is difficult: as it reflects genetic variance and breeding behavior, it usually has to be estimated.

One example of the use of the selection coefficient is that the MES theory can predict whether a novel allele with some selective advantage can survive the vagaries of genetic drift and become established within a population. Haldane (1927) showed that this could happen provided that its selection coefficient $\mathbf{s} > \mathbf{1/N_e}$, with the likelihood of it happening being ~**2s** (this is known as Haldane's sieve). Thus, the smaller the effective population size is for a given value of **s**, the more likely a novel gene will become part of the population's genotype. Kimura and Ohta (1969) then showed that the expected number of generations that it took for such a novel gene to become fixed within a population was $\mathbf{4N_e}$. Thus, as noted in the earlier discussion of coalescent theory (Chapter 8), the smaller the size of the breeding population, the faster and more likely it is that an advantageous new allele will spread through it. The importance of small population groups for encouraging speciation is discussed in the next chapter.

Balanced polymorphisms

The classic example of the effect of fitness on an allele is the mutation in the hemoglobin gene that confers malaria resistance (Jones, 1997). Most people bitten by a mosquito carrying *Plasmodium falciparum* will get malaria. People carrying one copy of a mutation that converts glutamic acid to valine in the β chain of hemoglobin have up to 90%

protection against infection, probably because the mutated protein can carry less oxygen than the normal one so that the blood becomes a less favorable environment for the plasmodium. One might suppose that this mutation would become the norm in areas where malaria was very common, but this is not the case: associated with single copies of the gene allele (heterozygotes) are a proportion of erythrocytes (red blood cells) that are rigid, sickle-shaped, and carry diminished amounts of oxygen; these predominate in homozygotes causing a series of disorders and premature death, which is, of course, associated with a very low likelihood of reproduction. This homozygous disadvantage leads to the recessive gene **S** continually being depleted.

Ridley (2004) has worked through the complete calculations to show how population frequencies of the malaria-insensitive heterozygote, the sickle cell homozygote, and the malaria-sensitive homozygote can be used to calculate the gene allele frequencies and fitness values. His analysis shows that the gene allele frequency ratio (A:S) is 0.877:0.123, and that the fitness coefficients for the AA, AS, and SS phenotypes are 0.14, 1.0, and 0.88, respectively. The different fitness values reflect the fact that someone with malaria is less likely to die before reproducing than someone with sickle cell anemia, and that both are less reproductively successful than the heterozygote.

The sickle cell mutation is described as a balanced polymorphism because, in regions where malaria is prevalent, the advantages of a single copy of the mutation are balanced by the disadvantage of two copies. Selection works both ways and the result is a dynamic and stable equilibrium with the single-gene model of the various phenotypes being able to explain the distribution of the phenotypes. Note, however, that this example reflects the effects of selection pressures on populations and not the sort of evolutionary change associated with speciation. Nevertheless, if populations form with new gene frequencies, such selection is a first step on the way to speciation.

There are a few other human phenotypes regulated by a single protein. An interesting example is the dual trait of dry earwax (cerumen) and minimal underarm (axillary) odor. Individuals carrying the rs17822931 single nucleotide polymorphism of ABCC11, an adenosine triphosphate transporter protein, have wet earwax and a strong axillary odor that can be sufficiently worrying for affected individuals to have their axillary apocrine glands removed from each armpit (Nakano et al., 2009). Note, however, that such genes modify a physiological property but are not responsible for it.

The strengths of the modern evolutionary synthesis

Within populations, there is an ever-shifting distribution of phenotypes. If a small group somehow becomes reproductively isolated from its parent population, it will have atypical phenotypes as a result of genetic drift and the Wahlund effect. If that subpopulation finds itself in a new environment with novel selection pressures within which it thrives, it may become first a subspecies with a distinct phenotype and eventually a full species in the sense that the former can breed with the parent line and the latter cannot. The phenomena that drive this progression are variation, selection, further mutation, and speciation, with selection of fit variants being the key step, as Darwin was the first to realize.

The strength of the modern evolutionary synthesis is that it provides a formal framework for analyzing the gene allele and phenotype or trait distributions in populations whose breeding and other behaviors can be explicitly defined, together with the effects of environmental pressures on that population. Here, the theory is coherent, rigorous, perceptive, and important. It can, for example, analyze the effects of selection acting on small populations where variants can flourish and gene or trait frequencies can shift. The significance of the small population is that, because of genetic drift, it is likely to have a genetic profile that is different from that of its founder population and a smaller pool of gene alleles. It is worth mentioning again here that new mutations arise far too rarely to play anything other than a very minor role in the short term.

EPG has other important uses. It can be used to make predictions about novel hypotheses; it provides a means of evaluating the relative contributions of heredity and environment to a trait and it is extremely useful in conducting and analyzing population data and in conducting breeding experiments where the evolutionary component is minimized. More recently, population genetics models have been integrated with sequence data to analyze the details of evolutionary events through the use of coalescence theory (Chapter 8), and this has made what might have seemed a rather traditional approach of direct relevance in the age of genomics. What is particularly interesting about coalescence theory in this context is that the analysis is not built on population distributions, but on the variants in long DNA sequences possessed by different individuals. It turns out that the significant parameter in the analysis is the number of genes being analyzed and that the number of gene variants used is of secondary importance. Coalescent analysis thus investigates a model of population behavior over time in the context of essentially unlimited amounts of data.

The limitations of the modern evolutionary synthesis

For all of its strengths, the classical EPG theory does have limitations, and the first derives from the fact that most phenotype variants, other than those directly traceable to a single-gene polymorphism, reflect a genetic complexity that was unimaginable to Fisher and his generation. Evolutionary population genetics is a theory describing how trait frequencies change in a population as a result of the hypothesized genes that underlie them, with these genes reflecting trait alleles rather than protein variants, a view that actually derives from the work of Mendel. Unless explicitly specified, the theory provides no links between these trait genes and what is generally understood as a normal gene, that is a functional unit coded in a DNA sequence. In the rest of this section, the genes of evolutionary population genetics will be referred to as trait genes to distinguish them from normal DNA genes.

Second, the application of the theory to the evidence on evolutionary change in a real situation is difficult as the main data that it has to work with are distributions of phenotypes across a population. A first step is partitioning of the variance (spread) of a phenotypic distribution into a component due to inheritance and a component due to the environment. The former is due to trait genes and, if the predicted phenotypes of a set of trait gene alleles is inadequate to explain the data, a more complex model of two or three trait genes and their interactions (epistasis) may be needed. The net result is that the phenotypic data have to be analyzed with considerable mathematical sophistication to identify the constants describing both the relative importance of the various trait genes and the strength of their interactions.

Third, the situation is further complicated mathematically if the genes are on the same chromosome: such linkage introduces yet further constants into the equations. The more constants that are required to describe the data for a given population and its associated trait distribution, the greater the standard error is in estimating each and the less precise the analysis becomes. Where a trait depends on more than three such trait genes, the theory becomes very unwieldy owing to the number of variables and arbitrary constants, and any predictions become too complex to test experimentally. Even where such constants can be obtained, it is hard to evaluate the meaning of the alleles that combine to produce that complex trait.

Quantitative traits and the difference between Mendelian and DNA-based genes

The MES has particular difficulty with analyzing complex traits that show continuous variation. It is however these phenotypes that are particularly interesting in the context of evolution because they include such species-defining properties as changes in growth, anatomical patterning, behavior, and physiological abilities. Apart from the difficulties in calculating parameters, there is a deeper problem. Any model that assumes a trait is based on a few trait genes and their interactions is a very weak approximation to reality. This is because many aspects of the phenotype derive from the functional outputs of networks composed of many interacting proteins. The quantitative theory, put together long before there was any understanding of molecular biology, simply cannot cope with such complexity because it includes no model that links mutation in normal genes to changes in traits.

In fact, and as the previous chapters of this book have emphasized, the phenotype of a complex trait is the result of a series of events at a succession of levels between the genome and the eventual trait (see Figure 12.4; **Table 14.1**). To increase the complexity of the situation, there can be further interactions between these levels. The trait genes underpinning a complex trait such as the features on a human face will reflect variants in the output properties of groups of networks, which include signaling, patterning, hormonal activity, and growth. The trait gene alleles of the evolutionary synthesis typically represent the limited range of options that protein polymorphisms generate in the mix of proteins and protein networks that underpin a trait. In terms of the hierarchy given in Table 14.1, they mainly represent level 4 objects. Rather than the two or three genes and their interactions that the theory can cope with, the trait is probably based on the activities of many hundreds of proteins. The available data from trait distributions in populations are just too thin to unravel this genomic reality.

For all of these criticisms, the theory can provide good analyses and illuminate many aspects of evolution, and it is important to understand why the few alleles of a trait gene can sometimes provide a good way of integrating the effects of large networks of proteins. There are two types of answer here. First, a fair amount of work in evolutionary population genetics is concerned with exploring gene flow, selection and drift; here, the analysis does

Table 14.1: Complex traits and their underlying levels

Level 5: complex trait	The observed phenotype for a feature (anatomical, physiological, or behavioral)
Level 4: network groups	Each are immediate generators of phenotypic traits (these are often trait-genes,[a] whose alleles reflect the range of possible phenotypes)
Level 3: protein networks	Each drives an aspect of the phenotypic trait (development: patterning, growth, differentiation; physiology: neuronal networks, muscle action, clocks)
Level 2: proteins	Each has a specific functional role either alone or in a network
Level 1: genes	Protein-coding DNA sequences and associated control sequences (for example, cis-acting sequences and methylation effects

[a]The level 4 trait genes are typical of those used in the modern evolutionary synthesis.

not need to specify the nature of the genes and the results are therefore independent of gene type. Second, and in cases where the phenotypes associated with trait gene alleles are being analyzed, it seems that the available options associated with phenotypes such as growth, specific behaviors, and physiological properties have ranges that are limited in extent. The net result is that, for all the complexity of their underlying networks, the outputs are restricted in range under normal conditions and all that mutation can do is to alter that range. The situation mirrors an example in physics: the temperature-dependent behavior of atomic systems is governed by the complex laws of statistical and quantum mechanics, but much of macroscopic behavior can be accurately described by the relatively simple laws of thermodynamics.

Finally, it is worth mentioning in this context a topic that has proved confusing and contentious, and this is the concept of the *selfish gene*. This idea comes from Dawkins (1976), who argued on the basis of the work of Hamilton (1964) and others that populations behave in practice as if selection acts on the phenotype of strong alleles to maximize their frequencies; such alleles can thus be viewed as if they were selfish, promoting themselves at the expense of alternatives. The sorts of genes that Dawkins was writing about included those that were particularly involved in social and altruistic behaviors, and thus involved in group cohesivity, but the argument ranged much more widely.

Not everyone has accepted Dawkins' views and this is because, from today's perspective particularly, there is a fallacy in the argument. Selfish genes are examples of what are referred to here as trait genes and are defined in terms of their complex phenotypes; the idea thus says little more than that alleles with beneficial phenotypes will be selected. Trait genes should not be thought of as normal genes as they can rarely if ever be reduced to a DNA sequence. Trait genes are high-level phenomena that represent a specific variant of the output of a set of proteins and protein networks (Table 14.1). The idea that such a trait gene drives change is, of course, correct because its phenotype will be more strongly selected than that of alternative alleles, but it is not a gene in the usual sense of the word. It is probably more helpful therefore to think about the selfish phenotype than the selfish gene. At the molecular level, of course, most proteins cooperate with others and are unselfish.

The multilevel view of how phenotypic change happens

At the end of each of the previous two chapters, the ways in which the effects of genomic mutation work their way up to phenotypic change across a population were discussed in terms of the graphical metaphor of a multidimensional, time- and space-dependent object, with each dimension reflecting a population distribution of some feature (see Figure 12.4; summarized in Table 14.1). In this metaphor, novel trait variants at the top, phenotype level, such as an abnormal anatomical feature or atypical innate behavior, can be represented as local bulges in the surface of one or another dimension in the general phenotype object that slowly change over time. The mapping of such bulges to the object at the level below reflects allele alternatives, each of which represents sets of proteins, many of which will be grouped in networks. Those alleles for anatomical traits will include networks for signaling, cell proliferation, differentiation, and the like, while those for physiological responses will include biochemical and neuronal pathways. Each dimension represents the population distribution of trait gene alleles for a specific trait, and unusual traits are located at the periphery of the distribution curve. Each of the possible alleles is connected to networks at the level below and further down to protein and gene objects. Producing phenotypes is complicated!

Selection involves those organisms in a subpopulation of some species that have rare features (seen as bulges) reproducing more successfully in a given environment than those that lack them. In this case, the mutated genes will slowly spread through the population, eventually becoming ubiquitous. At this point, the subpopulation has become a new, recognizable subspecies of the organism, still able to breed with the original parent population, and the multidimensional object will have changed to a new, temporarily stable form. For that subspecies to attain the status of being an independent species, its genome has to evolve further so that members of the group, even if they can, in principle, mate with the parent population, cannot produce fertile offspring. How this happens is the subject of the next chapter.

Key points

- There is an ever-shifting spread of phenotypes within a population due to genetic drift and mutation, and each is subject to selection on the basis of its fitness.
- Success here can derive from being reproductively successful in a given environment (natural selection), having an advantage in finding a mate (sexual selection), and evolving behaviours that benefit the group (kin selection).
- A variation that helps a population increase eventually becomes an adaptation.
- Organisms' breeding strategies lie on a spectrum. At one end is the K strategy, with few offspring but a high parental investment (for example, land vertebrates). At the other is the r strategy, with large numbers of fertile eggs but no parental care (for example, teleost fish).
- Selection in the wild is generally very weak and the acquisition of an adaptation so slow (thousands of generations) that it can usually only be observed or initiated under artificial or laboratory conditions.
- Under conditions of very strong, artificial selection, however, the acquisition of novel traits can be fast (~20 generations) in a population with considerable genetic diversity.
- Evolutionary population genetics is a mathematical model of the future behaviour of a population that describes how fitness, population size and behaviour, selection, and other such parameters can together lead to changes in the distribution of gene alleles over time.
- EPG has been important in explaining how parameters such as a small breeding population facilitate the rapid dispersal of high-fitness gene alleles through a population.
- Such alleles are, however, defined in terms of the trait or phenotype (the Mendelian view) rather than on DNA sequences. These alleles actually represent the behavior of groups of proteins and protein networks.

Further reading

Barton NH, Briggs DEG, Eisen JA, et al. (2007) Evolution. Laboratory Press.

Futuyma DJ (2013) Evolutionary Biology. Sinauer Press.

Hamilton MB (2009) Population Genetics. Wiley Blackwell.

Hedrick PW (2010) Genetics of Populations. Jones & Bartlett.

Jobling M, Hollox E, Hurles M, et al. (2013) Human Evolutionary Genetics. Garland Press.

CHAPTER 15
SPECIATION

Speciation not only reflects a temporary endpoint on the evolutionary path and an irreversible change in the local ecosystem, but it also represents an opportunity for future diversity. It is thus the key step in evolutionary change. In living species that reproduce sexually, full novel speciation can be recognized by reproductive isolation: a new species can, at best, produce sterile hybrids with its parent and other closely related species. However, for a separated daughter subpopulation to be unable to breed with its parent population, major genetic changes need to accumulate and these may take millions of years. This chapter discusses the reproductive isolation test of speciation and possible alternatives for when the breeding test cannot be done or is not applicable, and then reviews the mechanisms by which speciation can occur. This process always starts with a population subgroup becoming isolated from its parent population, finding itself in a novel environment in which some, at least, of the subgroup can adapt and flourish. Such progress is more likely in small than in large subpopulations because genetic drift and other population effects encourage the emergence of rare traits that may be more advantageous in the new than in the old environment. This separation initially leads to the accumulation of phenotypic and genetic differences in the subgroup that often result, first, in an unwillingness to mate with the parent population, and second the production of infertile hybrids and eventually to a failure in mating. These latter processes are slow, with the last typically taking perhaps a million generations; insight into the early stages of speciation has come from those few laboratory studies where strong selection has speeded them up.

The key events in evolution are those that create new species – they are the culmination of the processes of variation and selection. The separation of a new species from its parent species on the basis of reproductive incompatibility offers an irreversible step forward; it is a one-way ticket into the unknown and an opportunity for further new species to form in the future, so increasing diversity. If such separation only gives a subspecies, whose members are capable of mating with those from the parent species, then there is no incentive for novel adaptations to form and, indeed, that subspecies could be absorbed back into the parent species. It is clear on the basis of the title alone that, when Darwin published *On the Origin of Species* in 1859, he had already realized that the key to explaining evolution was understanding how new species formed and thought through many of the details by which it happened.

It is also worth noting that Darwin was the first person to focus on the significance of speciation and how it might happen. While others in the late nineteenth and early twentieth centuries discussed how speciation might happen (Berlocher, 1998), our current understanding can really be traced back to the work on the modern evolutionary synthesis in the 1930s and 1940s, when Mayr, in particular, emphasized that the first step on the way to new speciation was the separation of small populations of reproductively isolated organisms in distinct environments. Research by Haldane, Fisher, and others on the population genetics of such groups (Chapter 14), together with the work of Mayr, laid the foundations for our current understanding of the details of speciation.

Defining a species

As the differences between parent and daughter populations accumulate over time, they become ever more distinct. The first strong evidence of speciation is that, although organisms from a daughter population can mate with those from a parent one, the resulting hybrids are infertile. At this point, they are viewed as two distinct species. Then, as chromosome differences mount, such matings become unable to produce live offspring, or, indeed, any offspring at all. There is thus a simple biological meaning of a new species that is, in principle at least, functionally testable, and this is the *reproductive definition*:

- A species is a group of organisms with the property that, even if one of its members can successfully mate with an organism outside that group, any hybrid offspring will die or be sterile.

This definition is sometimes extended to include the idea that a particular species is associated with a specific habitat, but this is not a necessary addition.

One classic example here is the genus *Canis*, a family that currently includes wolves, dogs, golden jackals, coyotes, and the like, all of which have 78 chromosomes and can interbreed to give fertile offspring. This group diverged from the wider Canidae family some 1.7 Mya and can be viewed as a single species with a host of subspecies (Perini et al., 2010). The *Canis* family cannot, however, breed with other Canidae, such as the side-striped and black-backed jackals with 74 chromosomes, and the raccoon with 42 chromosomes. The situation is similar in the great cats: lions and tigers will interbreed and form hybrids that are fertile, even though they last shared a common ancestor several million years ago (Johnson et al., 2006).

A further, well-studied example is the family of Equidae, which provide more detailed examples of speciation and chromosomal alterations. In the past, the broad taxon of Equidae were widely dispersed but then went into decline. It seems that the last common ancestor of today's Equidae lived 4–4.5 Mya, with its offspring becoming geographically separated into the groups that are seen today; the last split being the domestic and Przewalski's horse separating only 72–38 Kya (Orlando et al., 2013). Details of the eight current species of Equidae are given in **Table 15.1**, which shows that each has its own distinct location and chromosome number.

Mules, which are hybrids between horses and donkeys, are easy to breed, presumably because the parents have very similar chromosome numbers, and the very occasional mule is fertile (Ryder et al., 1985; Yang et al., 2004). What is more unexpected, given that the parents have very different chromosome numbers, is that matings between the non-zebra and the various zebra species can also produce hybrids (for example, zebroids and zedonks; Bard, 1977), although all seem sterile. It seems that their genomes share sufficient genetic compatibility for meiosis and normal development to proceed, but that their

Table 15.1: The Equidae diploid chromosome numbers			
Species	Common name	Origin	Number of chromosomes
Equus ferus przewalski	Przewalski's horse	Mongolia	66
Equus caballus	Horse	North America	64
Equus africanus asinus	Donkey	East Africa	62
Equus hemious	Onager	Pakistan, India, Mongolia	56
Equus kiang	Wild ass	Tibet	52
Equus grevyi	Grévy's zebra	Kenya	46
Equus quagga	Plains zebra	Southern Africa	44
Equus zebra	Mountain zebra	South-West Africa	32

very different chromosome numbers and assignments preclude either meiosis or early development in any offspring that they can parent. One curiosity in horse–zebra hybrids is that they have more stripes than their zebra parent and this reflects a heterochronic postponement of stripe patterning in the embryonic epidermis (**Figure 15.1**).

The definition of a species based on reproductive isolation fails, of course, in a number of cases. One is where the distinction between strain and species occurs is fuzzy, such as in ring species (discussed below). A second occurs when organisms are extinct – the dead do not reproduce! A third is when the status of taxa, which may or may not be within the same species, is unclear because breeding experiments have not been or cannot be done for one or another reason (for example, the males and females may be unwilling to mate under laboratory conditions). The definition is also, of course, inapplicable to organisms that mainly reproduce asexually or parthenogenetically. It is also weak for many plant organisms whose reproductive tolerance is, compared with animals, unexpectedly broad.

Coyne and Orr (2004) discuss the various types of definition of speciation that there are in the literature and just how hard it is to provide one that matches all conditions. The most important definition for situations where the reproductive criterion cannot be used is the morphological definition:

- A species is a group of organisms that can be distinguished from other groups using standard techniques, particularly morphological ones.

Figure 15.1 An *Equus burchelli–Equus caballus* hybrid bred by Cossar Ewart in about 1895. The *E. burchelli* father (see Figure 13.6A) had about 25 stripes, but the hybrid offspring has about 70 stripes, with a striping morphology more typical of *Equus grevyi*.

This criterion works adequately for larger organisms, but smaller ones, particularly bacteria and very simple eukaryotes, require molecular assays as morphological ones are not sufficiently discriminatory. In both contexts, it is sometimes helpful to include a phylogenetic assay where a species includes some characters that can be seen as shared and derived within a monophyletic context and others that are unique. Indeed, sometimes a phylogenetic assay is the only one available (Baum, 1992).

Deciding whether, for example, a newly discovered fossil is a representative of a known species or whether it should be considered as a member of a new species is not easy. It requires detailed analysis of both morphology and other factors such as when and where it lived, before deciding in which clade it should be placed, and whether it should be viewed as a member of an existing species, a subspecies, or be assigned to a new species. This choice is hard because there is considerable variation within a species, some of which may reflect gender. One gets some sense of the difficulty of the problem when considering with no prior knowledge whether the skeleton of a modern Dutchman some 2 m high comes from the same species as the skeleton of a 1.5-m-high adult male Baka pygmy, given the major size and minor morphological differences. It would be easy to suggest on morphological and size grounds that they reflect different species, but one would first have to exclude the possibility that the differences reflected age, gender, strain variation, or diet. In fact, of course, they are simply subspecies or variants of *Homo sapiens*.

There is a further difficulty in considering species that are essentially defined by their ecosystem: these include examples such as host–parasite pairs and communities of species that have become ecologically dependent on one another through co-evolution. Here, the number of species in the group may be large and the interactions among them quite sophisticated (Gu et al., 2015, and see below). Decisions on such ecologically defined species can be quite difficult.

The population genetics of small groups

Mayr was the key figure to champion the idea that speciation starts in small groups that become reproductively isolated from their parent population; one reason for taking this view is that small groups of organisms have a series of properties that make them amenable to genetic change. The most important of these is that, because of genetic drift, their phenotypic and gene-frequency profiles will be different from that of the parent group, carrying only a small proportion of the genetic variance of the species - this is Mayr's founder effect. This may or may not be an advantage to a group that finds itself in a new and slightly different habitat: if the resulting trait profile is unsuitable, the group will be lost; if suitable, it will flourish.

Also important is the Wahlund effect (Chapter 14): because of genetic drift, there is likely to be a greater proportion of homozygotes in a small subpopulation with its asymmetric gene distribution than in its parent group. The net result of this is that there is likely to be a greater proportion of recessive traits seen in the subpopulation so that, if that recessive gene is advantageous, it's downstream trait stands a better chance of becoming the majority feature in the population and the frequency of that gene will hence increase in the population, eventually generating a trait that is part of the normal phenotype.

Perhaps the most important advantage that a small group has over a large population in changing its gene frequencies in response to selection comes from the time (measured in generations) that it takes for a trait novelty (which may be the result of new mutation or of an extant but rare gene variant) to become a dominant feature. Kimura and Ohta (1969) showed that, under reasonable conditions, this took $4N_e$ generations (where N_e is the effective breeding population, which is always less that the total population). In short, the smaller the population, the faster the change occurs.

Once, of course, the new and genetically isolated population is substantially different from its parent population in its genetic profile, further genetic changes in both populations will amplify those differences, while new subpopulations can form recently isolated subpopulations. Factors that are important here include new mutations, alterations to chromosome structures, and further changes in the local environment that again change selection pressures. It is these later changes that cumulatively drive genomic speciation.

Speciation is due to the accumulation of small genetic differences

The best and most direct evidence that speciation is due to the accumulated build up of genetic differences in populations that derive from a single founder population comes from the very few examples of ring species. This is a collection of related species that derive from a single founder population, subgroups of which extend clockwise and anticlockwise around an inhospitable central domain. This geography ensures that interbreeding is limited to neighbors, each of which can be seen as a subspecies. The criterion that defines a ring species is that, when the groups at the limits of the clockwise and anticlockwise migrations eventually meet up, they cannot breed and so meet the criterion of being different species (Irwin et al., 2001). Three well-known examples of ring species are the populations of *Larus* gulls surrounding the North Pole, song sparrows encircling the Sierra Nevada in California, and the greenish warbler (*Phylloscopus trochiloides*) populations around the Himalayas (**Figure 15.2**), with the last probably being the best known.

The small greenish warbler birds form a set of anatomically distinct populations that overlap at the margins of their territories and are mainly subspecies in that each can breed with members of its neighboring populations. There is, however, a discontinuity in breeding

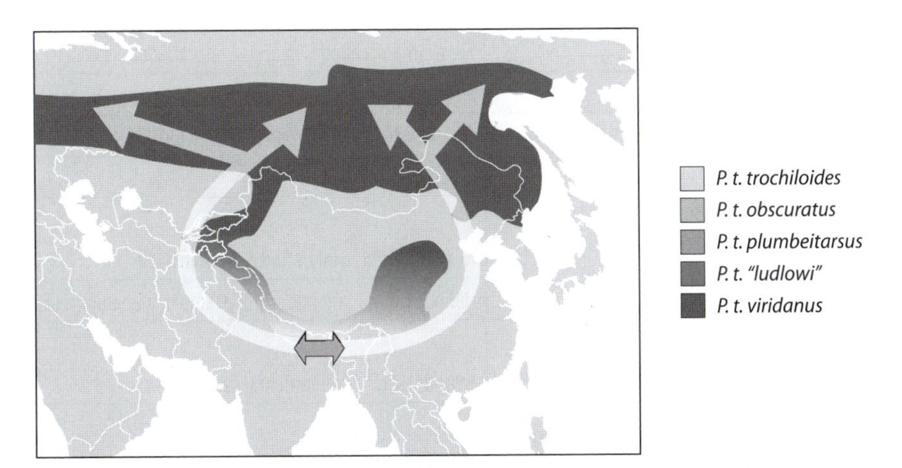

Figure 15.2 The ring of greenish warbler subspecies that spread out from the south to surround the Himalayas and meet up in Siberia where the overlapping populations no longer interbreed.

ability between the two adjacent groups at the northern part of the ring that derive from the easterly and westerly migrations. This started perhaps a few million years ago when a single population of birds in the southern Himalayas started to spread east and west. Groups formed a succession of minor variants as they colonized new habitats, each of which had its own ecological features, with the genetic differences between neighboring variants being sufficiently small that fertile hybrids could form at the borders of adjacent territories (this is the first stage of parapatric speciation (**Table 15.2**).

Further groups of birds continued the expansion in both directions, forming new variants as they encountered and then adapted to new habitats. When, however, the east-migrating and west-migrating populations eventually met up north of the Himalayas in central

Table 15.2: Types of speciation
(Note that some of these categories can be seen as overlapping. (From Coyne JA & Orr HA [2004] Speciation. Sinauer Press.))

Speciation type	Definition
Geographic speciation	
Allopatric speciation:	Subpopulation finds a new niche outside the parent territory
Peripatric speciation	A small group becomes separated from its parent population
Vicariant speciation	A population is split in two by a major geographic feature
Parapatric speciation	Subpopulation forms at border between two parent species
Sympatric speciation	Subpopulation finds a new niche within the parent territory
Ecological speciation:	Subpopulations become separated by ecological barriers
Habitat isolation	One species cannot use another's environment
Pollinator isolation	For example, pollen-carrying insects distinguish flower variants
Timing (allochronic) isolation	For example, different breeding or spawning times
Behavioral and nonecological isolation (reducing gene flow across populations)	
Mating system isolation	For example, self-fertilization
Behavioral isolation	For example, changes in appearance that affect courtship
Mechanical isolation	For example, incompatible reproductive apparatuses
Gametic (postmating, prezygotic) isolation	For example, sperm or pollen fails to fertilize
Postzygotic isolation	
Extrinsic isolation:	
Ecological inviability	Normal hybrids cannot find a suitable niche
Behavioral sterility	Hybrids cannot find mates (for example, inadequate courtship)
Intrinsic isolation:	
Hybrid inviability	Hybrids suffer developmental defects
Hybrid sterility:	
Physiological sterility	Hybrids have reproductive system defects
Behavioral sterility	Hybrids show inadequate courtship behavior

Siberia, these genetic differences had built up to the extent that the new neighbors could no longer produce offspring. The neighbors therefore have to be considered as a different species (more details on these birds, together with the songs of the various populations, are available at www.zoology.ubc.ca/~irwin/GreenishWarblers.html).

Genetic analysis of the greenish warblers has shown differences between groups that increase with distance of separation, although not linearly. There are, however, no major chromosomal distinctions between the subspecies, a result indicating the recent nature of the separations (Alcaide et al., 2014). It is, however, not yet clear either whether this interbreeding failure reflects an unwillingness to interbreed or a full genetic incompatibility. It is also not clear whether this set of variants is stable. Martins et al. (2013) expect that, within 10–50 Ky, each variant subspecies will become a distinct species, and that what is seen now represents only an intermediate stage on the way to dispersed speciation. The current family of greenish warbler variants not only highlights the difficulty in being precise about where the border lies between variants and species, but it also demonstrates how small differences, none of which alone block the formation of normal intergroup hybrids, can accumulate and lead to speciation.

Process of speciation

The essential process of speciation is clear: a subgroup of a population becomes reproductively isolated in some novel habitat, with that subgroup having an unrepresentative genetic profile due to genetic drift. What happens next depends on how the spectrum of phenotypes in a founder population handles the selection pressures associated with the new environment. This interaction cannot be predicted, but it is likely that many founder populations will fail to thrive and be lost. Those fortunate enough to do well will have trait variants enabling them to capitalize on the character of their new environment. Of particular interest is the possibility that local predation rates may be low, thus allowing the small group to expand rapidly both its numbers (the founder flush effect) and its range of phenotypes, to the extent that a distinct population emerges that differs significantly from the original population (Templeton, 2008).

Two factors then exacerbate any existing genetic differences and initiate new ones: first, the different selection pressures on reproductive success to which daughter populations may be subject; and, second, the genetic effects of drift and novel mutation, the latter being a much slower process. As a result and over time, the parent and daughter populations will, provided that they remain reproductively separated, become more and more genetically distant and phenotypically distinct as gene flow between the two is blocked. The point at which any potential hybrid offspring become nonfertile is not easy to predict. Nor is it clear when hybrids are no longer viable. A first stage is likely to be that embryos will form but die, as happens in almost all goat–sheep hybrids (the former has 60 and the latter 54 chromosomes). Eventually, however, meiosis fails, and this usually reflects *asynapsis*, the failure of homologous chromosomes to pair during meiosis, as was shown by Bhattacharyya et al. (2013) when they analyzed the infertility of hybrids between two mouse species.

The question that has received the most interest in this context is how subpopulations can become distinct. Coyne and Orr (2004) list the various possibilities, the effect of which is that organisms cannot mate or cross-pollinate, hybrids become less fertile, and environmental factors reduce intergroup fertility (see Table 15.2). Without doubt, however, the most common reason for group separation is spatial separation: a subgroup of a population becomes physically separated from its parent population and finds itself in a new environment where it is able to flourish. This called allopatric speciation, and is typified by the polar bear, which evolved from an ancient Irish brown bear species (Edwards et al., 2011). Here, a small population became isolated at the northern periphery of a large population (this type of allopatric speciation is known as peripatric speciation) and adapted to the icy, white conditions by losing melanin from its hairs.

The classic example of allopatric speciation, however, is the evolution of the monophyletic tribe of 15 or so species that comprise Darwin's finches (en.wikipedia.org/wiki/Darwin%27s_finches). It now seems likely that the parent population was a group of grassquit birds, *Tiaris obscura*, which originated in the Caribbean islands, spreading to central and South America, reaching the Galapagos Islands some 2.3 Mya (Sato et al., 2001). The various species then evolved in the different islands, reflecting vicariant speciation (see Table 15.2). The period required for speciation is not known; indeed, the process is not complete here as fertile hybrids can be bred between some species if the strong breed-specific mating preferences based on song can be altered through imprinting (Grant & Grant, 2008). It is, however, clear that the main driver of the adaptive radiation was food type (de León et al., 2014; Chapter 10). Ground finches have broad, strong beaks for tearing at cactus roots and eating insect larvae, while cactus finches have narrow beaks for punching holes in cactus leaves to access the pulp.

The other well-known class of separation is sympatric speciation: this occurs where a localized niche within the normal environment for some population becomes available for colonization by a subgroup. This is a likely source of speciation in asexual organisms where, for example, a subgroup can benefit from a highly localized food source. Sympatric speciation can also occur under laboratory conditions of strong selection and can happen in the wild if two groups physically overlap but survive in different ecological niches. Thus, one group of Darwin's finches with broad beaks can survive next to another with narrow beaks, with the two groups being sufficiently distinct that they choose not even to try to breed with one another (Huber et al., 2007). Here, beak size is an example of a magic trait, one which not only drives selection, but also has a role in mating choice. This is because beak size has a dual role in food choice and in affecting the sounds and calls that each group makes, which are part of the courting rituals.

Of particular interest is ecological separation because it can drive various types of speciation; here, a barrier develops in a local habitat and this eventually leads to subgroup separation and reproductive isolation (Hendry et al., 2007; Nosil, 2012). Such barriers include local changes in plant life that encourage different invertebrate subpopulations and changes in the mineral and fertilizer content of the local ground, which lead to different plants flourishing in distinct but nearby areas. Under such circumstances and where selection pressures are very strong, genetic difference in local but related plant, invertebrate, and vertebrate populations can be seen in as few as 14 generations (Hendry et al., 2007).

There are also population genetics reasons based on analyzing multilocus models of sympatric speciation for believing that that the early stages of speciation can take place in around 100 generations (Fry, 2003). That said, sympatric speciation does appear to be considerably less common than allopatric speciation.

A special case of sympatric speciation is co-evolution, where two organisms can evolve to become mutually interdependent to the extent that the one cannot exist without the other – they are one another's prime environment or habitat. Here, each organism defines the immediate component of the other's environment for the purposes of speciation. In the previous chapter, the topic of mutualism was discussed in the context of selection; co-evolution is the end result of this process (see Chapter 12). Co-evolution can extend beyond pairs of organisms and involve whole ecosystems to the extent that groups of organisms, both plants and animals, become so interdependent that loss of a single member of the system has wide and deleterious ramifications (Gu et al., 2015).

Species flocks

The dynamics of speciation have been illuminated by the study of species flocks. These are groups of geographically related but anatomically distinct species that probably evolved from single founder populations that colonized new areas in which a variety of distinct niches encouraged immediate allopatric speciation. The best-known case is, again, the Galapagos finches, but a far more recent example is the flock of cichlid species in Lake Victoria (Danley et al., 2012), which has 500–1000 anatomically distinct species. It is usually impossible to know with any precision when a process of speciation starts, but in this case the geological evidence shows that the lake only filled some 14.6 Kya (Seehausen, 2002). At this point, it seems that the lake was seeded by a small founder population of cichlids from adjacent streams with enough genetic variation to produce the considerable spectrum of anatomical traits seen today (Schluter & Conte, 2009).

It is reasonable to assume that the original population, wherever its origins, found itself in an environment in which there were few, if any, predators and was able to expand rapidly with few constraints on phenotypic and genetic variants, many of which found their own niches in the new lake. The assumption that these fish breed annually gives an upper bound of around 15,000 generations for novel anatomical features becoming established. This tallies with the figure of little more than 10 Ky and the same number of generations for the change in stickleback morphology that followed the evolution of freshwater variants from their marine forebears (see below and Chan et al., 2010).

These fish colonized a large and new lake with few, if any, predators; there was thus little to inhibit the evolution of a very large number of variant groups and competition between these groups would have been minimal, initially at least, so providing the conditions for both parapatric and sympatric speciation. Goldschmidt (1996) makes clear that there are many separate species in Lake Victoria that occupy similar functional niches but exist in geographically distinct locations: there are, for example 109 fish eaters, 29 insect eaters, and 21 zooplankton eaters (**Figure 15.3**). It would therefore be surprising if, as populations increase and move around the lake, these numbers did not decrease over the next few thousand years through competition between species currently occupying similar but distinct niches.

The anatomical variants seen today in the cichlids of Lake Victoria particularly include those tissues involved in food choice such as tooth number and shape, pharyngeal muscle size, and lower-jaw shape, and those involved in sexual selection, particularly male scale coloration and superficial ectodermal adornments (Miyagi & Terai, 2013). Goldschmidt (1996) has documented how anatomical adaptations, mainly to the oral cavity and teeth, have arisen that have permitted different species to feed off specific groups of organisms such as prawns, insects, other fish, algae, and snails (see Figure 15.3). Note that it is simply impractical to classify all these variants on the basis of interbreeding potential and more traditional morphological assays have had to suffice.

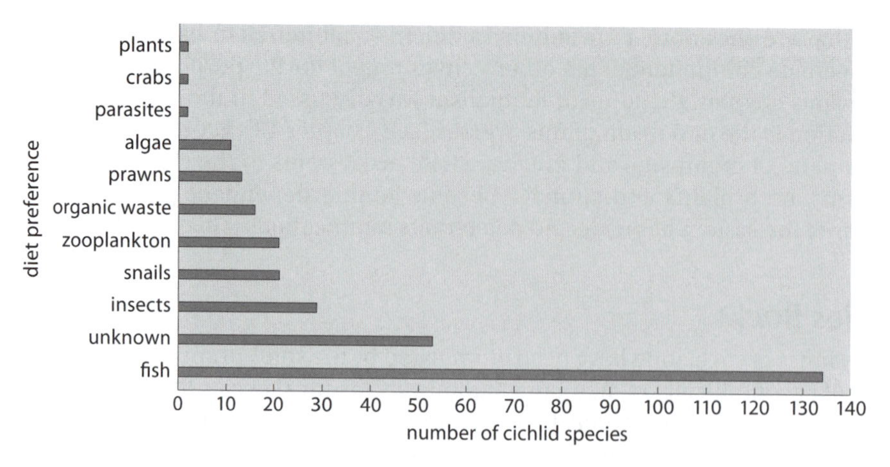

Figure 15.3 The numbers of cichlid species with particular diet preferences in a sample from Lake Victoria.

The evolution of ~600 distinct cichlid species that are claimed to exist in Lake Victoria on the basis of a mere 14 Ky of evolution from a small founder population gives a speciation time of a few thousand years, even allowing for single populations producing further subpopulations before full speciation has occurred. This period is so short that it demands some explanation. Part of this, at least, comes from the nature of the r selection strategy used by bony fish where thousands of eggs are produced as opposed to the very few produced by organisms using the K strategy. Normally, the many embryos and young fish (known as fry) produced as a result of the r strategy have a very low survival rate because they are eaten by predators. In a new habitat with few predators, however, r selection has the advantage that it allows very rapid growth in a population and, through normal recombination and mutation, the creation of a great number of variants. The argument against claiming that this mechanism is sufficient to achieve speciation in few thousand years (and probably generations) is that, while this figure may be enough for novel anatomical features to form, it is unlikely to have been enough for sufficient chromosomal alterations to accumulate for true speciation to occur. It would not therefore be surprising were most of these species still able to interbreed and, indeed, laboratory experiments have shown that hybrids can form (Keller et al., 2013; Turner et al., 2001).

There is another possibility: Lake Victoria is, for evolutionary purposes, much older than is claimed. The argument for its age being 14,600 years is that it was then that the deepest part was dry. Fryer (2001) has, however, pointed out that, for all the geological evidence, nothing is known about the various ponds and rivers that helped fill the new lake or about the various fish and other organisms that colonized it. While it seems that the cichlids in Lake Victoria originally came from one or two founder species, it is quite possible that the populations of fish that colonized the new lake were far more diverse than used to be thought and that these last common ancestors were present long before the lake filled. If this is so, then speciation will have been rather slower and more typical of other habitats than has been claimed, although still rapid (~135 Ky; Coyne & Orr, 2004).

Artificial speciation

The production of subspecies or variants by breeders is, of course, a well-established and fast way of producing organisms with specific features, be it pigeons with beautiful plumage (see Figure 12.1) or plants with high seed productivity. It does, however, seem unlikely that such breeding has actually led to a new species unable to produce fertile offspring with its parent population. This commercial work is complemented by a series of experimental explorations into the early stages of speciation, and most have used *Drosophila* as the model organism. This is because it breeds rapidly, its genetics are well understood, and it is cheap and easy to work with. The key feature of all of these experiments is that a single population is split or splits itself through habitat choice. Each group is then grown for many generations under its own choice of habitat and experimental breeding is conducted at regular intervals to see if the differential selection is enough to isolate the groups reproductively (Rice & Hostert, 1993). Such work sets a minimum generation number for at least the early stages of speciation to take place.

The best-known set of such experiments was performed by Rice and Salt (1990), who emulated sympatric events: they took a population of *Drosophila* pupae and placed them in a maze, with choices for accessing food that included light or dark (phototaxis), up or down (geotaxis), and acetaldehyde and acetone vapors (smell). They selected two groups of flies: the first emerged early from pupae and chose the dark, upwards, acetaldehyde routes to food, while the second emerged late and chose the light, downward, acetone route. After 25 generations of selective breeding, the two groups were unwilling to interbreed, and so had passed the first critical stage on the way to becoming two distinct species driven to separate under sympatric conditions. Such was the strength of the selection here that this figure is probably near the absolute minimum number of generations for the early, reproductive-isolation stages of speciation to occur on the basis of behavior.

A particularly clever aspect of these experiments was that they built on the considerably genetic diversity within a normal *Drosophila* population by selecting for several behavioral traits simultaneously. Rice and Salt were therefore able to capitalize on further opportunities for diversification that came from genetic drift in the small populations that chose four behavioral options - the number of generations involved in these experiments was almost certainly too low for mutation to play a role here (Waddington, 1961). It is worth noting that these experiments are very different from those of Waddington that were discussed in the last chapter. He attempted to breed into flies novel anatomical features that could normally only be experimentally induced, having no interest in whether these anatomically distinct flies could interbreed with their ancestral population. In fact, they still could, and Waddington later used this ability for genetic analysis of what had taken place as the lower thoracic region with halteres was replaced by the middle thoracic region with wings (Waddington, 1961).

The genetics of speciation

As daughter and parent populations become more different, their ability to interbreed declines and there are several stages in this process. The first is that the two separated populations fairly rapidly become less willing to mate, a behavioral process known as reinforcement, as was seen in the work of Rice and Salt (1990). The second step occurs as hybrids become increasing less fertile over time, eventually becoming sterile; mules, which are horse–donkey hybrids, have just reached that stage, while lion–tiger hybrids have not. Finally, mating fails because even if organisms from the two groups mate, meiosis or fertilization fails, or the embryo dies.

The precise genetic reasons for the slow decrease in hybrid fertility over time are not entirely clear but include increasing genetic incompatibility and decreasingly efficiency of meiosis. In vertebrates, it is usually the male that becomes infertile first, and this in accordance with Haldane's rule, which is based on his observation that "when in the offspring of two different animal races one sex is absent, rare, or sterile, that sex is more likely to be the heterozygous (heterogametic) sex" (Schilthuizen et al., 2011). This essentially means that where the sex chromosomes are different (usually males), any of its genes that acquire mutations will produce an abnormal phenotype as the mutant gene cannot be shielded by a wild-type gene. Further evidence on the nature of hybrid infertility came from inter-strain crosses of various species, with a particular focus on *Drosophila*: these demonstrated that hybrid sterility was most commonly associated with genes on sex chromosomes that particularly affected fertility (Coyne & Orr, 1989; for review, see Ridley, 2004).

There are also speciation genes, so-called because they have been genetically identified as playing a role in reproductive isolation and, in some cases, this role as been confirmed by genetic engineering, usually in *Drosophila* (Nosil & Feder, 2012). Examples of such genes here are Odysseus (*Odsh*), hybrid male rescue (*Hmr*), lethal hybrid rescue (*Lhr*), and nucleoporins 96 and 160 (*Nup96* and *Nup160*, respectively; Nosil & Schluter, 2011), but the reasons why they facilitate isolation remain unclear. There are, however, equivalent genes in mice, with six hybrid sterility (*Hst*) gene loci now identified, two of which are involved in sperm flagellar assembly (Pilder et al., 1993), and a possible reason for lowered fertility here is that interaction between the two distinct flagellar proteins of the hybrid sperm may lead to reduced motility.

In due course, a daughter population isolated from its parent population will become increasingly distinct as both acquire mutations that affect various aspects of interbreeding. Such changes can affect the geometry of the reproductive systems, the immune responses to sperm, and behavioral differences that further exacerbate mating preferences. Eventually, sufficient chromosome rearrangements such as inversions, duplications, coalescences, and splittings accumulate to the extent that meiosis becomes impossible through nondisjunction and the chromosomes from the two populations are unable to pair or separate properly during first meiosis and, later, perhaps mitosis. At this point, separation is complete and the original subpopulation has given rise to a new species.

The rate of speciation

On the basis of classical evolutionary theory and the fossil record, natural speciation was traditionally thought to have been a very slow and uniform process (this is phyletic gradualism), with individual species surviving for around 1–10 My before becoming extinct and perhaps being replaced by a sister taxon (Lawton & May, 2015). More recently, Coyne and Orr (2004) have reviewed the typical time required before an existing species gives rise to new ones (the biological speciation interval) and find similar figures. The major counter-argument to this idea that speciation was always slow came from Gould and Eldridge (1972, updated in 1993) who pointed out that punctuated equilibrium provided a better description of evolutionary change. Here, the long periods of slow evolutionary change could be interspersed with, for a variety of reasons, periods of relatively rapid speciation.

Such fast speciation certainly occurred in the periods following major extinctions such as at the ends of the Permian and Cretaceous Periods. The fossil record clearly shows there were major radiations over the next few million years as emptied habitats were recolonized by new taxa that evolved from previously minor taxa that had survived extinction.

After the Permian extinction, for instance, diapsid reptiles were replaced by synapsids as the dominant vertebrate taxon; they were, in turn, replaced by mammals after the Cretaceous (Chapter 6). Predation rates were obviously low in those early empty habitats and speciation was initially rapid; the process then settled down and became slower and more gradual. Eldridge and Gould were also able to find examples in the fossil record, where long periods of slow change were followed by bursts of rapid change that were not associated with major extinctions, but were in accordance with the predictions of punctuated equilibrium. Organisms for which longitudinal studies showed both types of change included trilobites, snails, bryozoans and dinosaurs (Eldredge & Gould, 1972; Horner et al., 1992). In most cases, however, the fossil record is too incomplete to establish with any certainty how long the process of speciation has usually taken.

Major changes can, indeed, be very rapid under experimental conditions of strong selectivity, as the work of Waddington (1953) and Rice and Salt (1990) has shown, but change is much slower in the wild. Even the fastest examples known, such as formation of the cichlid flock in Lake Victoria and the loss of pelvic spines in freshwater sticklebacks, required thousands of generations. This slowness is illustrated by a well-studied natural experiment. When anatomically modern humans colonized Europe some 40 Kya, they encountered groups of Neanderthals, a group that had left Africa some 400 Ky earlier. Assuming a reproduction time of about 20 years, these figures represent some 20,000 generations. As current Europeans include small amounts of Neanderthal DNA in their genome (Sankararaman et al., 2014), there was clearly no bar to intergroup mating. Not even the behavioral bar to reproduction had been reached (Chapter 16).

Although the morphological and behavioral changes that arise due to selection can be obvious, and can occasionally arise relatively fast on an evolutionary timescale, such changes say little about how long it takes for two groups that separate to become genetically distinct species, and so unable to breed, even in principle. One group about which we have some evidence is the Canidae. They appear to have separated from a common ancestor species almost 2 Mya, and about the same number of generations, but there is no evidence of hybrid infertility among this group or that of the great cats, which separated a little longer ago. Another group is the Equidae: the last common ancestor of today's eight species lived 4.5–4 Mya and almost all hybrids are sterile, except those between the domestic horse and Przewalski's horse, which separated 72–38 Kya (Lau et al., 2009; Orlando et al., 2013). Mules whose parents are horses and donkeys are infertile with the very occasional exception (Ryder et al., 1985). Given that mares first breed when they are 1–2 years old, it seems that hybrid infertility here takes 3–4 million years of species separation and perhaps half as many generations.

The figures for *Drosophila* are, at first sight, different, with an estimate of 150–350 Ky for novel speciation (Ridley, 2004). As the *Drosophila* reproductive cycle takes about 3 weeks in the wild, however, these figures suggest that it takes a minimum of several million generations for speciation to occur here. The situation for the evolution of *Homo sapiens* is unclear. The line that led to *Homo sapiens* separated from the clade of chimpanzees and bonobos some 7–8 Mya, with our taxon appearing about 200 Kya (Chapter 16). Given a reproductive cycle of about 25 years for hominids, these timings represent 300,000 and 8000 generations, although the question of the degree of speciation involved in the evolution of *H. sapiens* is left to the next chapter (the non-mating threshold has clearly been achieved). In short, it seems that full genetic speciation, in animals at least, usually requires a million or more generations under normal selection pressures.

The broader perspective

The forming of a new species that reproduces sexually is the culmination of a set of small-scale events that extends from the genome to the environment in its widest sense. At the organism level, these events initially involve mutations that affect protein function, protein networks, and their outputs. These molecular changes initially modify embryonic development and adult physiological function in a few individuals, with these effects slowly spreading across a population giving it a phenotype profile different from its parent population. At this point, selection pressures from the environment start to exert their effects and the net result is that some phenotypes will do better and others worse in the context of leaving fertile offspring. Over the long term, the former will survive and the latter will be lost as the features of the phenotype and the underlying genotype optimize themselves in the new environment. Over the even longer term, further mutations and additional changes to chromosome number and content ensure that this new population can no longer successively mate with the parental population. At this point, it has now met all the criteria for a new species.

When one looks at how well adapted a recent species is to its environment, it is easy to feel that they were made for one another, but a better metaphor is Dawkins' *The Blind Watchmaker* (1996), where the predominant driver of success is trial and error and eventually success, with each attempt being tested and to destruction if necessary. It is hard to get an intuitive feel for just how long all this takes: it generally needs more than a million generations for the mix of genetic drift, mutation, chromosome change, and selection within a population to effect full speciation – an interminably long time.

Working out the details of how speciation happens has thus been a triumph of scientific research because each step in the process is so hard to investigate experimentally. Some, such as mutation and selection, can, with difficulty, be modeled under laboratory conditions, but the later changes to chromosome are so slow and rare that their study is still a challenge. Such studies can usually only focus on one aspect of phenotypic change; the reality is that each aspect of an organism's phenotype is under continual pressure and, although change may be slow, it is ubiquitous across the genome and in the way that phenotypes change. Darwin's original broad ideas on the mechanism of evolution and the mode of speciation have mainly been proven right, but it has taken a century and a half to fill in most of the major details.

Key points

- A species is best defined as a group of organisms that is reproductively isolated from all other species. The evolutionary implication is that such a species cannot be absorbed back into its parent species but can only generate further diversity.
- Where this definition cannot be used, a species can usually be defined as a group of organisms that can readily be distinguished from other groups, usually on the basis of morphological criteria.
- Species derive from the splitting of lineages: a new one starts when a small founder subpopulation, becoming isolated in some way from its parent population, includes variants that thrive in their new environment.
- Isolation can occur for a range of reasons, the most important of which is geographic (allopatric speciation); other types include the availability of a range of behavioral and ecological niches in which a subpopulation can flourish.

- Population genetics shows that small founder populations are important partly because of genetic drift, so showing relatively more recessive homozygotes than large ones (the Wahlund effect), and partly because, the smaller the population, the more rapidly a novel mutant can become established.
- The initial cause of variation in a new population is through genetic drift, but, as new mutations affect the phenotypes, it starts to become sufficiently genetically distant from its parent population that any hybrid offspring are eventually infertile and, in due course, all reproduction becomes impossible.
- Once a population is reproductively isolated from its parents, it can give rise to further subpopulations before it becomes new species. Species divergence can be rapid.
- Selection pressures in the wild are normally low and speciation is very slow. Under controlled breeding, however, selection pressures can be very high, and subpopulations can become reproductively isolated, in the sense that they will not mate with one another, in as few as 25 generations.
- Full genetic speciation of a sexually reproducing eukaryote is unlikely to take less than a million generations.

Further reading

Barton NH, Briggs DEG, Eisen JA, Goldstein DB, Patel NH (2007) *Evolution*. Cold Spring Harbor Laboratory Press.

Coyne JA, Orr HA (2004) *Speciation*. Sinauer Press.

Ridley M (2004) *Evolution*. Wiley-Blackwell.

CHAPTER 16
HUMAN EVOLUTION

Human evolution is important for two linked reasons. First, it is about "us" and we have a natural interest in our biological history. Second, there is more evidence on the evolution of our species from fossils, from genomic analysis, and, uniquely, from artefacts than for any other. The main aims of this chapter are to (i) summarize the evidence on the evolutionary steps that led an early group of hominins descending from the African great apes eventually to give rise to modern man; (ii) discuss the various migrations from Africa through which modern humans colonized the world; and (iii) consider what light the anatomical and artefactual evidence casts on these events. In the wider context, this deep knowledge about the evolution of a single taxon helps answer broader questions about the processes of evolution.

No question in evolution has attracted as much attention as that of how humans evolved. The obvious reason is that we humans are unique: no other animal possesses our mental, speech, and creative capacities. Although creationists believe that producing such uniqueness required divine intervention (otherwise known as "intelligent design"), this idea is refuted by the evidence. A series of fossil skulls, together with a great many artefacts, has shown that, as hominins diverged from their hominid ancestors some 7 Mya (**Table 16.1**), the human cranium gradually tripled in volume and the genus *Homo* became increasingly more sophisticated in its capabilities.

It was, of course, Darwin who initiated serious thought on this subject through his book *The Descent of Man, and Selection in Relation to Sex*, which was written in 1871, at a time when very little as known about the subject. Since then, a very large amount of fossil, genetic, and artefactual evidence been discovered, while information has slowly accumulated on the many times that humans left Africa to colonize the world. This information has clarified the general lines of human evolution, the essential features of which are covered in this chapter; further detail is covered in the literature listed at the end of the chapter ("Further reading"). If, however, we stand back from our natural, personal interest in our origins and think of today's humans as just another organism that evolved over perhaps 7 Mya from rather different ancestors, it becomes clear that the enormous amount of evidence now available about *Homo sapiens* illuminates general questions about speciation, speed of anatomical variation, development of neuronal function, and selection.

It is not possible in a chapter as short as this to do more than indicate the richness of the evidence on human evolution now available and some of the problems that still remain – the area is, and probably always will be, contentious. One difficulty is classifying the number of distinct human taxa that there have been and deciding which should be assigned

Table 16.1: Terminology associated with human evolution		
Hominoidea		Lesser apes plus great apes
Hominidae		All modern and extinct great apes (today: humans, chimpanzees, gorillas, and orangutans)
Homininae		The Hominini and the Panini (chimpanzees and bonobos) clades
Hominini		All Homo taxa together with all fossil taxa more closely related to humans than to any other living taxon (Wood & Lonergan, 2008)
Ardipithecus	5.6–4.4 Mya	A very early hominin taxon adapted to life in trees
Australopithecus	4–2.4 Mya	A group of early hominin taxa that included not only gracile, but also robust members and was probably bipedal
Paranthropus	2.3–1.3 Mya	An early hominin taxon with a robust anatomy and large postcanine teeth (megadontia) probably related to *Australopithecus*
Homo	2.3 Mya to present	A group of bipedal hominin taxa characterized by increasing cranial capacity
Miocene Period	23–5.33 Mya	
Pliocene Period	5.33–2.588 Mya	Start of glacial period
Pleistocene Period	2.588 Mya–11.7 Kya	End of glacial period
Holocene Period	11.7 Kya–present	

species status. A second is eliciting the detail of the various migrations out of Africa and onwards around the world that were initiated by various early groups. A third is the lack of direct information on when the capacities for speech and associated cultural inheritance first appeared in *H. sapiens*. This last point is related to the evolution of the formidable human cerebral cortex and its properties, together with the other anatomical changes required for speech - an ability that chimpanzees lack, even though they have sufficient mental capacity to be taught the elements of language (Beran & Heimbauer, 2015).

The context of human evolution

The basic features of hominin evolution are unambiguously shown by the anatomical and genomic evidence (**Figure 16.1**). Some 7 Mya, a spur formed on the great ape clade and this eventually led to *H. sapiens*. It is now clear on the basis of overwhelming fossil evidence that all the essential features of human evolution took place in Africa alone. The characteristics such as pigmentation and facial morphology that define human subspecies in the rest of the world reflect later minor changes that appeared in populations that had left Africa. It is generally assumed that the most important populations associated with human evolution were those in the savannah areas of Central, Eastern, and Southern Africa, as this is where fossil and other evidence has been found. There is, however, a famous saying among human evolutionary biologists that "absence of evidence is not evidence of absence," and the question is still open as to whether some early humans also lived in the wetter and more forested areas of Africa, where any early evidence would be likely to have been degraded and, for this and geographical reasons, hard to find.

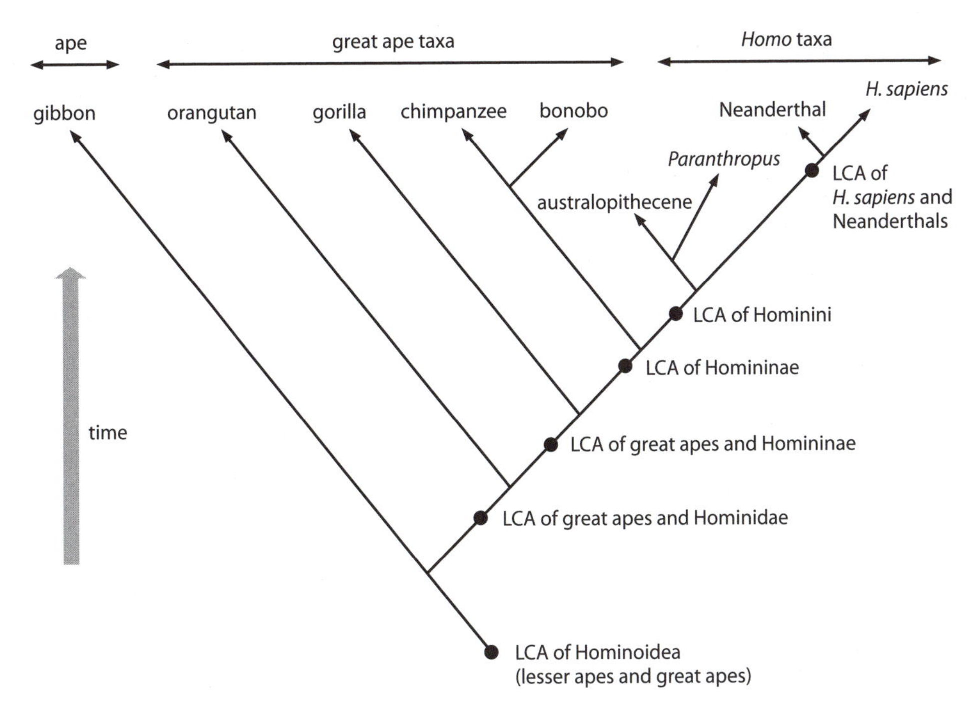

Figure 16.1 A cladogram showing the core line of descent of humans from monkeys over the last 7–5 My. Note: time is not to scale (LCA: last common ancestor).

The context within which the core aspects of human evolution took place in Africa is now clear. The climate and general faunal evidence suggests that the weather in the Pliocene (5.3–2.6 Mya) was perhaps 2–3°C warmer than today in Africa, and sea levels perhaps 25 m higher, with oscillating periods of wet and dry conditions (deMenocal, 2004). With the coming of the Pleistocene Period and the ice ages (2.6, 1.7, and 1.0 Mya) that so affected more northerly and southerly regions, there were times when the equatorial climate became much more arid and cooler, while sea levels dropped markedly. Over this whole period, there was a gradual change over nondesert Africa as forested areas became savannah.

The fossil evidence

It is generally agreed that the last common ancestor of human and chimpanzee lived ~7 Mya and was far more chimp-like than humanoid, albeit that we have relatively little Homininae skeletal material from that period. The major differences between chimpanzees and humans today are given in **Box 16.1**, and the most obvious skeletal changes reflect the progression from likely knuckle walking in a last common ancestor (like today's chimpanzees) to bipedal walking and running (see Lieberman, 2013, for a detailed analysis of this and other differences). Such differences probably reflect an adaptation to a savannah-based as opposed to an arboreal lifestyle.

It is interesting to see how the fossil evidence illustrates the timing of when the anatomical differences, at least, evolved. Eugene Dubois was the first to discover a hominin fossil that he named *Homo erectus* (because its skeleton was the first to show an erect posture). This discovery was made in 1891 in Java and led to the idea, now known to be wrong, that the

Box 16.1: Human–chimpanzee differences: 7 My of evolutionary change

Chimpanzees have 48 and humans have 46 chromosomes (human chromosome 2 includes much of chimpanzee chromosome 4 (Ijdo et al., 1991)).

Anatomical differences

Changes	Flattened face and enlarged nose
	Teeth and jaw change associated with a more varied diet
	Shortened and raised tongue, and lowered larynx associated with speech
	Enlarged cerebral cortex
	Fingers shortened and other hand changes
	Various modifications associated with bipedalism
	Coarse body hair becomes fine

Physiological differences

Gains	Sweating
	Obligate bipedal able to run long distances
	An improved diet that included meat and tubers

Behavioral and cultural differences

Gains	Speech
	Mental capacity
	Dexterity for tool making
	Art: body decorating, sculpture, painting etc.

origins of *H. sapiens* were not African. It was not until 1924 that Raymond Dart discovered the first human skeletal material in South Africa and this, now known as *Australopithecus africanus*, was far more chimpanzee-like than *H. erectus* and hence older. Since then, the skeletons of many other taxa have been discovered in Africa and around the world, and in increasing numbers as public interest has stimulated research and funding. Wood and Lonergan (2008; **Figure 16.2**) have ordered these fossils into six groups on the basis of anatomical diversity. It should, however, be emphasized that fossilized skeletons for many taxa are often incomplete and numbers are small. We generally have little idea of either the range of morphologies or of the extent of sexual dimorphism for a particular taxon, or, indeed, whether differences reflect group variation or true speciation.

Earliest hominins (7–4.5 Mya)

Sahelanthropus tchadensis, *Orrorin tugenensis*, and the later *Ardipithecus* hominins (*Ardipithecus* means "ground or floor ape"). The limited amount of skeletal material here comes from central and sub-Saharan Africa, and the features that mark out the *Ardipithecus* taxon, in particular, from chimpanzees are its more generalized teeth and a change to the foot and limb bones that would have facilitated a bipedal walking gait.

Archaic hominins (4.5–2.5 Mya)

The *Australopithecus* ("southern ape") taxa. These hominins were mainly of slender (or gracile) build, although a few taxa were robust, small (1.2–1.4 m high), and show several

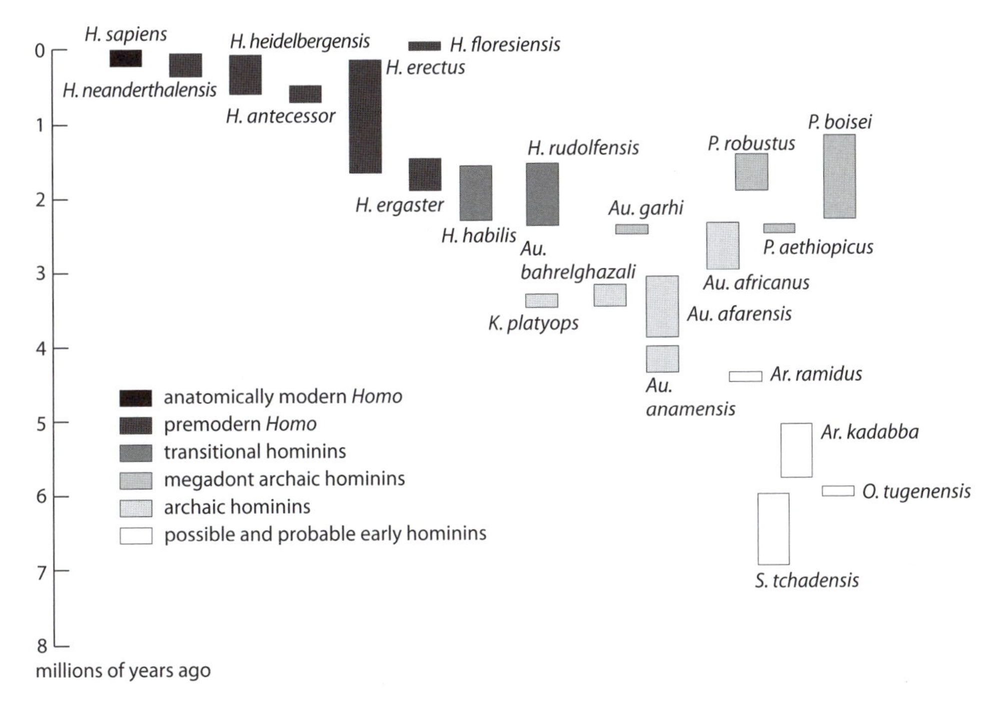

Figure 16.2 A broad ("speciose") taxonomy of human speciation. The columns reflect evidence of the earliest and latest identified fossils of each taxon but may be underestimates.

more advanced features than the earliest hominids. Their heads had become more human in orientation, the cranial base was shorter, their hands were more human-like (for example, the fingers had shortened relative to the thumb, so facilitating manual dexterity), and they had foot anatomy that was typical of bipedal organisms (Kimbel & Rak, 2010; Ward et al., 2012). Dennis et al. (2012) have argued that this was the first hominid to have a copy of *SRGAP2*, a gene involved in cortical development. This might account for the fact that the size of the *Australopithecus* brain is relatively large for its height compared with that of chimpanzees.

Archaic megadont hominins (2.5–1 Mya)

The *Paranthropus* ("near man") taxa. These were similar to and probably descended from the archaic australopithecines but distinguished from them by their robust skeletons, large jaws, and postcanine teeth (Wood & Constantino, 2007). Like all earlier hominins, their fossils are found in the arid areas of the east side of Africa in sub-Saharan and central areas – there is little chance of finding fossilized material in jungle areas. The *Paranthropus* taxa appear to have died out, leaving no successors.

Transitional hominins (~2.5–1.4 Mya; Table 16.2)

These taxa mark the beginnings of the *Homo* group, and include *H. habilis, H. rudolfensis,* and *H. gautengensis. H. habilis* had a short femur, long upper limbs, and other features marking it out as living a mixed arboreal and terrestrial existence (Curnoe, 2010). It also had an advanced cranial anatomy and seems to have been the first hominin who shaped stones to make tools (Plummer, 2004). While some chimps use stones to crack nuts,

Table 16.2: The fossil record of the various *Homo* "species"

Species	Known existence (Mya)	Location	Adult height (cm)	Adult weight (kg)	Cranial volume (cm^3)	Fossil record
Red Deer Cave people	0.0145–0.0115	China				Very few
Homo floresiensis	0.10–0.012	Indonesia	100	25	400	Seven individuals
Homo sapiens idaltu	0.16–0.15	Ethiopia			1450	Three craniums
Homo sapiens (modern *Homo*)[a]	0.2–present	Worldwide	150–190	50–100	950–1800	Still living
Homo rhodesiensis	0.3–0.12	Zambia			1300	Very few
Homo neanderthalensis	0.35–0.03	Europe, western Asia	170	55–70 (heavily built)	1200–1900	Many
Denisova hominin	0.35–0.03	Russia (Siberia)				A few bones; three individuals
Homo heidelbergensis	0.6–0.35	Europe, Africa, China	180	90	1100–1400	Many
Homo cepranensis	0.9–0.35	Italy			1000	One skull cap
Homo antecessor	1.2–0.8	Spain	175	90	1000	Two sites
Homo georgicus	1.85	Georgia			~560	Five skulls
Homo ergaster	1.9–1.4	Eastern and southern Africa	190		700–850	Many
Homo erectus (premodern *Homo*)[a]	1.9–0.2	Africa, Eurasia (Java, China, India, Caucasus)	180	60	850 (early)–1100 (late)	Many
Homo rudolfensis	1.9	Kenya			700	Two sites
Homo gautengensis	> 2–0.6	South Africa	100			A few crania
Homo habilis (transitional hominin)[a]	2.2–1.4	Africa	150	33–55	510–660	Many

[a] The key species for lumpers looking for taxon parsimony (Wood & Lonergan, 2008). Note: *Homo naledi* is not included here as the timing data are still incomplete (Berger et al., 2015).

H. habilis made hand choppers and flaked stones to give them sharp edges that were useful, for example, in cutting up meat. Lieberman (2013) has discussed the various ways in which such tools freed up members of the *Homo* taxa from spending almost all their time foraging and eating, as do modern chimpanzees.

The differences between the three taxa are minor, being based on cranial and tooth morphology, the most common skeletal remains. The long phalanges of the limbs associated with *H. habilis* suggest that it had the sort of grasping ability associated with at least a partly arboreal life. The longest-lived taxon was *H. gautengensis*, which may have invented fire: burnt animal bones are associated with some, at least, of their skeletal remains. In terms of progress towards modern man, novel features include an increased cranial capacity relative to the archaic hominins, a diminished snout (a less-protruding mouth), a reduction in tooth size that together with tooth specializations suggest an increasingly broad diet, and further anatomical changes that would have facilitated bipedalism.

Premodern Homo (1.9 Mya–30 Kya; see Table 16.2)

This class includes two early and several later taxa. The earliest was *H. ergaster* ("workman" - named for the stone tools associated with its skeletons) that was first found in Africa. It was followed by *H. erectus*, examples of which have been found in Africa, Java, and many other places. One tribe that has been much studied recently is *H. floresiensis*, a taxon recently found on the island of Flores in Indonesia that had undergone dwarfism. Its ancestry is unclear but may date back to the earliest emigrations of *H. erectus* from Africa (Aiello, 2010). *H. erectus* was followed by a series of taxa that include *H. antecessor*, *H. heidelbergensis*, *H. neanderthalensis*, the Denisovans, and, most recently discovered, *H. naledi* (Berger et al., 2015). Several anatomical features mark these taxa out as being very different from the transitional hominins: they possessed modern bipedal limbs; they had increased cranial capacities relative to their size; they had smaller teeth that were more like those of modern humans; and a more modern face with a nose and chin.

The current view is that the last common ancestor of *H. sapiens* and *H. neanderthalensis* was probably *H. heidelbergensis*. This taxon is sometimes identified with *H. rhodesiensis* who lived at the same time, and evolved some 600–500 Kya; the two taxa had very similar skeletons, but the latter has only been found in Africa (see Table 16.2). Some *H. heidelbergensis* individuals left Africa around 500 Kya and migrated first through the Middle East and later northwards, eventually inhabiting much of continental Europe. It is from their descendants that *H. neanderthalensis* evolved some 250–200 Kya, while *H. sapiens* evolved around 100 Ky later from those members of the taxon that had remained in Africa.

There seems little doubt that even early premodern *Homo* taxa were obligate bipeds, while their enlarged cranial capacity is certainly compatible with an enhanced intelligence. Moreover, their reduced snouts and enhanced noses would have facilitated the increased amount of breathing required for running as opposed to walking. If so, and as running requires an ability for the body to lose heat easily, this was probably the period during which body hair thinned to a light fuzz and sweat glands spread over the body. Lieberman (2013) has discussed how this ability would have made hunting much easier: chasing animals that cannot sweat eventually causes them to become exhausted. In short, the new abilities shown by the premodern hominins would have enabled them to eat better than their ancestors and so to flourish.

These early hominids seem to have been the first to leave Africa and evidence of their existence has been found across Asia (see below) dating back to about 1.85 Mya. The earliest non-African skeletal material currently known is from the Dmanisi cave in the Caucasus, and, while it is clearly related to one of the early hominin taxa (probably *H. erectus*), it is hard to be sure which because its brain volume is smaller than expected (546 rather than the 850–1100 cc typical of *H. erectus*). Thus, Lordkipanidze et al. (2013) consider that the morphology reflects within-species variation of *H. erectus* and the taxon should be named *H. erectus georgicus*. Schwartz et al. (2014) consider, however, that the morphology of skull and the mandible, in particular, is sufficiently distinct to confirm the older view that this hominin is a distinct species and should be called *H. georgicus*.

Modern *Homo* (from 0.2 Mya onwards)

Homo sapiens, also known as anatomically modern humans (AMH), evolved some 200 Kya in sub-Saharan Africa (McDougall et al., 2005; Cyran & Kimmel, 2010), and spread over all the continent. Over the last ~110–70 Ky, groups of AMH have emigrated in a series of dispersals to colonize the whole world (those who colonized Europe some 40 Kya used to be known as Cro-Magnons or stone-age men). Earlier groups included the Neanderthals, who left Africa 450 Kya and colonized much of Europe, and the Denisovans, who left a little earlier and reached as far as Siberia; both groups died out some 35–40 Kya. For all of the quantitative differences in size and pigmentation amongst AMH subgroups today, it is hard to pinpoint any absolute, qualitative differences amongst them; there is certainly no interbreeding difficulty. There is thus no reason to doubt that the rich diversification of the hominini has been reduced to a single species, albeit one with a considerable degree of genetic diversity. This diversity is, however, clearly less than that of the Canidae, the taxon that includes dogs, wolves, foxes, and jackals, all of which can interbreed and whose last common ancestor lived some 4 Mya.

It has to be said that the fossil evidence, for all the discoveries in the last 25 years, remains unsatisfactorily incomplete. Producing a reasonable cladogram of hominin phylogeny on the basis of anatomy requires the identification of enough shared and derived characteristics among the 20 or so taxa so far identified to produce a coherent story, and the data are not yet adequate for this. The fossil record does, however, make two clear points. First, there is a fairly clear evolutionary line between the last common ancestor of chimpanzees and anatomically modern humans, defined by increasing cranial capacity and an increasing tendency towards bipedality (**Figure 16.3**), and, second, there was, at all stages in the evolution of humans, a fair amount of variability in the hominin line: there was no simple ladder of ascent to humans but a branching bush of taxa.

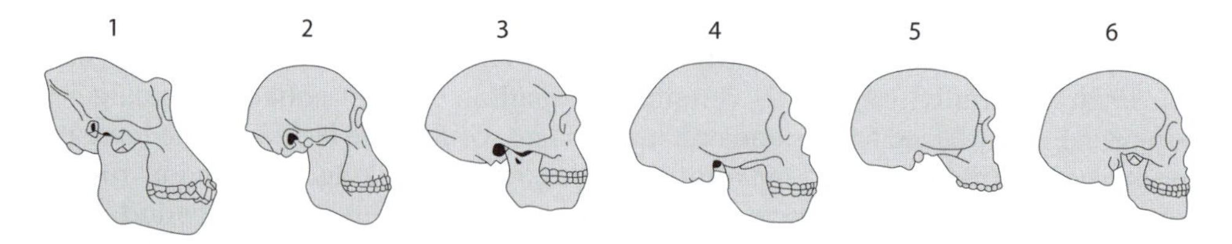

Figure 16.3 Skulls indicating cranial capacity. 1: A gorilla (c. 550 cc). 2: *Australopithecus* (c. 450 cc). 3: *Homo erectus* (c. 1000 cc). 4: Neanderthal (c. 1500 cc). 5: archaic human c. 0.3 Mya (c. 1100 cc). 6: modern human (c. 1400 cc). Note that the cranial capacities have not been adjusted to reflect body weight so that *Australopithecus*, has for its size (1.2–1.4 m), a larger brain than might have been expected.

The genetic evidence

The least ambiguous evidence on human evolution comes from analyzing DNA sequences, both genomic and mitochondrial, from the various *Homo* taxa (Oppenheimer, 2012). Fortunately, there is more of this than one might expect because DNA can decay slowly; under optimal conditions of dryness, it can last almost 500 Ky. The oldest example is mitochondrial DNA from Neanderthal ancestors (identified as late members of *H. heidelbergensis*) buried in a pit at Atapuerca (Spain) that date to around 400–300 Kya (Meyer et al., 2014). In addition, sufficient genomic and mitochondrial material for full sequencing of DNA has been retrieved from Neanderthals, Denisovans, and, of course, the many subgroups of *H. sapiens* (Veeramah & Hammer, 2014). The oldest full hominin genome so far sequenced comes from a Neanderthal woman who lived some 50 Kya in Siberia (Prüfer et al., 2014). Unfortunately, it has proven impossible so far to extract sequenceable DNA from the bones of the hominins found on the island of Flores and who might be descendants of *H. erectus*. Nevertheless, there are good prospects that other fossils will be discovered whose DNA will clarify our understanding of the lines of hominid evolution.

Detailed comparisons of genomic sequences from monkey, great ape, chimpanzee, and three human taxa, *H. sapiens*, the Denisovans, and *H. neanderthalensis* has now established the major lines of descent (**Figure 16.4**). Furthermore, there is clear evidence of interbreeding between *H. sapiens* and the Neanderthals and Denisovans that they encountered (Sankararaman et al., 2014; the arrows seen in Figure 16.4). AMH first seem to have left Africa around 100 Kya, and only met up with Neanderthals in Europe, perhaps 40 Kya. The sequencing evidence clearly shows that about 2% of the DNA of modern humans is Neanderthal-derived; it is thus obvious that the mutations and other genetic changes resulting from 400 Ky of separation were not enough even to send the two groups on the path to distinct speciation, under the criterion of a failure to mate. There is a similar story with the Denisovans: like *H. sapiens* and the Neanderthals, they were descendants of *H. heidelbergensis*. The Denisovan line emigrated from Africa before the Neanderthals, moving first to the Levant and then north to the Caucasus. In due course, they interbred with a later migration of AMH whose descendants now live in south-eastern Asia, particularly Melanesia, and Australasia (Reich et al., 2011) and even America (Qin & Stoneking, 2015). Sequencing shows that about 4% of Melanesian DNA is characteristic of Denisovan DNA (Reich et al., 2010). This and other computational evidence can be combined to show the likely lines of hominid evolution (see Figure 16.4).

Timing data on these populations comes from two sources: the age range and locations of the fossilized skeletons for each taxon and computational work on genomic and mitochondrial DNA sequences (Soares et al., 2009, 2012). In the latter case, the timings depend on two parameters, the generation time, which is generally taken to be 25 years or perhaps a little less, and the mutation rate, which is estimated to be between 1.1 and 2.5×10^{-8} per base pair per generation. In this context, analysis of the sequence changes among homologous genes in related organisms gives a figure of $1.3–1.8 \times 10^{-8}$ (Veeramah & Hammer, 2014). Coalescent analysis on mitochondrial DNA sequences from various groups, using a range of models for the population dynamics have been undertaken by Cyran and Kimmel (2010). Their results show that it matters relatively little as to which population model is chosen as each gives an expected coalescence time within the range 165–189 Kya, although the 95% confidence interval (CI) is somewhat wider. This figure represents the time when mitochondrial Eve, the last common ancestor of all anatomically modern females, lived. Similar calculations using Y chromosome sequences suggest first that Y-chromosome

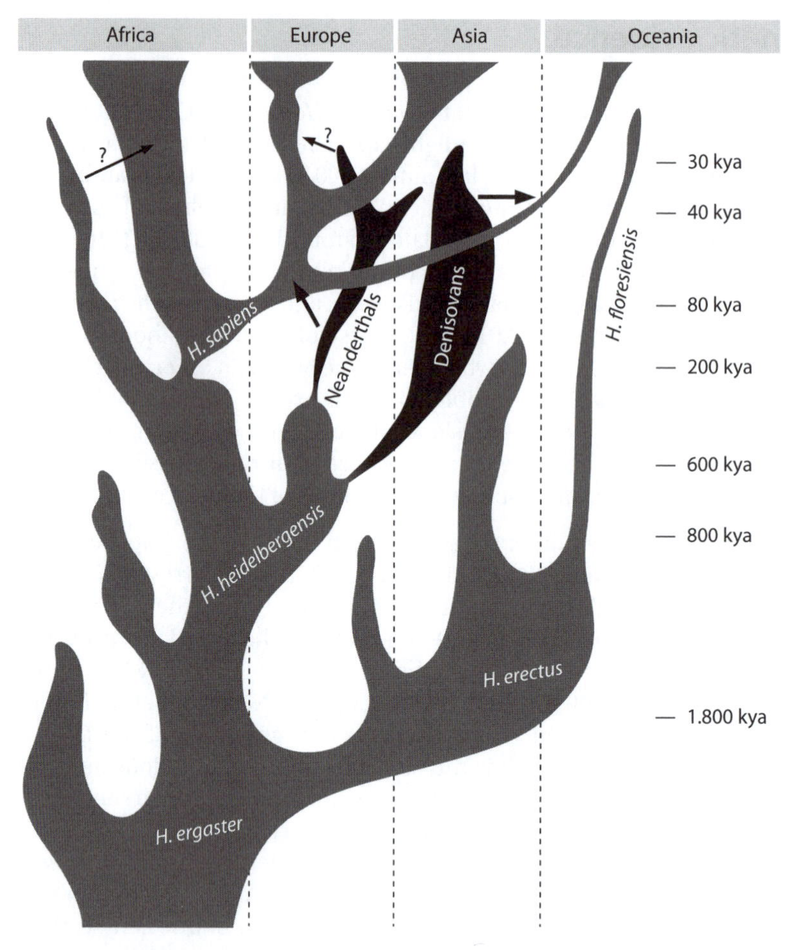

Figure 16.4 Diagram of likely evolutionary relationships within the later *Homo* taxa based on the fossil and genomic data. Arrows show where known interbreeding occurred, while arrows with queries show where possible interbreeding occurred. The width of a taxon at a given age is a measure of its estimated population size.

Adam lived some 254 (95% CI 192–307) Kya, and second that there were a cluster of small founder groups present at around 47–52 Kya (Karmin et al., 2015).

A further important area that DNA-sequence analysis is clarifying is the extent of DNA variation in different human populations and its implications. Analysis of 650,000 loci in 938 individuals from 53 populations distributed across Africa, Asia, and Europe has provided very strong evidence that humans evolved in sub-Saharan Africa, where there is far more genetic diversity than anywhere else and that populations elsewhere in the world derived from small groups that migrated out of Africa (Li et al., 2008). The degree of genomic diversity declines with the distance from Africa, and this provides strong evidence that genetic drift formed the basis of the differences among the various strains of AMH. More recently, similar such analyses done on Neanderthal DNA shows that they lived in small groups and had little genetic diversity (Henn et al., 2012; Castellano et al., 2014).

Such estimations of population size are based on coalescent analysis with Markov modeling (Li & Durbin, 2011; see Chapter 8): this uses computational informatics approaches to analyze the fine detail of the sequence differences between pairs of chromosomes from one or more individuals from a specific population, and to compare these results with those from individuals from different populations. The analysis, making reasonable estimates of generation time (25 years) and mutation rate (1.25×10^{-8} per site per generation), suggests that the modern human gene pool was essentially in place some 100 Kya, with genetic exchanges occurring until 20–40 Kya. More recent improvements in the technique have given estimates of how the population size varied over time (Sheehan et al., 2013). The analysis shows that the effective population of early *H. sapiens* that became European underwent a severe decline to only a few hundred some 70–100 Kya, before recovering to about 12,500 by 16 Kya. There was a similar but much less pronounced bottleneck in the within-Africa population, as well as in animal populations, and it is tempting to identify at least the former with the Toba supervolcanic eruption (Indonesia). This occurred 74 Kya and led to a global volcanic winter lasting up to a decade in some parts of the world (Lane et al., 2013).

It should be emphasized that the analysis of ancient DNA is a particularly active area with further samples constantly being discovered and new informatics techniques being invented to analyze their sequences. This work is likely to produce important new insights in the coming years.

Homo migrations out of Africa

It is now clear that there were several major migrations of *Homo* populations out of Africa to various parts of the world, and other minor ones (Deshpande et al., 2009). The earliest evidence of a non-African hominin is the ~1.85 Mya *H. erectus georgicus* skeletal material from the Dmanisi cave in Georgia, which is about 5000 km from central Africa. It is unrealistic to suppose that this route was taken for reasons other than by the slow dispersal of a growing population looking for new sources of food; reaching Georgia must have taken many thousands of years and it is likely that other exoduses from the same period led to other early hominin groups moving to the Middle East and then on to Pakistan, Java, and China (Zhu et al., 2003), with a late-surviving taxon being *H. floresiensis* that died out some 20 Kya in Indonesia. It does, however, seem likely that this early migration and any others before about 700 Kya ended in failure and death.

The next migration for which there is dated skeletal evidence occurred some 700–500 Kya and was probably when distinct tribes of *H. heidelbergensis* migrated out through the Levant to colonize both Europe and Asia. Here, mitochondrial DNA suggests that the last common ancestor of these tribes lived ~800 Kya (Lalueza-Fox & Gilbert, 2011), while fossil evidence suggests that the European branch evolved into the Neanderthals and the Asian branch into the Denisovans. Both died out some 30–40 Kya, after having interbred with members of *H. sapiens* (Higham et al., 2014). A third migration around 90 Kya led to *H. sapiens* dispersing across the Levant (Grün et al., 2005), but their descendants too seem to have died out.

The next known migration was some 70 Kya and resulted in a group of *H. sapiens* reaching as far as Australasia (Macaulay et al., 2005) (**Figure 16.5**); their descendants are the aborigines of Australia. The fourth and major migration of AMH from Africa occurred about 50 Kya and resulted first in groups moving east and then across Asia and later to Europe. It has generally been assumed that these migrations were from what is now Ethiopia across the Bab-el-Mandeb straights at the mouth of the Red Sea to the Arabian Peninsula. These would have been easier then because, as they were during an ice age, sea level

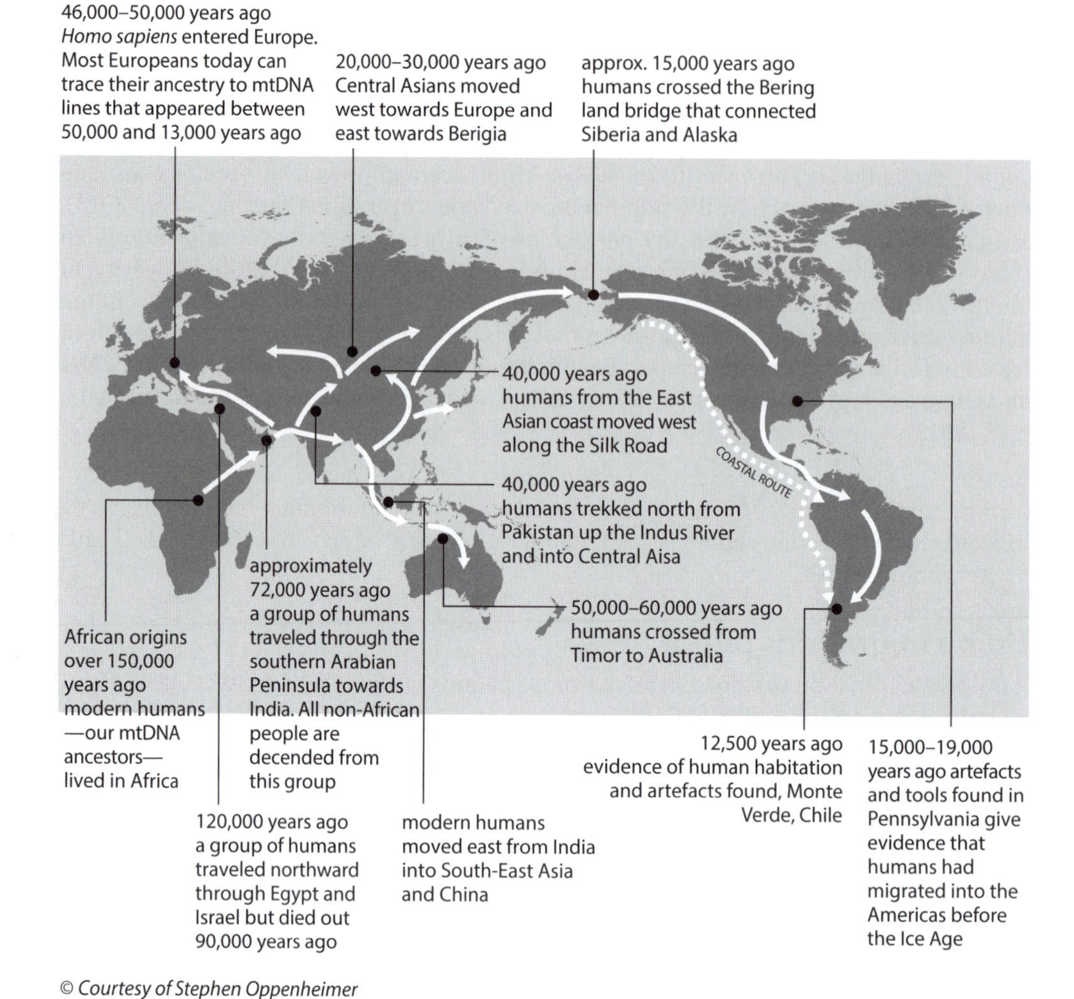
46,000–50,000 years ago
Homo sapiens entered Europe.
Most Europeans today can trace their ancestry to mtDNA lines that appeared between 50,000 and 13,000 years ago
20,000–30,000 years ago
Central Asians moved west towards Europe and east towards Berigia
approx. 15,000 years ago
humans crossed the Bering land bridge that connected Siberia and Alaska
40,000 years ago
humans from the East Asian coast moved west along the Silk Road
COASTAL ROUTE
40,000 years ago
humans trekked north from Pakistan up the Indus River and into Central Aisa
approximately 72,000 years ago
a group of humans traveled through the southern Arabian Peninsula towards India. All non-African people are decended from this group
50,000–60,000 years ago
humans crossed from Timor to Australia
African origins
over 150,000 years ago modern humans —our mtDNA ancestors— lived in Africa
12,500 years ago
evidence of human habitation and artefacts found, Monte Verde, Chile
15,000–19,000 years ago artefacts and tools found in Pennsylvania give evidence that humans had migrated into the Americas before the Ice Age
120,000 years ago
a group of humans traveled northward through Egypt and Israel but died out 90,000 years ago
modern humans moved east from India into South-East Asia and China

Figure 16.5 Generalized map of the modern human migration out of Africa (dates approximate; mtDNA: mitochondrial DNA; this figure omits the migration through Sinai ~50 Kya).

was lower and the boat trip was shorter than now. More recently, however, a series of genome comparisons across Egyptian, Ethiopian, sub-Saharan, West African, and non-African populations have given a slightly different picture (Pagani et al., 2015): their analysis shows that the splits between non-African populations and West African, Ethiopian, and Egyptian populations occurred around 75 Kya, 65 Kya, and 55 Kya, respectively (Pagani et al., 2015). Moreover, the Egyptian population was, on the basis of both haplotype and coalescent analysis, the closest to the non-African populations. This and other evidence suggest that the third emigration whose descendants first reached Australia was from Ethiopia (Tassi et al., 2015), while the fourth ~50 Kya was across the Sinai peninsula and into the Levant, with the founder population reflecting the local gene pool.

This latter population initially moved eastwards, first colonizing southern then eastern Asia. Different groups then separated, with some moving south, reaching Sahul (New Guinea, neighboring islands, and Australia; Pugach et al., 2013; Summerhayes et al.,

2010) and others moving northwards to colonize central Asia and Russia (see Figure 16.5). It was a small group of this latter population, perhaps 1500 in all, that eventually traversed the Bering land (now the Bering straits) some 20 Kya and whose descendants eventually colonized all of the Americas. It is, however, possible that there is a Polynesian contribution to the South American gene pool: there is fossil evidence in Brazil suggesting that its inhabitants reflect interbreeding between those who migrated from the north and a small number who crossed the southern Pacific from the Sahul (Neves et al., 2007; Malaspinas et al., 2014). The migration immediately northwards through Turkey to Europe was held up for about 10 Ky by the effects of the last ice age, the southern extent of which reached beyond the Caucasus. By 40 Kya, its effects had ameliorated and a group of these early humans were able to move north to Germany through the valleys of the Danube and its tributaries, then west across the Mediterranean towards Spain. Descendants of this group colonized all of Europe.

If done today, the migrations from Africa and onwards across the rest of the world would involve major sea crossings; then, however, such crossings were much shorter as the sea level was lower owing to the effects of the last ice age (110–12 Kya). This effect was associated with major periods of glaciation, the most severe being about 26–19 Kya. The effect of which was to lock enormous amounts of water as ice in the northern and southerly regions of the world and to lower the sea level elsewhere. The data are particularly good for the Australasian region, which coalesced into three areas of land connected by relatively short distances of water (Oppenheimer, 1998). The Sundaland land mass incorporated the Indonesian islands into the Asian continental mass, the Sahulland area included Australia, Tasmania, and New Guinea, while Wallacea was a group of intervening islands (**Figure 16.6**). It thus seems as if inter-landmass travel across the seas linking South-East Asia with Australia could be undertaken almost without losing sight of land (Irwin, 1994).

Genetic origin of the various human groups and their differences

The fossil history of the various hominin migrations does not alone explain how the various extant human groups acquired their differences, as seen in, for example, the degree of pigmentation and in facial features. Genetic and coalescent analysis of their different genomes has now clarified the basics if not the details of these differences. Many groups have now shown that the further a human subpopulation is from sub-Saharan Africa, the lower is its degree of genetic diversity. As already mentioned, the work of Li et al. (2008) on analyzing the genetic diversity in a wide range of African and non African populations has shown that the relationship between haplotype heterozygosity (roughly, the number of sequence alternatives for a particular gene) and distance from Africa was as expected for a population that started in sub-Saharan Africa, with each successive step in this migration being characterized by the majority of the local population remaining where they were, and only a minority moving on.

The human colonization of the world by AMH was thus initiated by a series of small founder populations of AMHs, each of which carried less genetic diversity than its parent population (Deshpande et al., 2009). Detailed genetic analyses supplement the fossil data and make it likely that there were at least two major migrations out of Africa to Western Asia (Tassi et al., 2015), with some of their descendants returning to Africa (Hodgson et al.,

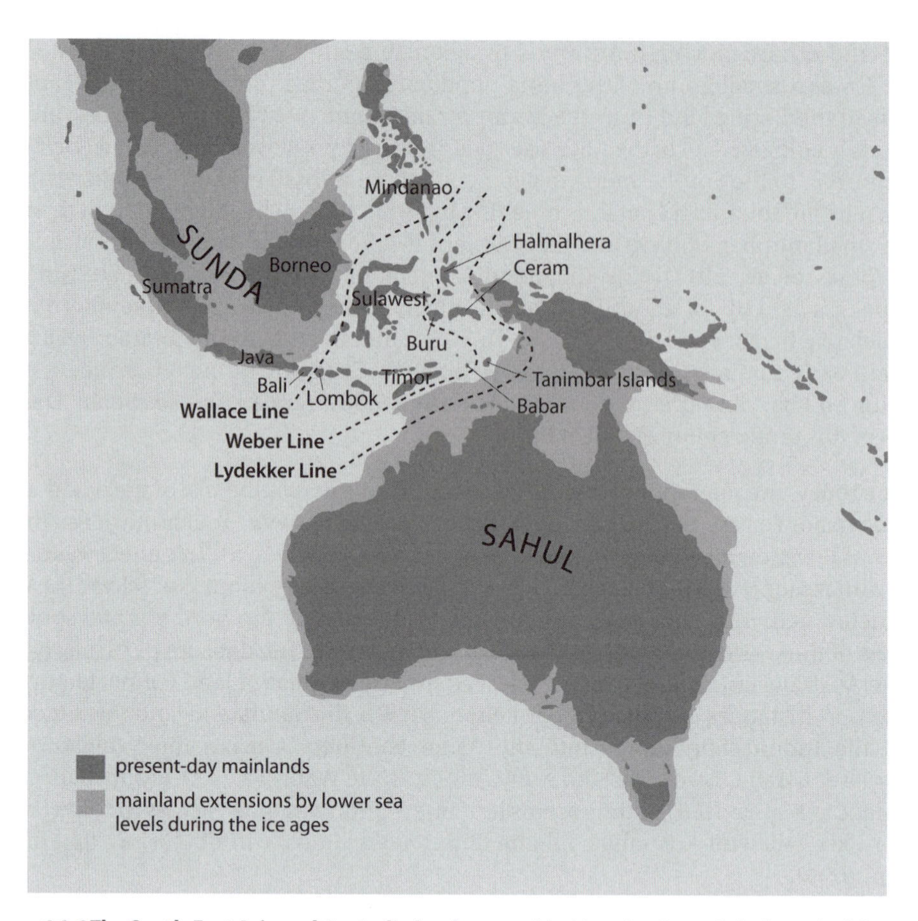

Figure 16.6 The South-East Asia and Australia landmasses (dark) at the time of the last glacial maximum compared with those today (see Oppenheimer, 1998).

2014) while others migrated further east. The general result from coalescent analyses is that all of these founder populations included at most a very few thousand breeding individuals.

It is now clear that the differences in today's various groups mainly derive from genetic drift, giving these small founder populations characteristics that were included within, but not necessarily typical of, their parent populations. In addition, one cannot exclude the possibility of small amounts of additional mutation over the past 2–3000 generations having an effect here. Nevertheless, drift is the main reason why there is much more genetic and phenotypic diversity in African populations than, for example, in Japanese or Aboriginal Australian populations. A further factor here was the effect of area-specific selection pressures on the genetic distribution. This, of course, reflected the environment in which the various populations came to live, with, for example, the local degree of ultraviolet radiation acting as a selection pressure that affected the amount of melanin pigmentation in the local population (Jablonski & Chaplin, 2010).

Since those early colonizations some 40 Kya, human populations have not been static: humans have explored every piece of habitable land, and other human groups have followed them. There have been invasions and wars that have been followed by interbreeding and genetic mixing, with the result that there are very few areas today whose populations are the sole descendants of those AMHs that first colonized them. Such migrations have been a continual part of human history, and genetic analysis is beginning to reveal ancient detail. Thus, whole-genome sequencing of the remains of individuals who lived 3–4000 years ago in Ireland has revealed that there were major migrations from the fertile crescent of the Levant to southern Russia and on to Ireland (Cassidy et al., 2016). The mass movements of slaves in the eighteenth and nineteenth centuries, and the current emigration from Syria are just recent examples of a human behavior that has a very long history.

Artefacts, art, and behavior

The skeletal evidence on human evolution is buttressed by ancillary artefacts associated with the various hominin taxa. Although chimpanzees, and even capuchin monkeys, can use stones to crack nuts and twigs to access insects, there is no evidence of either being able to make more sophisticated tools (Mangalam & Fragaszy, 2015). The importance of the artefacts associated with the various hominin taxa is that they provide clues as to when mental skills were acquired. The increase in sophistication of these artefacts over time helps understand the implications of increasing brain size as hominins evolved, one of which was, of course, cultural inheritance. It should be said that this is a contentious area and the treatment given here is intended to be conservative and do little more than capture the current consensus on when the various stone artefacts were first found and which species was responsible for them. These artefacts are grouped according to the degree of skill needed to make them, and it is clear that the sophistication of these lithic "industries" increased with time.

Stone tools

Working with stone to make useful tools has a long history, with discoveries being made over the last two centuries, mainly in Africa and Europe, and particularly in France (**Figure 16.7**; for review, see en.wikipedia.org/wiki/Stone_tool). The earliest evidence for tool-making comes from the grooved and fractured bones from Dikika (Ethiopia) and date to around 3.4 Mya, the Australopithecus Period. The stone tools made then are the simple (so-called mode 1) choppers found in Gona (Ethiopia) and created by hitting one stone with another. This technique was improved about 2.6 Mya when the predominant hominin was *H. habilis:* sharpened edges were created by removing flakes (knapping) that could be used for woodworking. The production of such tools, known as the Oldowan (Tanzania) industry because so many were found there, occurred widely across Africa.

The next level of tools (the Acheulean industry named after the site at St Acheul, France) was the production of hand axes for which a stone had been carefully shaped and flaked to produce an ovoid object with a blade and point at one end and a smoother, hand-holding area at the other (see Figure 16.7, mode 2). These date back to about 1.7 Mya, the time of premodern humans, and are associated with *H. ergaster*. Progress seems to have been slow until about 300 Kya, when Neanderthals in France and elsewhere improved the knapping procedure for removing fine flakes from stones (the Levallois technique), particularly flint, and were able to make slim knives and scrapers (see mode 3, 300–30 Kya - known as the Mousterian industry, as early examples were found in Le Moustier, France). In Africa, hand axes held in hafts for greater leverage were being made ~200 Kya or earlier (Rots & van Peer, 2006), and this may be the first innovation that can be associated with *H. sapiens* (McBrearty & Brooks, 2000; Henshilwood et al., 2001). By 70 Kya, stones were being heated over fires to facilitate knapping and the production of small tools.

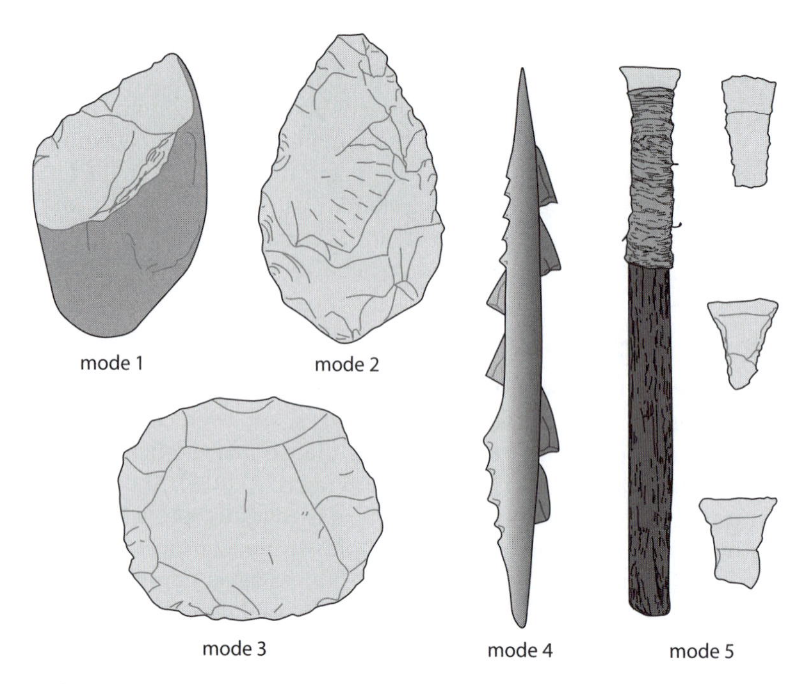

Figure 16.7 Paleolithic stone tools. Mode 1: Oldowan chopping tool (~5 cm across, 3.4–1.6 Mya; *Homo habilis*). Mode 2: Acheulan hand axe (~5 cm across, 1.7 Mya–300 Kya; *Homo erectus*). Mode 3: Mousterian scraper made using the Lavallois technique (~8 cm across, 300–30 Kya; Neanderthal). Mode 4: Aurignacian harpoon (~15 cm long, 50–10 Kya; *Homo sapiens*). Mode 5: microliths, with one tipping and strengthening an arrow (~3 cm long, ~40–10 Kya; *H. sapiens*).

With the later dispersals from Africa reaching Europe, there was an overlap of AMH with Neanderthals and various improvements in stone work have been noted. The final lithic work by Neanderthals was marked by toothed stone tools (the Châtelperronian industry, 35–29 Kya) and these may reflect interactions between the two taxa. The work of AMH in France was marked by more careful knapping to give long blades (the Aurignacian industry, 45–35 Kya; see mode 4), and by the start of polishing stone to give smooth-surfaced artefacts, which were first seen in Japan. Mode 5, the last in the staged series, covers the production of small stone tools that were used for delicate work and for arrow and spear tips (microliths). They are characteristic of the Gravettian (29–22 Kya) and Magdalenian (17–12 Kya) cultures, which were first identified in La Grevette and La Madelaine, sites near streams off the Dordogne river in France. Further improvements in stoneworking led to the production of better knives, finer axe heads, and grain-grinding tools.

Other physical evidence

There are other tantalizing clues about the increasing hominin mental abilities over the last million or so years. The oldest is probably the use of fire: there is now clear evidence from the Wonderwerk Cave in South Africa of burnt plants and bones that date to around 1.0 Mya and associated with *H. erectus*. Evidence of early hearths dated to 350–200 Kya comes from the Qesem caves in Israel, which were probably then inhabited by Neanderthals, while on the South African coast some 130–100 Kya, *H. sapiens* was using tools, fishing, and making fires (Will et al., 2013). The manufacture of kilns for firing clay is much later and associated with early modern humans: the earliest site is Dolní Věstonice in the Czech Republic, which dates back to 26 Kya.

Burying dead bodies is a more recent innovation than fire and the earliest evidence comes from a collection of Neanderthal bones dropped into a deep pit in the Sima del los Huesos (the pit of bones at Atapuerca, Spain) more than 300 Kya (Parés et al., 2000). The earliest known *H. sapiens* burial sites date to 135–100 Kya, and are in an area of the Levant where *H. sapiens* and Neanderthals co-existed (Grün et al., 2005).

Further evidence of early hominin activity comes from the nonstone material associated with their sites. The earliest comes from the wooden throwing spears from what is now a mine in Schöningen, Germany, where preservation conditions were remarkable, and date back to about 400 Kya. These are associated with *H. heidelbergensis* activity (Thieme, 1997). Perforated shell beads that date back some 82 Kya have been found in Morocco (Bouzouggar et al., 2007) and similar beads, colored with ochre, have been found in the Blombos Cave in South Africa (77 Kya; Henshilwood et al., 2002). Also found in this cave were examples of the earliest evidence for using bones to make tools such as awls and points. The earliest evidence of Neanderthal bone technology is of lissoirs (fine hide scrapers) from about 52 Kya in Mousterian camps next to the Dordogne River (Soressi et al., 2013).

Art

While there is no evidence to show that chimpanzees left to themselves will paint, they can certainly be taught to do so and apparently to enjoy the results. The oldest indications of a hominin interest here date back to 250–200 Kya, with the evidence suggesting that, as groups moved around, pieces of ochre, a pigmented stone with a color range of yellow to red, were carried with them. The earliest evidence of any hominin art is the presence of ochre associated with other hominin artefacts. Apart from its role in tanning, this pigmented stone has also been used for body painting, coloring human bones, and engraving. The earliest known engraved pieces of ochre were found in the Blombos Cave in South Africa (**Figure 16.8**A), which was inhabited some 77 Kya by early AMH (Henshilwood et al., 2002). In addition, pieces of ochre and black pigments (charcoal and manganese dioxide) have been found in Neanderthal sites in Europe (Roebroeks et al., 2012).

All of the early representative art, as we understand it today, comes from AMH sites (for review, see Morriss-Kay, 2010); there is minimal evidence that Neanderthal interests here extended beyond decorating shells and beads. The earliest, unequivocal piece of European figurative art found so far is the Venus of Hohle Fels (Venus is the generic name used for small female sculptures), a small, headless woman carved from a mammoth tusk found in the Hohle Fels cave and that may be a self-portrait of an AMH woman who has recently given birth (Figure 16.8B; Morriss-Kay, 2011). It was made around 40 Kya, much earlier than the earliest major examples of cave painting in France such as those of Lascaux, which date to 17.3 Kya, and Chauvet, originally dated to >32 Kya although analysis of its symbols suggests that be a little more recent (von Petzinger & Nowell, 2014). After seeing the paintings at Lascaux, Picasso is quoted as saying "we have learnt nothing."

It is clear that a fully developed sense of artistic creativity was present in the earliest members of *H. sapiens* to reach Europe. Given that there are traditions of representational art in the cultures of both early European humans and of early Indonesians (40 Kya; Aubert et al., 2014) and Australian aborigines, it is clear that these abilities were present in their ancestors who left Africa some 60 Kya, and probably earlier (Oppenheimer, 2003; Morriss-Kay, 2010), Thus far, however, the oldest known examples of African representational art

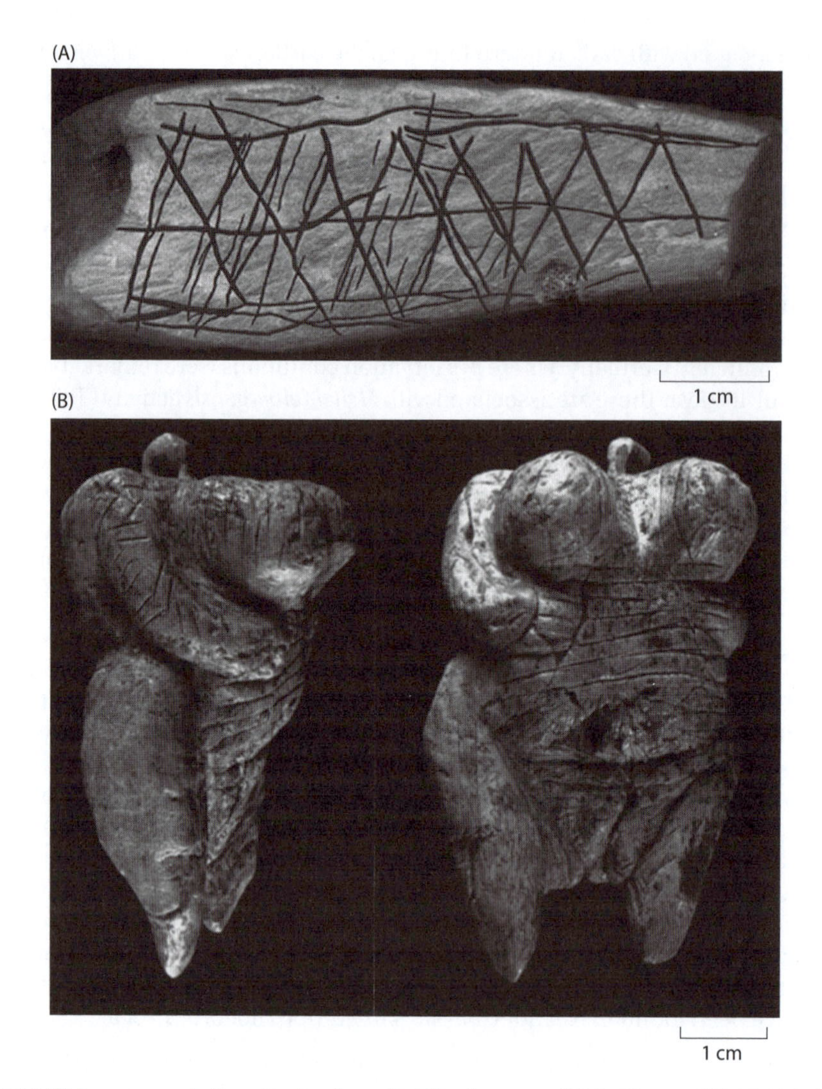

Figure 16.8 (A) An engraved piece of ochre from the Blombos cave. (B)The Venus of Hohle Fels (40 Kya).

are the bovid pictures near the Apollo 11 cave that date to about 25 Kya or a little older (Mason, 2006). It is unfortunate that no other art remains from those times: virtually all wooden artefacts from Africa would, by now, have decayed, with those from wetter regions being lost first.

Speech

The greatest gap in our knowledge of early hominins is, of course, when speech evolved. Apart from any mental abilities, the making of coherent sounds requires a set of anatomical changes in the soft tissue seen in chimpanzees. Lieberman (2011) has discussed the nature of these changes, which include the lowering of the larynx and vocal chords, the loss of laryngopharyngeal air sacs, the shortening and rounding of the tongue, and the shortening of the oral cavity. Most of these features reflect changes to soft tissues that are lost soon after death, so it is hard to know how these evolutionary changes occurred.

Recent evidence has suggested that Neanderthals may well have had at least some of the anatomical morphology required for speech. A key tissue here is the hyoid, an unjointed bone in the neck that plays a role in swallowing and tongue movement that, in modern humans, facilitates speech. A well-preserved hyoid bone from the skeleton of a Neanderthal discovered in a cave in Israel and dating to about 60 Kya has now been analyzed (D'Anastasio et al., 2013). It is similar in both internal architecture and biomechanical behavior to the hyoid bones of modern humans. As these features reflect use, it is quite possible that there was a capacity for speech in Neanderthals, although whether this reflects common descent or homoplasy is not known. If the former, then the possibility that *H. heidelbergensis* also had the appropriate physiology for speech cannot be excluded; it certainly had a brain capacity little different from that of modern man, and rather larger than its antecedents. Perhaps it was around then that the *Homo* taxon began to develop cultural inheritance, the ability to pass on immediately acquired knowledge to the next generation. This ability enabled group expertise to increase exponentially, with effects on selection that are unclear, but were almost certainly more immediate than those of genetic drift and mutation.

All of these artefactual data immediately raise the question of what was special about the AMH brain that distinguished it from those of earlier hominins and particularly of Neanderthals. There was certainly no obvious difference in gross size; if anything, the Neanderthal brain was slightly larger than that of AMH. Although it is normally difficult to infer brain detail from skull capacity, two distinctions between AMH and Neanderthals have been noted. First, there are slight differences to the skull vasculature: that of AMH is richer than that of Neanderthals, and this affords better protection for the brain against the cold (Rangel de Lázaro et al., 2016). Second, as the *Homo* genus evolved, the parietal region of the brain broadened, with that of AMH being noticeably larger in both absolute and relative size than that of Neanderthals (Bruner et al., 2011). The importance of this region is that it is associated with a mental self-representation, speech, visuospatial integration, body image, memory of experiences, and goal-related activity (Bruner et al., 2014). The increase in relative size of this region is thus one likely reason for the particular mental abilities that enabled AMH to be more successful than Neanderthals, once they were sharing the same territory; it is, however, unlikely to be the full answer.

Selection and speciation

At the beginning of this chapter, three questions were put. The first asked when did the members of the various hominin taxa first leave Africa and how many times did this happen? The answer is that groups of *H. erectus* left first, 2.0–2.5 Mya, and there were at least two more major dispersals by the descendants of *H. heidelbergensis* that led first to the Neanderthals and Denisovans and later to *H. sapiens* colonizing Europe and Asia; there were also many minor ones. This section looks at the other two questions on how many human species there were and when the descendant of *H. heidelbergensis* first became indistinguishable from modern human.

Lieberman (2013) has discussed in detail the evidence that the driving force for hominid natural selection was access to and ease of eating food: hominid variants were able to access food on land rather than fruit and nuts from trees, and the any ground-living animals were able to flourish. The ability to achieve this required hands that were more dexterous (and less well adapted to tree life) and a gait that was bipedal. In addition, more advanced taxa would have hunted and this required an ability to run over long distances;

this ability came with a loss of body hair and the ability to sweat, together with a flattened face and extended nose that facilitated breathing. There was also selection for enhanced mental ability, and this favored large brain size. We know very little of the molecular mechanisms underpinning these changes that took place over several million years and hence more than 100,000 generations.

Evidence from the artefacts and social behavior associated with the evolving taxa of hominins discussed above adds some detail to this picture. There is now evidence that australopithecines used simple rock tools, but the rate of tool development and use speeds up with the evolution of the *Homo* taxon and its increasing brain size. Here, it is interesting to compare Neanderthals and *H. sapiens* with their 400 Ky of separate evolution. The anatomical and genetic differences were small enough to permit interbreeding, but, on the basis of their respective artefacts, the mental differences were large enough for the AMH to replace Neanderthals. This is clear from their different abilities in making tools and works of art and, almost certainly, in their respective powers of speech abilities and hence cultural inheritance. The artistic abilities of *H. sapiens*, as seen in Europe 40–30 Kya were little different from our current abilities, and far in advance of the very few examples seen in Africa 77 Kya, although they may not reflect the full repertoire of what was being created then (again, absence of evidence is not evidence of absence). It has always been the case that analyses of hominin evolution need more data!

The discussion so far has avoided the question of how many species changes occurred during hominid evolution by using the more neutral term of taxon. Wood and Lonergan (2008) have discussed the different views of the lumpers who would like to have few species and more subspecies, and the splitters who would favor every anatomical variant having species status. The two views really reflect whether one views a species as being defined by a group with which other species cannot breed, or whether it is defined by a specific mix of shared-derived and novel anatomical characteristics (that is, synapomorphies and autapomorphies), and neither definition is satisfactory. Successful breeding is known on genomic grounds to have taken place between *H. sapiens* and *H. neanderthalensis* and between *H. sapiens* and the Denisovans some 50–40 Kya, when they met up in Europe. These taxa share a last common ancestor who lived more than 0.4 Mya in Africa and was probably *H. heidelbergensis*. Some of the descendants remained in Africa and evolved to give *H. sapiens,* while other groups left Africa and dispersed throughout Eurasia, eventually becoming Neanderthals and Denisovans, and all are, on the breeding criterion, members of the same species.

The genetic evidence comes down firmly on the side of the lumpers here, and probably supports the extreme position of there only ever having been a single hominin species, albeit one that has included a great deal of variation. The genomes of *H. sapiens* and chimpanzees are very close (Wildman et al., 2003), and the current human and chimpanzee chromosome numbers are 46 and 48, respectively, with human chromosome 2 resulting from the fusion of chromosomes 2 and much of 4 in a last common Homininae ancestor (Ijdo et al., 1991). Assuming that the Hominini line separated from the Homininae line 7.5 Mya and the reproductive cycle is 25 years or perhaps a little less, the separation occurred at least 300,000 generations ago, a number that seems rather low to produce full speciation (Chapter 15). It would not therefore be surprising if it were possible for human–chimpanzee hybrids to be produced, although, of course, such an experiment would never be approved on ethical grounds. It thus seems inappropriate to define the different human groups on anything other than morphological grounds and it is probably sensible to accept the conservative view that the hominins are a single species with subspecies and sub-subspecies (see Table 16.2).

More to the point and whether this view is right or wrong, the hypothesis cannot be tested. It is also singularly unhelpful: what matters is not whether one or another taxon has species or subspecies status but the details of the changes that led from one taxon to another and why they were successful. For this, the key criterion is that the definition of a taxon has to be broad enough to include quantitative variation (for example, in size) and narrow enough to exclude individuals with qualitative differences (for example, in the number of wrist bones). Given the paucity and fragmentation of the hominid fossil material, even meeting this criterion is not always straightforward (see Lordkipanidze, 2013, for example), particularly as one cannot exclude the possibility that similar features in different groups may reflect homoplasy (convergent change from different origins) rather than synapomorphy (shared, derived characteristics; see Wood & Harrison, 2011.) In principle, therefore, it is probably most helpful to take a speciose view of hominin evolution based on morphology as this will focus on small detail and accept that, in the light of future information, it will be possible to produce a clade of hominin taxa in which one can have confidence. In the meantime, it may be more appropriate to talk about *Homo* taxa rather than *Homo* species.

Perhaps a more interesting question is why, if the genomes of extant chimpanzees and humans are so close, the phenotypes are so distinctive? Minor differences in kinase domains have been observed in their respective genomes (Anamika et al., 2008) and in the patterns of gene expression in the two brains (Bauernfeind et al., 2015), but such observations do not really seem to get to the root of the problem. The question really reduces to how minor changes in sequencing produce major changes in phenotypes and, from the systems perspective taken here, the answer is that the mutations affect the networks that are particularly involved in growth, pigmentation, and brain development. Here, small changes in the kinetics of these networks can have important downstream effects not only on these properties, but also on gene expression. In this context, the differences in kinases observed by Anamika et al. (2008) may be significant as these proteins are known to play an important role in the growth pathways (Kosmidis et al., 2014).

Finally, it is worth asking whether the data give any clue as to when modern members of *H. sapiens*, with their full mental faculties, evolved. The existence of pictures on the walls of deep caves shows that the artists then were able to remember what they had seen and so had the ability to draw images from memory or imagination. This is now a ubiquitous ability of all human groups, though not of all individuals, and what Morriss-Kay (2010) has called the "mind's eye" must therefore have evolved before AMH left Africa. It seems that this ability evolved after the ancestors of Neanderthals left Africa, as there is no evidence that Neanderthals were capable of creating representational art. It is not possible to be precise about dates here, but it is hard to think of core faculties beyond language and a visual imagination that are needed to distinguish us from our premodern human ancestors. There is certainly no reason to doubt the hypothesis that anatomically modern humans, with the faculties that characterize this taxon today, were present in Africa soon after 100 Kya at the latest.

Key points

- The line that led to *H. sapiens* (the taxon of anatomically modern humans or AMHs) about 200 Kya originally split from the chimpanzees some 7 Mya in sub-Saharan Africa, and its subsequent history is of a complex shrub of taxa whose exact lines of descent are still not clear.

- The immediate ancestor of humans was probably *H. heidelbergensis*, whose non-African descendants also gave rise to Neanderthals in Europe and Denisovans in Asia some 4-500 Kya. When AMH left Africa some 50 Kya, interbreeding with both groups occurred.
- There were several important migrations of *Homo* taxa from Africa to colonize much of Asia and Europe, the earliest was about 1.5 Mya and the last about 50 Kya, coming from what is now Egypt and whose genotype reflected the local gene pool. It was a succession of small founder groups from this latter migration that mainly populated the world.
- Each successive founder population carried less genetic diversity than its parent population. This effect, combined with those of genetic drift and selection, was responsible for the different morphologies of the various human populations today.
- The various hominin taxa have produced a graded series of tools dating back over 1.5 My and pieces of art extending back over the last 77 Kya. The artefacts have given insight into the evolution of neurological capabilities and of cultural inheritance.
- As full speciation usually takes at least a million generations, it is likely that all hominin taxa are part of the same species, given the slow reproductive cycle of hominins and the likelihood that its taxon has only existed for about 7 Mya. It is, however, more helpful to view the many taxa as distinct in order to build up a picture of how the hominin phenotype evolved.

Further reading

Jobling M, Hollox E, Hurles M et al. (2013) Human Evolutionary Genetics, 2nd ed. Garland Press.

Lieberman D (2011) Evolution of the Human Head. Harvard University Press.

Lieberman D (2013) The Story of the Human Body: Evolution, Health and Disease. Pantheon Press.

Oppenheimer S (2012) Out-of-Africa, the peopling of continents and islands: tracing uniparental gene trees across the map. *Phil Trans R Soc* 367B:770–784.

Stringer C (2011) The Origin of our Species. Penguin Books.

Wood B, Elton S (eds) (2008) Human evolution: ancestors and relatives. *J Anat* 4:335–563.

CHAPTER 17 CONCLUSIONS

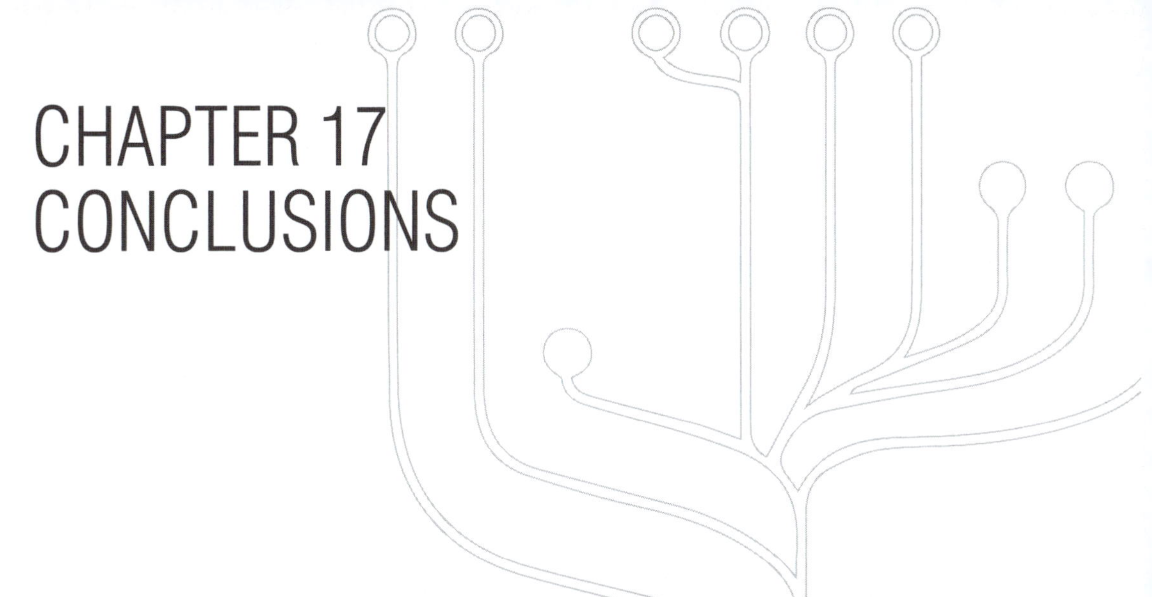

In 1859, Darwin ended *On the Origin of Species* by comparing the living world to a nineteenth-century English river bank with all of its flora and fauna in a state of flux.

> *It is interesting to contemplate an entangled bank, clothed with many plants of many kinds, with birds singing on the bushes, with various insects flitting about, and with worms crawling through the damp earth, and to reflect that these elaborately constructed forms, so different from each other, and dependent on each other in so complex a manner, have all been produced by laws acting around us. These laws, taken in the largest sense, being Growth with Reproduction; Inheritance which is almost implied by reproduction; Variability from the indirect and direct action of the external conditions of life, and from use and disuse; a Ratio of Increase so high as to lead to a Struggle for Life, and as a consequence to Natural Selection, entailing Divergence of Character and the Extinction of less-improved forms. Thus, from the war of nature, from famine and death, the most exalted object which we are capable of conceiving, namely, the production of the higher animals, directly follows. There is grandeur in this view of life, with its several powers, having been originally breathed into a few forms or into one; and that, whilst this planet has gone cycling on according to the fixed law of gravity, from so simple a beginning endless forms most beautiful and most wonderful have been, and are being, evolved.*

Apart from its lovely imagery, this paragraph is remarkable for two reasons. First is the comment in last sentence, that evolution never ends but is a natural part of life; at the time, this may have been the most revolutionary point in the whole book. Second is the middle part, which lays out the basic skeleton of the mechanism by which evolution works. More than 150 years of research has shown that Darwin was right in almost everything here, other than the comment on use and disuse. What generations of evolutionary biologists have done is to add tissues to the bare bones of Darwin's mechanism of evolutionary change.

From Darwin's perspective, perhaps the most surprising of the new observations would probably be the extent to which the dynamic of evolution is unhurried. Indeed, it is far slower than Darwin would ever have suspected, given his knowledge of selective breeding and his belief in pangenesis, a mechanism based on Lamarck's earlier ideas, by which acquired parental abilities could rapidly be passed on to offspring. Today, it is clear that variants take thousands of generations to predominate, while the process of speciation, in the sense of full reproductive isolation, may well not be complete after a million generations. Ligers, the hybrid offspring of lions and tigers are still fertile, as is the very occasional mule, the offspring of a horse and a donkey, even though each pair of parents shared a last common ancestor several million years ago.

The major advance since the time of Darwin, however, has certainly been the introduction of genetics to the science of evolution. Darwin, of course, knew nothing of the subject and never heard of Mendel's work, which was only publicized some 20 years after his death. The use of evolutionary population genetics in explaining how gene alleles are spread through a population and the particular importance of asymmetric allele distributions in subgroups of a population have been key to our understanding of how occasional phenotype variants can eventually become the normal feature of the descendants of that subpopulation. It has now been clear for around 80 years that the dance of gene variants through a sexually reproducing population across the generations slowly changes the distribution of proteins and their regulatory sites through the cumulative effects of mutation, recombination, genetic drift, and selection. The importance of populations in describing how evolutionary change happens is crucial, but it never occurred to Darwin - the word "population" will not be found in *On the Origin of Species*.

Today, a great deal more is also known about the processes of selection in its many manifestations and about the mechanisms by which speciation takes place. Here, the role of evolutionary population genetics has been the key tool to showing in a quantitative way the importance of mutation, genetic drift, selection pressures, and the particular properties of small populations in understanding evolutionary change. Curiously, from today's perspective, the whole subject, developed as it was in the 1940s and 1950s, essentially before the discovery of the genetic importance of DNA, is based on Mendel's nineteenth-century view of genes, that the various allelic alternatives for some trait could be defined by the phenotype that they elicited. Evolutionary population genetics was also implicitly based on the idea that the variants in phenotypic trait were underlain by the alleles of one or a few trait genes, as they are called in this book.

Today, it is known that the vertebrate genome typically includes some 25,000 protein-coding sequences, a number that is a substantial underestimate of the number of genes as it excludes alternate spliceforms and control sequences. Not only is this number very much greater than any phenotypic count on the number of traits, but it has also rarely been possible to match up trait genes with DNA sequence genes, other, of course, than for those few of the former that can be directly represented by the latter. Pax6 might be thought to be an example of a trait gene that was equivalent to the DNA sequence gene because of its basic role in eye development, but it is not a particularly convincing example because mutations in so many other genes can affect eye development. This book has tried to explain the profound difference between DNA sequence genes and trait genes on the basis that they reflect different levels of biological complexity and that there can be no simple matching up between them.

The main contribution to our understanding of life and evolution has come from advances in genomics and molecular genetics. The ability to sequence homologous genes from any organism and compare them computationally has not only sharpened our understanding of evolutionary descent through phylogenetics, but has also provided probes back into deep time for which there is not fossil record. Everything now known about the first three billion or so years of life comes from such work: it has shown, for example, how the first eukaryotic common ancestor evolved from the endosymbiosis of both eubacteria and archaebacteria, and how the early phyla formed from further endosymbiotic events, such as those that led to chloroplasts in plant and algal cells, together with the evolution of different ways of strengthening the cell coats of very early organisms.

While theoretical evolutionary work on genes and phenotypes is untouched by the exact meaning of what a gene is and does, this distinction becomes crucial in trying to analyze the evolution of a contemporary anatomical phenotype from its ancient, anatomically different ancestors. Evolutionary population genetics has nothing to say about this, and our knowledge of the subject has come from an area that was relatively quiescent between 1900 and 1980, the study of the development of embryos that are evolutionarily related (evo-devo). This field was given a new life by the realization that not only was it possible to elicit the molecular nature of the mechanisms that controlled development, but that one could also directly compare regulatory mechanisms from very different organisms. The results of this work have illuminated the history of molecular life in the time since the Cambrian Period, when the fossil record starts to provide substantial amounts of anatomical detail. It is now possible, for example, to understand not only the anatomy of urbilateria, the earliest bilaterian organism, but something of how the different animal phyla evolved from it.

None of this work explains either the generation of anatomical features or the way in which variants can arise from them. A major subtheme of this book comes from systems biology, which suggests that there are levels of complexity between DNA genes and trait genes, and that the key intervening level is that representing networks; these are composed of many proteins whose functional outputs (proliferation, differentiation, morphogenesis, and so on) drive anatomical change and ensure physiological stability. Under normal circumstances, such networks are buffered against change and a species is stable. Occasionally, however, a rare combination of protein variants will come about, often within a network, to change a trait and produce an anatomical variant with a heritable phenotype. It is on this that selection operates, often in ways that are still too complicated to understand. Such a process is, as evolutionary population genetics has shown, particularly efficient if that variant is part of a small population in a novel environment.

Contemporary challenges

As a result of a century and a half of research, there is now a clear picture of the history of species diversity that can be traced back to a first universal common ancestor (FUCA) and also of the principles that underlie the evolution of new species. Nevertheless, there are still many gaps in the details of our knowledge and understanding of evolution.

The fossil record

Knowledge about extinct organisms is always provisional and likely to be changed in the light of new fossils. Thus, for example, the time of the appearance of the earliest known angiosperm has recently been put back a few million years to 130–125 Mya with the recent discovery of *Montsechia vidalii*; this was a simple aquatic angiosperm, and almost certainly a derived form as it lacked roots (Gomez et al., 2015). New fossils are, of course, always needed to provide more information on the speed of evolutionary change and, in particular, whether there is evidence of saltatory or rapid morphological modifications to phenotypes. The most significant remaining gaps in evolutionary history are, however, the details of the early history of eukaryotic existence and the origins of the large organisms of the Ediacaran and Cambrian Periods. Future finds of these very rare fossils will depend on fortuitous discoveries of superbly preserved soft-bodied fossils in the rocks from these ancient periods.

Phylogenetics and computational biology

Much of the phylogenetic analysis so far carried out has focused on the comparative analysis of protein-coding sequences. Understanding the ways in which these proteins are mobilized will come from computational analysis of the sequences that regulate their activity. This may also involve analyzing the various classes of enhancer sites and regulatory RNAs (microRNA, small interfering RNA etc.). The other important gap that needs to be filled is the population history of species other than humans; it is likely that interesting work will emerge as coalescent analysis is applied, for example, to the great apes.

Evo-devo

While much is now known about how different types of embryos use similar molecular processes in generating structures and functions, almost nothing is known about the origins of the different cell types that are present in most of the bilateria, some of which were also present in the early radiata. Work here will probably depend on integrating evo-devo and computational phylogenetics, perhaps using very simple organisms that reflect the early stages of animal life. Of particular importance here will be information on the evolution of the nervous system which is based on neurons, a cell type whose activity is so subtle and whose use is so great that it is hard to believe that it evolved more than once, or changed much once it had formed.

Mechanisms of evolution

The evidence for natural selection is robust but inevitably sits on a smaller basis of knowledge than does evolutionary history. This is because it is much harder to provide the detail to support it: the fossil record suggests that full speciation usually takes a million or more generations and this is too slow a process to be studied experimentally in any easy way. Unpicking the evidence for how it happens has depended on (a) a mix of analyzing models of how populations change under selection, (b) laboratory studies where selection pressures can be speeded up, and (c) those occasional situations where natural events have illuminated facets of evolutionary behavior. Examples here are ring species, species flocks, and extinctions.

Variation

Little is still known about how mutation in a DNA sequence leads to a phenotypic variant. This book has argued that a key step here is the way that such a mutation affects the output of one or more protein networks, but understanding how this happens will involve an appreciation of first how mutation affects both protein and protein-regulatory sequences, and, second, how such changes affect network dynamics. This is a difficult problem, the solutions to which will involve substantial work by the systems biology and the evolutionary biology communities. This is partly because we know neither the rate constants of the various interactions nor how well buffering works within the networks, nor even how similar equivalent networks (for example, for proliferation) are in different tissues in a single organism. Perhaps most important, however, is our lack of understanding as to how mutation affects network outputs.

Selection

Under laboratory conditions, selection can be restricted to a single parameter, be it a physical or behavioral effect, a mating opportunity, or even just food access, and is thus straightforward to study. Under natural conditions, it is much harder to analyze selection pressures, and the cumulative effect of an animal's fitness is subsumed in one or two umbrella parameters whose values cannot be predicted but have to be measured. It will be interesting to see whether it is even possible to provide a means of estimating selection pressures. A second challenge will be to see if the theory of evolutionary population genetics, based as it is on a view of genes defined by phenotypes, can begin to incorporate the complexity of molecular genetics in ways other than through coalescent analysis. Even this approach only integrates the analysis of homologous DNA sequences with a model of population behavior and so avoids the complexities of gene function.

Speciation

The problems of predicting when the extent of the genomic differences between a parent and a daughter population are sufficient to block interbreeding are very hard to solve. This is because novel speciation depends partly on small-scale mutation that initially affects mating behavior and partly on major chromosomal changes. The former represent a slow and ongoing flux of change in DNA sequences, while the latter reflect those very rare major changes such as chromosome rearrangements that lead to a failure of hybrid fertility. Moreover, these latter changes need to have minimal effects on fitness: if they were in any way deleterious, they would be lost from the population. Both effects will be very hard to model. While the former step can be very rapid if populations are selected on the basis of different behaviors (Rice & Salt, 1990), it is hard to identify the changes to the anatomical and molecular phenotypes that underpin such behavioral changes. A further difficulty is that the occasional mutations that drive speciation need to spread through the population via genetic drift. In all, the process is so slow and multifaceted that it is hard to be optimistic that there will ever be a quantitative, predictive theory of speciation.

Human evolution

More is known about the evolution of *Homo sapiens* than any other species; it thus provides a measure of how much information is accessible and how far there is to go. Our understanding of the changes that have accompanied human evolution are now convincing, and more fossil information, particularly from Africa, would be expected to enrich this history and fill in evolutionary detail. Here, it is not a matter of being concerned with speciation details, but of obtaining an appreciation about anatomical changes and their timings so as to get insight into the selection pressures, the speed of anatomical change, and the associated physiological adaptations. Even more interesting, however, would be the discovery of further artefacts, particularly very early ones as they would provide insight into the most interesting aspect of anatomically modern human evolution, that of mental capacity.

Is anything missing?

The most difficult question to answer for the theory of evolution or indeed any other theory is whether or not a major part of the story is missing. The most obvious candidate here is the possibility that heritable change can be far more rapid than can be accounted for under normal genetic change, sometimes known as hard inheritance. Lamarck's solution, believed by Darwin and all other biologists until disproved in the late nineteenth century by Weissman's theory on the continuity of the germplasm, was that anatomical adaptations made by a parent in response to the environment were passed on to the next generation. This idea of soft inheritance has always lurked in the undergrowth of the evolutionary psyche and, with the discovery of various possible molecular mechanisms based on methylation, chromatin remodeling and RNA regulators of gene expression have taken on a new resonance (Gilbert & Epel, 2015).

The major reason for an interest in epigenetic mechanisms of inheritance (Chapter 12) is, of course, that they provide, in principle at least, a mechanism for generating rapid and heritable change, in contrast to standard mechanisms based on random mutation, recombination, genetic drift, and selection that are very, very slow. One concern with the idea of epigenetic change, however, is how it could become integrated into the genome. Acquired adaptation would normally be reflected in epigenetic changes in the genomes of cells in the tissues involved in the adaptation, and a mechanism is required to transfer these changes to the germ cells - but these are isolated from such tissues (hence Darwin's belief in pangenesis). Such a mechanism is required if the inheritance is to become permanent, but the very great majority of epigenetic changes that have so far been discovered are rapidly lost: of the 30 examples of epigenetic inheritance across 13 organisms cited by Gilbert and Epel (2015), very few persist beyond a few generations, mainly because the genome is wiped clean of methylation during early zygote development and any RNA populations associated with zygotes do not persist over the longer term.

It has to be said that there is little evidence in the fossil record or elsewhere that rapid generation-to-generation change is a key part of speciation. Nevertheless, if the requirement is only that change can be far more rapid that normally supposed, then there is good experimental evidence (for example, Waddington, 1953, 1961; Rice & Salt, 1990) that, if selection pressures are strong enough, the variation present within a population can be canalized to produce change extremely rapidly on the evolutionary timescale, in tens rather than millions of generations. Even under natural conditions, speciation can be surprisingly rapid on the evolutionary timescale: the evidence from species flocks (for example, the cichlids in Lake Victoria) suggest that the early stage of speciation where there is little selection and any small population with a selective advantage can find a niche can take only a few thousand generations.

In the context of this book, the obvious mechanism for fast, if not very fast, phenotypic change is through genetic drift acting within small populations under strong selection to bring together gene variants that have an effect on network dynamics. Because the very few outputs of networks depend on many genes, it reasonable to hypothesize that the same change in output can be effected through mutations to one of several genes. If so, phenotypic change in a network output will not only be faster than for changes to activities that depend on a single protein, but also has the advantage that it does not require the discovery of any novel mechanism of inheritance. The relative speed of changes to a network output compared with those in a single protein is, of course, a hypothesis and stands or falls on the basis of future work.

An area of evolutionary biology inadequately discussed in this book is evolutionary botany. An important reason for this is that plants are far more tolerant of interbreeding than animals, and it is therefore much harder to discover genetic constraints on breeding and evolutionary change. They are also much simpler than animals in the sense that they have many fewer tissue forms, cell types, and physiological properties. Unfortunately, they do not fossilize well and there are fewer form changes so that the details of their evolutionary history are harder to elicit, other than phylogenetically. It is interesting to note that plant physiology, biochemistry, and particularly development are very different from those of animals (Meyerowitz, 2002), and it will be interesting to see if there are lessons on evolutionary constraints to be drawn from further comparisons between the two.

The elephant in the room for all discussions of evolution is, of course, the origin of life and how a primitive bacterium known as the FUCA formed that was able to reproduce itself, the key criterion of life. As discussed in Chapter 9, there are indications that this bacterium arose from an RNA-based world, that energy for its reactions came from electric energy in the atmosphere and thermal energy from hot vents on the sea floor and that many of its organic constituents came from rocks originating from outside the earth, as well as from chemical activity within the oceans. But there are many more details to come.

Finally, this book has explored the highways and many of the byways of biology with the intention of capturing the contemporary zeitgeist of evolution. Studying the full breadth of evolution is difficult. This is partly because the main task is the discovery of the history of life on the basis of currently available material and partly because research in the area builds on the whole breadth of biological knowledge: evolution is both the pinnacle and the base of biology. We know a great deal but never quite enough. The general picture of how evolution works has been clear for more than 70 years. Two generations ago, the new problems were about genetics and DNA; for the last generation, the new areas were computational phylogenetics and evo-devo; for the next generation, one can hazard a guess that the problems will be in the areas of systems biology and perhaps epigenetics, with questions about evolutionary neurobiology and the origin of life likely to be for the longer term. The former will probably require a better understanding of neurophysiology than we have today, while the latter will require experimental genius. There is more to be done - there always is.

APPENDIX 1 SYSTEMS BIOLOGY

Systems biology is as much a way of thinking about biology as it is a technical subject. Most of what biologists do is involved in solving problems, answering questions, and clarifying detail – biology is a practical rather than a theoretical subject. It is, however, sometimes necessary to stand back and think about the biosphere as a whole and try to make sense of how the various levels of life, from DNA sequences to the behavior of populations, can be integrated into a single picture. This appendix provides the wider, systems-biology context for considering evolutionary change.

Physicists have always appreciated that their world involves two very different sorts of approach. First, events at an individual level have to be understood and formalized; second, the interactions between the different levels have to be clarified. The understanding of the effect of heat and energy on gases provides a good example: at the everyday level, there are laws such as Boyle's and Charles' that describe the effect of heat and pressure on gas volume; underneath these are the laws of thermodynamics which use the macroscopic parameters of energy, temperature, and entropy to provide a theoretical framework for the behavior of gases. Underpinning all of this is the thermal motion of gas molecules, which is described by the laws of statistical mechanics that, in turn, provide a theoretical basis for thermodynamics and hence for Boyle's and Charles' laws.

Physicists are fortunate here. First, they often deal with particles and waves that turn out to obey relatively simple laws that operate over all distances beyond those for which quantum effects are significant. Second, there is a large body of applied mathematics that can be used to help work out the implications of these laws under specific conditions. Biologists are unfortunate in this context. Mathematical descriptions are restricted to particular models that operate in very few areas (for example, population genetics, biomechanics, and some areas of physiology), and even then the treatments are not particularly simple. Moreover, there is only one single overarching theory and that is evolution, a subject that does not lend itself to mathematical formalization. This is partly because it represents the results of independent events at many levels and is thus intrinsically complex, and partly because evolutionary change depends on random events at the genomic and ecological levels.

Over the last decade or so, biology has had another layer of complexity added to an already complicated picture: it has become clear that many proteins do not operate alone but are part of complex networks that have an output, such as initiating growth, that could never have been predicted on the basis of their individual protein constituents or the structure of the networks. Systems biology is partly an attempt to understand how these protein networks operate (the narrow view) and partly an attempt to provide a framework for integrating the various levels of biological knowledge (the broader view).

The narrow view

This area of systems biology is primarily interested in understanding how complex molecular networks and pathways work. Over the last couple of decades, it has become clear that it is often not possible to predict the wider context of a particular protein's activity from its immediate interactions with other proteins. This is because that context is the end result of the interactions of up to a hundred or so proteins that cooperate within a single network or pathway. Well-known examples are the signal transduction pathways whose activities initiate the expression of new sets of genes, the epidermal growth factor (EGF) cycle which controls entry into mitosis (**Figure A1.1**), and the Krebs cycle that generates adenosine triphosphate (more than a hundred such networks can be seen at www.sabiosciences.com/pathwaycentral.php).

The reason why such networks are important is that they drive anatomical change in embryos and physiological function in adults. Their roles in embryos are particularly important in an evolutionary context because these roles include patterning, differentiation, cell proliferation, apoptosis, cell movement, and morphogenetic activities; they are thus the drivers of an organism's adult anatomical phenotype. Normal genetic variation together with novel mutations in their protein constituents lead to variation in the output of these networks and so to normal variation in the phenotype, and can occasionally even produce major changes. In this context, particularly important pathways are those for growth, patterning, and pigmentation (Chapter 12).

Understanding how such complex pathways work qualitatively and quantitatively underpins what can be seen as the *narrow view of systems biology*. Making numerical sense of these protein networks is, however, very difficult; this is partly because their normal mathematical formulations as sets of differential equations can only be done in a straightforward way where the law of mass action applies. Unfortunately, the concentrations of the participating proteins and substrates in cells are often so low that the law of mass action has to be replaced using stochastic methods (Kirkilionis, 2010). Two other difficulties are that such equations can only be readily solved near equilibrium, a condition that may not be obeyed, and that the rate constants for the equations are rarely, if ever, known. Such networks are thus very hard to model computationally. Some insight into their activity can be obtained by studying the effects of mutations on their constituent proteins, but there is a long way to go before we can elucidate the underlying principles of network dynamics or predict the effect of mutations.

The broader view

There is, however, a second aspect to systems biology and it comes from the fact that any event in the biosphere is underpinned by further events at any level of scale from the genome up to the environment, each of which has its own systems (**Table A1.1**). It is impossible to understand any rich aspect of biology without an appreciation that what is going on at one level depends on those levels above and below it. This is because there are interactions between levels that can be downwards and upwards. The following three key themes underlie this broader view of systems biology.

Events within each level are complex

Even within a level (see Table A1.1), events can be very complex. Obvious examples are protein networks: their function cannot be predicted on the basis of their composition.

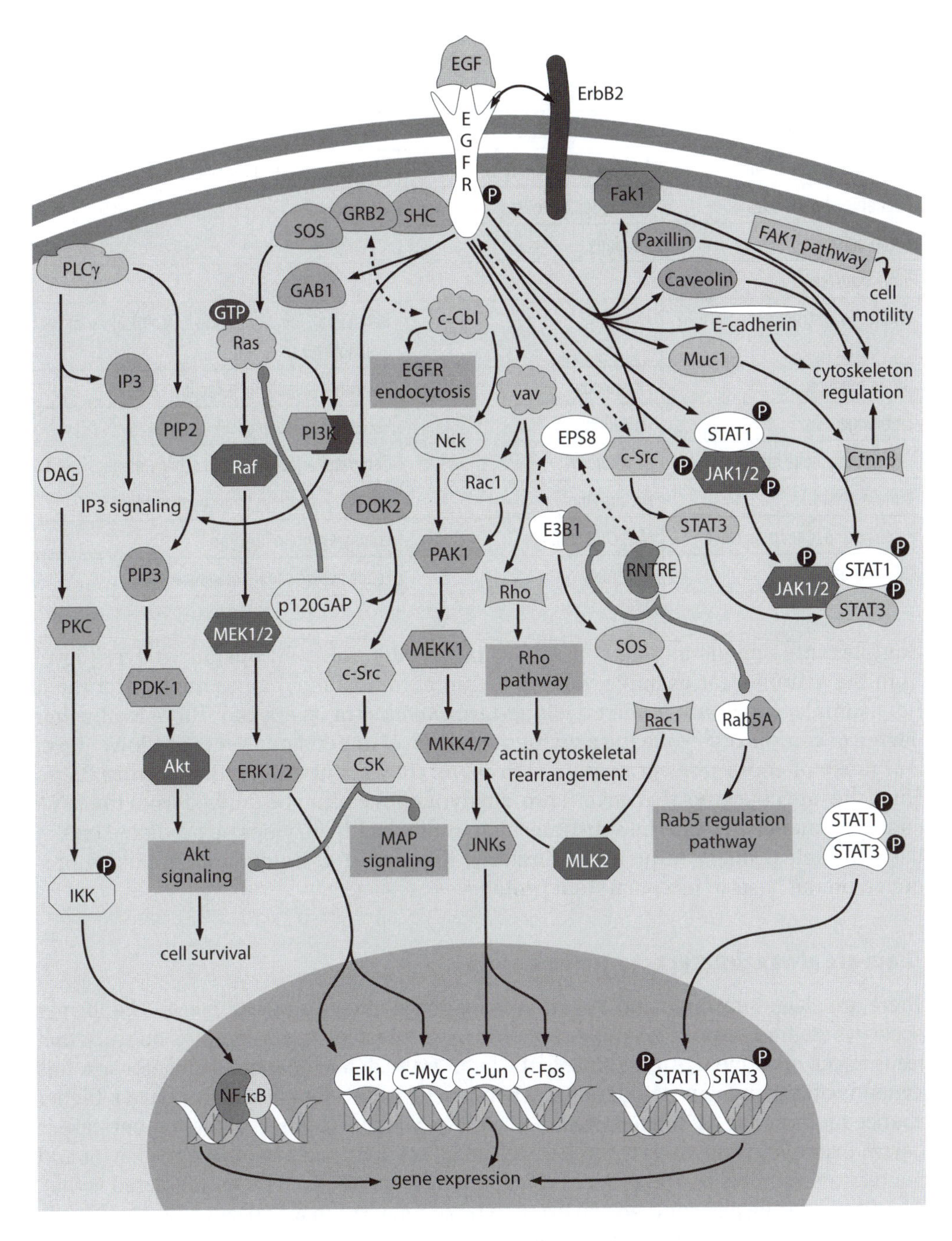

Figure A1.1 The mouse epidermal growth factor (EGF) signaling network. The net result of the EGF ligand binding to the receptor is that up to six transcription factors are activated, and these often lead to the initiation of the proliferation pathway. The details by which this happens are unclear; the extent to which all 60 or so proteins are needed in any particular situation is not known nor are the reasons why one or another internal pathway is activated (from www.sabiosciences.com/pathwaycentral.php).

Table A1.1: Effects of DNA mutation on key levels of biology, their subject names, and role in evolution

Population levels	Scientific domain	Evolutionary significance
Ecological environment	Ecology and geography	Each plays a role in the selection of variants
Phenotype frequencies	Population genetics	
Social behavior	Ethology	
Individual levels		
Whole-body structures	Anatomy	Behavioral, functional, and anatomical variants available for selection
Organ function	Physiology	Variants available for selection
Embryogenesis	Development	Variants available for selection
Cell and organelle function	Cell biology	Generates anatomical and other variants
Protein networks	Systems biology	Generates changes to network outputs
Proteins and other molecules	Molecular biology	Changes protein function
DNA mutation	Genomics	Leads to DNA sequence variants

Complex interactions among constituent factors do, however, operate at every level from the genome (for example, folding within chromosomes) up to the environment (for example, we cannot predict the long-term weather or its effects). There is a further degree of complexity: while the activities at one level depend on events at a lower level and must, of course, be compatible with them, they cannot be predicted from them. Thus, the rate constants that govern protein dynamics cannot be derived from the DNA coding sequence nor can the substrates that interact with enzymes be predicted on the basis of their primary sequences. Similarly, the output of a network cannot be predicted on the basis of its constituent proteins.

There are always interactions between levels

There are always upwards and downwards feedback effects between levels. The former occur when, for example, a simple signaling molecule activates a complex network that leads to cell movement. An example of the latter is the determination of alligator sex: this depends on the ambient temperature when its egg is developing - males have a higher chance of developing if the temperature is above 32°C and females if the temperature is below. Examples of upwards control are the effects of neuronal networks on behavior and population numbers on the greater environment. What needs to be remembered is that, when considering any complicated event, interactions within and between several levels may need to be taken into consideration.

Causality is distributed

The combination of complexity within levels and feedback both within and across them means that it can be very hard to work out exactly where causality lies. In the case of alligator sex determination, the immediate or *proximate* cause is ambient temperature, but there are other causes. The temperature-sensitive system presumably evolved because there was a selective advantage in having more males at higher temperatures, while the development of the early reproductive system depends on protein activity and novel gene expression. In general, events at many levels contribute to change, and the complexity of the feedback systems means that causality is distributed within and between levels - no level is privileged (Noble, 2012). It is almost always true that, where there is a claim for a single cause of a complex event, that cause is an oversimplification.

In short, complexity lies at the heart of life and understanding the biosphere requires us to grapple with and find ways of unpicking that complexity. Many areas of biology have always known this to be so; important examples from traditional areas include the complex physiological models of heart activity and nerve conduction. What is claimed today as novel systems biology actually has deep and old roots.

A note on systems terminology

The first step in being able to unravel complexity is to find a language for expressing it in such a way that the contribution of the individual facts can be seen as contributing to the whole story. One of the most straightforward ways of doing this is through formal or mathematical graphs; these do not describe data but are sets of linked statements, each of the general triadic form:

<A> *<links to>* <B>

or in mathematical terminology:

<node 1> *<edge>* <node 2>

If *A* or *B* each have other links (for example, <B> *<links to>* <C>, and so forth), then the group of such triads naturally forms a hierarchy or a network.

Mathematical graphs provide a natural language for capturing complex knowledge and seeing how such knowledge is built up from many simple, linked facts. This formalism is very powerful; thus, all of anatomy can be built on the *<part of>* link (albeit that *part of* can have several meanings). Classical taxonomy is based on the *<member of the class of>* (normally abbreviated to *<is a>*), while tissue lineage in embryogenesis is based on the *<develops from>* link. These examples are all hierarchies, but the framework also enables networks to be built: relationships such as *<is next to>* are the key for making maps of railway systems. More complicated are networks of interacting proteins where one can activate or repress another, with the effect, perhaps, depending on events elsewhere in the network and where there is a range of links (for example, activates, represses, and is a co-factor for).

Standard Linnaean taxonomy is based on class and the *<is a>* relationship. There, all groupings except the species or subspecies are idealizations based on common sets of properties. The taxonomy of the horse is one example:

<horse> *<is a>* <perissodactyl (odd-toed ungulate)> *<is a>* <eutherian (placental) mammal>

Although they are not usually seen as such, cladograms (hierarchies of evolutionary descent, usually based on anatomy) are also formal graphs based on the Darwinian relationship *<descends with modification from>* and thus capture complex evolutionary knowledge (Chapter 5). They are thus fundamentally different from a Linnaean taxonomic hierarchy, which is based on the <*is a*> relationship. It is worth noting that a cladogram describes descent but does not explicitly show the nature of the evidence on which the <*descends with modification*> link derives. It is, however, usually based on anatomical modifications, such as minor changes to a bone, with the evidence deriving from a close analysis of the fossil record and contemporary anatomy.

A similar hierarchy can be generated from analyzing related DNA sequences (this is known as a phylogram), but here it is not common ancestors, but ancestral sequences, that are parents. Thus:

<sequence Q of species B> <*descends through mutation from*> <sequence P>

and

<sequence R of species C> <*descends through mutation from*> <sequence P>

Further hierarchies can be constructed (though they rarely are) on the basis of protein function. It is sometimes claimed that the trouble with evolution as a science is that it does not make predictions, but this is simply not true. For example, it is a prediction that the groupings of living organisms based on these three very different and independent types of data will be the same if the data are rich enough, and, where the data for one or other area are incomplete, the weaker grouping should be a subset of the stronger one. This prediction has yet to be shown to be false. These ideas are further developed in Chapters 3, 5, 6 and 8.

The details of the evolution of a particular organism do, however, need far more than anatomical data: it requires information at levels from DNA mutation to selection of variants as a result of environmental pressures and rarely, if ever, do we have the full set of facts to model change graphically. It is, however, quite straightforward to include facts from and between several levels and types of information within a single graph, as an example from development illustrates (**Figure A1.2**).

In a growing vertebrate embryo, small tissues need a capillary blood supply once they are more than a millimeter or so in size. A growing tissue indicates this need through secreting a small protein, vascular endothelial growth factor (VEGF), that diffuses away from the tissue and eventually reaches the surface of a capillary where it binds to a receptor that activates a complex protein network whose downstream activity leads to new capillaries invading the original tissue. While the internal dynamics of such a single network may be very complex, its output behavior may be simple. Thus, the outcome of activating the EGF pathway (see Figure A1.1) may, for all the internal complexity of the network, simply be to direct the cell to enter mitosis.

The use of a graph to show how VEGF secretion by growing tissues leads to their vascularization is shown in Figure A1.2. The ligand diffuses away from the tissue and binds to the nearby endothelial cells of blood capillaries and causes the tips of the capillaries to proliferate up the VEGF gradient. The graph describing this is not simple, but it integrates a lot of experimental data to show how the overall behavior of the system operates. Of particular interest here is the way in which the graph incorporates information at three levels: the activity of molecules (for example, VEGF and notch), the activity of networks (for example, the VEGF and delta networks), and changes to whole tissues (capillary elongation and the provision of a blood supply to a tissue). Also interesting is the representation of time (from left to right), while local causality is represented through the form of the link that describes

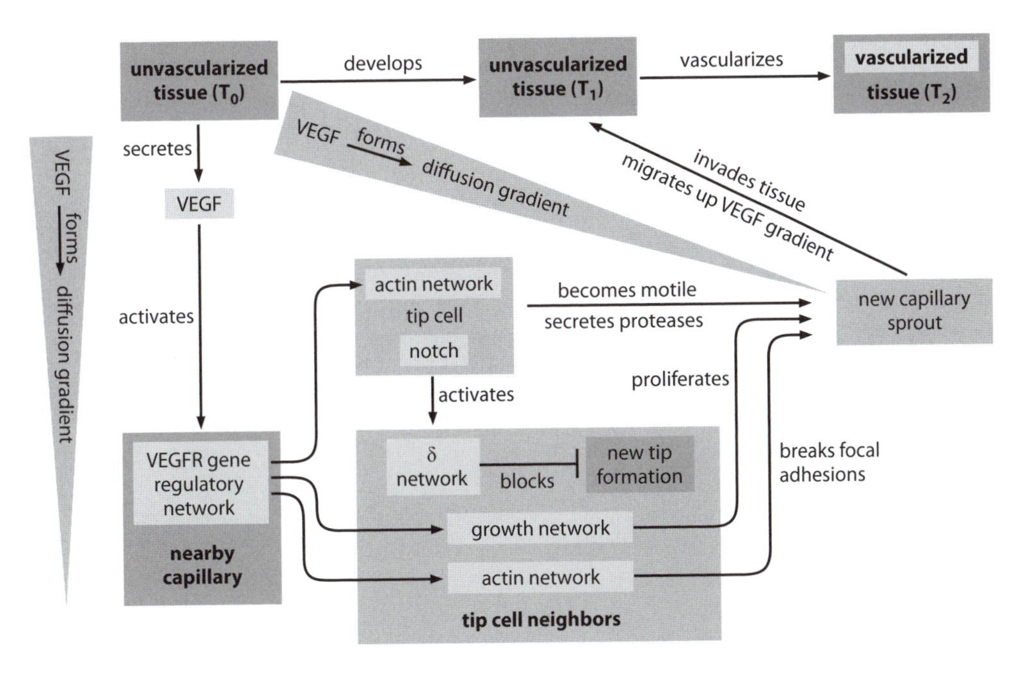

Figure A1.2 A graph describing some of the core events underpinning the way in which a developing tissue becomes vascularized. The unvascularized tissue at T_0 secretes vascular endothelial growth factor (VEGF), the signal that diffuses away from the tissue and activates the VEGF network in a nearby capillary cell causing it to extend at the tip and grow towards the source of the signal. The set of events that result in the tissue becoming vascularized is represented as a set of linked triad statements that together form a graph (VEGFR: VEGF receptor).

a change of state. The sheer number of such links emphasizes the extent to which causality is distributed:

<State A> <*verb*> <State B>

As this example illustrates, graphs are an important way of integrating sets of complex facts into a straightforward, formal picture. They are already used in the form of cladograms and phylograms, and should be helpful at in other biological contexts too.

Further reading

Bard J (2013) Systems biology – the broader perspective. *Cells* 19: 413–431.

Capra F, Luisi PL (2014) The Systems View of Life. Cambridge University Press.

Soya O (ed.) (2012) Evolutionary Systems Biology. Springer.

APPENDIX 2
A HISTORY OF EVOLUTIONARY THOUGHT

There is a view that history is irrelevant to science because today's knowledge makes yesterday's approaches redundant. It is, however, difficult to appreciate the complexity of modern views of evolution without a sense of the historical context in which they developed. It is also important to pay tribute to the many serious scientists who fought their contemporary orthodoxy to try to make sense of the story of life, and on whose shoulders we all stand. This chapter provides a summary of the history of evolutionary science from the creation myths to the molecular era. The reader who wants to know more should explore the references at the end of the chapter, particularly Mayr (1982).

Until relatively recently, the world of life seemed static: the sun rose in the east every morning, and the stars, moon, and planets appeared to circulate around us in a predictable way. Even when Galileo showed that the earth rotated around the sun, there was no reason to let these new views of the solar system affect ideas of natural history. The animals and plants of one's grandchildren were the same as the animals and plants of one's grandparents, and back as far as history went. The very idea of evolution seemed wrong for two reasons. First, animals were perfectly suited to their environments, all of which were populated, so there was no need for new creatures. Second, animals could only breed with members of their own kind, so new species could only be formed by *de novo* creation. The few who did think about such things realized that either the living world had been there forever or that, in the dim and distant past, it had been created. If the former, then there was nothing to be said; if the latter, then some great creative force had done it and the problems were who and how. So were borne the wonderful creation myths, of which Wikipedia currently details well over 50 glorious and very different stories from around the world (see en.wikipedia.org/wiki/List_of_creation_myths).

The best-known of these myths, in the Western world at least, are the two at the beginning of Genesis. The first (Genesis Chapter 1 and Chapter 2, verses 1–4) has God creating the universe in 6 days, starting with the physical world, then the living world, and, on the last day, Adam and Eve. All this work was followed by a day of well-deserved rest, the Sabbath. Modern textual analysis suggests that this story was not actually written by Moses in the thirteenth century BCE, but by one or more Hebrew priests around the sixth century BCE; it is still hard not to be overwhelmed by its wonderful sonorous phrases ("And it was the evening and the morning of the first day..."). The second version (the rest of Genesis Chapter 2) is much more mundane in tone: first the world and man were created, then the other living organisms, male and female, and, finally, and because there was no creature for man to mate with, Eve was created through God taking one of Adam's ribs and refashioning it as a woman. This version was probably written some 300 years earlier in Judea and includes the story of the Garden of Eden and the recognition of good and evil (see Friedman, 1987; and the wiki on the documentary hypothesis - en.wikipedia.org/wiki/Documentary_hypothesis). Some rabbis contend that the two versions are a single story, with the former concerned with the physical and the latter with the spiritual world. The tones and the language of the two stories, not to mention the timings, are so very different that one can only feel that this has to be seen as a sterling attempt to reconcile the irreconcilable - unless, of course, both are regarded as fables.

In an odd way, we can feel that those who invented these and the other creation stories were among the first scientists, looking for explanations of the complexities of the world. They were not, of course, real scientists because they never thought it necessary to test their ideas or to work through their implications. Getting to this stage was to take some thousands of years and it is only in the last 200 or so that we have begun to use an evidence-based approach to questions about where and how life arose, whether each of the species had always been as it is now or whether change had occurred and, particularly, how ideas might be tested.

What follows is a brief summary of that approach by biologists and philosophers, many of whom had to argue against those who believed that their own traditional myths were literal accounts of creation, irrespective of any facts to the contrary. Even today, when the evidence for evolution is overwhelming, there are still those who would prefer to believe ancient religious narratives. Appendix 4 considers the arguments of the two views.

The early days

Until about the eighteenth century, the very great majority of the people in the Western world accepted the biblical view that God created the world. Over the preceding centuries, a few people had had tried to make sense of the living world, mainly within this theistic framework. The earliest biologist about whom much is known was Aristotle (died 322 BCE). However, while he collected a great deal of biological knowledge and tried to systematize it and make sense of it, it seems unlikely that he ever considered that species were anything other than fixed. Likewise was his successor, Theophrastus (died 287 BCE) who wrote *Historia Plantarum* (*History of Plants*) in 10 volumes that was used as a resource for many centuries. The one early thinker to suggest that the biological world could change was Titus Lucretius Carus (died 55 BCE): in a six-part poem entitled *De Rerum Natura* (*On the Nature of Things*), he suggested (part five) that living things had a naturalistic, even evolutionary rather than a divine, origin. Thus, in Leonard's translation (2008):

How merited is that adopted name
Of earth - "The Mother!" - since from out the earth
Are all begotten.

Although various people interested in biology made comments on the possibility of evolution over the early years, the first substantial writings seem to have been by the Arabic writer Al-Jahiz (or "the goggle-eyed," so called because of an eye malformation) who lived in the ninth century in Baghdad and who was one of the very few who read the living world correctly (Bayrakdar, 1983). In his *Book of Animals*, he wrote, with brilliant insight:

> *Animals engage in a struggle for existence; for resources, to avoid being eaten and to breed. Environmental factors influence organisms to develop new characteristics to ensure survival, thus transforming into new species. Animals that survive to breed can pass on their successful characteristics to offspring.*

Stott (2012) makes it clear that Al-Jahiz thought that all this happened within a deistic context, but his ideas certainly became known in the following centuries, as did the suggestion of Ibn Khaldun (fourteenth century) that man derived from monkeys.

For the next five centuries, and in order to avoid religious controversies, such ideas were rarely discussed outside of the very small domain of academic medicine. Even Buffon (1707–88), the first major systematic biologist and director of the King's Garden (Jardin du Roi) in Paris, was careful not to publish his real views until he was very old (1778). His *Histoire Naturelle, générale et particulière* (39 volumes, 1749–89) was the first comprehensive discussion of biology and environment, and much of it was attacked by the church. Buffon's response, although he himself was unsure about evolution, was to apologize but to retract not a word! The next few generations of biologists were brought up on these texts.

It was perhaps the unlikely figure of a Swiss lawyer, who, in the 1770s, set the scene for modern thinking about evolution. Charles Bonnet (1720–93) was, however, not only a lawyer, but also a considerable experimental biologist who discovered parthenogenesis and plant and insect respiration, while also studying regeneration and mental disorders. His most interesting evolutionary idea was that man was the culmination of climbing up the 52 steps of the ladder of being that started with matter and progressed through plants, worms, insects, shellfish, fish, birds, and so on to its ultimate triumph, man. The idea that there is a hierarchy of being is one that goes back to Plato and perhaps to the first version of creation in the bible, but it seems to have been Bonnet who suggested that progression was possible. He also believed in preformationism, the idea that the sperm contained a small homunculus that could develop, once it met the fertile environment of the egg. Bonnet got a lot wrong but he made evolution a concept that serious people could consider.

In the last decades of the eighteenth century, ideas of biological evolution were clearly in the air, for all that the church tried to stop them. Notable authors who discussed evolution were the philosopher Denis Diderot (1713–84), who published an encyclopedia in France that included chapters on geology and evolution, James Burnett (Lord Monboddo, 1714–99) in Scotland, and Erasmus Darwin (1731–1802) in England. The last, a country doctor and Charles Darwin's grandfather, was also a serious naturalist who held original evolutionary views, especially on sexual selection. However, he remained publicly silent for many years until 1796, when he completed *Zoonomia*. This included the substantive sentence:

> *Would it be too bold to imagine, that in the great length of time, since the earth began to exist, perhaps millions of ages before the commencement of the history of mankind, would it be too bold to imagine, that all warm-blooded animals have arisen from one living filament, which THE GREAT FIRST CAUSE endued with animality, with the power of acquiring new parts, attended with new propensities, directed by irritations, sensations, volitions, and associations; and thus possessing the faculty of continuing to improve by its own inherent activity, and of delivering down those improvements by generation to its posterity, world without end!*

His final work on evolution, a poem entitled *The Origin of Society*, was published posthumously.

Although these authors all came to the conclusion that evolution had happened and each is historically important and interesting, none of them has anything scientific to say to us today; this is mainly because they never tested or properly worked through their ideas. Their significance was that they were at least prepared to think about rational as opposed to theistic explanations of life.

The move to evolutionary thinking

The turn of the nineteenth century was a pivotal time in the history of evolution and, before discussing what then happened, it is worth standing back and looking at the wider context. Four factors were pushing progress, two biological, one geological, and one philosophical. The first of the biological drivers was fossils. These had always seemed anomalous, particularly the presence of shells a long way from the sea. They had been commented on as far back as Xenophanes (c. 570–478 BCE) and Herodotus (484–425 BCE) in Greece, and Pliny the Elder (23–79 CE) in Rome. The standard explanation, promulgated by theologians for many centuries, had always been that they had been washed up onto land during Noah's flood.

Stott (2012) discusses how the problem of such fossils had worried Leonardo da Vinci (1452–1519), but it was not until 1508 that he noted fossil oyster shells present on the mountains around Milan and started to analyze them. He showed that the groups of shells included both young and old individuals, arranged much as they were found when alive, and concluded that they could neither have been washed there nor moved there on their own. He correctly deduced that, at some time in the dim and distant past, the Apennine mountain range had been under water. It was not only shells that were problematic; fossilized bones had been noted as curiosities as early as 2000 years ago in China. Examples were being found in Europe, and, by 1665, the subject of fossils was of sufficient interest for Robert Hooke to give a talk on them to the Royal Society in London. By the eighteenth century, bones much larger than those of living animals had been discovered in the Ohio valley and fossil collecting was becoming both fashionable and of serious geological interest.

The second biological driver was the appreciation of just how well adapted organisms were to the habitats in which they lived. Here, the 39 volumes of Buffon's *Histoire Naturelle, générale et particulière* published over the period 1748–89 were important because they included so much detail about so many animals. His almost exact contemporary Carl Linnaeus (1707–78) was equally important, as his systemization of some thousands of plants and animals was becoming a standard part of education. The question of why closely related animals were so well adapted to their niches was one that could sensibly be asked for the first time. The idea that God designed each animal for its correct place (the perfection of nature) was beginning to seem less credible.

The third was geology: the subject, as we know it today, started at the end of the sixteenth century with the geological investigation of rock strata and the religious need to find evidence for Noah's flood. The former proved more useful to science and, by the middle of the eighteenth century, was being popularized by Buffon in his *Histoire* and Diderot in his *Encyclopédie*. The key person, however, was James Hutton (1726–97), who worked out a great deal about layering and geological forces. He particularly suggested that the earth was not only very old, far older than Bishop Ussher's then-accepted biblical creation date of 4004 BCE, but also in a continuous state of flux. One reason for his taking this view was his discovery of ancient fish bones buried in the Salisbury Crags cliffs in the south of Edinburgh. By the 1790s, geologists, particularly William Smith in England, had realized that series of rock strata could be identified on the basis of the fossils that they contained, and a sense of the progression of organism complexity was becoming apparent. It seems odd today that the early geologists seem not to have taken a major part in the many discussions about how this progression occurred.

The final driver was philosophical: the seventeenth and eighteenth centuries had been the age of enlightenment when intellectuals set out to reform society on the basis of reason.

A key theme was to throw out ideas grounded in faith and tradition and start again. This was the environment in which the great physicists Isaac Newton and Gottfried Leibniz worked; it was impossible to be what we would now call a scientist and be untouched by these wider ideas. It was in this general context of inquiry that Bonnet, Erasmus Darwin, and a few others were expanding the boundaries of biological thinking. Not everyone, however, was at ease with such changes and the tensions were particularly apparent in the Museum of Natural History in Paris around 1800 and the following few years, when three major figures of biology fought out their ideas about whether evolution had happened.

The early nineteenth century

In 1793, the revolutionary government in Paris enlarged the Jardin du Roi to be the Museum of Natural History and appointed 12 professors, included among whom were Saint-Hilaire, Cuvier, and Lamarck. The fine detail of the natural world had only recently begun to be studied academically and each of these three focused, almost obsessively on a particular area. Étienne Geoffroy Saint-Hilaire (1772–1844) was the professor of zoology and an embryologist, paleontologist, and anatomist. His work on the diversity of vertebrates led him to the view that there was an underlying structural plan to animals and he was the first person to think seriously about anatomical similarities among unrelated organisms, such as the wings of bats and birds. He explored these ideas in the context of ontogeny (a term used in various ways but here meaning *the development of form*) and came to the view that evolution and species change (he used Lamarck's term, *transmutation*) probably occurred but involved major jumps. He was also of the view that structure determined function, unlike Cuvier, who took the view that it was the needs of function that determined structure.

George Cuvier (1769–1832) was the professor of natural history, and perhaps the greatest comparative anatomist and paleontologist of his day; his wonderful collection of tissues and fossils is maintained virtually unchanged today as part of the Gallery of Paleontology and Comparative Anatomy at the Jardin des Plantes in Paris, a place that remains a joy for any biologist to visit. Cuvier is known today partly for his work on taxonomy, but particularly for his discovery of extinctions. Here, he carefully analyzed the fossilized organisms in adjacent layers of rocks and found that the organisms in one layer often bore no resemblance to those in the layers above or below it (for instance, land animals could be covered by sea animals). He deduced that extinctions had actually occurred and that these had been followed by new bouts of creation, and hence that there was no evidence to support the idea of evolution. Although Cuvier was religious, he saw himself as a serious man of science whose views were grounded in knowledge; he therefore did not argue against evolution on grounds of faith, but on grounds of data; the trouble was that he had too few data and what he had, he misunderstood.

Jean-Baptist Lamarck (1744–1829), the third of the trio, was originally a soldier and then a botanist, but was made the professor of insects, worms, and microscopic animals (or invertebrates – a term that he coined) at the Museum. Lamarck's fame comes from his work on insect classifications and on evolution, with his views here being shaped over the period 1795–1809 and finally published in his *Philosophie zoologique* (1809). Gould (2000), in an article that is required reading for anyone interested in the history of evolutionary thought, has pointed out that the early chapters of this book see evolution as organisms climbing Bonnet's ladder of life. The picture in the appendix of the book is, however, very different: Lamarck had by then realized that, on the basis of anatomical geometry, earthworms and parasitic worms could not be in the same class since the anatomy of the former was very much more complex than that of the latter. It became clear to him that the only acceptable explanation of this problem, and, by extension, other types of anatomical change, was that evolution must have occurred not by climbing a ladder, but by descent and branching (**Figure A2.1**). Thus, Lamarck was the first person to put forward the idea of branching descent on the basis of evidence, and this is the foundation of all modern thinking on evolution.

Lamarck also grappled with the nature of anatomical change, which, he realized, underpinned evolutionary change. This was a problem that Darwin was not able to solve, merely commenting in Chapter 5 of *On the Origin of Species* that "Whatever the cause may be of each slight difference in the offspring from their parents - and a cause for each must exist..." Lamarck, as Gould (2000) has pointed out, came to the view that organisms had two abilities, to complexify and to adapt, and that the resulting changes were heritable, although he did not express his ideas very clearly. Lamarck's views were accepted by everyone, including Darwin (Chapter 12), until Weissman showed in the 1880s that they were incompatible with his theory of the continuity of the germplasm.

Lamarck today has a general reputation completely at odds with the importance of his work and the reason is that he antagonized Cuvier, who believed in change through successive acts of creation rather than heritable adaptation. After Lamarck died, Cuvier wrote an excoriating obituary in which he accused Lamarck of being a theoretician who went far beyond the data in his views on evolution. Gould (2000) quotes a particularly damning paragraph from this obituary.

> *These [evolutionary] principles once admitted, it will easily be perceived that nothing is wanting but time and circumstances to enable a monad [single-celled organism] or a polypus [sea anemone] gradually and indifferently to transform themselves into a frog, a stork, or an elephant...A system established on such foundations may amuse the imagination of a poet: a metaphysician may derive from it an entirely new series of systems; but it cannot for a moment bear the examination of anyone who has dissected a hand...or even a feather.*

Even at the time, the obituary was regarded as unfair because Lamarck was well known as a practical biologist, but Cuvier's views have prevailed in spite of the writings of everyone who has looked into the history more carefully. Today, after some 200 years, the work of Gould and others is restoring Lamarck's position as the first serious evolutionary biologist and as the key forerunner of Darwin, as Darwin himself eventually realized.

The other person in the first half of the nineteenth century who laid the groundwork for modern ideas of evolution was the Estonian, Karl Ernst von Baer (1792–1876; **Figure A2.2**), who was a major figure in early nineteenth-century biology, and one for whom Darwin had considerable respect. Although his work ranged across the sciences, it is as an embryologist that he is remembered today. Most of his work was technical: he discovered the blastula, the notochord, the human egg, and, with Christian Pander, the triploid structure of the gastrula. von Baer is important here because his laws, based on the examination of a great variety of embryos, provided the groundwork for integrating embryology and evolution, in spite of the fact that he never properly accepted evolution. His two key ideas were (i) that the general features of embryos in a group emerge before specific ones, and (ii) that embryos of "higher" forms do not resemble the adults of "lower" ones, only their embryos (Chapter 10).

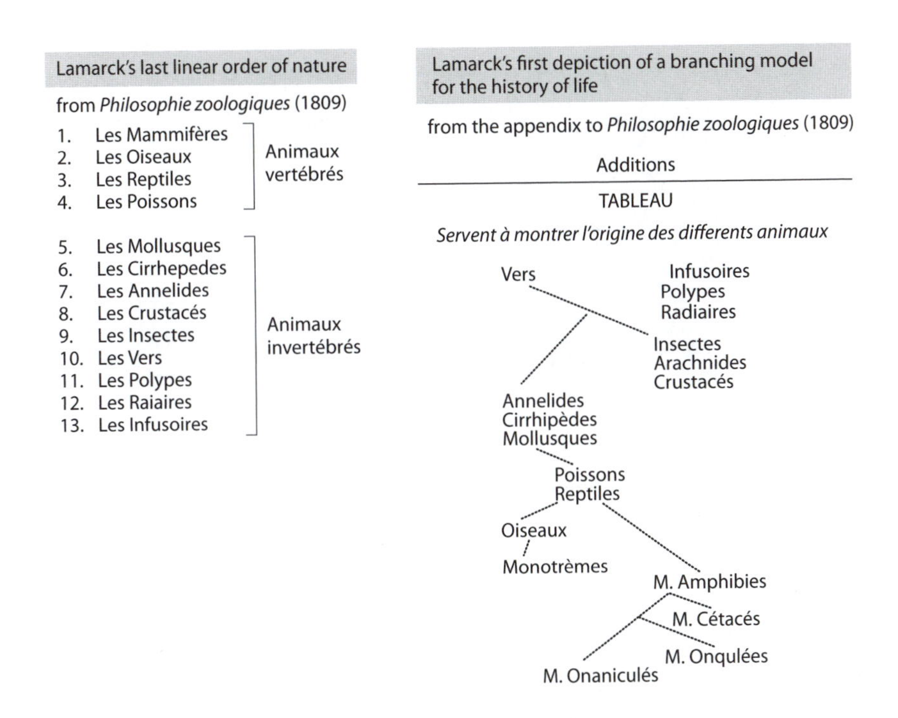

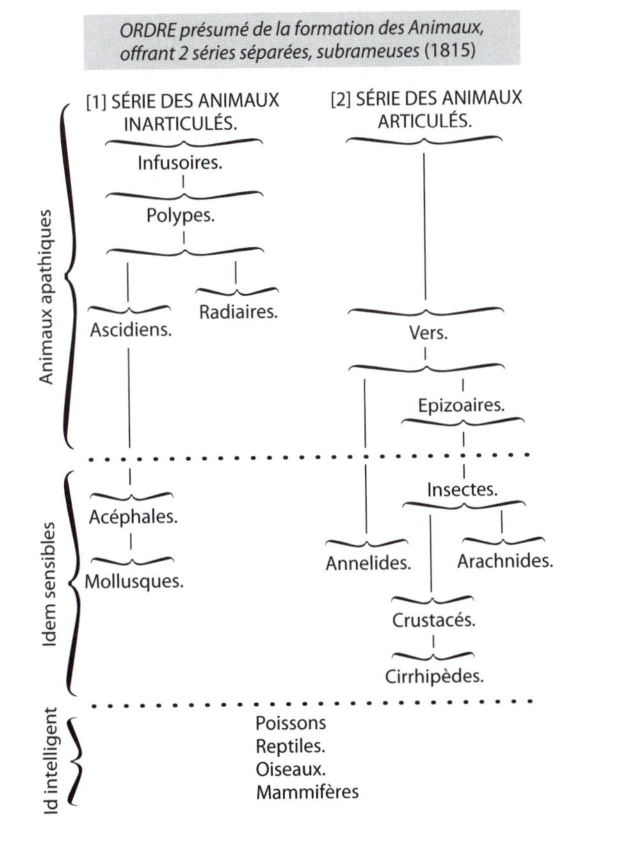

Figure A2.1 Lamarck's three views of phylogeny.

Figure A2.2 A picture of Ernst von Baer on an old Estonian 2 kroon banknote. von Baer is probably the only embryologist to have been honored in this way!

The era of Darwin

Although evolutionary thinking progressed little in the 50 or so years between Lamarck and the Darwin/Wallace era, it is clear that the idea of evolution was being widely discussed during this period. Robert Grant in England explored homologies, while Robert Chambers, a self-educated publisher, wrote *Vestiges of the Natural History of Creation*, which he published anonymously in 1844. This book, which proposed that the solar system and life had both evolved, was immediately controversial. While the public bought it in droves, churchmen and academics damned the book on grounds of religion and science, the latter because it went far beyond the evidence then available without producing any mechanisms as to how evolution, in particular, had happened. This risk of such public notoriety was one reason why Darwin held back from publishing his ideas on evolution for so long.

The crucial publication that eventually provided the key to understanding evolutionary change was Thomas Malthus' *An Essay on the Principle of Population*, which had actually been published some decades earlier (the most important version was printed in 1803). This argued that human populations tended to increase exponentially and were kept in check by natural constraints (sickness, famine, and war, for example). The inverse of this idea was taken up by both Wallace and Darwin: that a subpopulation of natural variants could escape these population constraints and flourish as a new species if it found an environment for which it was suited.

Alfred Russell Wallace (1823–1913) was a self-taught naturalist from a very modest background who became the first person to explore the distribution of animal species within their geographic environment while supporting himself by collecting tropical animals to be sold to collectors and museums in England. Wallace essentially invented the science of biogeography and is still remembered for identifying what is now known as Wallace's line, the boundary marking the western limit of Australasian fauna. Already convinced that evolution was a reality, Wallace became interested in why related species lived in different environments. The crucial insight came to him while he was recovering from a nasty bout of malaria in Borneo: he realized that Malthus' essay held the key to understanding how fitness determined why one species would flourish in a particular environment while another would not.

He wrote up his ideas in 1858 and sent the paper to Charles Darwin (1809–1882), the most distinguished biologist of the time, who had been quietly working on evolution for two decades and was shocked to think that he might lose the credit for all this effort. On the advice of his friends Charles Lyell and Joseph Hooker, Darwin abstracted a manuscript on the same theme that he had originally written in 1847 but not published and gave both articles to Lyell and Hooker, who sent them to the Linnaean Society, and they were published together in 1858. In one of the great misreadings of history, Thomas Bell, the president of the society, mentioned in his 1859 presidential address that the previous year had not been marked by any of those striking discoveries that at once revolutionize the department of science on which they bear (Guerrero, 2008).

Although Wallace was a remarkable scientist and deserves some of the credit for discovering natural selection, he rightly has a secondary role in the story. The first reason is on grounds of priority: although Wallace had been thinking about speciation for some years, his insight came in around 1858; Darwin had started his first notebook on evolution in 1837 and read Malthus' essay in 1838 (according to his autobiography), a couple of years after he had returned from his 5-year voyage around the world in the *Beagle*. By 1844, Darwin had reached Wallace's 1858 conclusion on the basis of thinking that had started when he had begun to consider the diversity of the Galapagos finches; indeed, all of Darwin's colleagues knew his views well before Wallace wrote his letter. The second reason is because of the depth of their respective research. Wallace's analysis derived from his knowledge of natural history and his academic interest in speciation. Darwin, however, had explored the evidence for evolution across the whole of biology because he realized that, to make the public case for evolution, he had to produce broad-based and convincing arguments. In particular, he had to show that variation was innate and commonplace. It was no coincidence that the first chapter of *On the Origin of Species* (1859) included information about breeding pigeon varieties.

The book itself focuses on a single question: what are the implications of the variation that any observer can see within a species? The answer, buttressed by a wide range of examples from across the then biological spectrum, starts with the fact that, left to themselves, populations grow exponentially but that this growth is constrained by the environment. Where variants arise, these may disappear or flourish (and may even replace their parent population) as selection pressures from that environment dictate (the *struggle for existence*). Such Malthusian competition (*natural selection* or *survival of the fittest*) acting on the nonuniform population leads to *descent with modification* (which is, of course, evolution), and these two ideas, which underpin the book, are the key to understanding the origin of new species.

It is also worth noting what is not in the book: although it is obviously about evolution, the term is barely mentioned in any explicit way, and one reason was that the word then had an embryological meaning. Instead, Darwin focused on the ideas and implications of branching descent and hence on the mechanisms by which new species could arise from existing ones, a problem that, as Mayr (1982) points out, had not previously been discussed in any substantive way. Perhaps surprisingly, Darwin's relatively brief discussion of the past is postponed until Chapter 10 (on the imperfections in the geological record) and Chapter 11 (on the fossil record); here, the information is used within the context of the theory rather than to assert that evolution had been acting since life had originated. Darwin clearly understood the evolutionary implications of his work but, at this stage, chose neither to emphasize them nor to consider the place of man. This was partly because he realized the likely public opprobrium to which he would be subjected - he did not want to be treated like the anonymous author of *Vestiges of the Natural History of Creation*. A second reason was that he did not want to upset his wife, who held traditional Christian views.

For all his hopes to the contrary, the book was extremely controversial, exciting public discussion and religious fury from some clerics but not others. Thus, the future Cardinal Newman (1801–1890) wrote to a colleague in 1868 that he was not really worried about how and when God created matter and laws nor was he disturbed by the idea of evolution, ending his letter by saying "I do not [see] that 'the accidental evolution of organic beings' is inconsistent with divine design - It is accidental to us, not to God." Others were, however, less tolerant. Darwin took little part in the controversy, but his views were stoutly defended by Thomas Huxley (1825–1895) who held a famous debate with "Soapy Sam" Wilberforce, Bishop of Oxford, in 1860 at the Natural History Museum in Oxford that seems to have ended in a draw. The result of all this controversy was that the subject of evolution was now firmly in the public domain. Over the next few years, Darwin brought out further editions of *Origins* in English and it was widely translated - it was perhaps the first international bestseller about science. He then worked on various other areas of biology, but it was not until 1871 that he produced his second book on evolution, *The Descent of Man, and Selection in Relation to Sex*.

The *Origin* had carefully avoided any discussion on the evolution of man apart from a cryptic comment towards the end saying that, in the distant future, "Light will be thrown on the origin of man and his history." In *The Descent of Man* and also in *The Expression of the Emotions in Man and Animals* (1872), Darwin sought to show how human emotions and psychology could have evolved from simpler aspects of animal behavior. He also used *The Descent* to explore the implications of beauty and courtship behavior in driving evolutionary change (sexual selection). His real conclusion was that man was just another animal and the rules that governed evolution elsewhere in the living world also applied to humans. While this work was inevitably controversial, the British public came to realize that Darwin, for all that he was reclusive, was a man of very great distinction. When he died in 1882, he was buried in Westminster Abbey, the only biologist to lie alongside royalty and the greatest of England.

It is sometimes claimed that, because Darwin was an upper-class and well-connected member of society, it was much easier for him to be a scientist than someone like Wallace, who was relatively poor and had no background advantages. This is true but only a small part of the story: Darwin was very rich because his father and wife were rich and, indeed, never had to do a day's work in his life. The reality, however, was that he had a restless mind and unstoppable energy for work (in spite of a long-term debilitating illness picked up during the *Beagle* voyage). Over a period of just over 40 years, he worked unceasingly and wrote a dozen books and many papers describing his experiments and theories on just about the whole range of biology as it then was. He was perhaps the most remarkable biologist of all time, and one of those few people about whom the more one reads, the more impressive that person seems.

This is not to say that Darwin got everything right or that he explained everything. In his search for causes of variation, he was sometimes less than critical in his reading of the data on the origins of variation. One example is his acceptance of telegony, the belief that, when a female who has already had offspring is impregnated by a second male, influences of the first male remain to affect subsequent offspring. Likewise, his theory of pangenesis to explain how variation in one generation became incorporated into the next was the crudest Lamarckism: he suggested that each domain of the adult body produced pangenes that reflected its anatomical details, and that these then travelled to the testis or ovary where they became incorporated into sperm and eggs. In this way, environment-induced change in the adult would be transmitted to the next generation. In an early example of testing a theory, Francis Galton (1822–1911), who was Darwin's cousin, disproved the predictions made by the pangene through a series of blood transfusion experiments on rabbits (for details, see Bulmer, 2003). Darwin also had little sense of quantitative biology or how population dynamics worked.

Such minor criticism is irrelevant to the importance of Darwin: before him, the idea of evolution was barely science; after him, no one who had read his writings with an open mind could doubt its reality or the way in which it happened. Through his ideas and the massive body of scientific data that he marshalled to support them, Darwin opened wide the door to acceptance of the ideas of evolution, and his successors flooded through it – his basic insights were right: he changed biology forever.

The nineteenth century after Darwin

The half century after Darwin saw a very large amount of work in all aspects of evolutionary biology across the scientific world (Mayr, 1982). For various reasons, England focused more on descent with modification than on natural selection; France and the USA remained uncomfortable with the idea of evolution for the next 20 or 30 years, and it was only in Russia and Germany that Darwin's views were completely accepted. The main reason for Germany initially accepting evolution was the effort put in by Ernst Haeckel (1834–1919), but he was so aggressive and so anti-Christian in his views that there was widespread rejection of the theory of evolution across Germany towards the end of the nineteenth century. Evolutionary research was particularly strong in Russia until the 1930s, when Trofim Lysenko (1898–1976), an agronomist, denounced Darwinian analysis as being incompatible with Marxist-Leninist thinking and claimed that evolution proceeded along simple Lamarckian lines. The net result of Stalin giving him power to implement his views was not only serious famine as his apparently evolved crops failed, but also the disgrace of many fine biologists and the blocking of serious evolutionary work in Russia for several decades.

Ernst Haeckel is better known today for his biogenetic law "Ontogeny recapitulates phylogeny" or, in more modern language, development repeats evolution. This idea, developed around 1870 and built on older ideas, holds that the development of later-evolved animals builds on the evolution of earlier ones so that adult features of early evolving animals are seen in the embryos of later ones. The classic exemplar is his claim that the branchial arches of all early vertebrate embryos reflect the presence of the gill slits of adult fish. Haeckel, as has been widely pointed out, was wrong for two reasons: in this specific case, the branchial arches of fish embryos are not gills, but give rise to them in fish, while the branchial arches of reptiles and their descendants form the tissues of the jaw and neck – there is no recapitulation of adult tissues. Indeed, Haeckel should have known better because von Baer had already published the correct view when he wrote that embryos of higher forms do not resemble the adults of lower examples, only their embryos (Gould, 1977).

For all that it was wrong, however, Haeckel's biogenetic law provided a framework for comparative embryology and stimulated a great deal of research in the latter part of the nineteenth century, particularly on comparative developmental anatomy (Gould, 1977). Important discoveries that resulted from this work were the relationship among the chordates and the distinction between protostomes and deuterostomes (the mouth forms from the first cavity in the blastula in the former, and from the second in the latter). By the end of the nineteenth century, however, the biogenetic law had rightly been abandoned and comparative embryology had been superseded by experimental embryology under the stimulus of Wilhelm Roux and Hans Driesch.

By then, the key evolutionary arguments were about the relative importance of hard inheritance, the effect of intrinsic variation, and of soft, sometimes called Lamarckian, inheritance, the result of environmentally determined change that became intrinsic, and whether variation proceeded by slow continuous change or substantial discontinuous jumps (saltations). The key figure here was August Weismann (1834–1914), who discovered the germ theory of inheritance (Weismann, 1889). This says that inheritance is carried by sperm and eggs, and that the crucial factor in generating variation is crossing-over of chromosomes during meiosis. Although Weisman was initially tolerant of soft inheritance, he decided in 1883 that it simply could not happen and such was his authority that, as Mayr (1982) points out, Weismann's neo-Darwinism set the scene for all discussion on evolution until well into the twentieth century.

It is interesting to see the extent to which the ideas of evolution became the stuff of popular culture, to the extent that the poet William Wordsworth's 1850 phrase "Nature, red in tooth and claw" came to be seen as representing natural selection, perhaps because it tallied so well with Victorian ideas of capitalism. An important counterbalance, initially ignored, came from the Russian scientist Pyotr Kropotkin, born a prince but who became a socialist and anarchist. His view, based on the work of the geographical survey that he ran in Siberia during the 1860s, was that much of evolutionary change depended on *Mutual Aid*, the title of his 1902 book, and still interesting to read. In this, he argued that evolutionary advances depended far more on what organisms could do for one another than on the struggle for existence. Today, we would say that Kropotkin's focus was more on the importance of ecological niches and kin selection than on competition.

The early twentieth century

It was in the context of trying to discover whether inheritance was discontinuous in around 1900 that Corren, Tschermak, and particularly de Vries independently realized the importance of Mendel's work on genes, which had been published in 1865; de Vries went on to emphasize the role of mutation in changing their effects. What Mendel showed was that inheritance was particulate and achieved through discrete genes, and hence that the blending of characteristics was not part of inheritance. This idea, that mutations in genes were responsible for variation, led to a split in the field. Those who took Mendel's work to heart were mainly experimentalists who became the first evolutionary geneticists, interested in genes and mechanisms, with many being originally trained in embryology (for example, Hugo de Vries, Thomas Hunt Morgan and, later, Conrad Waddington, universally know as Wad.) The other school was composed essentially of naturalists and their focus was on diversity, paleontology, evolutionary taxonomy, and, in particular, mechanisms of speciation. The schools of evolutionary biologists interested in genetic mechanisms (experimentalists and, later, mathematicians) and the naturalists who studied organisms moved so far apart in their interests that they barely communicated. It was not until 1937 when Dobzhansky (1900–1975) published *Genetics and the Origin of Species*, which integrated the theoretical work on genes and mutations with experimental work on animal populations, that they came back together.

Once this had happened, the biological community as a whole realized that there was a series of common problems that had to be solved if evolution was to be properly understood. Different groups then focused on specific areas. Naturalists asked questions about diversity, behavior, and selection within populations of organisms, and, in particular, looked at the relationship between species and environment. Paleontologists and comparative anatomists worked on evolutionary trends in anatomical structures. Taxonomists were interested in species and their relationships, while geneticists were interested in genes and characters (we would now say *genotypes* and the *phenotypes*) and how change occurred. A particular problem for more theoretically inclined biologists was the nature of variation. A key early scientist here was Sergei Chetverikov (1880–1959); he worked on *Drosophila* flies that he collected in the wild in the 1920s. He was able to show on the basis of genetic analysis that there was a great deal of hidden variation in a population.

One interesting question was whether genetic change through mutation occurred smoothly or in discrete jumps; the organism data suggested that this was smooth, while Mendelian genetics suggested it was discrete. This problem was solved by the great statistician, R.A. Fisher (1890–1962), who showed that the cumulative effect of sets of genes that each operated discretely would generate continuous change. He, together with J.B.S Haldane (1892–1964) and Sewall Wright (1889–1988), invented evolutionary population genetics. One of its triumphs was to show that the underlying cause of some aspect of continuous change, such as the spectrum of behavior or size seen in a population, could be partitioned into an inherited (genetic) component and an environmentally determined component. This was, of course, at a time when nothing was known of DNA, and genes were defined in terms of the phenotype that they underpinned.

This work did not properly explain how substantial anatomical variants could form during embryogenesis and whether they could be explained by Mendelian genetics. Goldschmidt (1940) felt that small changes, such as mutations in the sorts of genes then known, were inadequate and suggested that major anatomical changes such as were seen in the fossil record arose through very occasional mutations that had major effects on the phenotype. This was the saltation hypothesis, and the resulting organism would be what Goldschmidt called a "hopeful monster." Most would be lost, but the occasional mutant might breed and lead to a population of new variants that could survive and lead to new species. This suggestion was not met with much approval; indeed, it was widely ridiculed.

The only supporting evidence came from the work of C.H. Waddington (1905–1975), an early *Drosophila* geneticist and embryologist, as well as a major theoretical developmental biologist. He showed in 1953 that, provided selection was strong enough, the large amounts of variation seen in a normal *Drosophila* population was sufficient to produce genetically stable flies with four wings rather than the normal phenotype of two wings and two halteres, or balancing organs, an apparent example of a saltation, albeit one that reflected an earlier evolutionary state. He thus demonstrated that major anatomical changes could occur within a normal population, although it is only relatively recently that molecular explanations for such observations have become available (Chapter 14).

Oddly, all this work still did not provide an answer to Darwin's question of how variation actually led to a new species. This was not an easy question to answer as, left to themselves, gene frequencies remained constant in a population – this is the Hardy–Weinberg equilibrium (proved, or so it is claimed, by Hardy using a pencil on a dinner napkin in 1908). The answer, of course, was that selection pressures from an environment on a population changed the gene frequencies within it, and the dynamics of this could be modeled by the new evolutionary population genetics. The additional important insight came from Mayr and others in the 1940s: this was that novel speciation would happen most easily if a subpopulation with a gene distribution that differed from that of the original population was allowed to evolve in a novel environment with a degree of isolation.

This important work led to the modern evolutionary synthesis of the 1940s and 1950s, and was the work of a group of biologists that included Theodosius Dobzhansky, Julian Huxley, and Ernst Mayr. This was strengthened in the 1960s by Motoo Kimura, who realized that genetic drift, the distributing of genetic variants through random interbreeding, was an important component of variation. Together, they provided a framework that integrated population genetics, based on genes, variation, and mutation, with evolutionary thinking. Its key points were that:

- Within a population, there is slow accumulation of mutations that have small effects and these lead to genetic variation and a resultant expansion of phenotype variants.
- Through random interbreeding and genetic drift, mutations and allelic alternatives are distributed across the population, leading to normal variation.

One result of this work was formulating these ideas within a mathematical framework that could be used to analyze data (Chapter 14). Another was to provide a genetic basis for the answer to Darwin's question about how a new species can arise from an existing one. This can happen if a small group becomes reproductively isolated from the parent population. Through genetic drift, its genetic distribution will be skewed and may thus include a few phenotypic variants. If one of these is fitter than the others (more of their offspring survive to reproduce in their new environment), natural selection (Malthusian competition) will eventually lead to it dominating the population. Eventually, further mutation will lead to its being unable to breed with members of the original, parent population and the descendants of the variant will have become a new species.

It is also worth noting that this synthesis excludes or downplays several alternative ideas that had been previously thought important; among these were soft (or Lamarckian) inheritance and saltation (or mutations that have major effects on the phenotype). In the early 1970s, however, Eldridge and Gould suggested that the second of these points might not have been correctly interpreted and suggested that, rather than through the slow accumulation of small changes (or phyletic gradualism), evolution might proceed through periods of stasis alternating with short periods of rapid change, a process that they called punctuated equilibrium. This was obviously so after a major extinction (see Chapter 4), but there was considerable argument as to whether it might occur under more stable conditions.

The evidence that Eldridge and Gould cited to support their ideas came from longitudinal studies on trilobites and snails, while others have found a range of organisms from bryozoans to dinosaurs whose changes show punctuated equilibrium (Gould & Eldridge, 1972; Horner et al., 1992). This was not to say that change occurred through a single mutation that had a major effect (saltation) but that the weight of accumulated mutation became enough to reach a tipping point and so produce a rapid change. Analysis of the fossil record also showed that evolution in many species could be adequately described by the idea of phyletic gradualism. It is probably correct to say that the two views reflect the opposite ends of a spectrum, but the reasons why this spectrum exists are unclear. They will probably remain so until we have a deeper idea of how variation builds up in a population, and how and why specific phenotypes are selected.

Taxonomy

The other, more theoretical area that changed thinking on evolution in the twentieth century was the work of a line of biologists who had been concerned with how to produce a classification of organisms that reflected their evolutionary history. The standard taxonomy, based on Linnaeus' eighteenth-century naming scheme, grouped living organisms on the basis of shared and distinct features, usually morphological. Such a taxonomy is essentially based on the formal rule

<Organism A> *<is a member of the set of>* <Group B>

The result may well be to link groups with a common evolutionary history, but that requirement is not a necessary part of this type of classification. Such relationships can be linked to form a hierarchy, and this is an example of a formal graph (Appendix 1).

In the 1950s, on the basis of earlier thinking, Hennig put forward an alternative classification system that he called phylogenetic systematics (Hennig, 1966), and is now known as cladistics (the word clade derives from the Greek word for branch). This classification was based on Darwinian thinking whereby organisms were related if they shared derived characteristics, and that the more of these that they shared, the closer was the relationship. This classification was thus between individual species, rather than groupings of species, with the relationship between them being not *<is a member of the set of>* but *<derives with modification from>*. The information associated with satisfying this latter criterion allowed two species with a particular shared and derived character to trace their ancestry back to a last common ancestor that was the first to show such a character, and could be represented in tree form as a cladogram (Chapter 5).

It took some time before cladistics became accepted as the evolutionarily appropriate way of grouping species, one reason being that the cladistics terminology, while very precise, contained many new and complicated terms. For many years the two taxonomic approaches lived uneasily alongside one another until it became clear that the higher levels of Linnaean taxonomy had to be made evolutionally coherent and thus consistent with cladistics taxonomy. Thus, for example, the fossil evidence showed that today's ungulates and sea mammals shared a common ancestor, and this relationship had not been included in Linnaean taxonomy; a new group, the Cetartiodactyla, was therefore invented to include the two clades (Chapter 7). The net result is that cladistic and Linnaean taxonomies are, by and large, consistent, for all that they have different roles and structures.

A formalism based on inheritance turned out to be particularly suitable for the new molecular age: although cladistics views two organisms as being related if they share derived characteristics, it imposes no constraint on what these characteristics should be. The obvious choice is an anatomical feature (so producing a cladogram hierarchy), but it could equally be a DNA sequence that could be modified by mutation (this gives a phylogram hierarchy). Once sequencing became possible, informaticians were able to produce programs for analyzing sequence similarities and differences, and this work led to the new subject of phylogenetics, which provided a means of classifying and grouping species within an evolutionary hierarchy based on a sequence analysis. Because the analysis uses only contemporary sequences, its obvious roles are in examining the natural groupings of extant species, and in identifying ancient relationships where there are no fossil data. The methodology can thus, for example, identify those DNA sequences that were shared by a last common ancestor of mouse and *Drosophila* and those others that arose after the clades separated.

The molecular era

The Modern Evolutionary Synthesis dominated all thinking about the subject until the molecular era, which started in the 1950s, and is still an important component of evolutionary sciences. What seems remarkable today is the fact that the founders of this synthesis constructed a whole theory on the genetic basis of life without having much idea about what a gene was, how mutation worked, or what variation actually meant at the molecular level. Even more remarkable is how, given what we now know, that theory has stood the test of time and remained essentially unchanged. This is because the synthesis is a theory of phenotypes, organisms, and populations in which molecular details are essentially irrelevant and cannot be incorporated in any detail (Saylo et al., 2011). It was for this reason that the discovery of the structure of DNA and molecular biology in the 1950s initially had little effect on evolutionary thinking beyond explaining how DNA mutations altered proteins, although this provided the first mechanistic insight into the links between genotypic and phenotypic change.

Three subsequent molecular advances have immeasurably increased our understanding of the molecular basis of evolution and provide the context for much of evolutionary work today. The first comes from the insight that it is possible to use computer algorithms to analyze the DNA sequences of similar genes in different organisms (see Chapter 8). The second derives from the early work of Waddington on how evolutionary change reflects anatomical changes during embryogenesis that, in turn, reflects mutation-derived variation (see Chapter 13); this area of work is now part of evo-devo. The third comes from the more recent realization that many proteins not only work on their own in building anatomical structures in embryos, but also work cooperatively in complex networks; understanding how this happens is an important aspect of systems biology (Appendix 1). The evolutionary implications of systems biology for variation have yet to be worked through, but a preliminary analysis is considered in Chapter 13.

DNA sequence analysis

While cross-species comparison of DNA sequences has not changed our broad understanding of eukaryotic evolution, DNA sequencing has, since the 1970s, revolutionized our knowledge of the genotype and helped fill in many gaps in our understanding, particularly where there is no fossil evidence. It has allowed detailed clades of organism families to be reconstructed that are far more accurate than is possible using phenotypic characters of fossilized and living animals. It has also shown just how much variation there is in populations (a good example comes from the data generated by the "1000 Genomes Project" (2012)) and how this can be used to track population sizes back in time (coalescent theory; see Chapters 8 and 13).

Such work has allowed us to follow genetic changes over eons and, for instance, see how a specific human gene evolved from a bacterial sequence. It has also shown the importance of evolutionary opportunism in cases where a gene system that has been involved in one tissue can be reused in a new one: one classic example is the Hox set of proteins that was originally and widely used to pattern the anteroposterior axis, and was then reused to pattern vertebrate limbs. Of even greater significance may be its showing that very similar signal transduction pathways are used across a wide range of tissues in the development of a single organism with homologous pathways being used across organisms

(Gilbert, 2014). Perhaps the most spectacular achievement of cross-organism sequence analysis, however, has been to provide molecular evidence for the endosymbiotic theory: this states that the eukaryotic cell formed from the symbiosis of several prokaryotic (anuclear) organisms (Chapter 9).

Evo-devo

Although embryology had been an important evolutionary topic in the nineteenth century, evolutionary biologists seem to have lost interest in the area early in the twentieth century. It was not until the 1950s that things started to change. Waddington was the first major embryologist to explore how variants formed in embryos and how mutations that affected embryos could lead to evolutionary change (his classic experiments are discussed in Chapter 13). It was not until the early 1980s, however, that the field seems to have come to a collective realization that similar genes were responsible for similar aspects of development across a wide range of organisms to the extent that, for perhaps the first time, the vertebrate-invertebrate barrier was broken. This integration of development with evolution (the area informally known as evo-devo) has become a major area of research that bridges all of life from bacteria to plants and animals.

Systems biology

This area is still too new to have yet had a major influence on our understanding of evolution but probably will soon. It started to become clear in the 2000s that much of embryogenesis is driven by the outputs of complex protein networks that drive growth, differentiation, and morphogenesis, with the individual contributions of most individual proteins being subsumed within the properties of the whole network. The modern evolutionary synthesis deals in phenotypes underpinned by an ill-defined class of genes that actually reflect the action of such networks (for example, size and patterning) and it is clear that there is a degree of meshing between what the synthesis defines as genes and these networks (Chapter 14). Thus far, the main contribution of systems biology to evolution has been to show, in general terms, how mutation in a gene leads to quantitative variation in the output of a network. As the systems biology community works through the dynamics of these networks and their susceptibility to mutation, there will clearly be further and more detailed implications for our understanding of the relationship between genotypic and phenotypic change (see the latter pages of Chapters 12–15).

Further reading

A great deal has been written on the subject of this chapter and the key text is certainly:

Mayr E (1982) The Growth of Biological Thought. Belknap Press.

More easily accessible and of considerable value are the Wikipedia entries on the History of Evolutionary Thought (https://en.wikipedia.org/wiki/History_of_evolutionary_thought) and on the scientists mentioned in this chapter. Of particular interest is Darwin's autobiography:

Darwin F (1887) The Life and Letters of Charles Darwin, pp. 27–107. John Murray. http://darwin-online.org.uk/content/frameset?itemID=F1452.1&viewtype=side&pageseq=1

Other important texts worth consulting include:

Eldridge N, Gould SJ (1972) Punctuated equilibria: an alternative to phyletic gradualism. In Models in Paleobiology (Schopf TJM, ed.), pp. 82–115. Freeman Cooper. www.blackwellpublishing.com/ridley/classictexts/eldredge.pdf

Gould SJ (1977) Ontogeny and Phylogeny. Belknap Press.

Gould SJ (2000) A Tree Grows in Paris. Bucknell University. www.facstaff.bucknell.edu/sdjordan/PDFs/Gould_Lamarck'sTree.pdf

Guerrero G (2008) The session that did not shake the world (the Linnaean Society, 1st July 1858). *Int Microbiol* 11: 209–212.

Saylo MC, Escoton CC, Saylo MM (2011) Punctuated equilibrium vs. phyletic gradualism. www.sersc.org/journals/IJBSBT/vol3_no4/3.pdf

Stott R (2012) Darwin's Ghosts. Random House.

APPENDIX 3
ROCKS, DATES, AND FOSSILS

The history of evolution is to be found in the fossil record, remnants of organisms that have been fossilized and embedded in rock. This appendix examines how rocks are made, the major geological periods and the evolutionary events associated with them, the techniques by which rocks are dated, and the mechanisms of fossilization. It provides the background for Chapter 4, which discusses the fossil record, and Chapters 6 and 7, which analyze vertebrate evolution.

Rock types

The most common rock (~65%) in the 5–70 km-thick outer crust of the earth is igneous. It emerged from the earth's core by volcanic activity (for example, pumice and basalt) and by magma forced towards the surface, which slowly cooled to form great masses of, for example, granite, that are known as plutons. The other basic type of rock is sedimentary (~8%); this mainly forms from the deposition or sedimentation of particles that eventually become compressed into large masses. The particles are derived from other rocks eroded by weathering, tectonic forces, and the effects of vegetation, together with organic detritus such as the calcium-rich skeletons of corals and invertebrates. Occasionally, sediments include evaporates that are salts formed when mineral-rich solutions, such as seawater, dry out. Sediments subject to strong compressive forces form rocks that are often stratified, having layers that reflect the successive sets of sediments from which they have formed; these are classified by the size of the deposited particles and the mode of deposition. The third major class is metamorphosed rock (~27%), which is formed by extant rock that has been subjected to heat and pressure: calcareous sediments, for example, produce marble, while sedimentary shales and mudstones form slates and gneiss.

If, however, one looks only at surface rock, the proportions change and far more of this rock is sedimentary (~66%; Blatt & Jones, 1975). This is where fossils are mainly to be found: they represent the structures of dead organisms immortalized as stone. Occasionally, the fossilized material in the original sedimentary rock (such as shale and limestone) can still be seen after the rock has metamorphosed (for example, to slate or marble), but it will usually have been severely distorted. Rock type, on its own however, tells us nothing about evolution or about age, as rock is continually being made and changed. Making sense of the fossil record requires knowledge of the temporal, and occasionally the spatial, relationships among the fossil-bearing rock types. The current fossil record represents only a sample of what has been laid down, since the land surface of the world is mainly covered with vegetation and relatively little, other than desert and areas subject to weathering such as coastal cliffs, has easily accessible fossil-bearing rock.

Nicolas Steno (1638–1686) seems to have been the first to have discussed how strata (layers of rock with the same composition) are superimposed on one another, how they could have been laid down in different places at the same time but be different in appearance and could have become degraded and distorted, while detailed observational work was undertaken by James Hutton and a few others in the late eighteenth century. The link between rocks and fossils was made a little later by William Smith, when he realized that the various strata in a local coalmine in Somerset, UK, could be identified by the fossils that they contained. In 1799 he published a geological map marking the various rock strata in the area around the city of Bath in Somerset and, in 1815, a much larger map of most of England and Wales. This was a considerable achievement, as rocks of the same age and even the same basic type can look very different in different places because of their local mineral content and the various weathering and other forces to which they have been subjected over very long periods.

Smith's work stimulated others to analyze strata and identify them by their type and fossil content (where possible), and specific types were often named after the places where they were first recognized. It is ironic that Sedgwick, who identified Cambrian rock in the 1830s and called it after the Latin name for Wales, never accepted Darwin's ideas on evolution, even though it turned out that the fossils in Cambrian rocks represent the time when multicellular organisms from the great majority of the phyla evolved. A list of the major strata, their periods, and some major fossil features is given in **Table A3.1**.

Aging rocks (geochronology)

What was particularly interesting to early geologists about each layer was that that many of its constituent fossils bore little resemblance to those above or beneath it, and Cuvier (see Appendix 2) wrongly took this as evidence of major extinctions that were followed by periods of new creation. Cuvier did not appreciate two important factors. First, the nature of the surface in a particular region could change because of the effects of tectonic forces and evolutionary events: a region that was once sea could later become bare land that was, in due course, forested. Second, although the extinctions were always important and sometimes catastrophic, a proportion of organisms always survived and provided the base species for future diversity.

Common examples of invertebrate families with shells or robust exoskeletons that survived through at least several extinctions and became fossilized include ammonites, graptolites, and trilobites, although the last of these died out in the great Permian extinction (~252 Mya). Many such organisms evolved slowly, making small but clear incremental changes over time. Our knowledge of evolutionary history is now so good that we can link the detailed morphologies of particular organisms with the time intervals in which they lived. The presence of such index fossils in rocks is a strong indication of when the original organisms were alive and hence the minimum age of the rock.

How does one work out that one piece of rock is older than another in the absence of fossil data? The simplest way is to see if one came from a lower level than another and, in principle, one could just take a drill and remove a core of rock from a particular place and rank-order the various pieces on the basis of their depth from the surface: the further down, the older. Such relative measurements are, however, of very limited use, for three key reasons. First, even if one can identify distinct layers in the core, it is very hard to know how long each took to be laid down. Second, one cannot be sure that depth reflects age because the movements of tectonic plates and the enormous forces that occur when

Table A3.1 Geological time and events

Eon	Era	Period	Epoch	Starts at	Major biological events
	Cenozoic	Quaternary	Holocene	11.7 Kya	Starts after last ice age – modern times
			Pleistocene	2.59 Mya	Modern *Homo* species
		Neogene	Pliocene	5.33 Mya	*Homo habilis* (at end of epoch)
			Miocene	23.03 Mya	Modern mammals, first apes, modern birds
		Paleogene	Oligocene	33.9 Mya	Further mammals, anthropoid primates, plant diversification
			Eocene	56 Mya	More mammals, grasses appear
			Paleocene	66 Mya	Placental–mammal diversification, modern flowering plants
	Mesozoic	Cretaceous		145 Mya	Further dinosaur radiation, marsupial–placental mammal split
		Jurassic		201.3 Mya	Dinosaur radiation, first birds and mammals
		Triassic		252.2 Mya	Archosaur radiation, earliest small mammals, teleost fish, flowering plants
	Paleozoic	Permian		298.9 Mya	Mammal-like reptile radiation, therapsids, cycads, conifers
		Carboniferous		358.9 Mya	First land reptiles, amphibian radiation, winged insects, forests, gymnosperms
		Devonian		419.2 Mya	Jawed fish radiation, Amphibia, first trees, clubmosses, horsetails, land insects
		Silurian		443.4 Mya	Bony fish, first vascular plants on land, first land invertebrates
		Ordovician		485.4 Mya	Invertebrate diversification, jawless and jawed (late O) fish, first land plants and fungi
Phanerozoic		Cambrian		541 Mya	Explosion of phyla in the fossil record
	Neoproterozoic	Ediacaran		635 Mya	Ediacaran fauna and sponges, trace fossils of worms
		Cryogenian		850 Mya	Possible period of extreme cold – fossils are rare
		Ionian		1000 Mya	Trace fossils of multicellular organisms such as plankton (for example, dinoflagellates)
	Mesoprerozoic			1600 Mya	Green algae colonies
Proterozoic	Paleoproterozoic			2500 Mya	First eukaryotic cells and algae, oxygen from cyanobacteria change atmosphere
Archean				4000 Mya	Archaea, bacteria, stromatolites (blue–green, oxygen-producing algae)
Hadean				4567 Mya	No good evidence of life

plates collide can lead to major foldings in the rock, to the extent that older rock can be forced on top of younger. Third, such relative aging does not help when one wants to compare rocks from different places. The only useful way to age rocks is to use physical measuring techniques to date them; one can then compare this with information from the index fossils that they contain.

The best known approach to measuring absolute time is radiometric dating and is based on radioactive decay: many elements have several isotopes (the nuclei have the same number of charged protons, but each isotope has its own number of neutrons), with only one or two being stable. As a result of radioactive decay, unstable isotopes can lose one or more neutrons or protons and so decay to either another, lighter isotope of the element or a different element with a lower atomic weight, a process that is repeated until a stable isotope forms. Each such decay has a well-defined half-life (the time taken for half the original amount of an element to decay). In places where the amounts of both constituents can be determined (either by measuring radioactive emission or by direct counting of each atomic type using a mass spectrometer), it is now relatively straightforward to work out how long ago all of isotope or element two was isotope one.

The two most important radiometric pairs for dating older rocks are potassium/argon and uranium/lead. Potassium–argon dating depends on the fact that K^{40} decays to Ar^{40} with a half-life of 1.3 By, and the system depends on the fact that argon only starts to accumulate once the rock has solidified. The amount of argon is tiny, but, if the rock is heated to 400°C, it can be released and measured. The technique can be used for rocks aged more than 100 Ky. The decay of uranium to lead occurs in a series of steps that is often measured in the mineral zircon ($ZrSiO_4$): uranium, unlike lead, is soluble in this mineral so any lead present in zircon is a result of uranium decay. Uranium exists as two isotopes, U^{235} decays to Pb^{207} with a half-life of about 700 My, while U^{238} decays to Pb^{206} with a half-life of about 4.5 By, which is the approximate age of the earth. Because the different isotopes decay to lead via different routes, with each decay step having its own half-life, measurement of the relative amounts of all the elements and isotopes in the two decay sequences in a specimen give a very good measure of its age. In practice, this technique can be used to age any material that acquired its uranium from more than 500 My to less than 1 My ago.

These slow decay rates cannot be used for dating recent material, and one method frequently used for unfossilized organic material is radiocarbon dating: this makes use of the fact that cosmic radiation leads to normal carbon with an atomic weight of 12 (^{12}C) being activated at high altitudes to form radioactive ^{14}C; this reacts with oxygen to form $^{14}CO_2$, which can enter the food cycle. As ^{14}C breaks down to ^{12}C with a half-life of 5730 years, there is a continuous but slow cycling of carbon between its two forms and the net result is that about one billionth of the carbon in CO_2 is ^{14}C. This cycling breaks down if the carbon is ingested and incorporated into organisms as it becomes inaccessible to high-energy cosmic radiation; as a result, the ^{14}C slowly decays to ^{12}C. Although the practical details for measuring this accurately are complicated, modern techniques, which, for example, allow for variation in the cosmic ray background, can use the ratio of ^{14}C to ^{12}C to date a sample of organic material within a window of 60 Kya to quite recently. Grün (2006) has reviewed the methodologies of carbon dating, uranium-series dating, and other such techniques as they have been applied to human bones.

There is also a range of nonradiometric techniques that can be used to date rocks, particularly surface ones, with perhaps the most important being magnetostratigraphy. This capitalizes on the magnetic properties of a few magnetizable minerals, the most important of which is magnetite (Fe_3O_4), originally known as lodestone. This is found in many igneous and metamorphic rocks and has the property that, as the molten rock solidifies, its magnetic polarity is set along the local direction of the earth's magnetic field at that time. The rock then maintains this polarity, even when subject to later distortion or changes to the local magnetic field. Analysis of this magnetic remanence today thus gives the direction of the earth's magnetic field when the molten rock solidified. The resulting analysis has shown that, since the oldest magnetizable rocks were deposited, the polarity of the earth's magnetic field has changed direction many times, albeit for reasons that are still not properly understood. Such changes can be observed in the different layers of rock taken from a single location or in areas of slow rock deposition.

Perhaps the most striking examples of these reversals come from analyzing the seabed. Igneous rock is continually being extruded in a molten state at oceanic ridges (such as the mid-Atlantic ridge) and cools to form basalt rock, which moves laterally away from either side of the ridge under the pressure of further extrusions, so incidentally driving tectonic plate movement. As the rock solidifies, the magnetite within it aligns with the earth's current magnetic field. The most important studies here have been made using a magnetometer towed by a ship sailing directly away from the mid-Atlantic ridge that directly measured the inherent magnetism of the sea floor. This showed that, as one moves away from the ridge, sea-floor rock comprises a series of parallel stripes of variable width marked by successive reversals in magnetic polarity. As the rate of flow of rock from the ridge appears roughly constant, we now have a clock linking polarity with time that extends back to about 180 Mya, the longest period between sea-floor rock emerging from suboceanic ridges and being lost owing to subduction when it meets and then moves below the plates.

The next step was to correlate the magnetic analysis with absolute time using radiometric dating. Measurement showed that most chrons (the period between magnetic reversals) are in the range 0.1–1 My, with the time of the transition usually being around 1000–10,000 years, although it can be faster. Over the last five million years, there have been nine major periods during which the direction of the earth's magnetism has been reversed (two of them including short periods of reversal). Going back further in time, there have been two superchrons, or chrons lasting more than 10 My, the Cretaceous Normal from about 118–84 Mya and the Kiaman from about 362–212 Mya (the late Carboniferous Period to the late Permian Period). The net result of all this work is that we now have the geomagnetic polarity timescale that, for example, has 184 polarity reversals over the last 83 My.

What all this analysis says is that, given a piece of rock, we can, as a result of two centuries of detailed analysis, usually work out approximately when it was laid down. This is initially done on the basis of its inherent radioactivity and location, but we can then use its fossil content, residual magnetism, and other physical properties to confirm and improve the accuracy of the original estimate.

How fossils form

In most cases, dead organisms degrade rapidly because their tissues are eaten by vertebrates and invertebrates and degraded by their own enzymes (autolysis) and by microbes – the study of the events taking place after death is known as taphonomy. The net result is that, within a year or so, all that is left of even a large vertebrate are the bones and teeth, the hardest of its tissues. As bones are held together by softer connective tissue in joints and ligaments, these hard skeletal elements may soon separate and even disperse if subjected to weathering, scavenging, or water flow that breaks down connective tissue. Invertebrates, small organisms, and nonwoody plants will also decay rapidly, while even trees eventually break down.

What this means is that the various decomposition and dispersal forces have to be kept at bay if an organism is to become a fossil, and the key factors that need to be minimized are oxygen, water, and temperature, together with access by scavengers, both vertebrate and invertebrate. All organisms other than sulfate-reducing bacteria need oxygen and water to degrade tissues and the higher the temperature, the faster the breakdown of soft tissue occurs. The processes of fossilization and dispersal are thus least likely to take place if the dead organism is rapidly covered by sediment. This can occur at the bottom of the sea where it is also cool, or by mud in a river; under these conditions, the dead organism is protected from the effects of oxygen and small degrading organisms – it is no surprise that marine fossils are the most common. Other ways of keeping the decomposition forces at bay for larger organisms are being desiccated by the hot dry winds of a desert and for small invertebrates to become embedded in resin that eventually becomes amber. These constraints, of course, emphasize that the tissues most likely to become fossilized are hard tissues that break down very slowly. Examples are the mineralized bones and teeth of vertebrates, the chitin exoskeletons of invertebrates, and the complex polysaccharides that produce wood in plants and the outer walls of pollen.

Fossilization normally starts when dead tissue is rapidly buried in sediment such as mud or sand that will eventually become rock. First, soft tissue is usually lost through autolysis or scavenging; this allows mineral-rich water to permeate slowly through the small spaces in nondegraded tissues (for example, bone) that had originally contained tissue, water, or air, and eventually replaces the organic material, although hydroxyapatite, the mineral of bone, may well remain. Provided that the tissue structure does not break down before precipitation is complete, very fine detail can be preserved. In due course, the impregnated tissue and the surrounding sediment become rock, but, because of the way that the remains of the organism have constrained the penetration of minerals that may themselves be different from surrounding ones, its matrix remains recognizable as a fossil within a wider rock matrix. The actual time that this takes depends on the local environment. The minimum age for a fossil is defined as 10,000 years, but this may not be enough to complete the process: the bones of *Homo floresiensis*, an extinct species of the genus *Homo* and sometime known as the "hobbit" man, which were dated to ~18 Kya have not yet been fully fossilized (Brown et al., 2004).

The oldest fossils are stromatolites: these are layered mats of dense bacteria, some of which secreted carbonate, and trapped sediments; they date as far back as 3.5 Bya. Perhaps surprisingly, living stromatolites can still be seen today in Australian and Brazilian hypersaline lakes where few other organisms can survive. Even small algal and bacterial assemblages can be fossilized if they have a strong coat, and it is these that tell us about pre-Ediacaran life. An interesting and important class of early fossils are spores and pollen: these microfossils contain no free water, and often have a tough coat and no predators. They are often well preserved in ancient sediments and their morphology shows their source. Their particular importance is in dating early algal, fungal, and plant types.

Occasionally, soft tissue structures are maintained long enough to become fossilized and the result has been some of our most wonderful examples of preserved organisms. Many such examples of both vertebrates and invertebrates found in what are known as conservation fossil Lagerstätte (this German word means "storage place") are given in Bottjer et al (2002). Perhaps the oldest known example is a cyst of several cells that has been preserved in phosphate and dates back >1 By (Strother et al., 2011). Fossils from the Ediacaran period that are preserved in phosphate include those of the Doushantuo Formation, a site in Guizhou Province in China (570 Mya, Ediacaran), which includes phosphatized organisms of groups of a few cells. As the organisms seemed to show palintomic cleavage (cell division without growth), the fossils were originally thought to be of early invertebrate embryos, but more recent analysis has shown that they lack an external epithelium, a key feature of all early embryos, and for this and other reasons they are more likely to be non-Metazoan holozoans, and thus be more closely related to colonial unicellular organisms (Huldtgren et al., 2011).

Phosphate, because of its diffusion characteristics, only preserves tiny organisms, but larger soft-bodied organisms have also been fossilized. The classic example is the fauna of the Burgess Shale, a series of shale layers in rock that is now some 2500 m above sea level in the Rocky Mountains. Some 505 Mya (mid-Cambrian), these rocks were sand some 120 m below sea level, adjacent to a ledge some 10 m deep and at the base of a mountain. Mud and other debris would slide down the mountain and displace organisms living on the ledge, pushing them off and down to the sea floor below where they were rapidly covered in silt and so died of suffocation – these obrution conditions are ideal for fossilization. Recent work (Gaines et al., 2008) has shown that phosphate within the mid-gut glands of arthropods precipitated out replacing the soft tissue with mineral. The net result is that the Burgess Shale fossils (Chapter 4) show extraordinary detail of soft and hard tissues to the extent that it is possible to see gut contents. Other, earlier, soft-bodied fossils have been found in the Chengjiang Shales of China (Chapter 6). Such wonderful preservation is very rare in fossils later than the Cambrian Period, perhaps because the dead organisms were eaten by the many types of worm that had by then evolved (Brasier, 2010).

Fossils may form in other ways: the oxygen and nitrogen of plant tissues such as leaves may be lost leaving only carbon films (or compression fossils); small organisms such as insects and pollen can be trapped in resin (amber fossils), and organisms may degrade completely but slowly enough for the space they occupied to be replaced by minerals. These are cast fossils: the material that occupies the space is known as the part and the impression fossil in the surrounding rock as the counterpart; many of the Ediacaran organisms are represented today as such impression fossils. There are also trace fossils that represent evidence that an organism had been present: these include footprints, burrows, and coprolites (fossilized dung).

What all this implies is that, although a dead animal is unlikely to be preserved as a fossil, it stands a much better chance if it is marine, has either a tough exoskeleton or bone skeleton, and is rapidly covered with sediment after death, preferably as a result of suffocation. Perhaps the world's most common class of fossils comes from the calcareous skeletons of marine plankton (often from the Cretaceous, when global temperatures were high). These sunk to the sea floor, were compressed, and are now chalks, such as the white cliffs of Dover, and were available to become marble should the conditions for metamorphosis later occur. Plants stand a better chance of being preserved if they possess tough polysaccharide coats or woody tissues, as the world's supply of coal bears witness.

For vertebrates, a key tissue that preserves better than any other is teeth; indeed, they are the sole remnants of most Mesozoic mammals that were rodent-sized and whose small bones have not fossilized well. Enamel, the surface coating of teeth, is sufficiently tough that it can survive most degradative forces for a long time, even when in teeth from small animals. A very great deal of work has been done on fossil bones and teeth: their shape indicates their origin and function and, given this, their size is a measure of the original animal and paleontological age, while the degree of wear is a measure of the age and diet of the organism (Gill et al., 2014).

Further reading

Allègre CJ (2008) Isotope Geology. Cambridge University Press.

Benton J & Harper DAT (2009) Basic Palaeontology: Introduction to Paleobiology and the Fossil Record. 3rd ed. Blackwell.

McElhinny MW, McFadden PL (2000) Paleomagnetism: Continents and Oceans. Academic Press.

APPENDIX 4 EVOLUTION VERSUS CREATIONISM

Creationists who believe that the world was created by God or that, at the least, the production of the human brain required divine intervention (the so-called theory of intelligent design) serve a useful role for evolutionary biologists: they make claims and criticisms that have to be refuted, something that requires serious analysis and clear explanation. The first part of this chapter provides a sharp set of questions that creationists might ask to test evolution, and tries to answer them. The second part provides an equally serious set of questions about creationism together with possible answers. The aim is to explore the implications of the two sets of views and to test them. Note: very little literature is cited here: interested readers should explore and enjoy the wide variety of web resources on this debate.

All things bright and beautiful,
All creatures great and small,
All things wise and wonderful,
The Lord God made them all

This lovely traditional children's hymn summarizes in the simplest terms the theistic view of creation, although it excludes all things dull, ugly, stupid, and mundane. If one takes this view, all the evidence for evolution given in this book is irrelevant, as it does not include the participation of God. Some groups who do take this view, such as the Institute of Creation Research (www.icr.org) are mainly content to argue about the facts and ask hard questions about both theory and data; others feel obliged to attack evolution as it conflicts with the biblical view of creation. Such difficult questions are important because they force evolutionary biologists to articulate answers and test and even expand their understanding. Some questions, of course, cannot be answered fully yet because, as this book has emphasized, there are still gaps in knowledge about evolution; examples are details of the origins of life and the detailed evolution of the human brain over the last few million years. Nevertheless, it is to be regretted that creationist websites focus far more on querying the fine detail of evolutionary research than on explaining the data that support evolution or finding new data that would contradict the evidence for evolution and disprove the Darwinian mechanism of evolution.

Public argument between evolution and creationism has been going on since the famous 1860 debate in Oxford between Huxley and Wilberforce (see Appendix 2). Earlier evolutionary publications such as Lamarck's *Zoologie Philosophique* (1809) seem not to have led to major theological arguments, probably because they were viewed as interesting but technical, although Chamber's *Vestiges of the Natural History of Creation* (1844) was publicly criticized on both theological and scientific grounds. A legal example was the infamous Scopes trial in 1925 when John Scopes was prosecuted by the State of Tennessee for teaching evolution; he eventually won on a technicality rather than on the merits of the case.

More recent was the formal argument in 1949 between two New Zealanders, Dewar, a geologist, and Davies, a biologist, with Haldane, a major British evolutionary biologist and population geneticist. The former represented what they considered to be the "rationalist" antievolution approach, while the latter defended evolution. The debate took the form of question-and-answer letters and was published in the book *Is Evolution a Myth*? As this debate took place before anything was known about genomics and very little about developmental mechanisms, it inevitably focused on the fossil record, which was, of course, less detailed then than now. Haldane probably won but only on points; although he could not answer many questions fully because the information was lacking, the New Zealanders were far more interested in posing difficult questions about gaps in the fossil record than in providing detailed nonevolutionary answers to Haldane's questions.

The arguments always focus on two different questions: did evolution happen and, if it did, how? The first asks whether organisms arose from pre-existing ones or were independently created; the second asks whether natural selection, Darwin's mechanism for producing new species, is correct. Evolutionary biologists find overwhelming amounts of evidence that show that evolution happened with new organisms arising from old through descent with modification, while creationists deny that new organisms derive from old and believe that there is or was *de novo* creation by a deity, perhaps allowing for a little change in simple organisms through mutation. Similarly, evolutionary biologists have discovered a great deal of evidence to support the mechanism of natural selection and no evidence to contradict it, while creationists do not accept that this evidence is relevant to what happened during creation and particularly does not apply to humans.

It would be optimistic, putting it mildly, to believe that anything written here will have much effect on the argument. Nevertheless, and in the hope of at least clarifying the battleground and explaining in detail their different perspectives, this appendix takes the form of a question-and-answer session between the two sides. The first part asks critical questions of evolutionary science and the second asks critical questions of creationism. It is to be hoped that both sides will feel that the criticisms, if not the responses, are robust and go to the heart of the difficulties. **Box A4.1** gives the list of the two sets of criticisms.

The following sections present the evolution (E) versus creationism (C) debate in an interview format. For each side, the procedure is first to give a brief summary of the basis for its position, and then allow the other to pose a series of issues with some elaboration, if needed, followed by a paragraph or two for the response. Much of the content of these answers, of course, reflects points made earlier in the book. While each response might deserve further comeback comments from the other side, this would lead to long, tedious arguments and it is left to the reader to decide on the merits of the answers.

Evolution

The experimental data to support evolution come from three separate areas of research (Chapters 5–11). First, the fossil record: this gives clades of organisms going back in time based on species that differ from one another in specific ways; such cladograms show the relationships among today's organisms and their lines of descent. The validity of this

Box A4.1: Summary of criticisms of evolution and creationism

Some creationist criticisms of evolution

General criticisms

- Evolution is just a theory and has no more status than any other theory.
- The theory of evolution cannot make testable predictions and is therefore not a proper scientific theory.

The data

- As one can never prove that hypothesized mating behaviors actually occurred in the deep past, the computational analysis of the data reflects no more than speculation.
- The fossil evidence is inadequate: too few fossils bridge morphologies across species.
- The fossil data are better explained by alternating major extinctions and periods of creation that were followed by minor tinkering than by continuous speciation.
- The perfect adaptation of organisms to their environment is evidence for creationism.

The theory

- Species, by definition, cannot interbreed and the formation of new species is therefore impossible.
- The survival of the fittest is a meaningless tautology.
- Sophisticated organs are too complex to have evolved simply by selection acting on the effects of random mutation, particularly if they will not function until they have fully formed.
- Evolution can explain neither the human brain and its abilities nor social behavior.
- Evolution cannot explain the origin of life.

Some evolutionary criticisms of creationism

General criticism

- Creationism is a belief not a theory: it gives no explanations and makes no predictions.

The data

The creationist view cannot explain

- Fossil dating evidence and why dinosaur and human bones are not found together.
- The molecular data on mutation and its effects.
- The great but imperfect similarities between very different organisms (evo-devo).

The theory

- The bible, with its two very different accounts of creation, is inconsistent and just gives two more fables to add to a long list of world creation myths.
- Creationism cannot explain the evidence in favor of the mechanism of selection (variation, selection, and speciation).
- Creationism involves magic to explain the origin of life and its subsequent evolution.
- Human cognition is not fundamentally different from primates but builds on their abilities.
- Intelligent design or directing mutation at the single base level to microtune evolution is incompatible with quantum mechanics because of the uncertainty principle.
- Finally, why does creationism not accept the idea of natural theology where believers seek to explore the nature of God through understanding his work in the natural world?

hierarchy is independently buttressed by the chronological information derived from the geological record and radioactive dating. Second, the analysis of tens of thousands of DNA sequences in living animals based on comparisons between similar sequences with similar functions from different organisms: this gives a completely independent set of phylograms that show exactly the same relationships among living organisms as do cladograms, and, indeed, fills in many gaps. Third, protein function: comparisons of the molecular mechanisms underpinning the development of very different organisms show that, making allowances for mutation, they use homologs of many of the same proteins and networks for very similar purposes. Such work allows relationships among distantly related organisms (for example, mammals and insects) to be elucidated.

The theory of how evolutionary change happens (natural selection) is built on four successive stages, each supported by experimental evidence. (a) *Variation*: in any population of a single organism, each is distinct because its genome is a roughly random mix of the genes from its parents; the population thus reflects a spectrum of genotypes and phenotypes. (b) *Isolation*: a subgroup of an existing species becomes reproductively isolated in a novel environment; this small population will have, owing to random sampling, a spectrum of phenotypes and genotypes slightly different from that of the parent population that will, owing to genetic drift and further mutation, become amplified over time. As a result, modified proteins and networks with atypical properties will lead to the population having novel phenotypic variants. (c) *Selection*: some of these variants may do well in their new environment and, in due course, become the dominant and eventually the unique phenotype. (d) *Speciation*: further mutation combined with chromosomal change will eventually result in this population being unable to breed with the original population – it will then be recognized as a new species. As to timing, stages 1–3 can be relatively rapid, but the final stage is very slow. Thus, after 4 My of separation, dogs and wolves can still interbreed.

The evidence to support this framework comes from many sources but is more limited than that showing that evolutionary change has occurred. This is because there is no simple way of studying these events together. Experimentation is difficult here: the events of tens of thousands, even millions, of years have to be compressed into a few years through artificially amplifying mutation and selection pressures (Chapters 12–15).

The creationists' (C) criticisms of evolution (E)

General criticisms

C: Evolution is just a theory and has no more status than any other theory.

E: There are three points here. First, theories are not intrinsically inadequate. Anyone who flies bets his or her life on three very different theories being not only correct in principle, but also quantitatively right: the theory of mechanics, the theory of gravity, and the theory of thermodynamics. Second, the amount of evidence to support evolution (Chapters 6, 7, and 11) is so great as to be overwhelming – it is not theory but fact, and there is no evidence to support any other view of the origins of today's biodiversity. Third, while much of the evidence comes from analyzing descent with modification, actual timings depend on radioactive decay analysis. We know that this is correct because it is the same area of science that underpins atomic and nuclear bombs – and they work exactly as predicted. Perhaps more important here is that, even where there are gaps in the detail of our knowledge about evolution, no fact has emerged that contradicts the view that all of life evolved from very primitive beginnings more than 3.5 Bya and then diverged.

C: Neither the theory of evolution nor the mechanism of natural selection can make testable predictions, and neither is therefore a proper scientific theory.

E: This statement is just wrong. It is, for example, a simple prediction of evolution that, if we analyze very different types of data (for example, contemporary DNA sequences and the anatomy of ancient fossils), they should predict the same taxonomic groupings.

It is also a prediction that the cladistics analysis of fossil anatomy should predict the approximate timing order for when those organisms lived (making allowances for the approximate lengths of times that species persist and the intrinsic errors in dating methodologies - together these might add to about 5% of the number of years). In addition, it is a prediction that no organisms will be found whose cladistic status clashes with the established history of life (for example, there should be no teleost fish with a hand or gymnosperms with flowers). No such organisms have so far been found. All three predictions have so far been confirmed. The onus is on creationists to find a piece of data that contradicts the theory of evolution.

The second part of the criticism suggests that it is impossible to test the idea that speciation derives from the gradual accumulation of small differences. This prediction is tested by the natural experiment as a result of which ring species form: there are a few families of organisms that surround an inhospitable central area and whose adjacent relatives differ slightly from one another but eventually meet up and so the various species form a ring (Chapter 15). A nice example is the greenish warblers that form a ring of groups around the Himalayas. If one examines mating behaviors between neighbors almost every group can mate with both of its adjacent groups, apart from a region at the north where there is a discontinuity and the left group cannot breed with that on the right. By the definition of failed intergroup breeding, a new species has formed by a series of small changes in accordance with prediction. Other experimentation (Chapters 14 and 15) has shown that selection acting on isolated groups can lead to novel phenotypes and reproductive isolation, the key step on the way to speciation.

The data

C: The fossil data are better explained by alternating major extinctions and periods of creation that were followed by minor tinkering than by continuous evolution.

E: This idea, first put forward by Cuvier at the beginning of the nineteenth century, was wrong because it was based on inadequate data. The fossil evidence now shows that there was continuity of basic forms, and that, for example, the ancestors of today's mammals were the mammal-like reptiles of the late Paleozoic Era that were mainly lost in the Permian extinction: a few survived, becoming small rodent-like creatures that evolved to become full mammals during the Mesozoic Era, surviving the K-T extinction to radiate during the Cenozoic Era and so produce today's diversity.

C: The fossil evidence is inadequate: too few fossils bridge the morphology between species.

E: While it is true that the fossil record, depending as it does on dead organisms being preserved, is limited, it is still remarkably good and getting better. Indeed, in a few cases such as the Burgess Shale taxa, even soft tissue is preserved. As to the lack of transitional fossilized organisms, many are known. There are the feathered dinosaurs such as *Archaeopteryx* and the Dromaeosauridae, such as *Microraptor gui*, that bridge dinosaurs and birds, as well as a series of fossil fish showing how limbs form. Early examples have a humerus, radius, and ulna, while later ones have wrists and then digits (the details are given in Table 6.2). A full list of the several hundred currently known transitional fossils is readily available online at en.wikipedia.org/wiki/List_of_transitional_fossils.

C: As one can never prove that particular mating behaviors actually occurred in the deep past, the computational analysis of sequence data reflects no more than speculation.

E: The first part of the criticism is, of course, correct. The evidence for mating comes partly from analyzing anatomical relationships among groups of similar organisms to identify lines of descent and partly from showing that similar organisms have DNA sequences that clearly descended from an original ancestor sequence. Both types of evidence are only explicable on the basis of mating.

C: The perfect adaptation of organisms to their environment is evidence for creationism. It is hard to see how a new species can perfectly adapt to its environment in a relatively short time.

E: Unexpectedly, the perfection of adaptation is an argument in favor of evolution and against creationism because the latter does not allow for modification in a population as the environment changes. In practice, adaptation reflects an ongoing but slow selection process as minor variants in a population due to mutation and genetic drift are continually being subject to natural selection. This is because an environment is only rarely static: it is always altering owing to changes in the weather, food availability, or the types of predators. Hence, those with slightly fitter phenotypes, defined as those that leave slightly more fertile offspring in the changed environment, will flourish at the expense of others and will eventually become the new norm. The perfection in adaptation that we so admire reflects a dynamic process, continually tuning phenotype to the environment. Creationism, unlike evolution, does not allow for such dynamic tuning of adaptation.

There is a second point: the evidence shows that major changes do not reflect perfect *de novo* creation but build on pre-existing tissues and organs, and this process leads to far less "perfect" solutions than one might suppose. A classic example of how evolution works by a kludge or quick-and-dirty solution, rather than by the achievement of perfection, is provided by the panda's paw (Davis, 1964). Among the bears, and indeed most land-based vertebrates, the forelimb paw starts off in the embryo with five digits; only the panda has five digits together with an opposed thumb that it uses for shredding bamboo shoots. This extra thumb does not, however, form through a repatterning of the basic digit format. Instead, it forms by extension of one of the small wrist bones (Chapter 13).

There is an extra wrinkle here: although the panda only uses its front paw for eating bamboo, the alteration affects the hindpaw, too: the panda not only has six fingers, but also has six toes, even though the extra toe has no function. The mutation that causes this unusual morphology clearly leads to a minor change in the patterning of all limbs. The evidence shows that adaptations such as this (and the giraffe's recurrent laryngeal nerve. Chapter 7) have nothing to do with creation or intelligent design, but just represents an easy adaptation based on extra growth in a bone that is present in all mammalian limbs. It is hard to think of an example where the opportunistic nature of evolution is more decisively shown.

The reason why evolutionary change to tissues and organs can occur relatively rapidly is that in most cases mutation only affects the phenotype indirectly. The mechanisms responsible for producing tissues and organs in embryos are based on complex networks of proteins. While the details of how they are affected by mutation are not fully understood, it is clear that their network outputs can be altered by mutations to any of several of their constituent proteins rather than a single one. Change can thus happen far faster than used to be thought.

The theory

C: Species, by definition, cannot interbreed and the formation of new species is therefore impossible.

E: As detailed above, speciation occurs very slowly when a subgroup of an existing species becomes isolated. In due course, differential mutation and eventually chromosomal change in the two populations make them first only able to produce sterile hybrids and then unable to produce any offspring at all. Examples are particularly provided by the breeding behavior of the different species of *Equidae* (Chapter 15).

C: The survival of the fittest is a meaningless tautology.

E: If fittest meant strongest, it would be meaningless, but it does not. In the context of evolution, fittest has the specific meaning of leaving the most offspring capable of reproduction. Thus, if there is a subgroup of organisms whose offspring show better fitness in an environment than those of a parent group, the subgroup will eventually become the dominant group because they will leave more fertile offspring.

C: Sophisticated organs are too complex to have evolved simply by selection acting on the effects of random mutation, particularly if they will not function until they have fully formed.

E: The difficulty with discussing the evolution of sophisticated organs is that, because they are usually composed of soft tissues, they are very rarely preserved in fossils. Where they reflect hard tissues such as skeletal changes, the fossil record is often very good at demonstrating the steps of change. With regard to soft tissues, the evolution of the camera eye has, from Darwin onwards, been viewed as the classic problem in this context, particularly that of the squid, which is very similar to vertebrate eyes. The evidence comes from two very different areas. First, zoological work has shown that there is a range of mollusk eyes that range from little more than a curved dish of photoreceptors to the squid's fully formed camera eye (Salvini-Plawen & Mayr, 1977; Chapter 14; see Figure 14.5). It is thus easy to see how the camera eye could have formed through a long series of small steps from a few surface photoreceptors, with each having a small selective advantage. Second, an evolutionary genetics analysis of the evolution of the camera eye has been undertaken by Nilsson and Pelger (1994) on the assumption that a change in eye morphology that gave a 1% improvement in resolution would give a beneficial advantage to its host organism. Their analysis showed that it would take about 350,000 generations, and probably the same number of years, for a minimal eye based on a set of flat photoreceptors to become a full camera eye; this is a trivial amount of time in the scale of evolution or even speciation. Similarly, Clack (2012) has shown how the lung system of reptiles could evolve from the swim bladders of fish. There is no great conceptual problem in understanding how complex organs evolve.

C: Evolution can explain neither the human brain and its abilities nor social behavior.

E: The evolution of the unique features of the human brain, particularly the cerebral cortex, is still unclear. The increase in cranial capacity was not particularly rapid, taking perhaps two million years (~100,000 generations) to evolve from 600 cc in the case of *Homo habilis* to the 1500 cc for modern man. Even the major increase in the size of the cerebral cortex probably involves only about 2–3 extra neuronal cell divisions over this period.

More interesting is the increase in diversity of function, particularly with respect to its ability to process language, a topic about which we still know little. First, it is clear that human neuroanatomy builds on that seen in more neurologically simple vertebrates and, in particular, nonhuman primates (Bystron et al., 2008). The two unique features are the additional cortical volume, particularly in the parietal region, and the functional area associated with language. This does not seem to reflect an unusually fast degree of evolutionary change for the 6.5 My since the origin of the branch of primates leading to humans, particularly given that chimps, our closest relation here, can learn the basics of language if they are taught it.

As to social behavior, it is hard to see any major differences between human and other primates that cannot be explained by the acquisition of language and our remarkable facility for cultural inheritance. One can imagine how modern humans would appear if they grew up with no adults to teach them language and social behavior. Darwin (1871, 1872) devoted considerable time and effort to showing that human intellectual and social abilities have built on those to be found in the animal world, and modern research has confirmed this, particularly in chimpanzees (Matsuzawa, 2013; Melis, 2013). In short, while the details of the evolution of the human brain are not understood, its size and functions do not pose problems that are very much greater than understanding the brains of lesser primates

C: Evolution cannot explain the origin of either simple or complex life.

Evolution has two problems here: first it cannot account for the origin of what it claims was the original organism, probably a bacterium; second, it cannot account for the great richness of the Cambrian fauna that, it is claimed, suddenly appeared ~520 Mya. Such organisms, with their rich and novel complement of cell types and tissues have no antecedents and are obvious candidates for creation.

E: Evolution as a subject makes no claim to explain the origin of life and has never done so; it considers how life forms change. That said, clues as to how life began are beginning to be discovered: much is known about how the complex biochemical molecules that underpin proteins and nucleic acids, both RNA and DNA, can form from very simple molecules; experimentation has shown how lipid drops can form and can contain complex molecules (Chapter 9). Whatever its inadequacies, this explanation make far more sense than one based on either divine creation (how were all these complex biochemical networks assembled?) or through directed panspermia, the idea that basic life forms came from somewhere else in the universe and merely evolved here, a notion that does no more than relocate the origin of life.

As to the suggestion that the Cambrian life forms emerged, as it were, from nowhere, around 530 Mya, this is simply not so. There are early multicellular organisms dating back to more than 1 Bya, Ediacaran organisms (fronds and the like) dating to about ~580 Mya and fossil and trace evidence that sponges and cnidarians (for example, sea anemones and jellyfish), together with worm-like organisms such as *Dicksonia*, were all present well before the Cambrian (Chapter 4). Although the details of how the Cambrian organisms evolved remain opaque, mainly because small, soft-bodied organisms rarely make good fossils, the evidence of these earlier organisms shows that the Cambrian fauna did not just suddenly appear as if they had been newly created.

Creationism

The theory of creationism exists in two forms: the full version is that each organism was created separately (implicitly through divine activity) and either that there has been no change with time or that only minor changes can occur as, for example, in breeding experiments (such as for new pigeons or dog varieties). Either way, new species cannot form other than by creation, but extant ones can, of course, die out (and form fossils). A further, recent perspective here comes from Rutman (2014), who suggests that God created the laws of nature in such a way that life, and eventually humans, was inevitable. In other words, God's role in evolution was some 14 Bya and far earlier than others have suggested. The softer version of creationism says that Darwinian evolution can occur, but is inadequate to account for the evolution of sophisticated anatomical features such as the vertebrate eye and the human brain. Here, a divine helping hand was required in the form of intelligent design.

The mechanisms by which species were originally made is through God's undefined activity, which either created organisms *de novo* or just sometimes helped them on their way. The basis for this view comes from Genesis. The logic to support it comes from perceived inadequacies in the scientific data and theory.

The evolutionists' criticisms of creationism

General criticism

E: Creationism is a belief not a theory: it gives no explanations and makes no predictions.

C: Creationism is not a theory but a correct description of the origins of life and the creation of biological diversity, based on revealed truth. All of life could have formed by an alternative method such as evolution, but it did not.

The data

E: The creationist view cannot explain evidence of the dating of fossils nor why dinosaur and human bones are not found together.

C: There is no great problem with the very incomplete fossil evidence as this just consists of the remains of dead organisms. The presence of individual fossils gives no indication as to when their species was created or when it died out, while the fact that fossilized organisms that appear more evolved than those at lower levels merely reflects sampling problems. George Cuvier, the major anatomist of the early nineteenth century, who discovered the great extinctions (and we can include Noah's flood here) by analyzing layers of fossils, showed that each extinction was followed by a new bout of creation. Chateaubriand and Goss pointed out at around the same time that it was perfectly possible that God had created fossils, even coprolites (fossilized dung), much as, when he created man, he gave him a navel (the *omphalos* hypothesis). In short, the fossil record says nothing substantial, one way or the other, about evolution.

There is a further point here. There is no serious evidence of change over time, just differences among organisms. Indeed, there is no direct evidence of inheritance as the fossils reflect individuals not populations. As to timing, Genesis makes it clear that the sun was not created for several days after the "beginning", and what the bible refers to as days in those times obviously refers to undefined periods. The detailed times of creation given in the bible reflect God's timings not ours, and this explains the apparent isotope datings.

The reason why dinosaur and human bones are not found together is obvious: in those early days soon after creation, there were many more dinosaurs than humans because the former had been created earlier so the latter would have been much rarer and harder to find. Moreover, as they would not naturally have lived together, no reasonable person would expect their bones to be in the same place.

E: The creationist view cannot explain the molecular data on mutation and its effects.

C: Minor mutations are no problem. Consider the creation of the Burgess Shale fossils: all the basic phyla are there and, while some got lost, few new ones emerged before the

Cambrian extinction and the next round of creation. As to the similarities among the DNA sequence data, this merely means that God used a fairly standard template in creating related species. Any similarities reflect coincidences and the genomic data prove nothing.

E: The creationist view cannot explain the great but imperfect similarities between very different organisms (evo-devo).

C: Analysis of contemporary organisms just shows that there is a repertoire of possible developmental mechanisms, and appropriate ones were chosen for each organism; and if there has been a little mutation since their creation, it means nothing.

The theory

E: The bible, with its two very different accounts of creation, is inconsistent and just gives two more fables to add to a long list (see en.wikipedia.org/wiki/List_of_creation_myths).

There are two versions of the creation of the world in Genesis. The first (1:1–2:4) is based around the events of a week ("And it was the evening and the morning...") and has God creating Adam and Eve together on the sixth day. The second (2:5–25) starts with the creation of the world and of man, includes the Garden of Eden, and ends with the creation of woman from the rib of Adam. Version one creates man after mice and flies and version two before.

Detailed academic analysis, mainly by Christian scholars (Friedman, 1987), provides strong evidence that the second version of the creation was written in about the ninth century BCE in Judea, and the first version was written in about the sixth century BCE by the temple priests. Moses, himself, could not have written the Hebrew bible, as he died in the thirteenth century BCE, some 300 years before the Hebrew alphabet was invented and some 2500 years after the rabbis believed that the world was created (5776 years ago). All this suggests that the versions of creation told in the bible are just stories invented to make sense of a mystery.

The rabbis are interesting here. Rabbi Hertz, in his version of the five books of Moses, asserts strongly that, on the one hand, the two versions tell the same story and, on the other, that Moses should not have been expected to be up to date with modern science. A Hasidic view is that the former story refers to the earthly creation of man and the latter to his spiritual creation - and, implicitly, that the creation stories should therefore be seen as metaphor. Rabbi Lieber says in the Aitz Haim edition of the *Five Books of Moses* that "It is a book of morality not cosmology," and hence should not be taken literally. In general, the modern orthodox view is that evolution merely reflects the slow way that God drives the world, and is not particularly relevant to what happens on a day-to-day basis. Far more important to the rabbis are the laws that are contained in the five books.

C: The bible reports on what happened and there are, indeed, two versions of creation. The fact that they were written down much later merely reflects that difference. The bible does not, of course, say how God created each species and perhaps some scientific knowledge is helpful here, and one can see that this ambiguity is bound to excite differences in interpretation. We just do not know how much of these writings reflects signal and how much just noise (Silver, 2012).

E: The creationist view cannot explain the evidence in favor of natural selection (variation, selection, and speciation).

C: We do not need to do so as there is no reason to suppose that modern research has anything to say on what God did all those centuries ago.

E: The creationist view uses magic to explain the origin of life.

C: It is insulting to say that acts of God are just magic. Creationists have little problem with minor DNA mutation and tinkering that lead to variants (such as the various dog breeds) but believe that the major organism groupings were created at once. Here, it is worth noting that there are no antecedents to the Ediacaran fossils that vanished and to the later rich fauna of the Burgess Shale, with their complete range of organisms and cell types. Creating the diversity of flora and fauna is clearly complicated and there may well have been, as Cuvier first suggested two centuries ago, a series of catastrophic destructions that were followed by new acts of creation. The evidence is very thin and the jury is still out on this.

E: Human cognition is not fundamentally different from primates but builds on their abilities.

C: Of course human cognition is fundamentally different from that of apes. The latter cannot talk, and have no sense of history or culture, or, indeed, cultural inheritance. Most important, and what distinguishes us from animals is their complete lack of morality (Matsuzawa, 2013).

E: Intelligent design or directing mutation at the single-base level to microtune evolution (for example, to account for human cognition) to generate the power of the human brain is incompatible with quantum mechanics.

Genomic manipulation at the level of a few angstroms (10^{-8} cm) cannot be done with the necessary degree of precision because the uncertainty principle operates at such distances. In other words, intelligent design is impossible.

C: There are two points here. First, we do not know just how God works and there is no reason to suppose that the laws of physics apply to Him as they apply to us. But suppose you are right that there was a range of early primate species (Chapter 16); this does no more than prove that God needed several efforts before he got *Homo sapiens* "right" – no one thinks that this was an easy thing to do!

E: Finally, why is creationism so against science? There is a longstanding tradition of natural theology whereby believers seek to explore the nature of God through understanding his work in the natural world.

C: Of course this is correct, but any understanding of the world today cannot be extrapolated back to explain the creation.

Conclusions

Where does this question-and-answer session leave the two approaches? The evolutionary position remains extremely strong and nothing that the creationists say contradicts it. That said, there are still some gaps in our understanding of evolution: more information is still needed, for example, on how the many phyla of multicellular organisms with their various cell types evolved from unicellular and simple algae. Little is understood about the speed with which change can occur in new niches and after extinctions, although insights from systems biology on the effects of mutation on networks, together with the possibility of soft evolutionary inheritance (for example via RNA sequences; Chapter 12), are giving some insight here. It is also true that our understanding of how the human cerebral cortex evolved is inadequate, although this may not be as hard a problem to solve as once seemed (Chapter 16). The evidence to support the science of evolution is overwhelming but not complete. Perhaps more important, there is no evidence to contradict it.

The crux of the creationist case is that, although subspecies may form by mutation and selective breeding, the species themselves were created independently rather than from ancestors. The evidence shows that this view is scientifically untenable as it cannot explain how the very detailed analysis of the mutational differences in the DNA sequences of similar genes in related organisms shows heritable relationships. These data are so hard that it stretches credulity beyond breaking point, even to suggest that these relationships arose by chance. Arguments against the scientific data on timing are specious – unless one takes the view that, when God set all the atomic clocks ticking, He made zero time location-dependent. Similarly, the creationist criticisms of the fossil data have become progressively weaker as more have been discovered. As to the mechanisms by which speciation takes place, the lovely example of the greenish warbler ring species shows how small changes can lead to new species forming and directly counters creationism. In addition, systems biology approaches are beginning to show how complex tissues can form and change, and how evolutionary change can be faster than hitherto supposed.

There is, however, a different sort of problem with creationism, which is that, because it holds that God was responsible for all of life and that creation happened in the dim and distant past, its subscribers do not feel obliged to defend their views in any scientific way. It is enough to say that God created it all, even if He did it in several steps and perhaps allowed a little mutation to help Him; and, anyway, who are we to question Him? Any scientist who tries to explain creationist approaches inevitably feels like a defense lawyer desperately looking for something, indeed anything, to get their client off the hook while that client keeps saying "why bother, I know, and any reasonable jury who knew their bible would see that right is on my side." Nevertheless, it seems slightly sad to subject as rich an idea as God to this sort of scientific dissection when it is simply not necessary. It is wiser to view a deity as an ideal that guides our actions rather than a "God of the gaps," someone who micromanages nature and becomes decreasingly insignificant as the gaps that He is required to fill become ever smaller.

Any hard-nosed evolutionary biologist should enjoy the challenge of trying to work out how creationists think, and provide the evidence that counters their views. It is to be hoped that any creationists who read this will feel that their criticisms have been adequately represented. One might even hope that a few creationists might even change sides, but that is unlikely because the mindsets of the two communities are so different. One looks carefully at data and the other starts with the belief that the very different first and second chapters of Genesis reflect historical reality, even though they were clearly written more than 2000 years after the events that their authors believed they were describing. The former trust the well-proven methodology of science, while the latter feel that, wherever else it might explain, evolution is a special case.

GLOSSARY

adaptation A phenotypic feature whose evolution reflects the selective pressure exerted by a feature of the environment (in the widest sense).

Agnathan A taxon of early-evolving fish that lacked jaw bones.

allele An alternate form of a gene. For a gene defined by a DNA sequence, it represents a variant sequence. For a trait gene it represents the heritable component of a variant phenotype.

allopatric speciation Speciation that occurs when a subpopulation becomes geographically isolated from its parent population.

altruism A form of behavior whereby an organism performs some task that benefits its siblings or group at the expense of its own future reproductive capacity.

amniotic egg An egg for a vertebrate in which the embryo makes a set of protective membranes that ensures it does not need to develop in water.

Amphibia A taxon of early vertebrates that evolved from sarcopterygian fish, retaining a fish-like larval form. Their modern descendants are the lissamphibia.

anagenesis (otherwise known as phyletic transformation) The process whereby a species changes its form over time to the extent that it comes to be considered as a separate species.

Anapsids A taxon of sauropsid descendants that lost their temporal foramina and that includes turtles.

anatomically modern humans (AMH) This is the standard name for all members of *Homo sapiens*, a species that evolved about 200 Kya (early examples are also known as Cro-Magnon man or stone-age man).

ancient DNA DNA from organisms that died many tens of thousands of years ago but whose DNA is still sequenceable, in part at least.

anteroposterior patterning A key event in early embryonic development that establishes the head-to-tail organization of the tissues.

Anthropocene The (contemporary) epoch of time that began when humans started to have a significant effect on the earth's ecosystem through their numbers and other effects on the environment.

Archaebacteria The minor taxon of bacteria whose synapomorphies include cell membranes mainly made of L-glycerol ether lipids with pseudomurein coats, histones, and intron processing.

Archaeplastida A major supergroup of eukaryotic phyla whose taxa contain chloroplasts with a double membrane. The group comprises plants, red and green algae and glaucophytes (some single-celled eukaryotes).

Archosaurs A taxon of diapsid reptiles characterized by several synapomorphies that include a preorbital fossa, socketed teeth, and a fourth trochanter (femoral ridge), They included the dinosaurs, crocodiles, pterosaurs, and their descendants.

artificial speciation Attempts to produce subgroups of a species that show the properties of separate species. So far, advances have been restricted to an unwillingness to mate.

Avemetarsalia A taxon of archosaurs that included dinosaurs and birds. They all had a flexible ankle joint.

balanced polymorphism A rare genetic situation where the homozygotes of two alleles of a gene are each less fit than the heterozygote.

Bayesian tree construction This approach constructs a likely phylogenetic tree on the basis of as much data as possible and then uses Monte Carlo and Markov chain methods to generate more trees. After a million or so simulations, the program identifies the tree with the greatest probability of representing evolutionary history.

Bikont A very high-level taxon of phyla that carry two flagellae and chloroplasts and includes plants and algae.

Bilateria The taxon of animal phyla that show essential mirror symmetry, at least during their early stages (i.e. all phyla except the radiata which show rotational symmetry).

biramous limbs The double limb possessed by some arthropods. The outer limb is often for walking while the inner is a gill.

bootstrap resampling A means of validating phylogenetic analyses by using various subsets of the data to check the robustness of the analysis, but adding resamples of the original data so that the original sample size is unaffected.

bottleneck A dramatic reduction in the size of a breeding population as a result of a catastrophe or as a result of separation of a subpopulation from a larger one (*see* **founder population**). Protomammals in the early mesozoic era underwent a nocturnal bottleneck which ensured that only species able to live in the dark and underground survived.

Cambrian explosion The period (541–485 Mya) during which the great majority of contemporary animal phyla evolved.

Chordates A clade of deuterostome organisms characterized by possession, in the early developmental stage at least, of a long cartilaginous cylinder (or chord) ventral to the nerve cord.

cis-acting element A DNA regulatory element that controls gene expression and that works through direct interaction rather than through an RNA intermediary.

cladistics A type of tissue classification based on rules of the type <*Species P derives with modification from a last common ancestor Q*>.

cladogram A hierarchical tree based on cladistics analysis that details the evolutionary history of some taxon.

cleidoic egg An amniotic egg with a yolk, and external shell that allows it to develop in a dry environment after being laid. Mammalian eggs derived from the reptilian cleiodic egg.

co-evolution This occurs when two phenotypic features evolve to benefit one another. It can occur both in a single species, e.g. when a giraffe's neck elongated and its heart became modified to cope with the new pressure requirements, and in a pair of species, e.g. when a parasite and a host adapt to one another.

coalescent theory A stochastic approach to reconstructing the evolutionary history of an organism on the basis of sequence variants and a model of population genetics. Simulations are done to work out when in the past the different sequences would have coalesced back onto that of a last common ancestor. A strength of the approach is that it can also estimate the population history of the species and other such parameters.

coalescent time An estimate of the number of generations required for the backward running of mutation to result in two variant DNA sequences to coalesce onto a most recent common ancestor sequence.

competence A term describing the ability of an embryonic tissue to partake in a future developmental event. It usually means that the tissue has the appropriate cell-surface receptors to accept an initiating signal and transcription factors to respond to the signal by activating the requisite mRNA transcription.

Crurotarsi A taxon of archosaurs that includes crocodiles and is characterized by an inflexible ankle joint.

cultural inheritance The transmission of knowledge across generations by communication or imitation rather than by genetic inheritance.

Cyanobacterium A phylum of monoderm eubacteria that obtains energy through photosynthesis.

deep homologies A term invented by Shubin et al. (2009) and used in evo-devo work to describe contemporary molecular functional homologies that reflect events in ancient ancestors that typically lived before the protostome and deuterostome clades separated (c 540 Mya) and were thus shown by *Urbilateria*.

deuterostome An embryo whose initial gut cavity becomes the anus and whose second gut opening becomes the mouth (or that evolves from such an animal). This taxon includes all chordates and echinoderms.

Diapsids A major taxon of reptile species characterized by their skull having two temporal foramina posterior to the orbit on each side, or that descended from such an animals. The taxon includes all birds, archosaurs (dinosaurs, pterosaurs, etc.), and contemporary reptiles.

diderm A prokaryote with a cell membrane composed of two bilayers separated by polysaccharides.

Diploblast A taxon of animal organisms, the radiata, whose early embryo has two germ layers, an ectoderm and an endoderm, both of which mainly give rise to epithelia.

distance matrix A matrix whose components measure the genetic difference between specific pairs of regions in a set of homologous sequences.

ecological separation A form of speciation where the reproductive isolation derives from specific ecological conditons (e.g. water depth, foliage type, and saline concentration).

ectopic event An event in the wrong place, such as abnormal gene expression or an extra tissue. One example is the additional eye that forms on a *Drosphila* limb as a result of Pax6 expression being forced to occur there. Another is a sixth digit that occasionally forms spontaneously in humans.

Ediacaran Period The geological period (635–541 Mya) before the Cambrian Period during which the first substantial organisms evolved. These

Ediacaran taxa were almost completely displaced by the Cambrian taxa.

endocrine signaling A form of intercellular signaling in which signals are transported between two different cell types through the vascular system.

endosymbiosis The process whereby one organism engulfs another and, instead of breaking it down completely, develops a symbiotic relationship with it. The mechanism is thought to account for the origins in eukaryotes of nuclei, chloroplasts, mitochondria, and a great deal of molecular machinery.

epigenetic inheritance Inheritence that is not through a DNA sequence but through other processes such as methylation.

epistasis A term from population genetics that reflects the interactions needed between trait alleles to account for the observed phenotype distribution.

Eubacteria The major taxon of bacteria whose synapomorphies include cell membranes mainly made of R-glycerol ester lipids with murein coats.

eutherian mammals Those mammals with a full chorioallantoic placenta that give birth to newborn that are usually capable of walking.

evo-devo The subject of comparative embryology as interpreted through homologous proteins from very different organisms having similar functions.

evolutionary stable state A genetic state of a population that is stable to small changes.

evolutionary stable strategy A term that derives from game-theory simulations and represents a behavioral dynamic for a population that is more stable than any other over time.

exaptation A property that initially serves as an adaptation in one context and then as a different adaptation in another (e.g. feathers were first used for warmth and later for flying). Exaptation replaced the term pre-adaptation.

Excavata A major supergroup of unicellular eukaryotic phyla that have diminished mitochondria and phylum-dependent numbers of flagella. Some, such as *Dictyostelium discoideum*, a slime mold, include a social period in their life cycle.

FECA The first eukaryotic common ancestor; this was the first organism to acquire a nucleus by endosymbiosis and, under the standard assumption that this very rare event only happened once, the parent species of all other eukaryotic species.

fitness An evolutionary measure of reproductive success.

fossils Preserved remnants of dead organisms, normally meaning that the organic material has bee replaced by stone.

amber fossils Small organisms preserved in tree resin.

cast or impression fossils The shape of the organism is maintained in the surface of a rock, but all evidence of the organic matter has been lost.

compression fossils The organic matter of the organisms has been compressed to a thin film that has been replaced by rock.

index fossils Fossils of common species whose morphology and age are well known, Their presence can be used to date sediments.

soft fossils These are rare fossils whose soft tissues, as opposed to their skeletal and other hard tissues, have been preserved. For this to happen, the normal decay processes of physical destruction, oxygenation, and animal predation have to be severely inhibited (*see* **obrution conditions**).

trace fossils Evidence, normally in rock, of an organism's behavior. Examples include footprints, burrows, and copralites (fossil feces).

transitional fossil A fossil whose anatomy has features intermediate between those of an ancestor species (e.g. the fin of a fish) and a later, and very different taxon (e.g. a paw of a vertebrate reptile).

founder effect The loss of genetic variability when a subgroup from a parent population colonizes a new habitat.

founder flush effect The rapid expansion of a founder population that colonizes a new environment in which there are few if any predators.

founder population A subpopulation that becomes isolated from its parent population when it colonizes or expands into a new territory or environment.

FUCA The first universal common ancestor; this was the first organism capable of reproducing itself and so eventually gave rise to every other organism. It represents the origin of life.

gastrulation An early set of events movements in animal embryos that drives the change in geometry from a form that is essentially spherical to one in which there is a nerve cord, a gut, and appropriately distributed mesoderm.

gene A word with two very different meanings, neither of which is precise. The molecular definition is that it represents a DNA sequence with one of many possible functions. The Mendelian or traditional meaning is that it represents the heritable component of an observed phenotype. In this book, the latter is referred to as a *trait gene*, and is only rarely encoded in a single DNA sequence.

genetic assimilation The process whereby a feature that can be experimentally induced can, through selection, eventually come to be part of the heritable phenotype. This is not **Lamarckian inheritance**, but the result of the selection of several alleles that were distributed with low frequency across the wild-type genome and that together can generate the abnormal phenotype.

genetic background This is the complex set of particular allele variants that distinguishes one subspecies from another. It is important because the phenotype produced by one protein often depends on others and different backgrounds gives rise to subspecies-dependent genetic phenomena.

genetic drift The process by which gene alleles and hence gene frequencies are randomly distributed across a population through recombination in meiosis and random breeding. The smaller the breeding population, the faster the effect of drift.

girdle The ring of bones in a vertebrate that supports the forelimbs (pectoral girdle) and hindlimbs (pelvic girdle), enabling them, for example, to lift the body without splaying.

grade A group of taxa that shares features that were not acquired through common descent.

Hardy-Weinberg equilibrium The distribution across a population of allele or gene distributions, each of which has its own frequency. If breeding is random and there is no selection, these frequencies remain constant from generation to generation.

hermaphrodite An organism that produces both male and female zygotes. It may or may not be capable of self-fertilization.

heterochrony An evolutionary change in the timing at which homologous events happen in different species.

homeothermy The maintenance of a constant body temperature. This can result from biochemical activity or simply be a function of size: large organisms have relatively less surface area through which they can lose heat.

hominins (hominini) A group that includes all *Homo* taxa together with extinct taxa closer to humans than any other living vertebrate.

homologs Two genes or tissues are homologous if they derive through common descent from a common ancestor.

homoplasy A phenotypic similarity that arose separately in different taxa as a response to selection rather than through common descent.

hopeful monster An organism with a heritable mutation that led to a major change in anatomical phenotype that also increased fitness. Such a possibility was put forward by Goldschmidt (1940) to help explain complex evolutionary change. The nearest examples that have identified are mutations that affect embryonic patterning.

horizontal gene transfer A means of acquiring lengths of DNA through donation from another organism rather than through inheritance.

hybrid The offspring of mating between a male and female from different taxa. If the parents are from different species, the hybrid should be sterile, but if the species are on the species-separation boundary a hybrid may occasionally be fertile.

imperfect penetrance This term reflects the inability of an allele to display the expected Mendelian distribution of phenotypes: typically, there are too few heterozygotes and/or homozygotes. Molecular genetics views imperfect penetrance as deriving from secondary, genetic background effects on the relevant protein that may mask its effect.

intelligent design The belief that, in spite of evidence to the contrary, the human brain is so sophisticated and unique that it could only have evolved with divine help.

jackknife resampling A means of validating phylogenetic analysis by using various subsets of the data to check the robustness of the analysis.

juxtacrine signaling A form of intercellular signaling based on contact between two different cell types.

K strategy A reproductive strategy that emphasizes small numbers of offspring and considerable amounts of parental care.

kin selection Selection based on an individual's behavior that benefits its kin rather than itself.

lagerstätte A site in which superbly preserved fossils are found.

Lamarckian inheritance The inheritance of acquired adaptations; this reflected Lamarck's view (1809) that organisms had the heritable ability to become more complex and to adapt to their environment. It was generally accepted until

Weisman's work on the continuity of the germ-plasm at the end of the 19th century.

last common ancestor In a cladogram, this is the first species in which some synapomorphy was present. As it cannot be identified, it therefore a hypothetical species to be viewed as the ancestor of all subsequent species carrying that synapomorphy.

LECA The last eukaryotic common ancestor; this descendant of the FECA had acquired a mitochondrion by endosymbiosis, and was the parent species of all other eukaryotes, an early variant of which was a novel polysaccharide cell coat or the loss of any existing one.

lineage The term describes the developmental history of a tissue within an organism.

Linnaean taxonomy (classification) A means of classifying groups of species on the basis of shared properties, usually morphological. Higher classes incorporate lower classes on the basis of further and more basic (often developmentally earlier) shared properties. The classification is based on set membership, i.e. a rule of the form <*Taxon P is a member of the class of taxon B*>, a rule that does not include evolutionary data.

LUCA The last universal common ancestor; this organism is generally but not universally considered to be the parent organism of the eubacteria and archaebacteria.

magic trait A phenotypic trait that is involved in both adaptation to an environment and in selective (non-random) mating.

magnetic remanence The residual magnetism left in a ferromagnetic material once the original cause of the magnetism has been lost.

magnetostratigraphy The study of the geographical distribution of magnetic remanence.

mammaliaform The taxon of post-cynodont transitional synapsids of the early mesozoic era that later evolved into mammals.

mathematical graph A graphical representation of linked facts of the general form <term> <*relationship*> <term>. Such graphs include cladograms, phylograms, and networks. The most intuitively obvious example is a railways map where the facts are of the form <station A> < *is next to*> <station B>. More formally, it is a set of nodes (or vertices) connected by edges.

maximum likelihood A computational means of constructing phylogenetic trees on the basis of homologous sequences, using knowledge of the significance and probabilities of various mutational alternatives to construct all possible trees, and identifying that which is most likely.

maximum parsimony A computational means of constructing phylogenetic trees through examining all possible sequence trees and constructing the most likely, without using any additional information. This method is rarely used today.

metatherian mammals Marsupial mammals with an initial yolk-sac placenta that is later supplemented by a choriovitelline placenta and that give birth to offspring still in an early fetal state of development.

modern evolutionary synthesis A theory of evolution that built on Neo-Darwinism (the incorporaton of Mendelian genetics into Darwinian evolution) by including the mathematical field of population genetics as modified to handle first selection and later genetic drift.

molecular clock A means of measuring time through the amount of mutational change that has occurred. It is only approximately reliable, except when used to analyze small numbers of similar organisms over an evolutionary short time.

monoderm A prokaryote with a cell membrane composed of a single lipid bilayer with an external polysaccharide coat.

monophyletic group A taxon containing an ancestral group and all of its descendants.

morphogenesis The processes involving cell movement and cooperative cell behavior that give shape and structural detail to embryos.

morphospace A theoretical concept describing the multidimensional space within which every organism has a place. The interest in this concept derives from considering those parts of the space that are occupied and those that are not.

mutualism The evolutionary process by which two organisms come to increase one another's fitness.

natural extinction rate The typical period for which a species existed before going extinct. This is not easy to define because in general one cannot be certain when a taxon first appeared and when it was lost, unless it was at the time of a major extinction. The rate is typically 5–15 million years.

natural selection The pressures from the environment in its widest sense that determine whether an organism will be reproductively successful (or fit).

natural theology A means of investigating God through studying the evidence that comes from the natural world. In times past, it was an important justification for scientific investigation when religious authorities were nervous of scientific advances.

neighborhood joining A method for constructing phylogenetic trees on the basis of a distance matrix.

neo-Darwinism A theory of evolution developed in the 1930s that combined Darwinian natural selection with Mendelian genetics.

neoteny The slowing down of a developmental event so that full its development takes longer than expected.

neural crest cells (NCCs) A population of embryonic cells that migrate away from the neural crest, a region within the vertebrate ectoderm between future neuronal tissues and the future epithelium that will cover the organism. NCCs have a wide variety of fates that includes much of the peripheral nervous system, the face,the eye, the skull, pigment cells, and the adrenal medulla.

non-disjunction The inability of chromosomes to separate properly during mitosis or meiosis.

obrution conditions A term used to describe the rapid covering of a dead organism so as to protect it from decay.

Opisthokont or **Unikont** A supergroup of phyla that carry a single posterior flagellum; it includes the animals and the fungi as well as some unicellular phyla.

opportunism An informal term that reflects the ability of species to evolve new features that allow their descendants to colonize new habitats.

orthologous Homologous sequences are said to be orthologous if the sequences are from different organisms - *see* paralagous.

palynology The study of pollen spores that may be extant or fossilized.

paracrine signaling A form of intercellular signaling based on a molecular signal diffusing from one cell to another.

paralogous Homologous sequences are said to be paralagous if the sequences are from the same organism - *see* orthologous.

parthenogenesis The ability of an organism to reproduce itself from an unfertilized egg.

pattern formation The mechanism by which spatial patterns are set up within embryonic domains. It usually occurs through differential paracrine signaling.

pedomorphosis The retention in adults of an anatomical form of physiological ability normally seen in younger organisms a form of **heterochrony**).

phenocopy A property of a wild-type organism that can be experimentally or environmentally manipulated to mimic a mutant property.

phenotype The identifiable properties of an organism (typically traits or proteins.).

phyletic gradualism The idea that evolutionary change and novel speciation occur at a relatively slow and uniform rate.

phylogenetic tree (phylogram) A hierarchical tree (a mathematical graph) of a set of DNA or protein sequences, that connects genetic data of a group of existing sequences back in time to a series of shared last common ancestral sequences. Computational analysis is used to construct a set of likely last-common-ancestor sequences and mutational changes that led to the contemporary set of sequences. It is the genetic equivalent of a cladogram.

phylogenomics The evolutionary and comparative study of whole genomes so as to clarify the details of descent trees, horizontal gene transfer, and other such parameters.

plesiomorphy An ancestral innovation seen in all members of a clade.

prezygotic isolation An external pressure that inhibits interbreeding between two groups.

protein network A group of proteins that cooperate to produce an output such as proliferation or a physiological property such as a timing cycle. Their activity is normally initiated by a signal binding to the first protein, a receptor, whose response is to activate the network behavior.

Proto-mammals (Mammal-like reptiles) An early taxon of synapsid reptiles that were dominant in the Permian Period but became minor species during the Mesozoic era, the age of the archosaurs, during which they evolved into mammals.

protostome An embryo whose initial gut cavity becomes the mouth and whose second gut opening becomes the anus (or that evolved from such an animal).

prototherian mammals Those monotreme mammals that lay eggs and have an essentially reptilian reproductive system.

Pterosaurs A clade of flying archosaurs that differed from dinosaurs in many anatomical features, particularly their possession of the pteroid bone that was part of the wrist and helped support the wing membrane, and other adaptations for flight.

punctuated equilibrium The idea that, although evolutionary change and novel speciation often occur at a relatively slow and uniform rate, it can sometimes be surprisingly rapid.

r strategy A reproductive strategy that emphasizes large numbers of offspring and little if any parental care.

Radiata An animal taxon whose embryos and adult forms shows radial symmetry (e.g. jelly fish).

radiometric dating Dating rock on the basis of the state of radioactive decay of isotopes that were present when the rock formed.

reinforcement (the Wallace effect) The evolution of behaviors that discourage mating between populations that derive from a common ancestor population.

retrotransposons These are potentially mobile interspersed DNA elements that may be short or long (SINEs or LINEs) that can be viewed as synapomorphies for phylogenetic purposes.

ring species A group of species that spread from a founder population to surround an inhospitable domain and eventually met up. A property of the group is that neighbors can interbreed because there are only small genetic differences between them except at the join point. This is because the cumulative differences between the two populations there are too great for interbreeding to be possible.

saltation The view that evolutionary changes to aspects of the phenotype do not arise by a series of small changes but can occasionally arise through a major heritable change (*see* **hopeful monster**).

Sauropsid The earliest clade of reptiles with two temporal foramina (postorbital spaces in the cranium bones). Their descendants included the diapsids (e.g. snakes, reptiles, and archosaurs) and anapsids (e.g. turtles).

selection pressure A measure of the effect of a feature of the environment on the ability of members of a species to reproduce.

selective breeding A form of selection in which the experimentalist directs the deliberate breeding of organisms on the basis of their phenotypes.

sexual selection A mode of selection that depends on traits that facilitate mating and breeding.

somatic mutation A mutation that occurs in the DNA of a cell during mitosis. Although this mutation will then be carried by the subsequent lineage of that cell, it will not be transmitted to the next generation because it is not present in germ cells. An interesting example here is epigenetic change.

species A core taxon ideally defined by its inability to breed with other species. Where this test cannot be done, it is usually defined by a unique morphology.

species flock A group of species that formed from a founder population invading a new territory and that became diverse through the very different selection pressures in the various locales of that new territory.

stromatolite A mound comprising layers of bacteria and debris of sediment and biomaterials. Stromatolites are important because their fossils from c. 3.5 Bya are the oldest evidence of life.

survival of the fittest The idea that those organisms within a population that flourish are those that leave the most fertile offspring. *Fittest* is a technical word in evolution, and does **not** mean strongest.

sympatric speciation Speciation that occurs when a subpopulation becomes reproductively isolated in some way from its parent population while still occupying the same territory (e.g. through a difference in eating habits, or light preferences).

synapomorphy A term from cladistics which means a character possessed by a group of evolutionarily related organisms that is both shared and derived.

Synapsids A major taxon of reptiles characterized by their skull having a single temporal foramen posterior to each orbit. It includes all mammals and their Permian protomammalian ancestors.

systems biology An area of biology that partly tries to understand how complex protein networks operate and partly tries to make sense of biological complexity.

taphonomy The study of organism death and decay.

taxon A species or a group of species or any other rank, defined by common possession of some feature.

Therapsids A taxon of late Permian synapsids whose descendants evolved into cynodonts and eventually mammals.

tinkering (bricolage) A term introduced by Francois Jacob (1977) to describe the microevents downstream of mutation in the context of how evolutionary change occurs.

tissue module A structural unit that is repeated may times in an organism. Examples are sperm, fruits, vertebrae, leaves, and nephrons. Their synthesis is often activated by a signal that sets in train a standard set of genomic, network, and other events.

trait gene *See* gene

transcription factor A protein that is associated with a transcription complex that binds to a specific promoter site on the genome. Activation of the complex leads to the synthesis of mRNA downstream of the site and to the production of new proteins.

tree-searching A computational method of identifying the most likely phylogenetic tree for some data through exploring the likelihood of every possible tree.

triploblast A taxon of animal organisms whose early embryo has three germ layers, an ectoderm and an endoderm, both of which initially give rise to epithelia, and an intervening mesoderm which particularly gives rise to muscles and connective tissue. Triploblasty is a synapomorphy of the bilateria.

Urbilateria The last common ancestor of all multicellular organisms having an embryo with three germ layers and bilateral symmetry. It lived at the base of the Cambrian Period (541 Mya) or a little earlier.

Urmetazoa The last common ancestor of the radiata and bilateria, and the first animal. It lived well before the end of the Ediacaran Period and was probably a very primitive worm.

venus A term used for a figurine of a woman made over the period 40–10 Kya.

Wahlund effect This is the increase in homozygotes that is seen in a small population that derives from a larger one, and reflects the reduction in heterozygosity caused by the new gene distribution.

xenolog A DNA sequence in species A is a xenolog of a sequence in species B if an ancestor of species A originally acquired the sequence through horizontal gene transfer from an ancestor of species B.

REFERENCES

1000 Genomes Project consortium (2010) A map of human genome variation from population-scale sequencing. *Nature* 467:1061–1073 (doi: 10.1038/nature09708).

Abzhanov A (2010) Darwin's Galapagos finches in modern biology. *Philos Trans* 365B:1001–1007 (doi: 10.1111/j.1558-5646.2011.01385.x).

Abzhanov A, Kuo WP & Hartmann C (2006) The calmodulin pathway and evolution of elongated beak morphology in Darwin's finches. *Nature* 442:563–567.

Aerts P (1998) Vertical jumping in *Galago senegalensis*: the quest for an obligate mechanical power amplifier. *Philos Trans R Soc* 353:1607–1620.

Aiello LC (2010) Five years of *Homo floresiensis. Am J Phys Anthr* 142:167–169 (doi: 10.1002/ajpa.21255).

Albalat R, Martí-Solans J & Cañestro C (2012) DNA methylation in amphioxus: from ancestral functions to new roles in vertebrates. *Brief Funct Genomics* 11:142–155 (doi: 10.1093/bfgp/els009).

Alcaide M, Scordato ES, Price TD & Irwin DE (2014) Genomic divergence in a ring species complex. *Nature* 511:83–85 (doi: 10.1038/nature13285).

Al-Qattan MM & Abou Al-Shaar H (2015) Molecular basis of the clinical features of Holt-Oram syndrome resulting from missense and extended protein mutations of the TBX5 gene as well as TBX5 intragenic duplications. *Gene* 560:129–136 (doi: 10.1016/j.gene.2015.02.017).

Alibardi L (2003) Adaptation to the land: The skin of reptiles in comparison to that of amphibians and endotherm amniotes. *J Exp Zool B Mol Dev Evol* 298:12–41.

Alibardi L (2006) Structural and immunocytochemical characterization of keratinization in vertebrate epidermis and epidermal derivatives. *Int Rev Cytol* 253:177–259.

Alibardi L (2008) Microscopic analysis of lizard claw morphogenesis and hypothesis on its evolution. *Acta Zoo Morph Evo* 89:169–178.

Alibardi L (2012) Perspectives on hair evolution based on some comparative studies on vertebrate cornification. *J Exp Zool B (Mol Dev Evol)* 318:325–343 (doi: 10.1002/jez.b.22447).

Alibardi L, Dalla Valle L, Nardi A & Toni M (2009) Evolution of hard proteins in the sauropsid integument in relation to the cornification of skin derivatives in amniotes. *J Anat* 214:560-586 (doi: 10.1111/j.1469-7580.2009.01045.x).

Allègre CJ (2008) Isotope Geology. Cambridge University Press.

Amborella Genome Project (2013) The Amborella genome and the evolution of flowering plants. *Science* 342:1241089 (doi: 10.1126/science.1241089).

Anamika K, Martin J & Srinivasan N (2008) Comparative kinomics of human and chimpanzee reveal unique kinship and functional diversity generated by new domain combinations. *BMC Genomics* 9:625 (doi: 10.1186/1471-2164-9-625).

Anderson E & Hill RE (2014) Long range regulation of the sonic hedgehog gene. *Curr Opin Genet Dev* 27:54–59 (doi: 10.1016/j.gde.2014.03.011)

Anderson G, Beischlag TV, Vinciguerra M & Mazzoccoli G (2013) The circadian clock circuitry and the AHR signaling pathway in physiology and pathology. *Biochem Pharmacol* 85:1405–1416 (doi: 10.1016/j.bcp.2013.02.022).

Anthwal N, Joshi L & Tucker AS (2013) Evolution of the mammalian middle ear and jaw: adaptations and novel structures. *J Anat* 222:147–160 (doi: 10.1111/j.1469-7580.2012.01526.x).

Arendt D, Tessmar-Raible K & Snyman H (2004) Ciliary photoreceptors with a vertebrate-type opsin in an invertebrate brain. *Science* 306:869–871.

Atsumi T, McCarter L & Imae Y (1992) Polar and lateral flagellar motors of marine Vibrio are driven by different ion-motive forces. *Nature* 355:182–184.

Aubert M, Brumm A, Ramli M et al. (2014) Pleistocene cave art from Sulawesi Indonesia. *Nature* 514:223–227 (doi: 10.1038/nature13422).

Aulehla A & Pourquié O (2010) Signaling gradients during paraxial mesoderm development. *Cold Spring Harb Perspect Biol* 2:a000869 (doi: 10.1101/cshperspect.a000869).

Ayala FJ (1999) Molecular clock mirages. *BioEssays* 21:71–75.

Babbs C, Furniss D, Morriss-Kay GM & Wilkie AO (2008) Polydactyly in the mouse mutant Doublefoot involves altered Gli3 processing and is caused by a large deletion in cis to Indian hedgehog. *Mech Dev* 125:517–526 (doi: 10.1016/j.mod.2008.01.001).

Baehrecke EH (2000) How death shapes life during development. *Nat Rev Mol Cell Biol* 3:779–787.

Bajpai S, Thewissen JGM & Sahni A (2009) The origin and early evolution of whales: macroevolution documented on the Indian subcontinent *J Biosci* 34:673–686.

Bard JBL (1977) A unity underlying the different zebra striping patterns. *J Zoology* 183:527-539.

Bard JBL (1981) A model generating aspects of zebra and other mammalian coat patterns. *J Theor Biol* 93:363-385.

Bard J (2002) Growth and death in the developing mammalian kidney: signals, receptors and conversations. *Bioessays* 24:72–82.

Bard J (2011) A systems biology formulation of developmental anatomy. *J Anat* 218:591–599 (doi: 10.1111/j.1469-7580.2011.01371.x).

Bard J (2013a) Systems biology – the broader perspective. *Cells* 19:413–431 (doi: 10.1111/j.1469-7580.2011.01371.x).

Bard J (2013b) Driving developmental and evolutionary change: a systems biology view. *Prog Biophys Mol Biol* 11:83–91 (doi: 10.1016/j.pbiomolbio.2012.09.006).

Barrio RA, Varea C, Aragón JL & Maini PK (1999) A two-dimensional numerical study of spatial pattern formation in interacting Turing systems. *Bull Math Biol* 61:483–505.

Barton NH, Briggs DEG & Eisen JA (2007) Evolution. Cold Spring Harbor Laboratory Press.

Bauernfeind AL, Soderblom EJ, Turner ME et al. (2015) Evolutionary divergence of gene and protein expression in the brains of humans and chimpanzees. *Genome Biol Evol* 7:2276–2288 (doi: 10.1093/gbe/evv132).

Baum D (1992) Phylogenetic species concepts. *Trends Ecol Evol* 7:1-3.

Bayrakdar M (1983) Al-jahiz and the Rise of Biological Evolution. *Islamic Quart* 1983:3:149.

Beall CM (2013) Human adaptability studies at high altitude: research designs and major concepts during fifty years of discovery. *Am J Hum Biol* 25:141–147 (doi: 10.1002/ajhb.22355).

Bejder L & Hall BK (2002) Limbs in whales and limblessness in other vertebrates: mechanisms of evolutionary and developmental transformation and loss. *Evol Dev* 4:445–458.

Bengtson S, Belivanova V, Rasmussen B & Whitehouse M (2009) The controversial "Cambrian" fossils of the Vindhyan are real but more than a billion years older. *Proc Natl Acad Sci U S A* 106:7729–7734 (doi: 10.1073/pnas.0812460106).

Benton J (2005) Vertebrate Palaeontology. Wiley-Blackwell.

Benton J & Harper DAT (2009) Basic Palaeontology: Introduction to Paleobiology and the fossil record, 3rd ed. Blackwell.

Beppu H, Malhotra R & Beppu Y (2000) BMP type II receptor is required for gastrulation and early development of mouse embryos. *Dev Biol* 221:249–258.

Beran MJ & Heimbauer LA (2015) A longitudinal assessment of vocabulary retention in symbol-competent chimpanzees (*Pan troglodytes*). *PLoS One* 10:e0118408 (doi: 10.1371/journal.pone.0118408).

Berger LR, Hawks J, de Ruiter DJ et al. (2015) *Homo naledi*, a new species of the genus *Homo* from the Dinaledi Chamber, South Africa. *eLife* 4:e09560 (doi.org/10.7554/eLife.09560).

Berlocher SH (1998) Origins: a brief history of research on speciation. In Endless Forms: Species and Speciation (Howard DJ & Berlocher SH eds), pp 3–15. Oxford University Press.

Berlocher SH & Feder JL (2002) Sympatric speciation in phytophagous insects: moving beyond controversy? *Annu Rev Entomol* 47:773–815.

Berner RA (1999) Atmospheric oxygen over Phanerozoic time. *Science* 96:10955–10957.

Bernstein M (2006) Prebiotic materials from on and off the early Earth. *Philos Trans R Soc* 61B:1689–1700.

Bernt M, Braband A, Schierwater B & Stadler PF (2013) Genetic aspects of mitochondrial genome evolution. *Mol Phylogenet Evol* 69:328–338 (doi: 10.1016/j.ympev.2012.10.020).

Bhattacharyya T, Gregorova S & Mihola O (2013) Mechanistic basis of infertility of mouse inter-subspecific hybrids. *Proc Natl Acad Sci U S A* 110:468–477 (doi: 10.1073/pnas.1219126110).

Biscotti MA, Canapa A, Forconi M & Barucca M (2014) *Hox* and *ParaHox* genes: a review on molluscs. *Genesis* 52:935–945 (doi: 10.1002/dvg.22839).

Blackburn DG & Flemming AF (2012) Invasive implantation and intimate placental associations in a placentotrophic African lizard *Trachylepis ivensi* (scincidae). *J Morphol* 273:137–159 (doi: 10.1002/jmor.11011).

Blanco MJ, Misof BY & Wagner GP (1998) Heterochronic differences of Hoxa-11 expression in *Xenopus* fore and hind limb development: evidence for lower limb identity of the anuran ankle bones. *Dev Genes Evol* 208:175–187.

Blatt H & Jones RL (1975) Proportions of exposed igneous metamorphic and sedimentary rocks. *Geol Soc Am Bull* 86:1085–1088.

Boisvert CA, Joss JM & Ahlberg PE (2013) Comparative pelvic development of the axolotl (*Ambystoma mexicanum*) and the Australian lungfish (*Neoceratodus forsteri*): conservation and innovation across the fish-tetrapod transition. *Evo Devo* 4:3 (doi: 10.1186/2041-9139-4-3).

Boisvert CA, Mark-Kurik E & Ahlberg PE (2008) The pectoral fin of Panderichthys and the origin of digits. *Nature* 456:636–8 (doi: 10.1038/nature07339).

Bottjer DJ, Etter W, Hagadorn JW & Tang CM (eds) (2002) Exceptional Fossil Preservation: A Unique View on the Evolution of Marine Life. Columbia University Press.

Bourke AF (2014) Hamilton's rule and the causes of social evolution. *Philos Trans R Soc B* 69:20130362 (doi: 10.1098/rstb.2013.0362).

Bouzouggar A, Barton N & Vanhaeren M (2007) 82000-year-old shell beads from North Africa and implications for the origins of modern human behavior. *Proc Natl Acad Sci U S A* 104:9964–9969.

Brasier M (2010) Darwin's Lost World: the Hidden History of Animal Life. Oxford University Press.

Brazeau MD & Friedman M (2014) The characters of Palaeozoic jawed vertebrates. *Zool J Linn Soc* 170:779–821.

Brinkmann H, Venkatesh B, Brenner S & Meyer A (2004) Nuclear protein-coding genes support lungfish and not the coelacanth as the closest living relatives of land vertebrates. *Proc Natl Acad Sci* 101:4900–4905.

Brown P, Sutikna T & Morwood MJ (2004) A new small-bodied hominin from the Late Pleistocene of Flores Indonesia. *Nature* 431:1055–1061.

Bruce WA & Wrensch DL (1990) Reproductive potential sex ratio and mating efficiency of the straw itch mite (Acari: Pyemotidae). *J Econ Entomol* 83:384–391.

Bruner E, De La Cuétara JM & Holloway R (2011) A bivariate approach to the variation of the parietal curvature in the genus homo. *Anat Rec* 294:1548–1556 (doi: 10.1002/ar.21450).

Bruner E, Lozano M, Malafouris L et al. (2014) Extended mind and visuo-spatial integration: three hands for the Neandertal lineage. *J Anthropol Sci* 92:273–280 (doi: 10.4436/JASS.92009).

Brunet T, Bouclet A, Ahmadi P et al. (2013) Evolutionary conservation of early mesoderm specification by mechanotransduction in Bilateria. *Nat Commun* 4:2821 (doi: 10.1038/ncomms3821).

Buckley TR & Cunningham CW (2002) The effects of nucleotide substitution model assumptions on estimates of nonparametric bootstrap support. *Mol Biol Evol* 19:394–405.

Buckley M, Walker A, Ho SY et al. (2008) Comment on "Protein sequences from mastodon and *Tyrannosaurus rex* revealed by mass spectrometry". *Science* 319:33 (doi: 10.1126/science.1147046).

Budin I & Szostak JW (2010) Expanding roles for diverse physical phenomena during the origin of life. *Annu Rev Biophys* 39:245–263.

Bullara D & De Decker Y (2015) Pigment cell movement is not required for generation of Turing patterns in zebrafish skin. *Nat Commun* 6:6971 (doi: 10.1038/ncomms7971).

Bulmer MG (2003) Francis Galton: Pioneer of Heredity and Biometry. Johns Hopkins University Press.

Buneman P (1971) The recovery of trees from measures of dissimilarity. In Mathematics in the Archaeological and Historical Sciences (Hodson FR, Kendall DG & Tautu PT eds), pp 387–395. Edinburgh University Press.

Burke AC, Nelson CE, Morgan BA & Tabin C (1995) Hox genes and the evolution of vertebrate axial morphology. *Development* 12:333–346.

Burki F, Shalchian-Tabrizi K & Pawlowski J (2008) Phylogenomics reveals a new 'megagroup' including most photosynthetic eukaryotes. *Biol Lett* 4:366–369 (doi: 10.1098/rsbl.2008.0224).

Burrow CJ & Rudkin D (2014) Oldest near-complete acanthodian: the first vertebrate from the Silurian Bertie Formation Konservat-Lagerstätte, Ontario. *PLoS One* 9(8):e104171 (doi: 10.1371/journal.pone.0104171).

Butterfield NJ (2000) *Bangiomorpha pubescens* n gen n sp: implications for the evolution of sex multicellularity and the Mesoproterozoic/Neoproterozoic radiation of eukaryotes. *Paleobiology* 26:386–404.

Bystron I, Blakemore C & Rakic P (2008) Development of the human cerebral cortex: Boulder Committee revisited. *Nat Rev Neurosci* 9:110–122 (doi: 10.1038/nrn2252).

Caetano-Anollés G (ed.) (2010) Evolutionary Genomics and Systems Biology. John Wiley.

Capra F & Luisi PL (2014) The Systems View of Life. Cambridge University Press.

Carroll RL (1988) Vertebrate Paleontology and Evolution. WH Freeman & Co.

Cartwright RA, Lartillot N & Thorne JL (2011) History can matter: non-Markovian behavior of ancestral lineages. *Syst Biol* 60:276-90 (doi: 10.1093/sysbio/syr012).

Carvalho-Santos Z, Azimzadeh J, Pereira-Leal JB & Bettencourt-Dias M (2011) Evolution: Tracing the origins of centrioles, cilia, and flagella. *J Cell Biol* 194:165–175 (doi: 10.1083/jcb.201011152).

Cassidy LM, Martiniano R, Murphy EM et al. (2016) Neolithic and Bronze Age migration to Ireland and establishment of the insular Atlantic genome. *Proc Natl Acad Sci U S A* 113:368–73 (doi: 10.1073/pnas.1518445113).

Castellano S, Parra G, Sánchez-Quinto FA (2014) Patterns of coding variation in the complete exomes of three Neandertals. *Proc Natl Acad Sci U S A* 111:6666–6671 (doi: 10.1073/pnas.1405138111).

Cavalier-Smith T (2006) Rooting the tree of life by transition analyses. *Biol Direct* 1:19.

Cavalier-Smith T (2010) Origin of the cell nucleus mitosis and sex: roles of intracellular coevolution. *Biol Direct* 5:7 (doi: 10.1186/1745-6150-5-7).

Cavalier-Smith T (2014) The neomuran revolution and phagotrophic origin of eukaryotes and cilia in the light of intracellular coevolution and a revised tree of life. *Cold Spring Harb Perspect Biol* 6:a016006.

Cavalier-Smith T & Chao EE (2003) Molecular phylogeny of centrohelid heliozoa a novel lineage of bikont eukaryotes that arose by ciliary loss. *J Mol Evol* 56:387–396.

Chan YF, Marks ME & Jones FC (2010) Adaptive evolution of pelvic reduction in sticklebacks by recurrent deletion of a Pitx1 enhancer. *Science* 327:302–305 (doi: 10.1126/science.1182213).

Chaudhary R, Burleigh JG & Fernández-Baca D (2013) Inferring species trees from incongruent multi-copy gene trees using the Robinson-Fauld distance. *Alg Mol Biol* 8:28–37 (doi: 10.1186/1748-7188-8-28).

Chen CF, Foley J & Tang PC (2015) Development, regeneration, and evolution of feathers. *Annu Rev Anim Biosci* 3:169–195 (doi: 10.1146/annurev-animal-022513-114127).

Chen J-Y (2009) The sudden appearance of diverse animal body plans during the Cambrian explosion. *Int J Dev Biol* 53:733–751 (doi: 10.1387/ijdb.072513cj).

Chew KY, Shaw G & Yu H (2014) Heterochrony in the regulation of the developing marsupial limb. *Dev Dyn* 243:324–338 (doi: 10.1002/dvdy.24062).

Chow RL, Altmann CR, Lang RA & Hemmati-Brivanlou A (1999) Pax6 induces ectopic eyes in a vertebrate. *Development* 126:4213–4222.

Chown SL & Gaston KJ (2010) Body size variation in insects: a macroecological perspective. *Biol Rev Camb Philos Soc* 85:139–169 (doi: 10.1111/j.1469-185X.2009.00097.x).

Cisneros JC, Abdala F & Rubidge BS (2011) Dental occlusion in a 260-million-year-old therapsid with saber canines from the Permian of Brazil. *Science* 33:1603–1605 (doi: 10.1126/science.1200305).

Clack J (2012) Gaining Ground: The Origin and Evolution of the Tetrapods, 2nd ed. Indiana University Press.

Clack JA (2002) An early tetrapod from 'Romer's Gap'. *Nature* 418:72–76.

Clarkson E, Twitchet R & Smart C (2014) Invertebrate Palaeontology and Evolution. Wiley-Blackwell.

Coates MI (1994) The origin of vertebrate limbs. *Dev Suppl* 1994:169–80.

Cohn MJ & Tickle C (1999) Developmental basis of limblessness and axial patterning in snakes. *Nature* 399:474–479.

Collins D (1996) The "evolution" of *Anomalocaris* and its classification in the arthropod Class Dinocarida (nov) and Order Radiodonta (nov). *J Paleontol* 70:280–293.

Conway Morris S (1977) Fossil priapulid worms. Spec papers. Palaeontol Paleaeol Soc. No 20.

Cook LM & Saccheri IJ (2013) The peppered moth and industrial melanism: evolution of a natural selection case study. *Heredity* 10:207–212 (doi: 10.1038/hdy.2012.92).

Costanzo E, Trehin C & Vandenbussche M (2014) The role of WOX genes in flower development. *Ann Bot* 114:1545–1553 (doi: 10.1093/aob/mcu123).

Coyne JA & Orr HA (1989) Two rules of speciation. In Speciation and its Consequences (Otte D & Endler JA eds), pp 180–207. Sinauer Press.

Coyne JA & Orr HA (2004) Speciation. Sinauer Press.

Cresko WA, Armores A & Wilson C (2004) Parallel genetic basis for repeated evolution of armor loss in Alaskan threespine stickleback populations. *Proc Natl Acad Sci U S A* 101:6050–6055.

Crews D (2003) Sex determination: where environment and genetics meet. *Evol Dev* 5:50–55.

Cruz YP (1997) Mammals. In Embryology: Constructing the Organism (Gilbert SF & Raunio AM eds), pp 459–492. Sinauer Press.

Culyba MJ, Mo CY & Kohli RM (2015) Targets for combating the evolution of acquired antibiotic resistance. *Biochemistry* 54:3573–3582 (doi: 10.1021/acs.biochem.5b00109).

Curnoe D (2010) A review of early Homo in southern Africa focusing on cranial mandibular and dental remains with the description of a new species (*Homo gautengensis* sp nov). *J Comp Hum Biol* 61:151–177 (doi: 10.1016/j.jchb.2010.04.002).

Cyran KA & Kimmel M (2010) Alternatives to the Wright-Fisher model: the robustness of mitochondrial Eve dating. *Theor Popul Biol* 78:165–172 (doi: 10.1016/j.tpb.2010.06.001).

Dabney J, Knapp M & Glocke I (2013) Complete mitochondrial genome sequence of a middle Pleistocene cave bear reconstructed from ultrashort DNA fragments. *Proc Natl Acad Sci U S A* 110:15758–15763 (doi: 10.1073/pnas.1314445110).

D'Anastasio R, Wroe S, Tuniz C et al. (2013) Micro-biomechanics of the Kebara 2 hyoid and its implications for speech in Neanderthals. *PLoS One* 8:e82261 (doi: 10.1371/journal.pone.0082261).

Danley PD, Husemann M & Ding B (2012) The impact of the geologic history and paleoclimate on the diversification of East African cichlids. *Int J Evol Biol* 2012:574851 (doi: 10.1155/2012/574851).

Darwin E (1796) Zoonomia or The Laws of Organic Life, 2nd ed. J Johnson. http://www.gutenberg.org/files/15707/15707-h/15707-h.htm

Darwin CR (1859) On the Origin of Species by Means of Natural Selection or the Preservation of Favoured Races in the Struggle for Life. John Murray. http://darwin-onlineorguk/converted/pdf/1861_OriginNY_F382.pdf

Darwin CR (1871) The Descent of Man and Selection in Relation to Sex. John Murray.

Darwin CR (1872) The Expression of the Emotions in Man and Animals. John Murray.

Dave AJ & Godward MB (1982) Ultrastructural studies in the Rhodophyta I development of mitotic spindle poles in *Apoglossum ruscifolium Kylin*. *J Cell Sci* 58:345–62.

Davidson AJ (2009) Mouse kidney development. StemBook. www.ncbi.nlm.nih.gov/pubmed/20614633.

Davidson EH (2010) Emerging properties of animal gene regulatory networks. *Nature* 468:911–920 (doi: 10.1038/nature09645).

Davidson EH & Erwin DH (2009) An integrated view of precambrian eumetazoan evolution. *Cold Spring Harb Symp Quant Biol* 74:65–80 (doi: 10.1101/sqb.2009.74.042).

Davies JA & Bard J (1998) The development of the kidney. *Curr Top Dev Biol* 39:245–301.

Davis BM (2011) Evolution of the tribosphenic molar pattern in early mammals with comments on the "dual-origin" hypothesis. *J Mammal Evol* 18:227–244.

Davis DD (1964) The Giant Panda A Morphological Study of Evolutionary Mechanisms Fieldiana Zoology Memoirs, pp 120–124. Chicago Natural History Museum.

Davis MC, Shubin N & Daeschler EB (2004) A new specimen of *Sauripterus taylori* (*Sarcopterygii, Osteichthyes*) from the Fammenian Catskill Formation of North America. *J Vert Paleont* 24:26–40.

Dawkins R (1976) The Selfish Gene. Oxford University Press.

Dawkins R (1996) The Blind Watchmaker. W W Norton & Co.

de Bakker MA, Fowler DA & den Oude K (2013) Digit loss in archosaur evolution and the interplay between selection and constraints. *Nature* 500:445–448 (doi: 10.1038/nature12336).

de Herder WW (2012) Acromegalic gigantism, physicians and body snatching. Past or present? *Pituitary* 15:312–318 (doi: 10.1007/s11102-012-0389-5).

De León LF, Podos J, Gardezi T et al. (2014) Darwin's finches and their diet niches: the sympatric coexistence of imperfect generalists. *J Evol Biol* 27:1093–1104 (doi: 10.1111/jeb.12383).

De Robertis EM (2008) Evo-devo: variation on ancestral themes. *Cell* 132:185–195 (doi: 10.1016/j.cell.2008.01.003).

De Robertis EM & Sasai Y (1996) A common plan for dorsoventral patterning in Bilateria. *Nature* 380:37–40.

Decker RS, Koyama E & Pacifici M (2014) Genesis and morphogenesis of limb synovial joints and articular cartilage. *Matrix Biol* 39:5–10 (doi: 10.1016/j.matbio.2014.08.006).

deMenocal PB (2004) African climate change and faunal evolution during the Pliocene-Pleistocene. *Earth Planet Sci Lett* 220:3–24.

Denes AS, Jékely G & Steinmetz PR (2007) Molecular architecture of annelid nerve cord supports common origin of nervous system centralization in bilateria. *Cell* 129:277–288.

Dennis MY, Nuttle X & Sudmant PH (2012) Evolution of human-specific neural SRGAP2 genes by incomplete segmental duplication. *Cell* 149:912–922 (doi: 10.1016/j.cell.2012.03.033).

Denton D, Aung-Htut MT & Kumar S (2013) Developmentally programmed cell death in *Drosophila. Biochim Biophys Acta* 1833:3499–3506 (doi: 10.1016/j.bbamcr.2013.06.014).

Desalle R & Rosenfeld J (2012) Phylogenomics. Garland Press.

Deshpande O, Batzoglou S, Feldman MW & Cavalli-Sforza LL (2009) A serial founder effect model for human settlement out of Africa. *Proc Biol Sci* 276:291–300 (doi: 10.1098/rspb.2008.0750).

Dewar D, Merson Davies L & Haldane JBS (1949) Is Evolution a Myth. CA Watts & The Paternoster Press.

Dhouailly D (2009) A new scenario for the evolutionary origin of hair feather and avian scales. *J Anat* 214:587–606 (doi: 10.1111/j.1469-7580.2008.01041.x).

Dhouailly D & Sengel P (1973) Morphogenic interactions between reptilian epidermis and birds or mammalian dermis. *C R Acad Sci Hebd Seances Acad Sci D* 277:1221–1224.

Diogo R (2008) The Origin of Higher Clades: Osteology, Myology, Phylogeny and Evolution of Bony Fishes and the Rise of Tetrapods. CRC Press.

Dobzhansky T (1937) Genetics and the Origin of Species. Columbia University Press.

Dobzhansky T (1973) Nothing in biology makes sense except in the light of evolution. *Am Biol Teacher* 35:125–129.

Donoghue PC, Forey PL & Aldridge RJ (2000) Conodont affinity and chordate phylogeny. *Biol Rev Camb Philos Soc* 75:191–251.

Donoghue PC & Sansom IJ (2002) Origin and early evolution of vertebrate skeletonization. *Microsc Res Tech* 59:352–372.

Droser ML & Gehling JG (2015) The advent of animals: the view from the Ediacaran. *Proc Natl Acad Sci U S A* 112:4865–4870 (doi: 10.1073/pnas.1403669112).

Duboule D, Tarchini B, Zàkàny J & Kmita M (2007) Tinkering with constraints in the evolution of the vertebrate limb anterior-posterior polarity. *Novartis Found Symp* 284:130–141.

Dunning-Hotopp JC (2011) Horizontal gene transfer between bacteria and animals. *Trends Genet* 27:157–163 (doi: 10.1016/j.tig.2011.01.005).

Dziecio AJ & Mann S (2012) Designs for life: protocell models in the laboratory. *Chem Soc Rev* 41:79–85 (doi: 10.1039/c1cs15211d).

Edwards CJ, Suchard MA & Lemey P (2011) Ancient hybridization and an Irish origin for the modern polar bear matriline. *Curr Biol* 21:1251–1258 (doi: 10.1016/j.cub.2011.05.058).

El Albani A, Bengtson S, Canfield DE et al. (2010) Large colonial organisms with coordinated growth in oxygenated environments 2.1 Gyr ago. *Nature* 466:100–104 (doi: 10.1038/nature09166).

El Albani A, Bengtson S & Canfield DE (2014) The 2.1 Ga old Francevillian biota: biogenicity, taphonomy and biodiversity. *PLoS One* 9:e99438 (doi: 10.1371/journal.pone.0099438).

Eldredge N & Gould SJ (1972) Punctuated equilibria: an alternative to phyletic gradualism. In Models in Paleobiology (Schopf TJM ed.), pp 82–115. Freeman Cooper. http://www.nileseldredge.com/uploads/3/8/1/1/38114941/punctuated_equilibria_eldredge_gould_1972.pdf

Elzanowski A & Ostell J (2013) The genetic codes. www.ncbi.nlm.nih.gov/Taxonomy/Utils/wprintgc.cgi

Erzurumlu RS & Gaspar P (2012) Development and critical period plasticity of the barrel cortex. *Eur J Neurosci* 35:1540–1553 (doi: 10.1111/j.1460-9568.2012.08075.x).

Ewart JC (1897) A Critical Period in the Development of the Horse. A & C Black.

Ewart JC (1899) The Pennycuik Experiments. A & C Black.

Ezcurra MD (2014) The osteology of the basal archosauromorph Tasmaniosaurus triassicus from the Lower Triassic of Tasmania, Australia. *PLoS One* 9(1):e86864 (doi: 10.1371/journal.pone.008686).

Falcon-Lang HJ, Benton MJ & Stimson M (2007) Ecology of the earliest reptiles inferred from basal Pennsylvanian trackways. *J Geol Soc* 164:1113–1118.

Fedonkin MA & Waggoner BM (1997) The Late Precambrian fossil Kimberella is a mollusc-like bilaterian organism. *Nature* 388:868–871.

Fenchel T (2003) The Origin and Early Evolution of Life. Oxford University Press.

Ferris JP & Ertem G (1992) Oligomerization of ribonucleotides on montmorillonite: reaction of the 5-phosphorimidazolide of adenosine. *Science* 257:1387–1389.

Fisher RA (1930) The Genetical Theory of Natural Selection. Oxford Clarendon Press.

Flannery DT & Walter MR (2012) Archean tufted microbial mats and the Great Oxidation Event: new insights into an ancient problem. *Aust J Earth Sci* 59:1–11 (doi: 10.1080/08120099.2011.607849).

Fleming PA, Verburgt L, Scantlebury M et al. (2009) Jettisoning ballast or fuel? Caudal autotomy and locomotory energetics of the Cape dwarf gecko *Lygodactylus capensis* (Gekkonidae). *Physiol Biochem Zool* 82:756–765 (doi: 10.1086/605953).

Fortey R (2011) Survivors: the animals and plants that time left behind. Harper Press.

Fothergill T, Donahoo AL & Douglass A (2014) Netrin-DCC signaling regulates corpus callosum formation through attraction of pioneering axons and by modulating Slit2-mediated repulsion. *Cereb Cortex* 24:1138–1151 (doi: 10.1093/cercor/bhs395).

Frank SA (1995) George Price's contributions to evolutionary genetics. *J Theor Biol* 175:373–388.

Frank SA (2012) Natural selection IV: the Price equation. *J Evol Biol* 25:1002–1019 (doi: 10.1111/j.1420-9101.2012.02498.x).

Friedman JR & Nunnari J (2014) Mitochondrial form and function. *Nature* 505:335–343 (doi: 10.1038/nature12985).

Friedman RE (1987) Who Wrote the Bible? Summit Books.

Freyer C, Zeller U & Renfree MB (2003) The marsupial placenta: a phylogenetic analysis. *J Exp Zool* 299A:59–77.

Fry JD (2003) Multilocus models of sympatric speciation: Bush versus Rice versus Felsenstein. *Evolution* 57:1735–1746.

Fryer G (2001) On the age and origin of the species flock of haplochromine cichlid fishes of Lake Victoria. *Proc Biol Sci* 268:1147–1152.

Fuerst JA & Sagulenko E (2011) Beyond the bacterium: planctomycetes challenge our concepts of microbial structure and function. *Nat Rev Microbiol* 9:403–413 (doi: 10.1038/nrmicro2578).

Fuerst JA & Sagulenko E (2012) Keys to eukaryality: planctomycetes and ancestral evolution of cellular complexity. *Front Microbiol* 3:167 (doi: 10.3389/fmicb.2012.00167).

Futuyma DJ (2013) Evolutionary Biology. Sinauer Press.

Gage PJ, Rhoades W, Prucka SK & Hjalt T (2005) Fate maps of neural crest and mesoderm in the mammalian eye. *Invest Ophthalmol Vis Sci* 46:4200–4208.

Gaines RR, Briggs DEG & Yuanlong Z (2008) Cambrian Burgess Shale-type deposits share a common mode of fossilization. *Geology* 36:755–758.

Gañan Y, Macias D & Basco RD (1998) Morphological diversity of the avian foot is related with the pattern of msx gene expression in the developing autopod. *Dev Biol* 196:33–41.

Garm A & Ekström P (2010) Evidence for multiple photosystems in jellyfish. *Int Rev Cell Mol Biol* 280:41–78 (doi: 10.1016/S1937-6448(10)80002-4).

Gaston KJ & Spicer JI (2004) Biodiversity: An Introduction. Blackwell Science.

Gaunt SJ & Paul YL (2012) Changes in cis-regulatory elements during morphological evolution. *Biology (Basel)* 1:557–574 (doi: 10.3390/biology1030557).

Gauthier C & Peck LS (1999) Polar gigantism dictated by oxygen availability. *Nature* 399:114–115.

Gehring WJ (1996) The master control gene for morphogenesis and evolution of the eye. *Genes Cells* 1:11–15.

Gehring WJ, Kloter U & Suga H (2009) Evolution of the Hox gene complex from an evolutionary ground state. *Curr Top Dev Biol* 88:35–61 (doi: 10.1016/S0070-2153(09)88002-2).

Gerkema MP, Davies W, Foster RG et al. (2013) The nocturnal bottleneck and the evolution of activity patterns in mammals. *Proc Biol Sci* 280:20130508 (doi: 10.1098/rspb.2013.0508).

Gilbert SF (2014) Developmental Biology, 10th ed. Sinauer Press.

Gilbert SF & Epel D (2015) Ecological Developmental Biology, 2nd ed. Sinauer Press.

Giles S, Friedman M & Brazeau MD (2015) Osteichthyan-like cranial conditions in an Early Devonian stem gnathostome. *Nature* 520:82–5 (doi: 10.1038/nature14065).

Gill GG, Purnell MA & Crumpton N (2014) Dietary specializations and diversity in feeding ecology of the earliest stem mammals. *Nature* 512:303–305 (doi: 10.1038/nature13622).

Giribet G (2008) Assembling the lophotrochozoan (Spiralian) tree of life. *Philos Trans R Soc Lond B Biol Sci* 363B:1513–1522 (doi: 10.1098/rstb.2007.2241).

Giribet G & Edgecombe GD (2012) Reevaluating the arthropod tree of life. *Annu Rev Entomol* 57:167–186 (doi: 10.1146/annurev-ento-120710-100659).

Giribet G, Distel DL & Polz M (2000) Triploblastic relationships with emphasis on the acoelomates and the position of Gnathostomulida, Cycliophora, Plathelminthes, and Chaetognatha: a combined approach of 18S rDNA sequences and morphology. *Syst Biol* 49:539–562.

Gissis SB & Jablonka E (eds) (2011) Transformations of Lamarckism. MIT Press.

Glansdorff N, Xu Y & Labedan B (2008) The Last Universal Common Ancestor: emergence constitution and genetic legacy of an elusive forerunner. *Biol Direct* 3:29 (doi: 10.1186/1745-6150-3-29).

Godefroit P, Cau A, Hu D-Y, Escuillié F et al. (2013) A Jurassic avialan dinosaur from China resolves the early phylogenetic history of birds. *Nature* 498: 359–362. (doi:10.1038/nature12168).

Goldschmidt RB (1940) The Material Basis of Evolution. Yale University Press.

Goldschmidt T (1996) Darwin's Dreampond: Drama in Lake Victoria. MIT Press.

Gomez B, Daviero-Gomez V & Coiffard C (2015) *Montsechia*, an ancient aquatic angiosperm. *Proc Natl Acad Sci* 112:10985–10988 (doi: 10.1073/pnas.1509241112).

Gorrell JC, McAdam AG, Coltman DW et al. (2010) Adopting kin enhances inclusive fitness in asocial red squirrels. *Nat Commun* 1:1–4 (doi: 10.1038/ncomms1022).

Goswami A, Milne N & Wroe S (2011) Biting through constraints: cranial morphology disparity and convergence across living and fossil carnivorous mammals. *Proc R Soc* 278B:1831–1839 (doi: 10.1098/rspb.2010.2031).

Gould SJ (1977) Ontogeny and Phylogeny. Belknap Press.

Gould SJ (2000) A Tree Grows in Paris: Lamarck's Division of Worms and Revision of Nature. In The Lying Stones of Marrakech: Penultimate Reflections in Natural History, pp 115–114. Harmony Books. http://www.facstaff.bucknell.edu/sdjordan/PDFs/Gould_Lamarck'sTree.pdf

Gould SJ & Eldridge N (1972 [1993]) Punctuated equilibrium comes of age. *Nature* 366:223–227.

Grady JM, Enquist BJ, Dettweiler-Robinson E et al. (2014) Dinosaur physiology. Evidence for mesothermy in dinosaurs. *Science* 344:1268–1272 (doi: 10.1126/science).

Grant BR & Grant PR (2008) Fission and fusion of Darwin's finches populations. *Philos Trans R Soc Lond B Biol Sci* 63B:2821–2829 (doi: 10.1098/rstb.2008.0051).

Gray MW, Burger G & Lang BF (2001) The origin and early evolution of mitochondria. *Genome Biol* 2:1–5.

Green RE, Krause J, Brigs AW et al. (2010) A draft sequence of the neandertal genome. *Science* 328:710–722 (doi: 10.1126/science.1188021).

Greenfield A & Bard J (2015) The reproductive system. In Supplement to Kaufman's Atlas of Mouse Development (Baldock R, Bard J, Davidson D & Morriss-Kay G eds). pp 121–132 Academic Press.

Gribaldo S, Poole AM & Daubin V (2010) The origin of eukaryotes and their relationship with the Archaea: are we at a phylogenomic impasse? *Nat Rev Microbiol* 8:743–752 (doi: 10.1038/nrmicro2426).

Grimaldi D & Engel MS (2005) Evolution of Insects. Cambridge University Press.

Grindley JC, Hargett LK, Hill RE et al. (1997) Disruption of PAX6 function in mice homozygous for the Pax6Sey-1Neu mutation produces abnormalities in the early development and regionalization of the diencephalon. *Mech Dev* 64:111–126.

Gritli-Linde A (2007) Molecular control of secondary palate development. *Dev Biol* 301:309–326.

Groenewald GH, Welman J & Maceachern J A (2001) Vertebrate burrow complexes from the early Triassic Cynognathus Zone (Driekoppen Formation Beaufort Group) of the Karoo Basin South Africa. *Palaios* 16:148–160.

Grün R (2006) Direct dating of human fossils. *Am J Phys Anthr* 49:2–48.

Grün R, Stringer C & McDermott F (2005) U-series and ESR analyses of bones and teeth relating to the human burials from Skhul. *J Hum Evol* 49:316–334.

Gu H, Goodale E & Chen J (2015) Emerging directions in the study of the ecology and evolution of plant-animal mutualistic networks: a review. *Zoo Res* 36:65–71.

Guerrero G (2008) The session that did not shake the world (the Linnaean Society 1st July 1858). *Int Microbiol* 11:209–212.

Gupta RS (2011) Origin of diderm (Gram-negative) bacteria: antibiotic selection pressure rather than endosymbiosis likely led to the evolution of bacterial cells with two membranes. *Antonie Van Leeuwenhoek* 100:171–82 (doi: 10.1007/s10482-011-9616-8).

Hagen IJ, Donnellan SC & Bull CM (2012) Phylogeography of the prehensile-tailed skink Corucia zebrata on the Solomon. *Archipelago Ecol Evol* 2:1220–1234 (doi: 10.1002/ece3.84).

Hajkova Erhardt S & Lane N (2002) Epigenetic reprogramming in mouse primordial germ cells. *Mech Dev* 117:15–23.

Haldane JBS (1927) A mathematical theory of natural and artificial selection Part V: selection and mutation. *Proc Camb Philos Soc* 23:838–844.

Halder G, Callaerts P & Gehring WJ (1995) Induction of ectopic eyes by targeted expression of the eyeless gene in *Drosophila*. *Science* 267:1788–1792.

Hall BG (2004) Comparison of the accuracies of several phylogenetic methods using protein and DNA sequences. *Mol Biol Evol* 22:792–802.

Hamilton WD (1964) The genetical evolution of social behavior. I. *J Theoret Biol* 7:1–16.

Hamilton WD (1967) Extraordinary sex ratios. *Science* 156:477–488.

Hamilton MB (2009) Population Genetics. Wiley Blackwell.

Han M (1997) Gut reaction to Wnt signaling in worms. *Cell* 90:581–584.

Han TM & Runnegar B (1992) Megascopic eukaryotic algae from the 21-billion-year-old negaunee iron-formation Michigan. *Science* 257:232–235.

Hanson IM, Fletcher JM & Jordan T (1994) Mutations at the PAX6 locus are found in heterogeneous anterior segment malformations including Peters' anomaly. *Nat Genet* 6:168–173.

Hardy IC (ed.) (2002) Sex Ratios: Concepts and Research Methods. Cambridge University Press.

Harmston N & Lenhard B (2013) Chromatin and epigenetic features of long-range gene regulation. *Nucleic Acids Res* 41:7185–7199.

Harris WA (1997) Pax-6: where to be conserved is not conservative. *Proc Natl Acad Sci* 94:2098–2100.

Hedrick PW (2010) Genetics of Populations. Jones & Bartlett.

Heger TJ, Edgcomb VP & Kim E (2014) A resurgence in field research is essential to better understand the diversity, ecology, and evolution of microbial eukaryotes. *J Eukaryot Microbiol* 61:214–223 (doi: 10.1111/jeu.12095).

Hein J, Schierup M & Carten W (2005) Gene Genealogies, Variation and Evolution: A Primer in Coalescent Theory. Oxford University Press.

Held LI (2014) How the Snake Lost its Legs: Curious Tales from the Frontier of Evo-Devo. Cambridge University Press.

Hendry AP, Nosily P & Rieseberg LH (2007) The speed of ecological speciation. *Funct Ecol* 21:455–464.

Hennig W (1966) Phylogenetic Systematics. University of Illinois Press.

Henn BM, Cavalli-Sforza LL & Feldman MW (2012) The great human expansion. *Proc Natl Acad Sci U S A* 109:17758–17764 (doi: 10.1073/pnas.1212380109).

Henry AP, Nosily P & Reiseberg LH (2007) The speed of ecological speciation. *Funct Ecol* 21:455–464.

Henshilwood CS, d'Errico F & Marean CW (2001) An early bone tool industry from the Middle Stone Age at Blombos Cave, South Africa: implications for the origins of modern human behaviour, symbolism and language. *J Hum Evol* 41:631–678.

Henshilwood CS, d'Errico F & Yates R (2002) Emergence of modern human behavior: Middle Stone Age engravings from South Africa. *Science* 295:1278–1280.

Higgs PG & Lehman N (2015) The RNA World: molecular cooperation at the origins of life. *Nat Rev Genet* 6:7–17 (doi: 10.1038/nrg3841).

Higham T, Douka K & Wood R (2014) The timing and spatiotemporal patterning of Neanderthal disappearance. *Nature* 512:306–309 (doi: 10.1038/nature13621).

Ho SY & Duchêne S (2014) Molecular-clock methods for estimating evolutionary rates and timescales. *Mol Ecol* 23:5947–5965 (doi: 10.1111/mec.12953).

Hochuli PA & Feist-Burkhardt S (2013) Angiosperm-like pollen and Afropollis from the Middle Triassic (Anisian) of the Germanic Basin

(Northern Switzerland). *Front Plant Sci* 4:344 (doi: 10.3389/fpls.2013.00344).

Hodgson JA, Mulligan CJ, Al-Meeri A & Raaum RL (2014) Early back-to-Africa migration into the Horn of Africa. *PLoS Genet* 10:e1004393 (doi: 10.1371/journal.pgen.1004393).

Holland ND (2005) Chordates. *Curr Biol* 22:R911–R914.

Holland PWH (2013) Evolution of homeobox genes. *Wiley Interdiscip Rev Dev Biol* 2:31–45 (doi: 10.1002/wdev.78).

Holland PWH & Takahashi T (2005) The evolution of homeobox genes: implications for the study of brain development. *Brain Res Bull* 66:484–490.

Holleley CE, O'Meally D, Sarre SD et al. (2015) Sex reversal triggers the rapid transition from genetic to temperature-dependent sex. *Nature* 523:79–82 (doi: 10.1038/nature14574).

Holmes G (2012) The role of vertebrate models in understanding craniosynostosis. *Childs Nerv Syst* 28:1471–1481 (doi: 10.1007/s00381-012-1844-3).

Hornbruch A & Wolpert L (1970) Cell division in the early growth and morphogenesis of the chick limb. *Nature* 226:764–766.

Horner JR, Varricchio DJ & Goodwin MB (1992) Marine transgressions and the evolution of Cretaceous dinosaurs. *Nature* 358:59–61.

Hubaud A & Pourquié O (2014) Signalling dynamics in vertebrate segmentation. *Nat Rev Mol Cell Biol* 15:709–721 (doi: 10.1038/nrm3891).

Huber SK, De León LF & Hendry AP (2007) Reproductive isolation of sympatric morphs in a population of Darwin's finches. *Proc Biol Sci* 274:1709–1714.

Hulbert AJ & Else PL (1989) Evolution of mammalian endothermic metabolism: mitochondrial activity and cell composition. *Am J Physiol* 256:R63–R69.

Huldtgren T, Cunningham JA, Yin C & Stampanoni M (2011) Fossilized nuclei and germination structures identify Ediacaran "animal embryos" as encysting protists *Science* 334:1696–1699 (doi: 10.1126/science.1209537).

Hutchinson JR, Delmer C & Miller CE (2011) From flat foot to fat foot structure, ontogeny, function, and evolution of elephant sixth toes. *Science* 34:1699–1703 (doi: 10.1126/science.1211437).

Ijdo JW, Baldini A, Ward DC et al. (1991) Origin of human chromosome 2: an ancestral telomere-telomere fusion. *Proc Natl Acad Sci U S A* 88:9051–9055.

Iqbal K, Tran DA & Li AX (2015) Deleterious effects of endocrine disruptors are corrected in the mammalian germline by epigenome reprogramming. *Genome Biol* 16:59 (doi: 10.1186/s13059-015-0619-z).

Irwin DE, Irwin JH & Price TD (2001) Ring species as bridges between microevolution and speciation. *Genetica* 112–113:223–243.

Irwin G (1994) The Prehistoric Exploration and Colonisation of the Pacific. Cambridge University Press.

Jablonski NG & Chaplin G (2010) Human skin pigmentation as an adaptation to UV radiation. *Proc Natl Acad Sci U S A* 107(Suppl. 2):8962–8968 (doi: 10.1073/pnas.0914628107).

Jacob F (1977) Evolution and tinkering. *Science* 196:1161–1166.

Jandzik D, Garnett AT & Square TA (2015) Evolution of the new vertebrate head by co-option of an ancient chordate skeletal tissue. *Nature* 518:534–537 (doi: 10.1038/nature14000).

Jarman AP (2000) Developmental genetics: vertebrates and insects see eye to eye. *Curr Biol* 10:R857–8599.

Jarne P & Auld JR (2006) Animals mix it up too: the distribution of self-fertilization among hermaphroditic animals. *Evolution* 60:1816–1824.

Janvier P (1996) Early vertebrates. Clarendon Press.

Javaux EJ (2007) The early eukaryotic fossil record. *Adv Exp Med Biol* 607:1–19.

Jensen PE & Leister D (2014) Chloroplast evolution structure and functions. *F1000Prime Rep* 6:40 (doi: 10.12703/P6-40).

Ji QZ-X, Luo C-X, Yuan AR & Tabrum (2006) A swimming mammaliaform from the Middle Jurassic and ecomorphological diversification of early mammals. *Science* 311:1123–1127.

Jobling M, Hollox E, Hurles M et al. (2013) Human Evolutionary Genetics, 2nd ed. Garland Press.

Jodar M, Sendler E & Krawetz SA (2016) The protein and transcript profiles of human semen. *Cell Tiss Res* 363:85–96.

Johnson WE, Eizirik E, Pecon-Slattery J et al. (2006) The late Miocene radiation of modern Felidae: a genetic assessment. *Science* 311:73–77.

Jones TR (1997) Quantitative aspects of the relationship between the sickle-cell gene and malaria. *Parasitol Today* 13:107–111.

Jorde LB, Fineman RM & Martin RA (1983) Epidemiology and genetics of neural tube defects: an application of the Utah Genealogical Data Base. *Am J Phys Anthropol* 62:23–31.

Just J, Kristensen RM & Olesen J (2014) Dendrogramma, new genus, with two new non-bilaterian species from the marine bathyal of southeastern Australia (Animalia, Metazoa incertae sedis) – with similarities to some medusoids from the Precambrian Ediacara. *PLoS One* 9:e102976 (doi: 10.1371/journal.pone.0102976).

Kaas JH (2011) Neocortex in early mammals and its subsequent variations. *Ann N Y Acad Sci* 1225:1225–1236 (doi: 10.1111/j.1749-6632.2011.05981.x).

Kanehisa M, Goto S, Sato Y et al. (2012) KEGG for integration and interpretation of large-scale molecular data sets. *Nucleic Acids Res* 40: D109–D114 (doi: 10.1093/nar/gkr988).

Kardong KV (2014) Vertebrates: Comparative Anatomy, Function, Evolution, 7th ed. McGraw-Hill.

Karmin M, Saag L, Vicente M et al. (2015) A recent bottleneck of Y chromosome diversity coincides with a global change in culture. *Genome Res* 25:459–66 (doi: 10.1101/gr.186684.114).

Katz A (2012) Origin and diversification of eukaryotes. *Annu Rev Microbiol* 66:411–427 (doi: 10.1146/annurev-micro-090110-102808).

Kawakami Y, Capdevila J & Büscher D (2001) WNT signals control FGF-dependent limb initiation and AER induction in the chick embryo. *Cell* 104:891–900 (doi:10.1016/S0092-8674(01)00285-9).

Keeling PJ (2013) The number, speed, and impact of plastid endosymbioses in eukaryotic evolution. *Annu Rev Plant Biol* 64:583–607 (doi: 10.1146/annurev-arplant-050312-120144).

Keeling PK & Palmer JD (2008) Horizontal gene transfer in eukaryotic evolution. *Nat Rev Genet* 9:605–618 (doi: 10.1038/nrg2386).

Keith A (1910) Abnormal ossification of Meckel's cartilage. *J Anat Physiol* 44:151–152.

Keller I, Wagner CE & Greuter L (2013) Population genomic signatures of divergent adaptation gene flow and hybrid speciation in the rapid radiation of Lake Victoria cichlid fishes. *Mol Ecol* 22:2848–2863 (doi: 10.1111/mec.12083).

Kemp TS (2005) The Origin and Evolution of the Mammals. Oxford University Press.

Keyte AL & Smith KS (2010) Developmental origins of precocial forelimbs in marsupial neonates. *Development* 137:4283–4294 (doi: 10.1242/dev.049445).

Kielan-Jaworowska Z, Cifelli RL & Luo Z-X (2004) Mammals from the Age of Dinosaurs: Origins, Evolution, and Structure. Columbia University Press.

Kim KM & Caetano-Anollés G (2011) The proteomic complexity and rise of the primordial ancestor of diversified life. *BMC Evol Biol* 11:140 (doi: 10.1186/1471-2148-11-140).

Kimbel WH & Rak Y (2010) The cranial base of *Australopithecus afarensis*: new insights from the female skull. *Philos Trans R Soc* 365B:3365–3376 (doi: 10.1098/rstb.2010.0070).

King N, Hittinger CT & Carroll SB (2003) Evolution of key cell signaling and adhesion protein families predates animal origins. *Science* 301:361–363.

Kingman JFC (1982) The coalescent. *Genetics* 156:1461–1463.

Kingman JF (2000) Origins of the coalescent 1974–1982. *Genetics* 156:1461–1463.

Kimura M & Ohta T (1969) The average number of generations until fixation of a mutant gene in a finite population. *Genetics* 61:763–771.

Kirkilionis M (2010) Exploration of cellular reaction systems. *Brief Bioinform* 11:153–178 (doi: 10.1093/bib/bbp062).

Kitching IJ, Forey PL, Humphries CJ & Williams D (1998) Cladistics: Theory and Practice of Parsimony Analysis, 2nd ed. Oxford University Press.

Kleisner K, Ivell R & Flegr J (2010) The evolutionary history of testicular externalization and the origin of the scrotum. *J Biosci* 35:27–37.

Knoll AH (2004) Life on a Young Planet: The First Three Billion Years of Evolution on Earth. Princeton University Press.

Knoll AH (2015) Paleobiological perspectives on early eukaryotic evolution. *Cold Spring Harb Perspect Biol* 7: a018093 (doi: 10.1101/cshperspect.a018093).

Knoll AH, Javaux EJ, Hewitt D & Cohen P (2006) Eukaryotic organisms in proterozoic oceans. *Philos Trans R Soc B* 361:1023–1038.

Koonin EV & Galperin MY (2003) Sequence – Evolution – Function: Computational Approaches in Comparative Genomics. Kluwer Academic. http://www.ncbi.nlm.nih.gov/books/NBK20260/

Kosmidis EK, Moschou V, Ziogas G et al. (2014) Functional aspects of the EGF-induced MAP kinase cascade: a complex self-organizing system approach. *PLoS One* 9:e111612 (doi: 10.1371/journal.pone.0111612).

Koumandou VL, Wickstead W & Ginger ML (2013) Molecular paleontology and complexity in the last eukaryotic common ancestor. *Crit Rev Biochem Mol Biol* 48:373–396 (doi: 10.3109/10409238.2013.821444).

Kozmik Z, Daube M & Frei E (2003) Role of Pax genes in eye evolution: a cnidarian PaxB gene uniting Pax2 and Pax6 functions. *Dev Cell* 5:773–785.

Krabbenhoft KM & Fallon JF (1989) The formation of leg or wing specific structures by leg bud cells grafted to the wing bud is influenced by proximity to the apical ridge. *Dev Biol* 131:373–82.

Kramerov DA & Vassetzky NS (2011) SINEs. *Wiley Interdiscip Rev RNA* 2:772–786 (doi: 10.1002/wrna.91).

Krause WJ & Cutts JH (1985) Placentation in the opossum, *Didelphis virginiana. Acta Anat* 123:156–171.

Krause WJ & Cutts JH (1986) Scanning electron microscopic observations on developing opossum embryos: days 9 through 12. *Anat Anz* 161:11–21.

Kropotkin P (1902 [2009]) Mutual Aid: A Factor of Evolution. New York University Press. http://www.complementarycurrency.org/ccLibrary/Mutual_Aid-A_Factor_of_Evolution-Peter_Kropotkin.pdf

Kumar JP & Moses K (2001) Eye specification in Drosophila: perspectives and implications. *Sem Cell Dev Biol* 12:469–474.

Kumar S (2005) Molecular clocks: four decades of evolution. *Nat Rev Genet* 6:654–662.

Kun Á, Szilágyi A & Könnyű B (2015) The dynamics of the RNA world: insights and challenges. *Ann N Y Acad Sci* 1341:75–95 (doi: 10.1111/nyas.12700).

Kusuhashi N (2013) A new Early Cretaceous eutherian mammal from the Sasayama Group Hyogo Japan. *Proc Biol Sci* 280:20130142 (doi: 10.1098/rspb.2013.0142).

Lalueza-Fox C & Gilbert MTP (2011) Paleogenomics of archaic hominins. *Curr Biol* 21:R1002–R1009 (doi: 10.1016/j.cub.2011.11.021).

Lane CS, Chorn BT & Johnson TC (2013) Ash from the Toba supereruption in Lake Malawi shows no volcanic winter in East Africa at 75 ka, *Proc Natl Acad Sci U S A* 110: 8025–8029 (doi: 10.1073/pnas.1301474110).

Langeland JA, Holland LZ, Chastain RA & Holland ND (2006) An amphioxus LIM-homeobox gene AmphiLim1/5 expressed early in the invaginating organizer region and later in differentiating cells of the kidney and central nervous system. *Int J Biol Sci* 2:110–116.

Lau AN, Peng L, Goto H et al. (2009) Horse domestication and conservation genetics of Przewalski's horse inferred from sex chromosomal and autosomal sequences. *Mol Biol Evol* 26:199–208 (doi:10.1093/molbev/msn239).

Lawton JH & May RM (2005) Extinction Rates. Oxford University Press.

Le PT, Pontarotti P & Raoult D (2014) Alphaproteobacteria species as a source and target of lateral sequence transfers. *Trends Microbiol* 22:147–156 (doi: 10.1016/j.tim.2013).

Lebedev OA & Coates MI (1995) The postcranial skeleton of the Devonian tetrapod Tulerpeton curtum Lebedev. *Zoo J Linn Soc* 114 (3): 307–348 (doi:10.1111/j.1096-3642.1995.tb00119.x).

Lee MS, Cau A, Naish D & Dyke GJ (2014) Dinosaur evolution. Sustained miniaturization and anatomical innovation in the dinosaurian ancestors of birds. *Science* 345:662–566 (doi: 10.1126/science.1252243).

Lee RT, Thiery JP & Carney TJ (2013) Dermal fin rays and scales derive from mesoderm not neural crest. *Curr Biol* 6:R336–R337 (doi: 10.1016/j.cub.2013.02.055).

Lemey P, Salemi M & Vandamme A-M (eds) (2009) The Phylogenetic Handbook: A Practical Approach to Phylogenetic Analysis and Hypothesis Testing. Cambridge University Press.

Leonard WE (2008) On the Nature of Things (Part 5): Translation of: Titus Lucretius Carus (d. 55 BCE) *De Rerum Natura.* http://www.gutenberg.org/files/785/785-h/785-h.htm#link2H_4_0027

Li H & Durbin R (2011) Inference of human population history from individual whole-genome sequences. *Nature* 475:493–496 (doi: 10.1038/nature10231).

Li JZ, Absher DM & Tang H (2008) Worldwide human relationships inferred from genome-wide patterns of variation. *Science* 319:1100–1104 (doi: 10.1126/science.1153717).

Li Q, Gao KQ & Vinther J (2010) Plumage color patterns of an extinct dinosaur. *Science* 327:1369–1372 (doi: 10.1126/science.1186290).

Lichtneckert R & Reichert H (2005) Insights into the urbilaterian brain: conserved genetic patterning mechanisms in insect and vertebrate brain development. *Heredity* 94:465–477.

Lieberman D (2011) Evolution of the Human Head. Harvard University Press.

Lieberman D (2013) The Story of the Human Body: Evolution Health and Disease. Pantheon Press.

Liebers R, Rassoulzadegan M & Lyko F (2014) Epigenetic regulation by heritable RNA. *PLoS Genet* 10:e100429 (doi: 10.1371/journal.pgen.1004296).

Ligrone R, Duckett JG & Renzaglia KS (2012) Major transitions in the evolution of early land plants: a bryological perspective. *Ann Bot* 109:851–871 (doi: 10.1093/aob/mcs017).

Lipscombe D (1998) Basics of cladistics analysis. http://www.gwu.edu/~clade/faculty/lipscomb/Cladistics.pdf

Liu L, Yu L & Kubatko L (2009) Coalescent methods for estimating phylogenetic trees. *Mol Phylogenet Evol* 53:320–328.

Liu Z, Lavine KJ, Hung IH & Ornitz DM (2007) FGF18 is required for early chondrocyte proliferation hypertrophy and vascular invasion of the growth plate. *Dev Biol* 302:80–91.

Livi CB & Davidson EH (2006) Expression and function of blimp1/krox, an alternatively transcribed regulatory gene of the sea urchin endomesoderm network. *Dev Biol* 293:513–525.

Long JA, Young GC, Holland T et al. (2006) An exceptional Devonian fish from Australia sheds light on tetrapod origins. *Nature* 444:199–202.

Long JA, Trinajstic K, Young GC & Senden T (2008) Live birth in the Devonian period. *Nature* 453:650–652 (doi: 10.1038/nature06966).

Long M, VanKuren NW, Chen S & Vibranovski MD (2013) New gene evolution: little did we know. *Annu Rev Genet* 47:307–333 (doi: 10.1146/annurev-genet-111212-133301).

Longrich NR, Vinther J, Meng Q et al. (2012) Primitive wing feather arrangement in *Archaeopteryx lithographica* and *Anchiornis huxleyi*. *Curr Biol* 22:2262–2267 (doi: 10.1016/j.cub.2012.09.052).

Loomis WF (2015) Genetic control of morphogenesis in *Dictyostelium*. *Dev Biol* 402:146–161 (doi: 10.1016/j.ydbio.2015.03.016).

Lordkipanidze D, Ponce de León MS & Margvelashvili A (2013) A complete skull from Dmanisi Georgia and the evolutionary biology of early Homo. *Science* 342:326–331 (doi: 10.1126/science.1238484).

Louchart A & Viriot L (2011) From snout to beak: the loss of teeth in birds. *Trends Ecol Evol* 26:663–673 (doi: 10.1016/j.tree.2011.09.004).

Lowe CJ, Clarke DN & Medeiros DM (2015) The deuterostome context of chordate origins. *Nature* 520:456–465 (doi: 10.1038/nature14434).

Luo ZX (2007) Transformation and diversification in early mammal evolution. *Nature* 450:1011–1019.

Luo Z-X, Ji Q, Wibble JR & Yuan C-X (2003) An early Cretaceous tribosphenic mammal and metatherian evolution. *Science* 302:1934–1940.

Luo ZX, Ruf I, Schultz JA & Martin T (2011a) Fossil evidence on evolution of inner ear cochlea in Jurassic mammals. *Proc Biol Sci* 278:28–34 (doi: 10.1098/rspb.2010.1148).

Luo ZX, Yuan CX, Meng QJ & Ji Q (2011b) A Jurassic eutherian mammal and divergence of marsupials and placentals. *Nature* 476:442–445 (doi: 10.1038/nature10291).

Lyson TR, Sperling EA & Heimberg AM (2012) MicroRNAs support a turtle + lizard clade. *Biol Lett* 8:104–107 (doi: 10.1098/rsbl.2011.0477).

Lyson TR, Bever GS & Scheyer TM (2013) Evolutionary origin of the turtle shell. *Curr Biol* 23:1113–1119 (doi: 10.1016/j.cub.2013.05.003).

Macaulay V, Hill C & Achilli A (2005) Single rapid coastal settlement of Asia revealed by analysis of complete mitochondrial genomes. *Science* 308:1034–1036.

Marsicano CA, Irmis RB, Mancuso AC et al. (2016) The precise temporal calibration of dinosaur origins. *Proc Natl Acad Sci* 113:509–513 (doi: 10.1073/pnas.1512541112).

McBrearty S & Brooks AS (2000) The revolution that wasn't: a new interpretation of the origin of modern human behaviour. *J Hum Evol* 39:453–563.

McCourt RM, Delwiche CF & Karol KG (2004) Charophyte algae and land plant origins. *Trends Ecol Evol* 19:661–666.

McDougall I, Brown FH & Fleagle JG (2005) Stratigraphic placement and age of modern humans from Kibish, Ethiopia. *Nature* 433:733–736.

McElhinny MW & McFadden PL (2000) Paleomagnetism: Continents and Oceans. Academic Press.

McFadden GI (2001) Primary and secondary endosymbiosis and the origin of plastids. *J Phycol* 37:951–959.

McGhee G (2011) Convergent Evolution. MIT Press.

McGuire JA & Dudley R (2011) The biology of gliding in flying lizards (genus *Draco*) and their fossil and extant analogs. *Integr Comp Biol* 51:983–990 (doi: 10.1093/icb/icr090).

McKenna MC (1975) Toward a phylogenetic classification of the Mammalia. In Phylogeny of the Primates (Luckett WP & Szalay ES eds), pp 21–46. Plenum Press.

MacNaughton RB, Cole JM & Dalrymple RW (2002) First steps on land: arthropod trackways in Cambrian-Ordovician eolian sandstone southeastern Ontario Canada. *Geology* 30:391–394.

Maddison WP (1997) Gene trees in species trees. *Syst Biol* 46:523–536.

Malaspinas AS, Lao O & Schroeder H (2014) Two ancient human genomes reveal Polynesian ancestry among the indigenous Botocudos of Brazil. *Curr Biol* 24:R1035–R1037 (doi: 10.1016/j.cub.2014.09.078).

Malfait F & De Paepe A (2014) The Ehlers-Danlos syndrome. *Adv Exp Med Biol* 802:129–143 (doi: 10.1007/978-94-007-7893-1_9).

Malik S (2013) Polydactyly phenotypes genetics and classification. *Clin Genet* 85B:203–212 (doi: 10.1111/cge.12276).

Mallarino R, Grant PR, Grant BR et al. (2011) Two developmental modules establish 3D beak-shape variation in Darwin's finches. *Proc Natl Acad Sci U S A* 108:4057–4062 (doi: 10.1073/pnas.1011480108).

Mangalam M & Fragaszy DM (2015) Wild bearded capuchin monkeys crack nuts dexterously. *Curr Biol* 25:1334–1339 (doi: 10.1016/j.cub.2015.03.035).

Mansy SS, Schrum JP & Krishnamurthy M (2008) Template-directed synthesis of a genetic polymer in a model protocell. *Nature* 454:122–125 (doi: 10.1038/nature07018).

Margolin W (2014) Sculpting the bacterial cell. *Curr Biol* 19:R812–R822 (doi: 10.1016/j.cub.2009.06.033).

Marjanović D & Laurin M (2007) Fossils molecules divergence times and the origin of lissamphibians. *Syst Biol* 56:369–388.

Marjanović D & Laurin M (2009) The origin(s) of modern amphibians: a commentary. *Evol Biol* 36:336–338.

Marsicano CA, Irmis RB, Mancuso AC et al. (2016) The precise temporal calibration of dinosaur origins. *Proc Natl Acad Sci* 113:509–513 (doi: 10.1073/pnas.1512541112).

Martill DM, Tischlinger H & Longrich NR (2015) A four-legged snake from the Early Cretaceous of Gondwana. *Science* 349:416–419 (doi: 10.1126/science.aaa9208).

Martin LD & Czerkas SA (2000) The fossil record of feather evolution in the Mesozoic. *Am Zool* 40:687–694.

Martin MW, Grazhdankin DV & Bowring SA (2000) Age of Neoproterozoic bilatarian body and trace fossils, White Sea, Russia: implications for metazoan evolution. *Science* 288:841–845.

Martinez RN, Sereno PC, Alcober OA et al. (2011 A basal dinosaur from the dawn of the dinosaur era in southwestern Pangaea. *Science* 331:206–210 (doi: 10.1126/science.1198467).

Martínez-Abadías N, Motch SM & Pankratz TL (2013) Tissue-specific responses to aberrant FGF signaling in complex head phenotypes. *Dev Dyn* 242:80–94 (doi: 10.1002/dvdy.23903).

Martins AB, de Aguiar MA & Bar-Yam Y (2013) Evolution and stability of ring species. *Proc Natl Acad Sci U S A* 110:5080–5084 (doi: 10.1073/pnas.1217034110).

Masel J (2011) Genetic drift - a quick guide. *Curr Biol* 21:R837–R838 (doi: 10.1016/j.cub.2011.08.007).

Mason J (2006) Apollo 11 cave in southwest Namibia: some observations on the site and its rock art. *S African Archeol Bull* 61:76–89.

Mat W-K, Hong Xue H & Wong JT-F (2008) The genomics of LUCA. *Front Biosci* 13:5605–5613.

Mateo JM (2002) Kin-recognition abilities and nepotism as a function of sociality. *Proc R Soc Lond B* 269:721–727.

Matsui T, Yamamoto T & Wyder S (2009) Expression profiles of urbilaterian genes uniquely shared between honey bee and vertebrates. *BMC Genomics* 10:17 (doi: 10.1186/1471-2164-10-17).

Matsuzawa T (2013) Evolution of the brain and social behavior in chimpanzees. *Curr Opin Neurobiol* 23:443–449 (doi: 10.1016/j.conb.2013.01.012).

Maynard Smith J, Burian R & Kauffman S (1985) Developmental constraints and evolution. *Q Rev Biol* 60:265–287.

Mayr E (1982) The Growth of Biological Thought. Belknap Press.

Medina L & Abellán A (2009) Development and evolution of the pallium. *Semin Cell Dev Biol* 20:698–711 (doi: 10.1016/j.semcdb.2009.04.008).

Medland SE, Nyholt DR, Painter JN et al. (2009) Common variants in the trichohyalin gene are associated with straight hair in Europeans. *Am J Hum Genet* 85:750–755 (doi: 10.1016/j.ajhg.2009.10.009).

Meinhardt H (1984) Models for positional signalling the threefold subdivision of segments and the pigmentation pattern of molluscs. *J Embryol Exp Morphol* 83(Suppl.):289–311.

Melis AP (2013) The evolutionary roots of human collaboration: coordination and sharing of resources. *Ann N Y Acad Sci* 1299:68–76 (doi: 10.1111/nyas.12263).

Meng J, Hu Y, Wang Y et al. (2006) A Mesozoic gliding mammal from northeastern China. *Nature* 444:889–893.

Meuner FJ & Laurin M (2012) A microanatomical and histological study of the fin long bones of the Devonian sarcopterygian Eusthenopteron foordi. *Acta Zool* 93:88–97 (do1: 10.1111/j.1463-6395.2010.00489.x).

Meyer M, Fu Q & Aximu-Petri A (2014) A mitochondrial genome sequence of a hominin from Sima de los Huesos. *Nature* 505:403–406 (doi: 10.1038/nature12788).

Meyerowitz EM (2002) Plants compared to animals: the broadest comparative study of development. *Science* 295:1482–1485.

Mikkelsen MD, Harholt J & Ullvskov P (2014) Evidence for land plant cell wall biosynthetic mechanisms in charophyte green algae. *Ann Bot* 114:1217–1236 (doi: 10.1093/aob/mcu171).

Miller RF, Cloutier R & Turner S (2003) The oldest articulated chondrichthyan from the Early Devonian period. *Nature*. 425:501–4.

Miller SL (1953) Production of amino acids under possible primitive earth conditions. *Science* 117:528–529.

Mirabeau O & Joly J-S (2013) Molecular evolution of peptidergic signalling systems in bilaterians. *Proc Natl Acad Sci U S A* 110:E2028–E2037 (doi: 10.1073/pnas.1219956110).

Misof B, Liu S & Meusemann KB (2014) Phylogenomics resolves the timing and pattern of insect evolution. *Science* 346:763–767 (doi: 10.1126/science.1257570).

Mitchell FL & Lasswell J (2005) A dazzle of dragonflies. Texas A&M University Press.

Mitchell G & Skinner JD (2009) An allometric analysis of the giraffe cardiovascular system. *Comp Biochem Physiol A Mol Integr Physiol* 154:523–529 (doi: 10.1016/j.cbpa.2009.08.013).

Mitgutsch C, Richardson MK & Jiménez R (2012) Circumventing the polydactyly 'constraint': the mole's 'thumb'. *Biol Lett* 8:74–77 (doi: 10.1098/rsbl.2011.0494).

Miyagi R & Terai Y (2013) The diversity of male nuptial coloration leads to species diversity in Lake Victoria cichlids. *Genes Genet Syst* 88:145–153.

Molnár Z, Kaas JH, de Carlos JA et al. (2014) Evolution and development of the mammalian cerebral cortex. *Brain Behav Evol* 83:126–139 (doi: 10.1159/000357753).

Mokkonen M & Lindstedt C (2015) The evolutionary ecology of deception. *Biol Rev Camb Philos Soc* In Press. (doi: 10.1111/brv.12208).

Monk D (2015) Germline-derived DNA methylation and early embryo epigenetic reprogramming: The selected survival of imprints. *Int J Biochem Cell Biol* 67:128–138 (doi: 10.1016/j.biocel.2015.04.014).

Moore PB & Steiz TA (2005) The role of RNA in the synthesis of proteins. In The RNA World, 3rd ed. (Gesteland RF Cech TR & Atkins JF eds), pp 257–285. Cold Spring Harbor Laboratory Press.

Mora C, Tittensor DP & Adl S (2011) How many species are there on Earth and in the ocean? *PLoS Biol* 9:e1001127 (doi: 10.1371/journal.pbio.1001127).

Morris SC & Caron JB (2012) *Pikaia gracilens Walcott,* a stem-group chordate from the Middle Cambrian of British Columbia. *Biol Rev Camb Philos Soc* 87:480–512 (doi: 10.1111/j.1469-185X.2012.00220.x).

Morriss GM (1975) Placental evolution and embryonic nutrition. In *Comparative placentation* (Dawes GS ed). pp. 87–107, Academic Press.

Morriss-Kay GM (2010) The evolution of human artistic creativity. *J Anat* 216:158–176 (doi: 10.1111/j.1469-7580.2009.01160.x).

Morriss-Kay GM (2011) A new hypothesis on the creation of the Hohle Fels "Venus" figurine. In L'art pléicistocène dans le monde (Clotte J ed.), pp 65–66. Société préhistorique de l'Ariège.

Morriss-Kay GM & Wilkie AO (2005) Growth of the normal skull vault and its alteration in craniosynostosis: insights from human genetics and experimental studies. *J Anat* 207:637–653.

Müller J & Reisz RB (2005) An early captorhinid reptile (Amniota Eureptilia) from the Upper Carboniferous of Hamilton Kansas. *J Vert Pal* 25:561–568.

Munns SL, Owerkowicz T, Andrewartha SJ & Frappell PB (2012) The accessory role of the diaphragmaticus muscle in lung ventilation in the estuarine crocodile *Crocodylus porosus*. *J Exp Biol* 215:845–852 (doi: 10.1242/jeb.061952).

Nakano M, Miwa N, Hirano A et al. (2009) A strong association of axillary osmidrosis with the wet earwax type determined by genotyping of the ABCC11 gene. *BMC Genet* 10:42 (doi: 10.1186/1471-2156-10-42).

Needham J (1931) Chemical Embryology, Cambridge University Press.

Nesbitt SJ (2011) The early evolution of archosaurs: relationships and the origin of major clades. *Bull Am Mus Nat Hist* 352:1–292.

Neves WA, Hubbe M & Piló LB (2007) Early Holocene human skeletal remains from Sumidouro Cave, Lagoa Santa, Brazil: history of discoveries, geological and chronological context, and comparative cranial morphology. *J Hum Evol* 52:16–30.

Newman JH (1973) Letter to J. Walker of Scarborough, May 22, 1868. In The Letters and Diaries of John Henry Newman, pp 77–78. Clarendon Press.

Nikaido M, Rooney AP & Okada N (1999) Phylogenetic relationships among cetartiodactyls based on insertions of short and long interspersed elements: Hippopotamuses are the closest extant relatives of whales. *Proc Natl Acad Sci U S A* 96:10261–10266.

Nilsson DE & Pelger S (1994) A pessimistic estimate of the time required for an eye to evolve. *Proc Biol Sci* 256B:53–58.

Nisbet EG & Nisbet REN (2008) Methane oxygen photosynthesis rubisco and the regulation of the air through time. *Proc R Soc B* 363:2745–2754 (doi: 10.1098/rstb.2008.0057).

Nochomovitz YD & Li H (2006) Highly designable phenotypes and mutational buffers emerge from a systematic mapping between network topology and dynamic output. *Proc Natl Acad Sci U S A* 103:4180–4185.

Noble D (2008) The Music of Life: Biology Beyond Genes. Oxford University Press.

Noble D (2011) Successes and failures in modeling heart cell electrophysiology. *Heart Rhythm* 8:1798–1803 (doi: 10.1016/j.hrthm.2011.06.014).

Noble D (2012) A theory of biological relativity: no privileged level of causation. *Interface Focus* 2:55–64 (doi: 10.1098/rsfs.2011.0067).

Nobles DR, Romanovicz DK & Brown RM (2001) Cellulose in cyanobacteria. Origin of vascular plant cellulose synthase? *Plant Physiol* 127:529–542.

Noffke N, Christian D, Wacey D & Hazen RM (2013) Microbially induced sedimentary structures recording an ancient ecosystem in the ca. 3.48 billion-year-old Dresser Formation, Pilbara, Western Australia. *Astrobiology* 13:1103–1124 (doi: 10.1089/ast.2013.1030).

Nord AS (2015) Learning about mammalian gene regulation from functional enhancer assays in the mouse. *Genomics* 106:178–184 (doi: 10.1016/j.ygeno.2015.06.008).

Nosil P (2012) Ecological Speciation. Oxford University Press.

Nosil P & Feder JL (2012) Genomic divergence during speciation: causes and consequences. *Philos Trans R Soc* 367B:332–342 (doi: 10.1098/rstb.2011.0263).

Nosil P & Schluter D (2011) The genes underlying the process of speciation. *Trends Ecol Evol* 26:160–167 (doi: 10.1016/j.tree.2011.01.001).

Novacek MJ, Rougier GW & Wible JR (1997) Epipubic bones in eutherian mammals from the late Cretaceous of Mongolia. *Nature* 389:483–486.

Nowack EC, Melkonian M & Glöckner G (2008) Chromatophore genome sequence of Paulinella sheds light on acquisition of photosynthesis by eukaryotes. *Curr Biol* 18:410–418 (doi: 10.1016/j.cub.2008.02.051).

Nowitzki U, Flechner A & Kellermann J (1998) Eubacterial origin of nuclear genes for chloroplast and cytosolic glucose-6-phosphate isomerase from spinach: sampling eubacterial gene diversity in eukaryotic chromosomes through symbiosis. *Gene* 214:205–213.

Nozaki H, Matsuzaki M & Takahara M (2003) The phylogenetic position of red algae revealed by multiple nuclear genes from mitochondria-containing eukaryotes and an alternative hypothesis on the origin of plastids. *J Mol Evol* 56:485–497.

Oftedal OT (2002) The mammary gland and its origin during synapsid evolution. *J Mammary Gland Biol Neoplasia* 7:225–252.

Oftedal OT (2012) The evolution of milk secretion and its ancient origins. *Animal* 6:355–368 (doi: 10.1017/S1751731111001935).

Ohtomo Y, Kakegawa T & Ishida A (2014) Evidence for biogenic graphite in early Archaean Isua metasedimentary rocks. *Nature Geosci* 7:25–28.

Oppenheimer S (1998) Eden in the East: the Drowned Continent of Southeast Asia. Phoenix Press.

Oppenheimer S (2003) Out of Eden: the Peopling of the World. Constable.

Oppenheimer S (2009) The great arc of dispersal of modern humans: Africa to Australia. *Quat Int* 202:2–13.

Oppenheimer S (2012) Out-of-Africa: the peopling of continents and islands: tracing uniparental gene trees across the map. *Philos Trans R Soc Lond B Biol Sci* 367B:770–784 (doi: 10.1098/rstb.2011.0306).

Oren A & Garrity GM (2014) Then and now: a systematic review of the systematics of prokaryotes in the last 80 years. *Antonie Van Leeuwenhoek* 106:43–56 (doi: 10.1007/s10482-013-0084-1).

Orlando L, Ginolhac A & Zhang G (2013) Recalibrating *Equus* evolution using the genome sequence of an early middle Pleistocene horse. *Nature* 499:74–78 (doi: 10.1038/nature12323).

Oster G & Wang H (2003) Rotary protein motors. *Trends Cell Biol* 13:114–121 (doi:10.1016/S0962-8924(03)00004-7).

Ota KG, Fujimoto S, Oisi Y & Kuratani S (2011) Identification of vertebra-like elements and their possible differentiation from sclerotomes in the hagfish. *Nat Commun* 2:373 (doi: 10.1038/ncomms1355).

Otsuna H, Shinomiya K & Ito K (2014) Parallel neural pathways in higher visual centers of the Drosophila brain that mediate wavelength-specific behavior. *Front Neural Circuits* 8:8 (doi: 10.3389/fncir.2014.00008).

Pagani L, Schiffels S, Gurdasani D et al. (2015) Tracing the route of modern humans out of Africa by using 225 human genome sequences from Ethiopians and Egyptians. *Am J Hum Genet* 96:986–991 (doi: 10.1016/j.ajhg.2015.04.019).

Page RDM & Holmes EC (1998) Molecular Evolution: A Phylogenetic Approach. Wiley-Blackwell.

Parés JM, Pérez-González A, Weil AB & Arsuaga JL (2000) On the age of the hominid fossils at the Sima de los Huesos Sierra de Atapuerca Spain: paleomagnetic evidence. *Am J Phys Anthropol* 111:451–461.

Parfrey LW, Lahr DJ, Knoll AH & Katz LA (2011) Estimating the timing of early eukaryotic diversification with multigene molecular clocks. *Proc Natl Acad Sci U S A* 108:13624–13629 (doi: 10.1073/pnas.1110633108).

Pascual-Anaya J, D'Aniello S, Kuratani S & Garcia-Fernàndez J (2013) Evolution of Hox gene clusters in deuterostomes. *BMC Dev Biol* 13:26 (doi: 10.1186/1471-213X-13-26).

Paton RL, Smithson TR & Clack JA (1999) An amniote-like skeleton from the Early Carboniferous of Scotland. *Nature* 398:508–513 (doi:10.1038/19071).

Pavlopoulos A & Akam M (2011) Hox gene Ultrabithorax regulates distinct sets of target genes at successive stages of *Drosophila* haltere morphogenesis. *Proc Natl Acad Sci U S A* 108:2855–2860 (doi: 10.1073/pnas.1015077108).

Pereira CF, Lemischka IR & Moore K (2012) Reprogramming cell fates: insights from combinatorial approaches. *Ann N Y Acad Sci* 1266:7–17 (doi: 10.1111/j.1749-6632.2012.06508.x).

Pereira J, Johnson WE & O'Brien SJ (2014) Evolutionary genomics and adaptive evolution of the Hedgehog gene family (Shh Ihh and Dhh) in vertebrates. *PLoS One* 9:e74132 (doi: 10.1371/journal.pone.0074132).

Perini, FA, Russo CAM & Schrago CG (2010) The evolution of South American endemic canids: a history of rapid diversification and morphological parallelism. *J Evol Biol* 23:311–322 (doi:10.1111/j.1420-9101.2009.01901.x).

Petersen KK, Hørlyck A & Ostergaard KH (2013) Protection against high intravascular pressure in giraffe legs. *Am J Physiol Regul Integr Comp Physiol* 305:R1021–R1030 (doi: 10.1152/ajpregu.00025.2013).

Peterson KJ & Eernisse DJ (2001) Animal phylogeny and the ancestry of bilaterians: inferences from morphology and 18S rDNA gene sequences. *Evol Dev* 3:170–205.

Petroutsos D, Amiar S & Abida H (2014) Evolution of galactoglycerolipid biosynthetic pathways—from cyanobacteria to primary plastids and from primary to secondary plastids. *Prog Lipid Res* 54:68–85 (doi: 10.1016/j.plipres.2014.02.001).

Pfennig DW & Mullen SP (2010) Mimics without models: causes and consequences of allopatry in Batesian mimicry complexes. *Proc Biol Sci* 277:2577–2585 (doi: 10.1098/rspb.2010.0586).

Philippe H, Brinkmann H & Copley RR (2011) Acoelomorph flatworms are deuterostomes related to Xenoturbella. *Nature* 470:255–258 (doi: 10.1038/nature09676).

Pierce SE, Clack JA & Hutchinson JR (2012) Three-dimensional limb joint mobility in the early tetrapod Ichthyostega. *Nature* 486:523–526 (doi: 10.1038/nature11124).

Pilder SH, Olds-Clarke P, Phillips DM & Silver LM (1993) Hybrid sterility-6: a mouse t complex locus controlling sperm flagellar assembly and movement. *Dev Biol* 159:631–642.

Pisani D, Poling LL, Lyons-Weiler M & Hedges SB (2004) The colonization of land by animals: molecular phylogeny and divergence times among arthropods. *BMC Biol* 2:1.

Plummer T (2004) Flaked stones and old bones: biological and cultural evolution at the dawn of technology. *Am J Phys Anthropol* 39(Suppl.):118–164.

Polaszek A (ed) (2010) Systema Naturae 250—The Linnaean Ark. CRC Press.

Pradel A, Sansom IJ, Gagnier PY et al. (2007) The tail of the Ordovician fish *Sacabambaspis*. *Biol Lett* 3:72–75.

Prinzinger R, Preßmar A & Schleucher E (1991) Body temperature in birds. *Comp Biochem Phys* 99A:499–506.

Prüfer K, Racimo F, Patterson N et al. (2014) The complete genome sequence of a Neanderthal from the Altai Mountains. *Nature* 505:43–49 (doi: 10.1038/nature12886).

Prum RO & Brush AH (2002) The evolutionary origin and diversification of feathers. *Q Rev Biol* 77:261–295.

Pugach I, Delfin F & Gunnarsdóttir E (2013) Genome-wide data substantiate Holocene gene flow from India to Australia. *Proc Natl Acad Sci U S A* 110:1803–1808 (doi: 10.1073/pnas.1211927110).

Pyron RA (2011) Divergence time estimation using fossils as terminal taxa and the origins of Lissamphibia. *Syst Biol* 60:466–481 (doi: 10.1093/sysbio/syr047).

Qin P & Stoneking M (2015) Denisovan ancestry in East Eurasian and Native American populations. *Mol Biol Evol* 32:2665–2674 (doi: 10.1093/molbev/msv141).

Rands CM, Darling A & Fujita M (2013) Insights into the evolution of Darwin's finches from comparative analysis of the Geospiza magnirostris genome sequence. *BMC Genomics* 12:1495 (doi: 10.1186/1471-2164-14-95).

Rangel de Lázaro G, de la Cuétara JM, Píšová H et al. (2016) Diploic vessels and computed tomography: segmentation and comparison in modern humans and fossil hominids. *Am J Phys Anthropol* 159:313–324 (doi: 10.1002/ajpa.22878).

Rauhut OW, Foth C, Tischlinger H & Norell MA (2012) Exceptionally preserved juvenile megalosauroid theropod dinosaur with filamentous integument from the Late Jurassic of Germany. *Proc Natl Acad Sci U S A* 109:11746–11751 (doi: 10.1073/pnas.1203238109).

Raven JA & Edwards D (2014) Roots: evolutionary origins and biogeochemical significance. *J Exp Bot* 52(Suppl.):381–401.

Reich D, Green RE & Kircher M (2010) Genetic history of an archaic hominin group from Denisova Cave in Siberia. *Nature* 468:1053–1060 (doi: 10.1038/nature09710).

Reich D, Patterson N, Kircher M et al. (2011) Denisova admixture and the first modern human dispersals into Southeast Asia and Oceania. *Am J Hum Genet* 89:516–528 (doi: 10.1016/j.ajhg.2011.09.005).

Reidenberg JS (2007) Anatomical adaptations of aquatic mammals. *Anat Rec (Hoboken)* 290:507–513.

Reisz RR (1977) Petrolacosaurus, the oldest known diapsid reptile. *Science* 196:1091–1093.

Renfree MB (2010) Marsupials: placental mammals with a difference. *Placenta* 31:S21–S26 (doi: 10.1016/j.placenta.2009.12.023).

Retallack GJ (2013) Ediacaran life on land. *Nature* 493:89–92 (doi: 10.1038/nature11777).

Rice WR & Hostert EE (1993) Laboratory experiments on speciation: what have we learned in forty years? *Evolution* 47:1637–1653.

Rice WR & Salt GW (1990) The evolution of reproductive isolation as a correlated character under sympatric conditions: experimental evidence. *Evolution* 44:1140–1152.

Richardson L, Venkataraman S, Stevenson P et al. (2014) EMAGE mouse embryo spatial gene expression database: (2014 update). *Nucleic Acids Res* 42:D835–D844 (doi: 10.1093/nar/gkt1155).

Richter DJ & King N (2013) The genomic and cellular foundations of animal origins. *Ann Rev Genet* 47:509–537 (doi: 10.1146/annurev-genet-111212-133456).

Ridley M (2004) Evolution. Wiley-Blackwell.

Riedel-Kruse IH, Müller C & Oates AC (2007) Synchrony dynamics during initiation failure and rescue of the segmentation clock. *Science* 317:1911–1915.

Rinehart LF & Lucas SG (2013) Tooth form and function in temnospondyl amphibians: relationship of shape to applied stress. *New Mexico Mus Nat Hist Sci Bull* 61:533–542.

Roebroeks W, Sier MJ & Nielsen TK (2012) Use of red ochre by early Neandertals. *Proc Natl Acad Sci U S A* 109:1889–1894 (doi: 10.1073/pnas.1112261109).

Romanes GJR (1910) Darwin and after Darwin. Open Court Publishing.

Rota-Stabelli O, Daley AC & Pisani D (2013) Molecular timetrees reveal a cambrian colonization of land and a new scenario for ecdysozoan evolution. *Curr Biol* 23:392–398 (doi: 10.1016/j.cub.2013.01.026).

Rots V & Van Peer P (2006) Early evidence of complexity in lithic economy: core-axe production hafting and use at Late Middle Pleistocene site 8-B-11 Sai Island (Sudan). *J Archaeol Sci* 33:360–371.

Rowe TB, Macrini TE & Luo ZX (2011) Fossil evidence on origin of the mammalian brain. *Science* 332:955–957 (doi: 10.1126/science.1203117).

Rowntree LG, Clark TH & Hanson AM (1935) Accruing acceleration in growth and development in five successive generations of rats under continuous treatment with thymus extract. *Arch Int Med* 56:1–29.

Rücklin M, Donoghue PC, Johanson Z, Trinajstic K et al. (2012) Development of teeth and jaws in the earliest jawed vertebrates. *Nature* 491:748–751 (doi: 10.1038/nature11555).

Ruse M & Travis J (eds) (2009) Evolution: The First Four Billion Years. Harvard University Press.

Rutman JY (2014) Why evolution matters: a Jewish view. Valentine Mitchell.

Ryan JF & Baxevanis AD (2007) Hox Wnt and the evolution of the primary body axis: insights from the early-divergent phyla. *Biol Direct* 2:37.

Ryder OA, Chemnick LG, Bowling AT & Benirschke K (1985) Male mule foal qualifies as the offspring of a female mule and jack donkey. *J Hered* 76:379–381.

Sagulenko E, Morgan GP & Webb RI (2014) Structural studies of planctomycete *Gemmata obscuriglobus* support cell compartmentalisation in a bacterium. *PLoS One* 9:e91344 (doi: 10.1371/journal.pone.0091344).

Saitou N (2013) Introduction to Evolutionary Genetics. Springer.

Sakamaki K, Imai K, Tomii K & Miller DJ (2015) Evolutionary analyses of caspase-8 and its paralogs: deep origins of the apoptotic signaling pathways. *Bioessays* 37:767–776 (doi: 10.1002/bies.201500010).

Salvini-Plawen LV & Mayr E (1977) On the evolution of photoreceptors and eyes. *Evol Biol* 10:207–263.

San Mauro D (2010) A multilocus timescale for the origin of extant amphibians. *Mol Phylogenet Evol* 56:554–561 (doi: 10.1016/j.ympev.2010.04.019).

Sankararaman S, Mallick S, Dannemann M et al. (2014) The genomic landscape of Neanderthal ancestry in present-day humans. *Nature* 507:354–357 (doi: 10.1038/nature12961).

Sansom IJ, Donoghue PCJ & Albanesi G (2005) Histology and affinity of the earliest armoured vertebrate. *Biol Lett* 1:446–449 (doi:10.1098/rsbl.2005.0349).

Santarella-Mellwig R, Franke J & Jaedicke A (2010) The compartmentalized bacteria of the Planctomycetes-Verrucomicrobia-Chlamydiae superphylum have membrane coat-like proteins. *PLoS Biol* 8:e1000281 (doi: 10.1371/journal.pbio.1000281).

Sanz-Ezquerro JJ & Tickle C (2003) Digital development and morphogenesis. *J Anat* 202:51–58.

Sartorius GA & Nieschlag E (2010) Paternal age and reproduction. *Hum Reprod Update* 16:65–79 (doi: 10.1093/humupd/dmp027).

Sassa T (2013) The role of human-specific gene duplications during brain development and evolution. *J Neurogenet* 27:86–96.

Sato A, Tichy H & O'hUigin C (2001) On the origin of Darwin's finches. *Mol Biol Evo*l 18:299–311.

Saylo MC, Escoton CC & Saylo MM (2011) Punctuated equilibrium vs. phyletic gradualism. *Int J Biosci Biotech* 3:27–42. www.sersc.org/journals/IJBSBT/vol3_no4/3.pdf.

Schilthuizen M, Giesbers MC & Beukeboom LW (2011) Haldane's rule in the 21st century. *Heredity* 107:95–102 (doi: 10.1038/hdy.2010.170).

Schluter D & Conte GL (2009) Genetics and ecological speciation. *Proc Natl Acad Sci U S A* 106 (Suppl. 1):9955–9962 (doi: 10.1073/pnas.0901264106).

Schneider NY (2011) The development of the olfactory organs in newly hatched monotremes and neonate marsupials. *J Anat* 219:229–242.

Schneider H & Sampaio I (2013) The systematics and evolution of New World primates—a review. *Mol Phylogenet Evol* 82B:348–357 (doi: 10.1016/j.ympev.2013.10.017).

Schraiber JG & Akey JM (2015) Methods and models for unravelling human evolutionary history. *Nat Rev Genet* 16:727–740 (doi: 10.1038/nrg4005).

Schubbert S, Bollag G & Shannon K (2007) Deregulated Ras signaling in developmental disorders: new tricks for an old dog. *Curr Opin Genet Dev* 17:15–22.

Schultz JA & Martin T (2014) Function of pretribosphenic and tribosphenic mammalian molars inferred from 3D animation. *Naturwissen* 101:771–781 (doi: 10.1007/s00114-014-1214-y).

Schwartz JH, Tattersall I & Chi Z (2014) Comment on "A complete skull from Dmanisi, Georgia, and the evolutionary biology of early Homo". *Science* 344:360 (doi: 10.1126/science.1250056).

Secor DH (2015) Marine Ecology of Marine Fishes. Johns Hopkins University Press.

Seehausen O (2002) Patterns in fish radiation are compatible with Pleistocene desiccation of Lake Victoria and 14600 year history for its cichlid species flock. *Proc Biol Sci* 269:491–497.

Seilacher A, Bose PK & Pfluger F (1998) Triploblastic animals more than 1 billion years ago: trace fossil evidence from India. *Science* 282:80–83.

Seipel K & Schmid V (2005) Evolution of striated muscle: jellyfish and the origin of triploblasty. *Dev Biol* 282:14–26.

Selwood L & Johnson MH (2006) Trophoblast and hypoblast in the monotreme marsupial and eutherian mammal: evolution and origins. *BioEssays* 28:128–145.

Semon R (1901) Book - Normal Plates of the Development of Vertebrates 3. https://embryology.med.unsw.edu.au/embryology/index.php/Book_-_Normal_Plates_of_the_Development_of_Vertebrates_3

Seymour RS, Bennett-Stamper CL, Johnston SD et al. (2004) Evidence for endothermic ancestors of crocodiles at the stem of archosaur evolution. *Physiol Biochem Zool* 77:1051–1067.

Shapiro JA (2011) Evolution: A View from the 21st Century. FT Press.

Shattock SG (1880) A new bone in human anatomy together with an investigation into the morphological significance of the so-called internal lateral ligament of the human lower jaw. *J Anat Physiol* 14:201–204.

Sheehan S, Harris K & Song YS (2013) Estimating variable effective population sizes from multiple genomes: a sequentially markov conditional sampling distribution approach. *Genetics* 194:647–662 (doi: 10.1534/genetics.112.149096).

Sherman PW (1977) Nepotism and the evolution of alarm calls. *Science* 97:1246–1253.

Shi Y & Yokoyama S (2003) Molecular analysis of the evolutionary significance of ultraviolet vision in vertebrates. *Proc Natl Acad Sci U S A* 100:8308–8313.

Shichida Y & Matsuyama T (2009) Evolution of opsins and phototransduction. *Philos Trans R Soc* 364B:2881–2895 (doi: 10.1098/rstb.2009.0051).

Shih PM & Matzke NJ (2013) Primary endosymbiosis events date to the later Proterozoic with cross-calibrated phylogenetic dating of duplicated ATPase proteins. *Proc Natl Acad Sci U S A* 110:12355–12360 (doi: 10.1073/pnas.1305813110).

Shimada H & Yamagishi A (2011) Stability of heterochiral hybrid membrane made of bacterial sn-G3P lipids and archaeal sn-G1P lipids. *Biochemistry* 50:4114–4120R (doi: 10.1021/bi200172d).

Shu D, Morris SC & Zhang ZF (2003) A new species of Yunnanozoan with implications for deuterostome evolution. *Science* 299:1380–1384.

Shu D, Zhang X & Chen L (1996) Reinterpretation of Yunnanozoon as the earliest known hemichordate. *Nature* 380:428–430.

Shubin N (2007) Your Inner Fish. Random House.

Shubin NH, Daeschler EB & Jenkins FA Jr (2014) Pelvic girdle and fin of *Tiktaalik roseae*. *Proc Natl Acad Sci U S A* 111:893–899 (doi: 10.1073/pnas.1322559111).

Shubin N, Tabin C & Carroll S (2009) Deep homology and the origins of evolutionary novelty. *Nature* 457:818–823 (doi: 10.1038/nature07891).

Silver N (2012) The Signal and the Noise. Penguin Press.

Simons RS & Brainerd EL (1999) Morphological variation of hypaxial musculature in salamanders (Lissamphibia: caudata). *J Morphol* 241:153–164.

Smith MR & Ortega-Hernández J (2014) Hallucigenia's onychophoran-like claws and the case for Tactopoda. *Nature* 514:363–366 (doi: 10.1038/nature13576).

Soares P, Ermini L & Thomson N (2009) Correcting for purifying selection: an improved human mitochondrial molecular clock. *Am J Hum Genet* 84:740–759 (doi: 10.1016/j.ajhg.2009.05.001).

Soares P, Alshamali F & Pereira V (2012) The expansion of mtDNA haplogroup L3 within and out of Africa. *Mol Biol Evol* 29:915–927 (doi: 10.1093/molbev/msr245).

Soltis DE, Bell CD, Kim S & Soltis PS (2008) Origin and early evolution of angiosperms. *Ann N Y Acad Sci* 1133:3–25 (doi: 10.1196/annals.1438.005).

Soressi M, McPherron SP, Lenoir M et al. (2013) Neandertals made the first specialized bone tools in Europe. *Proc Natl Acad Sci U S A* 110:14186–14190 (doi: 10.1073/pnas.1302730110).

Soya O (ed.) (2012) Evolutionary Systems Biology. Springer.

Spang A, Saw JH & Jørgensen SL (2015) Complex archaea that bridge the gap between prokaryotes and eukaryotes. *Nature* 521:173–179 (doi: 10.1038/nature14447).

Stringer C (2011) The Origin of our Species. Penguin Books.

Stott R (2012) Darwin's ghosts: in search of the first evolutionists. Bloomsbury.

Strother PK, Battison L, Brasier MD & Wellman CH (2011) Earth's earliest non-marine eukaryotes. *Nature* 473:505–509 (doi: 10.1038/nature09943).

Suárez R, Gobius I & Richards LJ (2014) Evolution and development of interhemispheric connections in the vertebrate forebrain. *Front Hum Neurosci* 8:497 (doi: 10.3389/fnhum.2014.00497).

Summerhayes GR, Leavesley M, Fairbairn (2010) Human adaptation and plant use in highland New Guinea 49000 to 44000 years ago. *Science* 330:78–81 (doi: 10.1126/science.1193130).

Szöllősi GJ, Tannier E, Daubin V & Boussau B (2015) The inference of gene trees with species trees. *Syst Biol* 64:e42–e62 (doi: 10.1093/sysbio/syu048).

Tabin CJ, Carroll SB & Panganiban G (1999) Out on a limb: parallels in vertebrate and invertebrate limb patterning and the origin of appendages. *Am Zool* 39:650–666.

Tajsharghi H & Oldfors A (2013) Myosinopathies: pathology and mechanisms. *Acta Neuropathol* 125:3–18 (doi: 10.1007/s00401-012-1024-2).

Takarada T, Hinoi E & Nakazato R (2013) An analysis of skeletal development in osteoblast-specific and chondrocyte-specific runt-related transcription factor-2 (Runx2) knockout mice. *J Bone Miner Res* 28:2064–2069 (doi: 10.1002/jbmr.1945).

Takechi M & Kuratani S (2010) History of studies on mammalian middle ear evolution: a comparative morphological and developmental biology perspective. *J Exp Zool Mol Dev Evol* 314:417–433.

Tapaltsyan V, Charles C & Hu J (2016) Identification of novel Fgf enhancers and their role in dental evolution. *Evol Dev* 18:31–40 (doi: 10.1111/ede.12132).

Tassi F, Ghirotto S & Mezzavilla M (2015) Early modern human dispersal from Africa: genomic evidence for multiple waves of migration. *Investig Genet* 6:13 (doi: 10.1186/s13323-015-0030-2).

Teeling EC, Springer MS & Madsen O (2005) A molecular phylogeny for bats illuminates biogeography and the fossil record. *Science* 307:580–584.

Templeton AR (2008) The reality and importance of founder speciation in evolution. *BioEssays* 30:470–479 (doi: 10.1002/bies.20745).

Thieme H (1997) Lower Palaeolithic hunting spears from Germany. *Nature* 385:807–810.

Thornton JW (2004) Resurrecting ancient genes: experimental analysis of extinct molecules. *Nat Rev Genet* 5:366–375.

Thulborn RA (1980) The ankle joints of archosaurs. *Aust J Palaeont* 4:241–261 (doi: 10.1080/03115518008558970).

Tickle C (2006) Developmental cell biology: making digit patterns in the vertebrate limb. *Nat Rev Mol Cell Biol* 7:45–53.

Tissir F & Goffinet AM (2013) Shaping the nervous system: role of the core planar cell polarity genes. *Nat Rev Neurosci* 14:525–535 (doi: 10.1038/nrn3525).

Towers M, Mahood R, Yin Y & Tickle C (2008) Integration of growth and specification in chick wing digit-patterning. *Nature* 452:882–886 (doi: 10.1038/nature06718).

Turing AM (1952) The chemical basis of morphogenesis. *Philos Trans R Soc* B 237:37–72.

Turner GF, Seehausen O & Knight ME (2001) How many species of cichlid fishes are there in African lakes? *Mol Ecol* 10:793–806.

Tyndale-Biscoe CH & Renfree MB (1987) Reproductive Physiology of Marsupials. Cambridge University Press.

Uhen MD (2007) Evolution of marine mammals: back to the sea after 300 million years. *Anat Rec* 290:514–522.

Ungar PS (2010) Mammal Teeth: Origin Evolution and Diversity. Johns Hopkins Press.

Valas RE & Bourne PE (2011) The origin of a derived superkingdom: how a Gram-positive bacterium crossed the desert to become an archaeon. *Biol Direct* 6:16 (doi: 10.1186/1745-6150-6-16).

Van Tuinen M & Blair Hedges S (2001) Calibration of avian molecular clocks. *Mol Biol Evol* 18:206–213.

Van Valkenburgh B (2007) Deja vu: the evolution of feeding morphologies in the Carnivora. *Integr Comp Biol* 47:147–163 (doi: 10.1093/icb/icm016).

Varriale A (2014) DNA methylation epigenetics and evolution in vertebrates: facts and challenges. *Int J Evol Biol* 2014:475981 (doi: 10.1155/2014/475981).

Veeramah KR & Hammer MF (2014) The impact of whole-genome sequencing on the reconstruction of human population history. *Nat Rev Genet* 15:149–162 (doi: 10.1038/nrg3625).

Ventura GT, Kenig F & Reddy CM (2007) Molecular evidence of Late Archaen archaea and the presence of a subsurface hydrothermal biosphere. *Proc Natl Acad Sci U S A* 104:14260–14265.

Verbrugge LM (1982) Sex differentials in health. *Public Health Rep* 97:417–437.

Vereecken NJ, Wilson CA, Hötling S et al. (2012) Pre-adaptations and the evolution of pollination by sexual deception: Cope's rule of specialization revisited. *Proc Biol Sci* 279:4786–4794 (doi: 10.1098/rspb.2012.1804).

von Koenigswald WV (2000) Two different strategies in enamel differentiation: Marsupialia versus Eutheria. In Development Function and Evolution of Teeth (Teaford MF, Meredith-Smith M, Ferguson MWJ eds), pp 107–118. Cambridge University Press.

von Petzinger G & Nowell A (2014) A place in time: situating Chauvet within the long chronology of symbolic behavioral development. *J Hum Evol* 74:37–54 (doi: 10.1016/j.jhevol.2014.02.022).

Waddington CH (1953) Genetic assimilation of the bithorax phenotype. *Evolution* 10:1–13.

Waddington CH (1961) Genetic assimilation. *Adv Genet* 10:257–293.

Wakeley J (2010) Natural selection and coalescent theory. In Evolution Since Darwin: The First 150 Years (Bell MA, Futuyma DJ, Eanes WF & Levinton JS eds), pp 119–149. Sinauer Press.

Wall JD & Slatkin M (2012) Paleopopulation genetics. *Annu Rev Genet* 46:635–649.

Wang X, Nudds RL, Palmer C & Dyke GJ (2012) Size scaling and stiffness of avian primary feathers: implications for the flight of Mesozoic birds. *J Evol Biol* 25:547–555 (doi: 10.1111/j.1420-9101.2011.02449.x).

Ward CV, Kimbel WH, Harmon EH & Johanson DC (2012) New postcranial fossils of *Australopithecus afarensis* from Hadar Ethiopia (1990–2007). *J Hum Evol* 65:1–51 (doi: 10.1016/j.jhevol.2011.11.012).

Wayne L (1988) International Committee on Systematic Bacteriology: Announcement of the report of the ad hoc Committee on Reconciliation of Approaches to Bacterial Systematics. *Zentralb Bakteriol Hyg A* 268:433–434.

Weir JT & Schluter D (2008) Calibrating the avian molecular clock. *Mol Ecol* 17:2321–2328 (doi: 10.1111/j.1365-294X.2008.03742.x).

Weaver TD (2012) Did a discrete event 200,000–100,000 years ago produce modern humans? *J Hum Evol* 63:121–126 (doi: 10.1016/j.jhevol.2012.04.003).

Weismann A (1889) The continuity of the germ-plasm as the foundation of a theory of heredity. In Essays on Heredity. Oxford Clarendon Press. www.esp.org/books/weismann/essays/facsimile/

Westneat MW, Betz O & Blob RW (2003) Tracheal respiration in insects visualized with synchrotron X-ray imaging. *Science* 299:558–560.

Wildman DE, Uddin M, Liu G et al. (2003) Implications of natural selection in shaping 99.4% nonsynonymous DNA identity between humans and chimpanzees: enlarging genus Homo. *Proc Natl Acad Sci U S A* 100:7181–7788.

Wiles AM, Doderer M & Ruan J (2010) Building and analyzing protein interactome networks by cross-species comparisons. *BMC Syst Biol* 4:36 (doi: 10.1186/1752-0509-4-36).

Wilkins AS (2002) The Evolution of Developmental Pathways. Sinauer Press.

Wilkins AS (2007) Genetic networks as transmitting and amplifying devices for natural genetic tinkering. *Novartis Found Symp* 284:71–86.

Will M, Parkington JE, Kandel AW & Conard NJ (2013) Coastal adaptations and the Middle Stone Age lithic assemblages from Hoedjiespunt 1 in the Western Cape South Africa. *J Hum Evol* 64:518–537 (doi: 10.1016/j.jhevol.2013.02.012).

Wille M, Nägler TF & Lehmann B (2008) Hydrogen sulphide release to surface waters at the Precambrian/Cambrian boundary. *Nature* 453:767–769 (doi: 10.1038/nature07072).

Williams TA, Foster PG & Nye TMW (2012) A congruent phylogenomic signal places eukaryotes within the Archaea. *Proc Biol Sci* 279:4870–4879 (doi: 10.1098/rspb.2012.1795).

Williams TA, Foster PG, Cox CJ & Embley TM (2013) An archaeal origin of eukaryotes supports only two primary domains of life. *Nature* 304:231–236 (doi: 10.1038/nature12779).

Williamson TE, Brusatte SL & Wilson GP (2014) The origin and early evolution of metatherian mammals: the Cretaceous record. *Zookeys* 465:1–76 (doi: 10.3897/zookeys.465.8178).

Willis K & McElwain J (2013) The Evolution of Plants. Oxford University Press.

Wilson EO (2000) Sociobiology: The New Synthesis. Belknap Press.

Witmer LM (1987) The nature of the antorbital fossa of archosaurs: shifting the null hypothesis. In Fourth Symposium on Mesozoic Terrestrial Ecosystesms (Currie PJ & Koster EH eds), pp 230–235. http://www.ohio.edu/people/witmerl/Downloads/1987_Witmer_Archosaur_antorbital_cavity.pdf

Woese CR, Kandler O & Wheelis ML (1990) Towards a natural system of organisms: proposal for the domains Archaea Bacteria and Eucarya. *Proc Natl Acad Sci U S A* 87:4576–4579.

Wolpert L, Tickle C & Jessell T (2015) Principles of Development, 5th ed. Oxford University Press.

Wood B & Constantino P (2007) *Paranthropus boisei*: fifty years of evidence and analysis. *Am J Phys Anthropol* 45(Suppl.):106–132.

Wood B & Elton S (eds) (2008) Human evolution: ancestors and relatives. *J Anat* 4:335–562. (This is a set of symposium papers).

Wood B & Harrison T (2011) The evolutionary context of the first hominins. *Nature* 470:347–352 (doi: 10.1038/nature09709).

Wood B & Lonergan N (2008) The hominin fossil record: taxa, grades and clades. *J Anat* 212:354–376 (doi: 10.1111/j.1469-7580.2008.00871.x).

Wray GA (1997) Echinoderms. In Embryology: Constructing the Organism (Gilbert SF & Raunio AM eds), pp 309–330. Sinauer Press.

Xu X (2012) A gigantic feathered dinosaur from the Lower Cretaceous of China. *Nature* 484:92–95 (doi: 10.1038/nature10906).

Xu X, Zhou Z, Dudley R et al. (2014) An integrative approach to understanding bird origins. *Science* 346:1253293 (doi: 10.1126/science.1253293).

Xue H, Tong KL, Marck C et al. (2003) Transfer RNA paralogs: evidence for genetic code-amino acid biosynthesis coevolution and an archaeal root of life. *Gene* 310:59–66.

Yan W (2014) Potential roles of noncoding RNAs in environmental epigenetic transgenerational inheritance. *Mol Cell Endocrinol* 398:24–30. (doi: 10.1016/j.mce.2014.09.008).

Yang F, Fu B & O'Brien PC (2004) Refined genome-wide comparative map of the domestic horse, donkey and human based on cross-species chromosome painting: insight into the occasional fertility of mules. *Chromosome Res* 12:65–76.

Yang Y, Guillot P, Boyd Y et al. (1998) Evidence that preaxial polydactyly in the doublefoot mutant is due to ectopic Indian Hedgehog signalling. *Development* 125:3123–3132.

Yehuda R, Daskalakis NP & Bierer LM (2015) Holocaust exposure induced intergenerational effects on FKBP5 methylation. *Biol Psychiatry* pii: S0006-3223(15)00652-6 (doi: 10.1016/j.biopsych.2015.08.005).

Yoshida MA & Ogura A (2011) Genetic mechanisms involved in the evolution of the cephalopod camera eye revealed by transcriptomic and developmental studies. *BMC Evol Biol* 11:180 (doi: 10.1186/1471-2148-11-180).

Zardoya R & Meyer A (1998) Complete mitochondrial genome suggests diapsid affinities of turtles. *Proc Natl Acad Sci* 95:14226–14231.

Zhang F, Zhou Z & Xu X (2008) A bizarre Jurassic maniraptoran from China with elongate ribbon-like feathers. *Nature* 455:1105–1108 (doi: 10.1038/nature07447).

Zhao F, Bottjer DJ & Hu S (2013) Complexity and diversity of eyes in Early Cambrian ecosystems. *Nat Sci Rep* 3:2751 (doi: 10.1038/srep02751).

Zhao B, Tumaneng K & Guan KL (2011) The Hippo pathway in organ size control tissue regeneration and stem cell self-renewal. *Nat Cell Biol* 13:877–883 (doi: 10.1038/ncb2303).

Zheng X-T, You HL, Xu X & Dong ZH (2009) An Early Cretaceous heterodontosaurid dinosaur with filamentous integumentary structures. *Nature* 458:333–336 (doi: 10.1038/nature07856).

Zmasek CM, Zhang Q, Ye Y & Godzik A (2007) Surprising complexity of the ancestral apoptosis network. *Genome Biol* 8:R226.

Zhu M & Ahlberg PE (2004) The origin of the internal nostril of tetrapods. *Nature* 432:94–7.

Zhu M, Yu X & Ahlberg PE (2013) A Silurian placoderm with osteichthyan-like marginal jaw bones. *Nature* 502:188–193 (doi: 10.1038/nature12617).

Zhu R, Zhisheng AZ, Potts R & Hoffmand KA (2003) Magnetostratigraphic dating of early humans in China. *Earth Sci Rev* 61:341–359.

FIGURE ACKNOWLEDGMENTS

Figure Number	Acknowledgment
3.1	Courtesy of Oswego City School District.
4.1a	Courtesy of Verisimilus published under CC BY-SA 3.0.
4.1b	Published under CC BY-SA 3.0
4.2	From Just J, Kristensen RM & Olesen J (2014) *PLoS ONE* 9:e102976. With permission from Public Library of Science.
4.3a	Courtesy of Simon Conway Morris, University of Cambridge.
4.3c	Courtesy of Martin R. Smith.
4.3d	Courtesy of Keith Schengili-Roberts published under CC BY-SA 3.0.
4.4	Courtesy of Hcrepin published under CC BY-SA 3.0.
4.5	From Chen J-Y (2009) *Int. J. Dev. Biol.* 53:733–751. With permission from J-Y Chen.
4.8a	Courtesy of Eduard Solà published under CC BY-SA 3.0.
4.8b	Courtesy of H. Raab published under CC BY-SA 3.0.
5.4	From Harris WA (1997) *PNAS* 94:2098–2100. With permission from National Academy of Sciences, USA (© 1997).
5.6	Courtesy of Peter V. Sengbusch.
5.8	From Cohn MJ & Tickle C (1999) *Nature* 399:474–479. With permission from Nature Publishing Group.
6.2	From Donoghue PCJ & Sansom IJ (2002) *Microsc. Res. Tech.* 59:352–372. With permission from John Wiley and Sons.
6.3a, b	From Larsen WJ (1997) Human Embryology 2nd Ed. WB Saunders Company. With permission from Elsevier.
6.3c	Courtesy of Nobu Tamura published under CC BY-SA 3.0.
6.5	From Schneider I, Aneas I, Gehrke AR et al. (2011) *PNAS* 108:12782–12786. With permission from The National Academy of Sciences.
6.7	Courtesy of Steven M Carr.
6.9a	From Peters D (1991) From The Beginning – The Story Of Human Evolution. With permission from David Peters.
6.9b	Courtesy of DiBgd published under CC BY-SA 3.0.
6.11	Kardong K (2002) Vertebrates: Comparative Anatomy, Function, Evolution 3rd Ed. McGraw-Hill Education. With permission from McGraw-Hill Education ©.
6.13	Courtesy of Debivort published under CC BY-SA 3.0.

Figure Number	Acknowledgment
7.2	From Takechi M & Kuratani S (2010) *J. Exp. Zool. B. Mol. Dev. Evol.* 314B:417–433. With permission from John Wiley & Sons.
7.3	From Carroll RL (1988) Vertebrate Paleontology and Evolution. W.H. Freeman & Company.
7.4	Beddard FE (1902) The Cambridge Natural History. Cambridge University Press.
7.6	From Nikaido M, Rooney AP & Okada N (1999) *PNAS* 96:10261–10266. With permission from The National Academy of Sciences.
7.7	From Bejder L & Hall BK (2002) *Evol Dev* 4:445–458. With permission from John Wiley and Sons.
8.1	From Gilbert SF (2014) Developmental Biology (10th Ed). Sinauer Press. With permission from Sinauer Press.
8.3	From Langeland J, Holland LZ, Chastain RA & Holland ND (2006) *Int. J. Biol. Sci.* 2:110–116. With permission from Ivyspring.
8.4	From Li J, Chen X, Kang, B & Liu M (2015) *PLoS ONE* 10:e0123894. With permission from Public Library of Science.
8.5	From Weaver J (2012) *J. Hum. Evol.* 63:121–6. With permission from Elsevier.
9.1	Courtesy of Susan Offner.
9.2	From Kim KM & Caetano-Anollés G (2011) *BMC Evol. Biol.* 11:140 published under CC BY 2.0.
9.3	From Fuerst JA & Sagulenko E (2012) *Front. Microbio.* 3:167. With permission from JA Fuerst.
9.6	From El Albani, A Bengtson, S Canfield, DE et al. (2010) *Nature* 466:100-104 (doi: 10.1038/nature09166). With permission from Public Library of Science.
9.7	From Fey P, Kowal AS, Gaudet P et al. (2007) *Nature Protocols* 2:1307–1316. With permission from Nature Publishing Group.
10.2 #1	Kominami T & Takata H (2004) *Dev Growth Differ* 46:309–326. With permission from John Wiley and Sons.
10.2 #2	Courtesy of University of South Carolina Press.
10.4	From www.sdbonline.org/sites/fly/atlas/52.htm. Courtesy of Volker Hartenstein.
10.5	Courtesy of EMAP eMouse Atlas Project (http://www.emouseatlas.org). With reference to Richardson L, Venkataraman K, Stevenson P, et al. (2014) *Nucleic Acids Res.* 42:D835–44.
10.9	From http://www.sabiosciences.com/pathway.php?sn=Rho_family_GTPase. Courtesy of Qiagen.
10.11	From Davies JA & Bard J (1998) *Curr. Top Dev. Biol.* 39:245–301. With permission from Elsevier.
10.12	Krabbenhoft KM & Fallon JF (1989) *Dev. Biol.* 131:373–382. With permission from Elsevier.
11.1a	Courtesy of Lewis Held.
11.1b	From Harris WA (1997) *PNAS* 94:2098–2100. With permission from The National Academy of Sciences.
11.1c, d	Halder G, Callaerts P & Gehring WJ (1995) *Science* 267:1788–1792. With permission from The American Association for the Advancement of Science.

Figure Number	Acknowledgment
11.2 & 11.3	From Gilbert SF (2014) Developmental Biology (10th Ed). Sinauer Press. With permission from Sinauer Press.
11.4	From Lichtneckert R & Reichert H (2005) *Heredity* 94:465–477. With permission from Nature Publishing Group.
11.5a	From Kanehisa M, Goto S, Sato Y et al. (2014) *Nucleic Acids Res.* 42:D199–D205. http://www.kegg.jp/pathway/dme04330. With permission from KEGG.
11.5b	From Kanehisa M, Goto S, Sato Y et al. (2014) *Nucleic Acids Res.* 42:D199–D205. http://www.kegg.jp/pathway/mmu04330. With permission from KEGG.
11.6	From Wiles AM, Doderer M, Ruan J et al. (2010) *BMC Syst Biol.* 4:36 published under CC BY 2.0.
12.1a, b, d, e, f	Courtesy of Jim Gifford published under CC BY-SA 2.0.
12.1c	Courtesy of Graham Manning published under CC BY-SA 3.0.
12.2	From Bard JBL (1981) *J. Theor. Biol.* 93:363–385. With permission from Elsevier.
12.3a, b, c	From Morriss-Kay GM & Wilkie AOM (2005) *J. Anat.* 207:637–653. With permission from John Wiley and Sons.
13.1	From Hayes C, Lyon MF & Morriss-Kay GM (2002) *J. Anat.* 193:81–91. With permission from John Wiley and Sons.
13.2a	Courtesy of S. Blair Hedges.
13.2b	Courtesy of Dr Tim Vickers.
13.4	Courtesy of Lewis Held. From Held LI (2014) How the Snake Lost Its Legs: Curious Tales from the Frontier of Evo-Devo, 1st Ed. Cambridge University Press.
13.5a	From Mitgutsch C, Richardson MK, Jiménez R et al. (2012) *Biol. Lett.* 8:74–77. With permission from The Royal Society.
13.5c	From Miller CE, Basu C, Fritsch T et al. (2008) *J. R. Soc. Interface* 5:465–475. With permission from The Royal Society.
13.6 & 13.7	From Bard JBL (1977) *J. Zool.* 183:527–539. With permission from John Wiley and Sons.
14.1	Courtesy of Olaf Leillinger published under CC BY-SA 3.0.
14.2c	Courtesy of William Warby published under CC BY 2.0.
14.2d	Courtesy of PiccoloNamek published under CC BY-SA 3.0.
14.3a	Courtesy of Nicolas Gompel ©.
14.3b	Courtesy of EB Lewis.
14.4	From Waddington CH (1956) *Evolution* 10:1–13. With permission from John Wiley and Sons.
14.5 (top)	From Nilsson DE & Pelger S (1994) *Proc. Roy. Soc.* 256B:53–8. With permission from The Royal Society.
14.5 (bottom)	From Salvini-Plawen LV & Mayr E (1977) *Evol. Biol.* 10:207–63. With permission from Springer Science+Business Media.
15.1	From Ewart JC (1899) The Pennycuik Experiments. Adam & Charles Black.

Figure Number	Acknowledgment
16.2	From Wood B & Lonergan N (2008) *J. Anat.* 212:354–376. With permission from John Wiley and Sons.
16.3	Courtesy of PersianEvolution published under CC BY-SA 3.0.
16.4	From Lalueza-Fox C & Gilbert MTP (2011) *Curr. Biol.* 21:R1002-R1009. With permission from John Wiley and Sons.
16.5	Adapted from Oppenheimer S (2012). Out-of-Africa, the peopling of continents and islands: tracing uniparental gene trees across the map. *Phil. Trans. R. Soc.* 367:770–784. Courtesy of Stephen Oppenheimer ©.
16.6	Courtesy of Maximilian Dörrbecker published under CC BY-SA 3.0.
16.7 (mode 4)	Courtesy of José-Manuel Benito published under CC BY-SA 2.5.
16.7 (mode 5)	From Clark JGD (1936) The Mesolithic Settlement of Northern Europe. Cambridge University Press.
16.8a	Courtesy of Prof Christopher Henshilwood, University of Bergen.
16.8b	From Conard NJ (2009) *Nature* 459:248–252. With permission from Nature Publishing Group.
A1.1	From http://www.sabiosciences.com/pathway.php?sn=EGF_Pathway. Courtesy of Qiagen.
A1.2	From Bard J (2011) *J. Anat.* 218:591–599. With permission from John Wiley and Sons.

INDEX

C

D

I

J

K

L

M

N

Q

R

S

T